ACKNOWLEDGEMENTS

My grateful thanks are due to those who read through parts of the manuscript of this book, and made valuable comments, all of which have been followed – though any remaining omissions and errors are purely my own responsibility:

Professor Sir Francis Graham-Smith (Astronomer Royal),
Dr Paul Murdin, OBE (Royal Greenwich Observatory),
Dr Gilbert Fielder (University of Lancaster),
Dr Peter Cattermole (University of Sheffield),
Dr John Mason,
Iain Nicolson (Hatfield Polytechnic Observatory).

I am most grateful for invaluable help in proof reading to Barney D'Abbs and John English; also to Paul Doherty for the maps on pages 55, 82–89, 96–103, 109–113, 164–252, and of course to the publishers, particularly Béatrice Frei.

PATRICK MOORE

Selsey, Sussex

METRIC CONVERSION

In this book I have followed the current practice of giving lengths in metric units rather than the familiar imperial ones. To help in avoiding confusion, the following table may be found useful.

centimetres		inches	kilometres		miles
2·54	1	0.39	1·61	1	0·62
5·08	2	0·79	3·22	2	1·24
7·62	3	1·18	4·83	3	1·86
10·16	4	1·58	6·44	4	2·49
12·70	5	1·97	8·05	5	3·11
15·24	6	2·36	9·66	6	3·73
17·78	7	2·76	11·27	7	4·35
20·32	8	3·15	12·88	8	4·97
22·86	9	3·54	14·48	9	5·59
25·40	10	3·94	16·09	10	6·21
50·80	20	7·87	32·19	20	12·43
76·20	30	11·81	48·28	30	18·64
101·6	40	15·75	64·37	40	24·86
127·0	50	19·69	80·47	50	31·07
152·4	60	23·62	96·56	60	37·28
177·8	70	27·56	112·7	70	43·50
203·2	80	31·50	128·7	80	49·71
228·6	90	35·43	144·8	90	55·92
254·0	100	39·37	160·9	100	62·14

© Patrick Moore and Guinness Publishing Ltd, 1988

First published in 1979
Second edition 1983

Editor: Béatrice Frei
Picture Editor: Alex Goldberg
Design and Layout: Michael Morey

Front cover picture: Largest solar telescope, Kitt Peak Observatory, Tucson, Arizona. (**Colorific/Black Star/James Sugar**)

Published in Great Britain by
Guinness Publishing Ltd, 33 London Road, Enfield, Middlesex

Typeset by Fakenham Photosetting Ltd, Fakenham, Norfolk
Printed and bound in Great Britain by Hazell Watson and Viney Ltd, Member of BPCC plc, Aylesbury, Bucks.

British Library Cataloguing in Publication Data

Moore, Patrick
    The Guinness book of astronomy facts & feats.–3rd ed.
    1. Astronomy
    I. Title
    520   QB43.2

    ISBN 0–85112–375–9

'Guinness' is a registered trade mark of Guinness Superlatives Ltd

CONTENTS

THE SOLAR SYSTEM

THE SUN

The Sun is by far the nearest star; it is 270 000 times closer than α Centauri. It is therefore the only star which may be studied in detail. It is a normal Main Sequence star.

The first known estimate of the distance of the Sun was made by Aristarchus of Samos, about 270 BC. His value, derived from observations of the angle between the Sun and the exact half-moon, was approximately 4 800 000 km. Ptolemy (circa AD 150) increased this to 8 000 000 km, but about AD 1543 Copernicus reverted to only 3 200 000 km. Kepler, in 1618, gave a value of 22 500 000 km.

The first reasonably accurate estimate of the mean distance of the Sun was made by G. D. Cassini in 1672. He gave a value of 138 370 000 km. Successive determinations have been as follows:

Year	Authority	Method	Distance, km
1672	Cassini	Parallax of Mars	138 370 000
1770	Euler	Transit of Venus, 1769	151 225 000
1771	Lalande	Transit of Venus, 1769	154 198 000
1814	Delambre	Transit of Venus, 1769	153 841 000
1823	Encke	Transits of Venus, 1761 and 1769	153 303 000
1862	Foucault	Velocity of light	147 459 000
1867	Newcomb	Parallax of Mars	148 626 000
1872	Le Verrier	Masses of the planets	148 459 000
1875	Galle	Parallax of asteroid Flora	148 290 000
1877	Airy	Transit of Venus, 1874	150 152 000
1878	Stone	Transit of Venus, 1874	148 125 000
1881	Puiseux	Transit of Venus, 1874	146 475 000
1931	Spencer Jones	Parallax of Eros	149 645 000
1976	various	Radar to Venus	149 597 000

The first suggestion of measuring the Sun's distance (astronomical unit) by using transits of Venus was made by J. Gregory in 1663, and was extended by Edmond Halley in 1678. The method was sound in theory, but was affected by the 'Black Drop' – the apparent effect of Venus drawing a strip of blackness after it has passed on to the Sun's disk, thus making exact timings difficult. (Captain Cook's famous voyage, during which he discovered Australia, was made in order to take the astronomer Green to a suitable position from which to observe the transit of 1769.) Parallax measurements of the planets and asteroids were more accurate, but Spencer Jones' value as derived from the close approach of Eros in 1931 was too high, and was later revised by Rabe to 149 493 000 km. The modern method – radar–was introduced in the early 1960s in the United States. The present accepted value for the astronomical unit is accurate to a tiny fraction of one per cent.

The first comments upon the Sun's rotation were made by Galileo, following his observations of sunspots from 1610. He gave a value of rather less than one month.

The discovery of the Sun's differential rotation, i.e. that the Sun does not rotate as a solid body would do, the equatorial rotation being shorter than the polar – was made by Richard Carrington in 1863. In order to help in identifying specific rotations of the Sun, Carrington had introduced a numbering system, beginning with Rotation No 1 on 9 November 1853. Rotation No 1500 began on 19 October 1965. **The first observations of Doppler shifts at opposite limbs due to the solar rotation** were made by H. C. Vogel in 1871.

Synodic rotation periods for features at various heliographic latitudes are as follows:

Latitude	Average synodic rotation period, days	Latitude	Average synodic rotation period, days
0	24·6	50	29·2
10	24·9	60	30·9
20	25·2	70	32·4
30	25·8	80	33·7
40	27·5	90	34·0

The first serious attempt to measure the solar constant was made by Sir John Herschel in 1837–8, using an actinometer (basically a bowl of water; the estimate was made by seeing the rate at which the bowl was heated). He gave a value which is about half the actual figure. The solar constant may be defined as the amount of energy in the form of solar radiation which is normally received on unit area at the top of the Earth's atmosphere; it is roughly equal

---

DATA

Mean distance from the Earth:
149 597 900 km (= 1 astronomical unit, a.u.)

Maximum distance from the Earth:
152 100 000 km

Minimum distance from the Earth:
147 100 000 km

Mean parallax: 8″·794

Distance from centre of Galaxy:
25 000 light-years

Period of revolution round centre of Galaxy: about 225 000 000 years (= 1 'cosmic year')

Velocity round centre of Galaxy:
2150 km/s

Velocity toward solar apex: 19·5 km/s

Apparent diameter: max 32′35″, mean 32′01″, min 31′31″

Equatorial diameter: 1 392 000 km

Density (water = 1): 1·409

Mass (Earth = 1): 332 946

Mass: 2×10^{27} tonnes (99 per cent of the mass of the entire Solar System)

Volume (Earth = 1): 1 303 600

Surface gravity (Earth = 1): 27·90

Escape velocity: 617·5 km/s

Mean apparent magnitude: −26·8 (= 600 000 times as brilliant as the full moon)

Absolute magnitude: +4·83

Spectrum: G2

Surface temperature: 5500 °C

Core temperature: 14 000 000 °C*

Rotation period,
sidereal: mean 25·380 days
synodic: mean 27·275 days

Time taken for light from the Sun to reach the Earth: mean
499·012 sec = 8·3 minutes

*Some authorities prefer a value of over 15 000 000 °C.

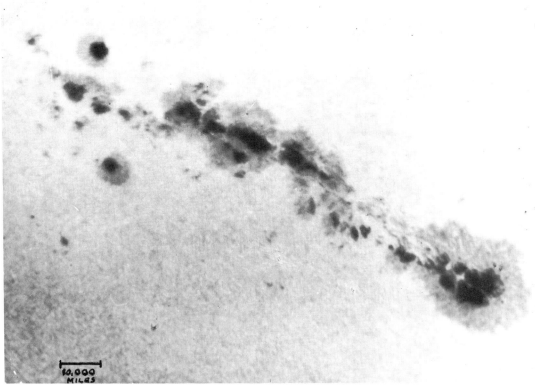

Gigantic sunspot group photographed on 28 February 1967 (W. M. Baxter)

to the amount of energy reaching ground level on a clear day. The modern value is 1·95 calories per square cm per minute.

The first photograph of the Sun (a Daguerreotype) seems to have been taken by Lerebours, in France, in 1842. However, the first good Daguerreotype was taken by Fizeau and Foucault, also in France, on 2 April 1845, at the request of F. Arago. In 1854 J. B. Reade used a dry collodion plate to show mottling on the Sun. **The first systematic series of solar photographs** was taken at Kew (London) from 1858 to 1872, using equipment designed by the British astronomer Warren de la Rue. Nowadays the Sun is photographed daily at many observatories throughout the world.

SUNSPOTS AND ASSOCIATED PHENOMENA

The discovery of sunspots was made in 1610–11. (Naked-eye spots had been previously recorded, but had not been explained; one given in a Chinese re-cord of 28 BC is described as 'a black vapour as large as a coin'). The first observer to publish telescopic observations of them was J. Fabricius, from Holland, in 1611, and though his drawings are undated he probably saw the spots toward the end of 1610. C. Scheiner at Ingolstädt recorded spots in March 1611, with his pupil C. B. Cysat. Schein-er wrote a tract, which came to the notice of Galileo, who claimed to have been observing sunspots since November 1610. No doubt all these observers recorded spots telescopically at about the same time (the period was close to solar maximum). However, interpretations differed. Galileo's explanation was basically correct; Scheiner regarded the spots as dark bodies moving round the Sun at a distance close to the solar surface; Cassini, later, regarded them as mountains protruding through the bright surface!

The first to describe the projection method of observing sunspots may have been Galileo's pupil B. Castelli. Galileo himself certainly used the method, and said (correctly) that it was 'the method that any sensible person will use'. (This seems to dispose of the legend that he ruined his eyesight by looking directly at the Sun through a telescope.)

The Wilson effect was announced by A. Wilson, of Glasgow, in 1774. He observed that with a regular spot, the penumbra to the limbward side seemed to become broadened, as against the opposite side, as the spot neared the edge of the disk. From this, Wilson deduced that the spots must be hollows. The original observations were made in 1769.

The largest spot-group on record was that of April 1947; it covered an area of $18\,130\,000\,000\ \text{km}^2$, reaching its maximum on 8 April. To be visible with the naked eye, a spot-group must cover 500 millionths of the visible hemisphere. (One millionth of the hemisphere is equal to $3\,000\,000\ \text{km}^2$.)

The longest-lived spot group lasted for 200 days, between June and December 1943. Very small spots (pores) may have lifetimes of less than an hour.

The first suggestion of a solar cycle seems to have come from the

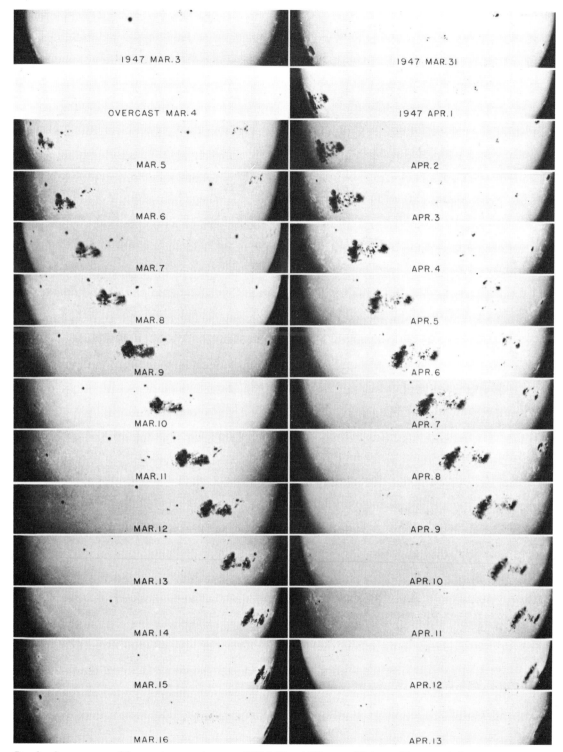

Day-by-day progress of the great sunspot group of 1947 (Mount Wilson and Palomar Observatories)

Danish astronomer Horrebow in 1775–6, but his work was not published until 1859, by which time the cycle had been definitely identified.

The 11-year solar cycle was dis-covered by H. Schwabe, a Dessau pharmacist, who began observing the Sun regularly in 1826 – mainly to see whether he could observe the transit of an intra-Mercurian planet. In 1851 his findings were popularized by Humboldt. A connection between solar activity and terrestrial magnetic phenomena was found by E. Sabine in 1852, and in 1870 E. Loomis, at Yale, established the link between the solar cycle and the frequency of auroræ.

The mean value of the length of the solar cycle since 1715 is 11·04 years. **The longest interval between successive maxima** has been 17·1 years (1788 to 1805); **the shortest interval** has been 7·3 years (1829·9 to 1837). Since 1715, when reasonably accurate records began, the **most energetic maximum** has been that of 1957·9. The **least energetic maximum** was that of 1816.

The 'Maunder Minimum' was discovered, from examination of old records, by the British astronomer E. W. Maunder in 1890. (It had been noted independently by F. G. W. Spörer.) He found that between 1645 and 1715 there were virtually no spots at all, so that the solar cycle was suspended; more recent research indicates that the corona may also have been virtually absent. It may be significant that there was freak weather in England during the 1680s, when the Thames froze regularly and 'frost fairs' were held on it. Auroræ, too were lacking; Halley noted that he saw his first aurora only in 1716, after forty years of watching. There may have been an earlier spotless period of the same type from 1400 to 1510, though the records are very incomplete. Further evidence of these prolonged minima comes from tree-ring studies, such as that carried out by F. Vercelli, who examined a tree which had lived between about 275 BC to AD 1914. Tree-rings are affected by events on the Sun; the solar cycle effects are well marked, and it is clear that conditions during the Maunder Minimum were decidedly abnormal.

Maxima

1718·2	1805·2	1894·1
1727·5	1816·4	1907·0
1738·7	1829·9	1917·6
1750·5	1837·2	1928·4
1761·5	1848·1	1937·4
1769·7	1860·1	1947·5
1778·4	1870·6	1957·9
1788·1	1883·9	1968·9
		1979·9

Minima

1723·5	1810·6	1901·7
1734·0	1823·3	1913·6
1745·0	1833·9	1923·6
1755·2	1843·5	1933·8
1766·5	1856·0	1944·2
1775·5	1867·2	1954·3
1784·7	1878·9	1964·7
1798·3	1889·6	1976·5
		1986·8

There is strong evidence for a longer cycle superimposed on the 11-year one.

The law relating to the latitudes of sunspots (Spörer's Law) was discovered by the German amateur F. G. W. Spörer in 1861. At the start of a new cycle after minimum, the first spots appear at latitudes between 30 and 45 degrees north or south. As the cycle progresses, spots appear closer to the equator, until at maximum the average latitude of the groups is only about 15 degrees north or south. After maximum the spots become less common, but the approach to the equator continues, reaching only about 7 degrees north or south. The spots of the old cycle then die out (before reaching the equator), but even before they have completely disappeared the first spots of the new cycle are seen at the higher latitudes.

The first 'butterfly diagram', showing the effects of Spörer's Law, was drawn by Maunder in 1904.

The Wolf or Zürich sunspot number for any given day, indicating the state of the Sun at that time, was worked out by R. Wolf of Zürich in 1852. The formula is: $R = k(10g + f)$, where R is the Wolf number, g is the number of groups seen, f is the total number of individual spots seen, and k is a constant depending on the equipment and site of the observer. (k is not usually far from unity.) The Wolf number may range from 0 for a clear disk up to over 200. A spot less than about 2500 km in diameter is officially known as a **pore**.

The magnetic fields associated with sunspots were discovered by G. E. Hale, in the United States, in 1908. This resulted from the Zeeman effect (discovered in 1896 by the Dutch physicist P. Zeeman), according to which the spectral lines of a light source are split into two components if the source is associated with a magnetic field. Hale also found that in a spot-group the leader and the follower are of opposite polarity – and that the conditions are the same over a complete hemisphere of the Sun, though reversed in the opposite hemisphere. At the end of each cycle the whole situation is reversed, and there are grounds for supposing that the true cycle is 22 years in length rather than 11. Thus in the present cycle (beginning in 1986)

the leader of a northern-hemisphere group is a south-polarity spot, the follower a north-polarity; in the southern hemisphere of the Sun, it is the leader which has north polarity. The magnetic fields of spots are very powerful, and may exceed 4000 gauss. **The strongest field observed** was with a group seen in 1967; the field was 5000 gauss, according to Steshenlio.

Plages are bright, active regions in the Sun's atmosphere, usually seen around sunspot groups; when observed in monochromatic light (usually hydrogen or calcium) they are termed **flocculi**. The brightest features of this type are seen in integrated light as **faculæ**.

The discovery of faculæ was made by C. Scheiner, probably about 1611. Faculæ (Latin, 'torches') are clouds of incandescent gases lying above the brilliant surface; they are composed largely of hydrogen, and are best seen near the limb, where the photosphere is less bright than at the centre of the disk (in fact, the limb has only two-thirds of the brilliance of the centre, because at the centre we are looking down more directly into the hotter material). Faculæ may last for over two months, though their average lifetime is about 15 days. Also, faculæ often appear in areas where a spot-group is about to appear, and persist after the group has disappeared.

Polar faculæ are different from those of the more central regions, and are much less easy to observe; they are most common near sunspot minimum, and have latitudes higher than 65 degrees, with lifetimes ranging from a few days to no more than 12 minutes. They may well be associated with coronal plumes.

Flares are violent, short-lived outbursts, usually occurring above active spot-groups. They emit charged particles as well as radiation, ranging from very short gamma-rays up to long-wavelength radio waves. They are most energetic in the X-ray and EUV (extreme ultra-violet) regions of the electromagnetic spectrum. They produce shock-waves in the corona and chromosphere, and considerable quantities of plasma may be expelled from the Sun. A typical flare lasts for around 20

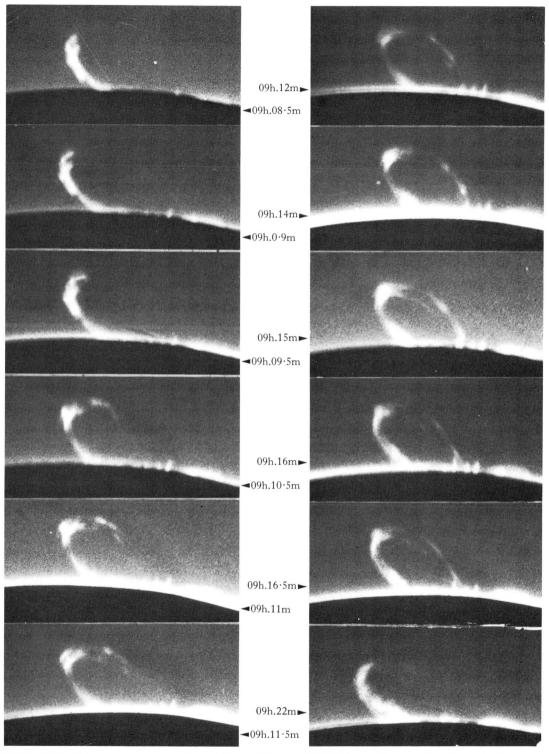

09h.12m▶

◀09h.08·5m

09h.14m▶

◀09h.0·9m

09h.15m▶

◀09h.09·5m

09h.16m▶

◀09h.10·5m

09h.16·5m▶

◀09h.11m

09h.22m▶

◀09h.11·5m

Development of a loop prominence on the east limb on 29 August 1970 (R. Lane, Shiremoor Solar Observatory)

minutes, though extreme cases may extend over a few hours, and they are phenomena of the chromosphere and lower corona.

Flares studied in the light of H (hydrogen) are classified according to their area. The classification is as follows:

Area square degrees	Classification
Over 24·7	4
12·5–24·7	3
5·2–12·4	2
2·0–5·1	1
Less than 2	s

F = faint, N = normal, B = bright. Thus the most important flares are classified as 4B.

Flares are thought to be due to the sudden release of magnetic energy which has been built up in the magnetic fields of complicated active regions, but it is true to say that their mechanism is not yet fully understood. The **first flare** to be observed was that of 1 September 1859, by R. Carrington, in white light, but very few have since been seen except with spectroscopic equipment or the equivalent.

The modern sunspot theory was proposed by H. Babcock in 1961. It may be assumed that the solar magnetic lines of force run from one magnetic pole to the other below the bright surface. The differential rotation means that the lines are distorted and drawn out into loops. Over a period of years, the lines are coiled right round the Sun and are bunched in mid-latitudes, creating knots. Eventually a loop of magnetic energy erupts through the surface, steadying and cooling it, and producing the two spots characteristic of a group – with, predictably, opposite polarities. After about 11 years, the knots have become so complex that they break. The Sun 'snaps back' to its original state, but 'overshoots', so that the polarities in the two hemispheres are reversed.

Every spot-group has its own characteristics, but in general an 'average' two-spot group begins as two tiny pores at the limit of visibility. The pores develop into proper spots, growing and separating in longitude. Within two weeks the group has reached its maximum length, with a fairly regular leading spot together with a less regular

follower of opposite polarity – and, of course, various minor spots and clusters. The darkest part of a spot – the umbra – still has a temperature of about 4500 °C, while the surrounding penumbra is at about 5000 °C; this means that a spot is by no means black, and if it could be seen shining on its own the surface brilliancy would be greater than that of an arc-lamp. After the group has reached its peak a slower decline sets in. The leader is generally the last survivor. Roughly 75 per cent of groups fit into this general pattern, but of the remainder some do not conform, and there are also frequent single spots.

Even in non-spot zones, the solar surface is not calm. The brilliant surface or photosphere has a granular structure; each granule is about 1000 km in diameter (1″·3) and has a life of about 8 minutes. It is estimated that the whole surface includes about 4 000 000 granules at any one time; they represent upcurrents, and the general situation has been compared with 'boiling' of a liquid, though the photosphere is of course entirely gaseous. The granular structure is easy to observe, though the first really good pictures of it were obtained from a balloon (*Stratoscope II*). Rising from the surface are **spicules**, not visible in integrated light; the average size of a spicule is 1000 km, and a typical height is 7000 km. The average lifetime of a spicule is no more than 5 minutes.

SPECTRUM AND COMPOSITION OF THE SUN

The first solar spectrum was obtained by Isaac Newton in 1666, but he never took his investigations much further, though he did of course prove the complex nature of sunlight.

The discovery of dark lines in the solar spectrum was made in England by W. H. Wollaston, in 1802. However, Wollaston merely took the lines for the boundaries between different colours of the rainbow spectrum.

The first systematic studies of the dark lines were carried out in Germany by J. von Fraunhofer, from 1814. Fraunhofer realized that the lines were permanent; he recorded 574 of them, and mapped 324.

The first explanation of the dark 'Fraunhofer lines' was given by G. Kirchhoff in 1859 (initially working with R. Bunsen). Kirchhoff found that the photosphere yields a rainbow or continuous spectrum; the overlying gases produce a line spectrum, but since these lines are seen against the rainbow background they appear dark instead of bright. Since their positions and intensities are not affected, each line may be tracked down to a particular element or group of elements. In 1861–2 Kirchhoff produced **the first detailed map of the solar spectrum**. (His eyesight was affected, and the work was actually finished by his assistant, K. Hofmann.) In 1869 A. Ångström, the Swedish physicist, studied the solar spectrum by using a grating instead of a prism, and in 1889 Rowland produced a detailed photographic map of the spectrum.

The most prominent Fraunhofer lines in the visible spectrum are:

Letter	Wavelength Ångströms	Identification	Letter	Wavelength Ångströms	Identification
A	7593	O_2			
a	7183	H_2O			
B	6867	O_2			

(These three are telluric lines – due to the Earth's intervening atmosphere.)

Letter	Wavelength Ångströms	Identification	Letter	Wavelength Ångströms	Identification
C (Hα)	6563	H	b_4	5167	Mg
D_1	5896		F (Hβ)	4861	H
D_2	5890	Na	f (Hγ)	4340	H
E	5270	Ca, Fe	G	4308	Fe, Ti
E	5269	Fe	g	4227	Ca
b_1	5183	Mg	h (Hδ)	4102	H
b_2	5173	Mg	H	3967	Ca^{II}
b_3	5169	Fe	K	3933	

(One Ångström is equal to one hundred-millionth part of a centimetre; it is named in honour of Anders Ångström. The diameter of a human hair is roughly 500 000 Å.)

By now many of the chemical elements have been identified in the Sun. The list of elements which have and have not been identified is as follows:

THE CHEMICAL ELEMENTS, AND THEIR OCCURRENCE IN THE SUN

The following is a list of elements 1 to 103
* = detected in the Sun.
R = included in H. A. Rowland's list published in 1891.

Atomic No.	Name	Atomic Weight	Occurrence in the Sun
1 H	Hydrogen	1·008	* R
2 He	Helium	4·003	*
3 Li	Lithium	6·939	* (in sunspots)
4 Be	Beryllium	9·013	* R

Atomic No.	Name	Atomic Weight	Occurrence in the Sun
5 B	Boron	10·812	* (in compound)
6 C	Carbon	12·012	* R
7 N	Nitrogen	14·007	*
8 O	Oxygen	16·000	*
9 F	Fluorine	18·999	* (in compound)
10 Ne	Neon	20·184	*
11 Na	Sodium	22·991	* R
12 Mg	Magnesium	24·313	* R
13 Al	Aluminium	26·982	* R
14 Si	Silicon	28·090	* R
15 P	Phosphorus	30·975	*
16 S	Sulphur	32·066	*
17 Cl	Chlorine	35·434	
18 A	Argon	39·949	* (in corona)
19 K	Potassium	39·103	* R
20 Ca	Calcium	40·080	* R
21 Sc	Scandium	44·958	* R
22 Ti	Titanium	47·900	* R
23 V	Vanadium	50·944	* R
24 Cr	Chromium	52·00	* R
25 Mn	Manganese	52·94	* R
26 Fe	Iron	55·85	* R
27 Co	Cobalt	58·94	* R
28 Ni	Nickel	58·71	* R
29 Cu	Copper	63·55	* R
30 Zn	Zinc	65·37	* R
31 Ga	Gallium	69·72	*
32 Ge	Germanium	72·60	* R
33 As	Arsenic	74·92	
34 Se	Selenium	78·96	
35 Br	Bromine	79·91	
36 Kr	Krypton	83·80	
37 Rb	Rubidium	85·48	* (in spots)
38 Sr	Strontium	87·63	* R
39 Y	Yttrium	88·91	* R
40 Zr	Zirconium	91·22	* R
41 Nb	Niobium	92·91	* R
42 Mo	Molybdenum	95·95	* R
43 Tc	Technetium	99	
44 Ru	Ruthenium	101·07	*
45 Rh	Rhodium	102·91	* R
46 Pd	Palladium	106·5	* R
47 Ag	Silver	107·87	* R
48 Cd	Cadmium	112·41	* R
49 In	Indium	114·82	* (in spots)
50 Sn	Tin	118·70	* R
51 Sb	Antimony	121·76	*
52 Te	Tellurium	127·61	
53 I	Iodine	126·91	
54 Xe	Xenon	131·30	
55 Cs	Cæsium	132·91	
56 Ba	Barium	137·35	* R
57 La	Lanthanum	138·92	* R
58 Ce	Cerium	140·13	* R
59 Pr	Praseodymium	140·91	*
60 Nd	Neodymium	144·25	* R
61 Pm	Promethium	147	
62 Sm	Samarium	150·36	*
63 Eu	Europium	151·96	*
64 Gd	Gadolinium	157·25	*
65 Tb	Terbium	158·93	*
66 Dy	Dysprosium	162·50	*
67 Ho	Holmium	164·94	
68 Er	Erbium	167·27	* R
69 Tm	Thulium	168·94	*
70 Yb	Ytterbium	173·04	*
71 Lu	Lutecium	174·98	*
72 Hf	Hafnium	178·50	*
73 Ta	Tantalum	180·96	*
74 W	Tungsten	183·86	*
75 Re	Rhenium	186·3	
76 Os	Osmium	190·2	*
77 Ir	Iridium	192·2	*
78 Pt	Platinum	195·1	*

Atomic No.	Name	Atomic Weight	Occurrence in the Sun
79 Au	Gold	197·0	*
80 Hg	Mercury	200·6	
81 Tl	Thallium	204·4	
82 Pb	Lead	207·2	* R
83 Bi	Bismuth	209·0	
84 Po	Polonium	210	
85 At	Astatine	211	
86 Rn	Radon	222	
87 Fr	Francium	223	
88 Ra	Radium	226	
89 Ac	Actinium	227	
90 Th	Thorium	232	*
91 Pa	Protoactinium	231	
92 U	Uranium	238	

The remaining elements are 'transuranic' and radioactive, and have not been detected in the Sun. They are:

Atomic No.	Name	Atomic Weight	Occurrence in the Sun
93 Np	Neptunium	237	
94 Pu	Plutonium	239	
95 Am	Americium	241	
96 Cm	Curium	242	
97 Bk	Berkelium	243	
98 Cf	Californium	244	
99 Es	Einsteinium	253	
100 Fm	Fermium	254	
101 Md	Mendelevium	254	
102 No	Nobelium	254	
103 Lw	Lawrencium	257	
104 Rf	Rutherfordium	–	
105 Ha	Hahnium	–	

For elements 43, 61, 85–89, 91, 93–103 the mass number is that of the most stable isotope.

The fact that the remaining elements have not been identified in the Sun does not necessarily indicate that they are completely absent. They may be present, though no doubt in very small amounts.

So far as relative mass is concerned, the most abundant element by far is hydrogen (71 per cent). It is followed by helium (27 per cent). All the others combined make up only 2 per cent.

Helium was identified in the Sun (by Sir Norman Lockyer, in 1868) before being found on Earth. Lockyer named it after the Greek ἥλιος, the Sun. It was detected on Earth in 1894, by Sir William Ramsay, as a gas occluded in cleveite. For a time it was believed that the corona contained another element unknown on Earth, and it was even given a name – Coronium – but the lines, described initially by Harkness and Young at the eclipse of 1869, proved to be due to elements already known. The brilliant photosphere is surprisingly thin – only about 300 km thick. **The limb darkening was first explained** by K. Schwarzschild in 1906; as already noted, it is due to the fact that when we look at the centre of the disk we are seeing into deeper and hotter layers. Old ideas about the Sun sound strange today. Sir William Herschel believed that below the bright surface there was a cool layer, which could well be inhabited; up to the time of his death, in 1822, he maintained this view. As recently as 1952 a German lawyer, Godfried Büren, stated that the Sun had a vegetation-covered inner globe, and offered a prize of 25 000 marks to anyone who could prove him wrong. The leading German astronomical society took up the challenge, and won a court case. (Whether the prize was actually paid does not seem to be on record!)

The spectroheliograph, enabling the Sun to be photographed in the light of one element only, was invented by G. E. Hale in 1892. The visual equivalent, the **spectrohelioscope**, was invented in 1923, also by Hale. In 1933 B. Lyot, of France, developed the **Lyot filter**, which is less versatile but more convenient, and also allows the Sun to be studied in the light of one element only.

The meteoritic theory of solar energy was discussed by J. R. Mayer in 1848. Mayer found that a globe of hot gas the size of the Sun would cool down in 5000 years or so if there were no other energy source, while a Sun made of coal, and burning furiously enough to produce as much heat as the Sun actually does, would last for only 4600 years. He therefore assumed that the energy was produced by meteorites striking the solar surface.

The contraction theory was proposed in 1834 by H. von Helmholtz. He calculated that if the Sun contracted by 60 m per year, the energy produced would suffice for 15 000 000 years. This theory was supported later by the great British physicist Lord Kelvin. However, it had to be abandoned when astronomers concluded that the Sun is certainly at least as old as the Earth (about 4700 million years) and probably older.

The nuclear transformation theory was worked out by H. Bethe in 1938, during a train journey from Washington to Cornell University. Hydrogen is being converted to helium, so that energy is released and mass is lost: the decrease in mass amounts to

4 000 000 tonnes per second. Bethe assumed that carbon and nitrogen were used as catalysts, but C. Critchfield, also in America, subsequently showed that in solar-type stars the proton-proton reaction is dominant. Eventually the Sun will enter the red giant stage, with a probable diameter of about 300 000 000 km, before collapsing into a white dwarf. Fortunately, no dramatic changes in the Sun are likely for at least 5000 million years in the future, though slight variations may occur, and some authorities maintain that it is these minor changes which have produced the Ice Ages which have affected the Earth now and then throughout its history.

It now seems that the solar core, in which thermonuclear reactions are taking place, extends out to 0·25 of the solar radius from the centre of the globe. Out to 0·83 of the radius, energy is transported outward by radiative diffusion, while in the outer layers it is convection which is the transporting agency.

The first indications of a solar oscillation were obtained in 1960. The period was found to be 5 minutes, and was thought to be a surface ripple affecting not more than the outermost 10 000 km of the Sun's globe. **Solar vibrations were discovered in 1973** by R. H. Dicke, who was attempting to make measurements of the polar and equatorial diameters of the Sun to see whether there was any observable flattening. Dicke found that the Sun is 'quivering like a jelly', so that the equator bulges as the poles are flattened, but the maximum amplitude is only about 5 km and the velocity about 10 m/s, so that the observations are immensely difficult and delicate. There are also some indications that the diameter of the Sun is not absolutely constant, though any changes at the present epoch must be very minor and probably periodical.

There are other, longer-term oscillations; one of 50 minutes, one of 2 hours 40 minutes discovered in 1976 by a Russian team together with a team from Birmingham, and so on. These may represent global pulsations of the Sun as a whole, and by now the study of what has been called 'solar seismology'

has become of immense importance.

Solar neutrinos are presenting astronomers with many problems. Neutrinos are particles with no mass and no electric charge, so that they are extremely difficult to detect. Theoretical considerations indicate that the Sun should emit quantities of them, and in 1966 efforts to detect them were begun by a team from the Brookhaven National Laboratory in the USA, led by R. Davis. The 'telescope' is located in the Homestake Gold Mine in South Dakota, inside a deep mine-shaft, and consists of a tank of 454 600 litres of cleaning fluid (tetrachloroethylene). Only neutrinos can penetrate so far below ground level, and on the rare occasions when a chlorine atom happened to be struck by a neutrino, radioactive argon would be produced; this could therefore provide a key to the number of neutrinos. However, the observed flux was much smaller than that which had been predicted, and the same has been found by a team from the USSR, using 100 tonnes of liquid scintillator and 144 photodetectors in a mine in the Donetsk Basin; these experiments were started in 1978. It is true that these experiments could detect only certain types of neutrinos, and other types of detectors are being planned; the Kamiokande water Čerenkov detector in Japan is one – it uses sensitive light detectors on the walls of a water tank holding some 3000 tonnes of water.

Why the neutrino flux is so low is not clear. It has been suggested that the core temperature of the Sun may be rather lower than is generally believed, but this would raise other theoretical problems, and at present (1988) the 'neutrino problem' remains unsolved.

The chromosphere of the Sun lies above the photosphere. The temperature increases from 4200 °C in the low chromosphere to 8000 °C at an altitude of 1500 km, and then increases rapidly until the chromosphere merges with the corona. The dark lines in the solar spectrum are produced in the upper photosphere and lower chromosphere, often called the **reversing layer**. The chromosphere is 2000 to 10 000 km deep.

The corona, visible with the naked eye only during a total solar eclipse, lies above the chromosphere. The mean

temperature is nearly 2 000 000 °C, but the density is so low – less than one million millionth of the Earth's air at sea-level – that there is very little 'heat'. (The density of the Earth's air at sea-level is about 10^{19} particles per cubic centimetre; of the Sun's photosphere, 10^{17}; of the corona, only 10^5 particles per cubic centimetre.) The corona sends out only one-millionth as much light as the photosphere. It was once believed that the high temperature was due to sound waves, produced by turbulence in the photosphere, but it now seems more likely that magnetic phenomena are responsible. Areas in the corona where the temperatures and pressures are much lower than average are known as **coronal holes**. The corona has no definite boundary, but simply thins out until its density has become no greater than that of the interplanetary medium.

The first coronagraph was built by B. Lyot in 1930, and tested at the Pic du Midi Observatory (altitude 2870 m). It depends upon producing an 'artificial eclipse' instrumentally. With it, Lyot was able to study the inner corona and its spectrum.

Solar Wind is the radial, continuous outflow of charged particles from the Sun. It was first predicted in 1951 by Biermann, following a study of comets' tails, though indications go back to a 1900 paper by Sir Oliver Lodge. The solar wind, consisting of 'plasma', is made up of protons, helium nuclei (alpha-particles) and electrons; the velocity away from the Sun ranges between 200 and 900 km/s, with the average velocity past the Earth given as about 600 km/s. Low-velocity streams come from the loops in the corona, and high-velocity streams from coronal holes. The solar wind affects the Earth's magnetosphere, and during periods of great activity on the Sun, when the wind is enhanced, the particles overload the Van Allen zones surrounding the Earth; particles cascade down into the lower part of the atmosphere, producing auroræ. The region in space where the solar wind ceases to be appreciable is known as the **heliopause**. It is hoped to track some current space-vehicles (Pioneers 10 and 11, and Voyagers 1 and 2) until they

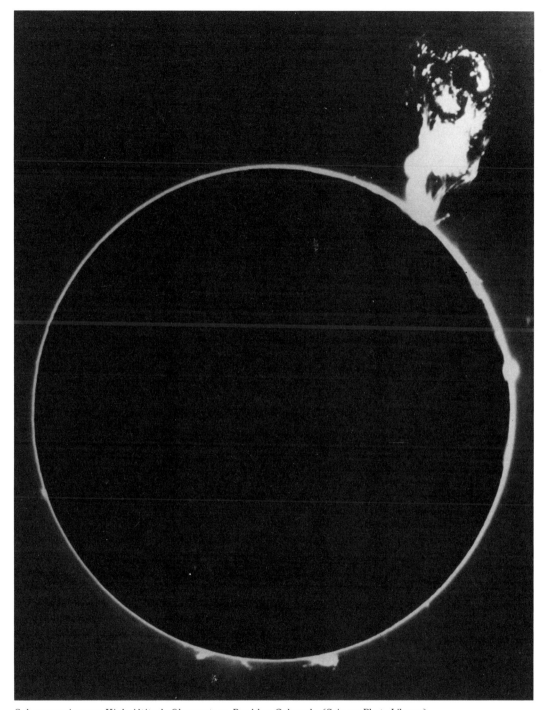

Solar prominence; High-Altitude Observatory, Boulder, Colorado. (Science Photo Library)

reach the heliopause.

Cosmic rays from the Sun were detected by Forbush in 1942. In 1954 Forbush established that cosmic-ray intensity decreases when solar activity increases (Forbush effect).

Prominences were first described in detail by the Swedish observer Vas-senius at the total eclipse of 1733, though he believed that they belonged to the Moon rather than the Sun. (They may have been seen earlier – by Stannyan in 1706, from Berne.) It was only after the eclipse of 1842 that astronomers became certain that they were solar rather than lunar. They were formerly termed 'red flames', but are in fact regions of hot hydrogen gas, reddish in colour. **Quiescent** prominences may persist for months; **eruptive** prominences show violent motions, and may attain several hundred thousand km.

Prominences are visible with the naked eye only during a total eclipse.

However, following the eclipse of 19 August 1868, J. Janssen (France) and Sir Norman Lockyer (England) developed the method of observing them spectroscopically without an eclipse. By observing at hydrogen wavelengths, prominences may be seen against the bright disk of the Sun as dark filaments, sometimes termed flocculi. (Bright flocculi are due to calcium.)

The discovery of ultra-violet radiation from the Sun was made in 1801 by J. Ritter, using a prism to produce a solar spectrum and noting that paper soaked in NaCl was darkened if held in a region beyond the violet end of the visible spectrum.

The discovery of radio emission from the Sun was due to J. S. Hey and his team in 1942 (27–8 February). Initially, the effect was thought to be due to German jamming of the radar transmitters! Various types of emission are now known, and there are 'bursts', some of which are associated with flares. In June every year the 'radio sun' occults the Crab Nebula, itself a radio source, and the phenomenon yields valuable information about the corona. The first observations of this kind were carried out in June 1952.

The first radar contact with the Sun was made in 1959, by Eshleman and his colleagues at the Stanford Research Institute USA.

Sunrise is defined as the moment when the Sun's upper limb appears above the horizon. As refraction reduces the apparent zenith distance of the Sun by 34 minutes of arc at the horizon, and as the Sun's semi-diameter is about 16′, the moment of sunrise is defined as being the instant when the Sun has a zenith distance of 90 degrees 50 minutes. **Sunset** is the moment when the upper limb of the Sun disappears below the horizon: the zenith distance is again 90 degrees 50 minutes. Because of refraction effects, it has sometimes been possible to see the Sun and the full Moon visible simultaneously above opposite horizons.

The apex of the Sun's way (that is to say, the point in the sky toward which the Sun is at present moving) is in Hercules, at R.A. 18 h, declination 34 °N. The **antapex** is in Columba (R.A. 6 h, declination 34 °S).

ECLIPSES OF THE SUN

An eclipse of the Sun occurs when the Moon passes in front of the Sun; strictly speaking, the phenomenon is an occultation of the Sun by the Moon. Solar eclipses may be total (when the whole of the photosphere is hidden), partial, or annular (when the Moon's apparent diameter is smaller than that of the Sun, so that a ring of light is left round the lunar disk; Latin **annulus**, a ring.)

The greatest number of eclipses possible in one year is 7; thus in 1935 there were 5 solar and 2 lunar eclipses, and in 1982 there were 4 solar and 3 lunar.

The least number of eclipses possible in one year is 2, both of which must be solar, as in 1984.

The longest possible duration of totality for a solar eclipse is 7 m 31 s. This has never been observed, but at the eclipse of 20 June 1955 totality over the Philippine Islands lasted for 7 m 8 s.

The shortest possible duration of totality may be a fraction of a second. This happened at the eclipse of 3 October 1986, which was annular along most of the central track, but was total for about a tenth of a second over a restricted area in the North Atlantic Ocean.

The longest possible duration of the annular phase of an eclipse is 12 m 24 s.

The longest totality ever observed was during the eclipse of 30 June 1973. A Concorde aircraft, specially equipped for the purpose, flew underneath the Moon's shadow and kept pace with it, so that the scientists on board (including the British astronomer John Beckman) saw a totality lasting for 72 minutes! They were carrying out observations at millimetre wavelengths, and at their height of 55 000 feet were above most of the water vapour in our atmosphere, which normally hampers such observations. They were also able to see definite changes in the corona and prominences over the full period. The Moon's shadow moves over the Earth at over 3000 km/h.

The widest track for totality across the Earth's surface is 272 km. In most total eclipses, of course, the width is considerably less than this; and a partial eclipse is seen to either side of the track of totality.

The first eclipse predictions were made by studies of the Saros period. This is the period after which the Sun, Moon and node arrive back at almost the same relative positions; it amounts to 6585·321 solar days, or approximately 18 years 11 days. Therefore, an eclipse tends to be followed by another eclipse in the same Saros series 18 years 11 days later, though conditions are not identical, and the Saros is at best a reasonable guide. **The first known prediction** was made by the Greek philosopher Thales, who forecast

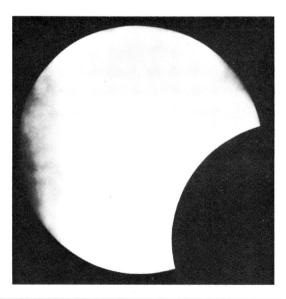

Partial eclipse of the Sun on 29 April 1976 at 10 h 35 m UT
(K. Kennedy)

the eclipse of 25 May 585 BC. This occurred near sunset in the Mediterranean area, and is said to have put an end to a battle between the forces of King Alyattes of the Lydians and King Cyaxares of the Medes; the combatants were so alarmed by the sudden darkness that they concluded a hasty peace. One Saros series lasts for 1150 years; it includes 64 eclipses, of which 43 or 44 are total, while the rest are partial eclipses seen from the polar zones of the Earth.

The first recorded solar eclipse seems to have been that of 2136 BC (22 October), seen in China during the reign of the Emperor Chung K'ang. The next observed eclipse which may be dated with any certainty is that of 1375 BC, described on a clay tablet found at Ugarit (Syria).

The first solar eclipse recorded in Britain was that of 15 February 538, described in the Anglo-Saxon Chronicle; it occurred four years after the death of Cerdic, first king of the West Saxons. The Sun was two-thirds eclipsed in London. English total eclipses over the past thousand years have been those of 1140, 1715, 1724 and 1927. The eclipse of 3 May 1715 was observed by Halley from London and by Flamsteed from Greenwich; that of 22 May 1724 was observed only by Dr. Stukeley from Haradon Hill, near Salisbury, because of generally cloudy conditions. Totality was, however, seen from France. The eclipse of 29 June 1927 was total over part of north England; the conditions were generally poor, and the best photographs were obtained from Giggleswick – totality lasted for less than half a minute. The eclipse of 30 June 1954 was total over parts of Norway and Sweden; the extreme southern edge of the track just brushed the Shetland Isles. The next English total eclipse will be that of 11 August 1999, when the track will cross Cornwall; it will be followed by the total eclipses of 14 June 2051 (total in London), 23 September 2090 and 7 October 2135. No total eclipse was seen from London between 875 and 1715, though those of 968 and 1140 were nearly total there. The 1927 eclipse was 95 per cent total in London.

There were total eclipses over parts of Scotland in 1133, and on 17 June 1433, when the length of totality reached 4·5 minutes at Inverness and caused what became known as 'Black Hour'. The eclipses of 1598 and 1652 were also total over parts of Scotland, and that of 1699 nearly so.

The maximum theoretical length for a British total eclipse is 5·5 minutes. That of 15 June 885 lasted for almost 5 minutes, and the same will be true for the Scottish total eclipse of 20 July 2381.

The first total eclipse recorded in the United States was that of 24 June 1778, when the track passed from Lower California to New England. **The first official American total eclipse expedition** was that of 21 October 1790, when a party went to Penobscot, Maine; it was led by S. Williams of Harvard, and it was given 'free passage' by the British forces, but unfortunately a mistake in the calculations meant that the party remained outside the track of totality! **The first American eclipse expedition to Europe** was that of 28 July 1851, when G. P. Bond led a team to Scandinavia.

The only emperor to have died of fright because of an eclipse was Louis of Bavaria, in 840 (his three sons then proceeded to indulge in a ruinous war over the succession).

The only astronomer to have escaped from a besieged city in a balloon to study a total eclipse was Jules Janssen. The eclipse was that of 22 December 1870, and Janssen flew out from Paris, which was surrounded by the German forces. He arrived safely at Oran, but clouds prevented him from making any observations.

The first mention of the corona may have been due to Plutarch, who lived from about AD 46 to 120. Plutarch's book 'On the Face in the Orb of the Moon' contains a reference to 'a certain splendour' round the eclipsed Sun which could well have been the corona. The corona was definitely recorded from Corfu during the eclipse of 22 December 968. The astronomer Clavius saw it at the eclipse of 9 April 1567, but regarded it as merely the uncovered edge of the Sun; Kepler showed that this could not be so, and attributed it to a lunar atmosphere. After observ-

ing the eclipse of 16 June 1806 from Kindehook, New York, the Spanish astronomer Don José Joaquin de Ferrer pointed out that if the corona were due to a lunar atmosphere, then the height of this atmosphere would have to be 50 times greater than that of the Earth, which was clearly unreasonable. However, it was only after careful studies of the eclipses of 1842 and 1851 that the corona and the prominences were shown unmistakably to belong to the Sun rather than to the Moon.

There is some evidence that during eclipses which occurred during the 'Maunder Minimum' (1645–1715) the corona was absent, though it is impossible to be sure. Certainly the shape of the corona at spot-maximum is more symmetrical than at spot-minimum, when there are 'polar streamers'. This was first recognized after studies of the eclipses of 1871 and 1872.

The first observation of Baily's Beads seems to have been made by Halley during the eclipse of 1715. They were also seen by Maclaurin, from Edinburgh, at the annular eclipse of 1 March 1737. They were described in detail by the English astronomer Francis Baily (after whom they are named) in 1836 during the annular eclipse of 15 May. The brilliant bead-like effect is caused by the Sun's rays shining through valleys on the lunar limb immediately before and immediately after totality. They were **first photographed** at the eclipse of 7 August 1869 by C. F. Hines and members of the Philadelphia Photographic Corps, observing from Ottumwa, Iowa.

The first observation of the shadow-bands was made by H. Goldschmidt in 1820. These are wavy lines seen across terrestrial features before and after totality; they are, of course, produced in the Earth's atmosphere.

The first attempt to photograph a total eclipse was made by the Austrian astronomer Majocci on 8 July 1842. He failed to record totality, though he did succeed in photographing the partial phase.

The first successful photograph of a total eclipse, showing the corona and prominences, was taken by Berkowski on 28 July 1851, using the 6·25 in Königsberg heliometer and giving an exposure time of 24 s.

The first photograph of the flash spectrum was taken by the American astronomer C. Young on 22 December 1870. (The flash spectrum is the sudden, brief change in the Fraunhofer lines from dark to bright, when the Moon blots out the photosphere in the background, and the chromosphere is left shining 'on its own'). The flash spectrum was first observed during an annular eclipse by Pogson in 1872.

The only cricket match to have been interrupted by an eclipse of the Sun was the Jubilee Test between India and England in February 1980. A solar eclipse was due that afternoon, and the Indian Board, in agreement with the English team, did not want the responsibility of a crowd of 50 000 damaging their eyes by looking at the Sun when the eclipse began. So that day was declared a 'rest day' – and the Test Match continued on the next morning!

The first attempt to show a total eclipse on television from several stations along the track was made by the BBC at the eclipse of 15 February 1961. The attempt was successful, and totality was shown successively from stations in France, Italy and Yugoslavia.

SOLAR ECLIPSES, 1923–1977

T=total, P=partial, A=annular

Date	Type	Area
1923 March 16	A	S. Africa
1923 Sept. 10	T	California, Mexico
1924 March 5	P	S. Africa
1924 July 31	P	Antarctic
1924 Aug. 29	P	Iceland, N. Russia, Japan
1925 Jan. 24	T	North-eastern USA
1925 July 20/1	A	New Zealand, Australia
1926 Jan. 14	T	E. Africa, Indian Ocean, Borneo
1926 July 9/10	A	Pacific
1927 Jan. 3	A	New Zealand, S. America
1927 June 29	T	England, Scandinavia
1927 Dec. 24	P	Polar zone
1928 May 19	T	S. Atlantic
1928 June 17	P	N. Siberia
1928 Nov. 12	P	England to India
1929 May 9	T	Indian Ocean, Philippines
1929 Nov. 1	A	Newfoundland, C. Africa, Indian Ocean
1930 April 28	T	Pacific
1930 Oct. 21/2	T	S. Pacific to S. America
1931 April 17/18	P	Arctic
1931 Sept. 12	P	Alaska, N. Pacific
1931 Oct. 11	P	S. America, S. Pacific, Antarctic
1932 March 7	A	Antarctic
1932 Aug. 31	T	USA
1933 Feb. 24	A	S. America, C. Africa
1933 Aug. 21	A	Iran, India, N. Australia
1934 Feb. 13/14	P	Pacific
1934 Aug. 10	A	S. Africa
1935 Jan. 5	P	No land surface
1935 Feb. 3	P	N. America
1935 June 30	P	Britain
1935 July 30	P	No land surface
1935 Dec. 25	A	New Zealand, south S. America
1936 June 19	T	Greece, Turkey, Siberia, Japan
1936 Dec. 13/14	A	Australia, New Zealand
1937 June 8	T	Pacific, Chile
1937 Dec. 2/3	A	Pacific
1938 May 29	T	S. Atlantic
1938 Nov. 21/2	P	E. Asia, Pacific coast of N. America
1939 April 19	A	Alaska, Arctic
1939 Oct. 12	T	Antarctic
1940 April 7	A	USA, Pacific
1940 Oct. 1	T	Brazil, S. Atlantic, S. Africa
1941 March 27	A	S. Pacific, S. America
1941 Sept. 21	T	China, Pacific
1942 March 16/17	P	S. Pacific, Antarctic
1942 Aug. 12	P	Invisible in Britain
1942 Sept. 10	P	Britain
1943 Feb. 4/5	T	Japan, Alaska
1943 Aug. 1	A	Pacific
1944 Jan. 25	T	Brazil, Atlantic, Sudan
1944 July 20	A	India, New Guinea
1945 Jan. 14	A	Australia, New Zealand
1945 July 9	T	Canada, Greenland, N. Europe
1946 Jan. 3	P	Invisible in Britain
1946 May 30	P	S. Pacific
1946 June 29	P	Arctic
1946 Nov. 23	P	N. America
1947 May 20	T	Pacific, Equatorial Africa, Kenya
1947 Nov. 12	A	Pacific
1948 May 8/9	A	E. Asia
1948 Nov. 1	T	Kenya, Pacific
1949 April 28	P	Britain
1949 Oct. 21	P	New Zealand, Australia
1950 March 18	A	S. Atlantic
1950 Sept. 12	T	Siberia, N. Pacific
1951 March 7	A	Pacific
1951 Sept. 1	A	Eastern USA, C. and S. Africa
1952 Feb. 25	T	Africa, Arabia, Russia
1952 Aug. 20	A	S. America
1953 Feb. 13/14	P	E. Asia
1953 July 11	P	Arctic
1953 Aug. 9	P	Pacific
1954 Jan. 5	A	Antarctic
1954 June 30	T	Iceland, Norway, Sweden, Russia, India
1954 Dec. 25	A	S. Africa, S. Indian Ocean
1955 June 20	T	S. Asia, Pacific, Philippines
1955 Dec. 14	A	Sudan, Indian Ocean, China
1956 June 8	T	S. Pacific
1956 Dec. 2	P	Europe, Asia
1957 April 29/30	A	Arctic
1957 Oct. 23	T	Antarctica
1958 April 19	A	Indian Ocean, Pacific
1958 Oct. 12	T	Pacific
1959 April 8	A	S. Indian Ocean, Pacific
1959 Oct. 2	T	N. Atlantic, N. Africa
1960 March 27	P	Australia, Antarctica
1960 Sept. 20/1	P	N. America, E. Siberia
1961 Feb. 15	T	France, Italy, Greece, Yugoslavia, Russia
1961 Aug. 11	A	S. Atlantic, Antarctica
1962 Feb. 4/5	T	Pacific
1962 July 31	A	S. America, C. Africa
1963 Jan. 25	A	Pacific, S. Africa
1963 July 20	T	Japan, north N. America, Pacific
1964 Jan. 14	P	Tasmania, Antarctica
1964 July 9	P	N. Canada, Arctic
1964 Dec. 3/4	P	N.E. Asia, Alaska, Pacific
1965 May 30	T	Pacific, New Zealand, Peruvian coast
1965 Nov. 23	A	Russia, Tibet, E. Indies
1966 May 20	A	Greece, Russia
1966 Nov. 12	T	S. America, Atlantic
1967 May 9	P	N. America, Iceland, Scandinavia
1967 Nov. 2	T	S. Atlantic
1968 March 28/9	P	Pacific, Antarctica
1968 Sept. 22	T	Arctic, Mongolia, Siberia
1969 March 18	A	Indian Ocean, Pacific
1969 Sept. 11	A	Peru, Bolivia
1970 March 7	T	Mexico, USA, Canada
1970 Aug. 3/Sept. 1	T	East Indies, Pacific
1971 Feb. 25	P	Europe, N.W. Africa
1971 July 22	P	Alaska, Arctic
1971 Aug. 20	P	Australasia, S. Pacific
1972 Jan. 16	A	Antarctica
1972 July 10	T	Alaska, Canada
1973 Jan. 4	A	Pacific, S. Atlantic
1973 June 30	T	Atlantic, N. Africa, Kenya, Indian Ocean
1973 Dec. 24	A	Brazil, Atlantic, N. Africa
1974 June 20	T	Indian Ocean
1974 Dec. 13	P	N. and C. America
1975 May 11	P	Europe, N. Asia, Arctic
1975 Nov. 3	P	Antarctic
1976 April 29	A	N.W. Africa, Turkey, China
1976 Oct. 23	T	Tanzania, Indian Ocean, Australia

SOLAR ECLIPSES 1977–1999

Date	Area	Type	Notes
1977 April 18	Atlantic, S.W. Africa, Indian Ocean	Annular	
1977 October 12	Pacific, Peru, Brazil	Total	Max. length 2 m 37 s
1978 April 7	Antarctic	Partial	79% eclipsed
1978 October 2	Arctic	Partial	69% eclipsed
1979 February 26	Pacific, USA, Canada, Greenland	Total	Max. length 2 m 48 s
1979 August 22	Pacific, Antarctica	Annular	
1980 February 16	Atlantic, Kenya, India, China	Total	Max. length 4 m 8 s
1980 August 10	S. Pacific, Brazil	Annular	
1981 February 4	Pacific, S. Australia and New Zealand	Annular	
1981 July 31	Russia, N. Pacific	Total	Max. length 2 m 3 s
1982 January 25	Antarctic	Partial	57% eclipsed
1982 June 21	Antarctic	Partial	62% eclipsed
1982 July 20	Arctic	Partial	46% eclipsed
1982 December 15	Arctic	Partial	74% eclipsed
1983 June 11	Indian Ocean, E. Indies, Pacific	Total	Max. length 5 m 11 s
1983 December 4	Atlantic, Equatorial Africa	Annular	
1984 May 30	Pacific, Mexico, USA, Atlantic, N. Africa	Annular	
1984 November 22/3	E. Indies, S. Pacific	Total	Max. length 1 m 59 s
1985 May 19		Partial	84% eclipsed
1985 November 12	S. Pacific, Antarctica	Total	Max. length 1 m 55 s
1986 April 9	Antarctic	Partial	82% eclipsed
1986 October 3	N. Atlantic	Total	Max. length 0 m 1 s Annular along most of track
1987 March 29	Argentina, Atlantic, Central Africa, Indian Ocean	Total	Max. length 56 s, in Atlantic. Annular along most of track
1987 September 23	Russia, China, Pacific	Annular	
1988 March 18	Indian Ocean, E. Indies, Pacific	Total	Max. length 3 m 46 s
1989 March 7	Arctic	Partial	83% eclipsed
1989 August 31	Antarctic	Partial	63% eclipsed
1990 January 26	Antarctic	Annular	
1990 July 22	Finland, Russia, Pacific	Total	Max. length 2 m 33 s
1991 January 15/16	Australia, N. Zealand, Pacific	Annular	
1991 July 11	Pacific, Mexico, Brazil	Total	Max. length 6 m 54 s
1992 January 4/5	Central Pacific	Annular	
1992 June 30	S. Atlantic	Total	Max. length 5 m 20 s
1992 December 24	Arctic	Partial	84% eclipsed
1993 May 21	Arctic	Partial	74% eclipsed
1993 November 13	Antarctic	Partial	93% eclipsed
1994 May 10	Pacific, Mexico, USA, Canada	Annular	
1994 November 3	Peru, Brazil, S. Atlantic	Total	Max. length 4 m 23 s
1995 April 29	S. Pacific, Peru, S. Atlantic	Annular	
1995 October 24	Iran, India, E. Indies, Pacific	Total	Max. length 2 m 5 s
1996 April 17	Antarctic	Partial	88% eclipsed
1996 October 12	Arctic	Partial	76% eclipsed
1997 March 9	Russia, Arctic	Total	Max. length 2 m 50 s
1997 September 2	Antarctic	Partial	90% eclipsed
1998 February 26	Pacific, Atlantic	Total	Max. length 3 m 56 s
1998 August 22	Indian Ocean, E. Indies, Pacific	Annular	
1999 February 16	Indian Ocean, Australia, Pacific	Annular	
1999 August 11	Atlantic, England (Cornwall), France, Turkey, India	Total	Max. length 2 m 23 s

Progress of an annular eclipse on 29 April 1976 (Patrick Moore)

SOLAR PROBES

The Sun emits not only visible light, but also radiations in other parts of the electromagnetic spectrum – from very long radio waves through to ultra-short X-rays. **The first suggestion of a 'solar wind'**, made up of streams of low-energy atomic particles sent out constantly in all directions, was made on theoretical grounds by S. Chapman and V. C. A. Ferraro in 1931. In 1958 E. Parker showed that the corona must be in a state of constant expansion, so that particles will escape all the time and by-pass the Earth at from 400 to 800 km/s. The existence of the solar wind was proved soon after the start of the Space Age, by the early Russian vehicles of 1959–61 and by the US Pioneer and Explorer probes.

The first attempt at solar observations from high altitude was made in 1914 by Charles Abbott, using an automated pyrheliometer launched from Omaha by hydrogen-filled rubber balloons. The altitude reached was 24·4 km. In 1935 a balloon, Explorer II, took a two-man crew to the same height.

The first attempt at solar research by a modern-type rocket was made in 1946, when a captured and converted German V2 was launched from White Sands, New Mexico. It reached 55 km, and recorded the solar spectrum down to 2400 Ångströms. **The first X-ray solar flares** were recorded in 1956, from balloon-launched rockets, though solar X-rays had been detected as early as 1948.

The closest deliberate approach to the Sun of a solar probe was that of the German-built Helios 2, in 1976. The minimum distance from the Sun was 45 000 000 km.

Many solar probes have now been launched. Most of them carry out general studies of the Sun as well as specialized programmes. The following list includes only some of the more important vehicles, with specialized investigations given. * indicates an American probe; ** a Soviet one.

Name	Launch date	Remarks on specialized programmes, etc.
*Pioneer 4	3 Mar 1959	Flare studies; Earth's magnetic field.
*Vanguard 3	18 Sep 1959	Solar X-rays.
*Pioneer 5	11 Mar 1960	Flares; solar wind. In solar orbit, 0·806×0·995 astronomical units. Operated till 26 Jun 1960, when it was 37 000 000 km from Earth.
*OSO 1	7 Mar 1962	First Orbiting Solar Observatory (there were 8 altogether in the series). Sent back data on 75 flares before losing contact on 6 Aug 1963.
**Cosmos 3	24 Apr 1962	Solar and cosmic radiation.
**Cosmos 7	28 Jul 1962	Monitoring of flares during flights of Vostoks 3 and 4.
*Explorer 18	26 Nov 1963	IMP – Interplanetary Monitoring Platform. In Earth orbit. Provision for flare warnings during manned space missions.
*OGO 1	4 Sep 1964	Orbiting Geophysical Observatory 1. Earth-Sun relationships. The first of a series of OGO probes.
*Explorer 30	18 Nov 1965	'Solrad'. Solar radiation in connection with the International Year of the Quiet Sun (IQSY).
*OSO 4	18 Oct 1967	Very short-wave solar ultra-violet.
*OGO 5	4 Mar 1967	Earth's magnetic field. Hydrogen cloud round Bennett's Comet.
*HEOS 1	5 Dec 1968	In Earth orbit, at 418×112 440 km. With HEOS 2, covered 7 years of the 11-year cycle.
**Cosmos 262	26 Dec 1968	Earth orbit, 262×965 km. Solar X-rays and ultra-violet.
Shinsei SS1	28 Sep 1971	First Japanese solar probe, launched from Kagoshima. In Earth orbit, 870×1870 km. Operated for 4 months.
*HEOS 2	31 Jan 1972	In Earth orbit. Studies of high-energy particles in conjunction with HEOS 1.
**Prognoz 1	14 Apr 1972	First of a series. Solar wind and X-radiation.
Intercosmos 9	19 Apr 1973	'Copernicus'; international, launched from the USSR. In Earth orbit, 202×1551 km.
*Skylab	14 May 1973	Manned station, carrying out important studies of the Sun; manned by 3 successive crews, 1973–4. Decayed in the Earth's atmosphere on 11 Jul 1979.
*Explorer 52	3 Jun 1974	'Injun'. Solar wind and general activity.
Helios 1	10 Dec 1974	German-built, American-launched. On 15 Mar 1975 it approached the Sun to 48 000 000 km at a velocity of 238 000 km/h, spinning at 1 rev per sec. Close-range studies of solar wind and the Sun's surface.
Aryabhāta	19 Apr 1975	Indian-built, Russian launched. Solar neutrons and gamma radiation.
Helios 2	15 Jan 1976	German-built, American-launched. Approached the Sun to 45 000 000 km; same programmes as Helios 1.
**Prognoz 6	22 Sep 1977	Effects of solar X- and gamma-rays on Earth's magnetic field; studies of ultra-violet, X- and gamma-radiation from the Galaxy.
*Solar Maximum Mission	14 Feb 1980	Detailed studies of the Sun near maximum of the solar cycle. Faults developed in it, and in 1984 it was brought back into the Space Shuttle, repaired and re-launched. Instruments on it confirmed the variability of the **solar constant**.

Total solar eclipse, showing structure of the outer corona: Sheldon M. Smith and Leonard B. Weinstein (NASA)

THE MOON

The Moon is officially ranked as the Earth's satellite. Relative to its primary, it is however extremely large and massive, and it might well be better to regard the Earth-Moon system as a double planet.

The first suggestion that the Moon is mountainous was made by the Greek philosopher Democritus (460–370 BC). Earlier, Xenophanes (c. 450 BC) had supposed that there were many suns and moons according to the regions, divisions and zones of the Earth!

The first map of the Moon was drawn by W. Gilbert, probably around 1600 (Gilbert died in 1603) though not published until 1651. This was compiled with the naked eye, as telescopes had not then come into use.

The first telescopic map of the Moon was drawn by Thomas Harriot in July 1609. It shows a number of identifiable features, and was much more accurate than the drawings made by Galileo from 1610.

The first British telescopic observations of the Moon following those of Harriot seem to have been made by Sir William Lower, from Wales, from about 1611. Lower's drawings have not survived, but he compared the appearance of the Moon with a tart that his cook had made!

The first serious attempts to measure the heights of lunar mountains were made by Galileo, from 1610. He measured the heights of the Lunar Apennines, and though he overestimated the altitudes his results were of the right order. Much better results were obtained by J. H. Schröter, from 1778.

The first explanation of the faint visibility of the 'night' side of the Moon, when the Moon is in the crescent stage, was given by Leonardo da Vinci (1452–1519). Leonardo correctly stated that the phenomenon is due to light reflected on to the Moon from the Earth.

The first systems of lunar nomenclature were introduced in 1645 by Langrenus, and in 1647 by

DATA	
Distance from Earth, centre to centre: **mean**: 384 400 km; **closest (perigee)**: 356 410 km; **furthest (apogee)**: 406 697 km	
Distance from Earth, surface to surface: **mean**: 376 284 km; **closest (perigee)**: 348 294 km; **furthest (apogee)**: 398 581 km	
Revolution period: 27·321 661 days	
Axial rotation period: 27·321 661 days	
Synodic period: 29 d 12 h 44 m 2·9 s	
Mean orbital velocity: 3680 km/h	
Axial inclination of equator, referred to ecliptic: 1°32′	
Orbital inclination: 5°09′	
Orbital eccentricity: 0·0549	
Diameter: 3475·6 km	
Apparent diameter seen from Earth: max 33′31″, mean 31′5″, min 29′22″	
Reciprocal mass, Earth = 1: 81·3	
Density, water = 1: 3·342	
Mass, Earth = 1: 0·0123	
Volume, Earth = 1: 0·0203	
Escape velocity: 2·38 km/s	
Surface gravity, Earth = 1: 0·1653	
Albedo: 0·07	
Mean magnitude, at full: − 12·7	

Hevelius; but these have not survived. (For instance, the crater now called Plato was called by Hevelius 'the Greater Black Lake'.) The modern system was introduced in 1651 by G. Riccioli, who named the features in honour of scientists – plus a few others. He was not impartial; for instance he allotted a major formation to himself and another to his pupil Grimaldi, but he did not believe in the 'Copernican' theory that the Earth moves round the Sun, and so he 'flung Copernicus into the Ocean of Storms'. Riccioli's principle has been followed since, though clearly all the major craters were 'used up' and later distinguished scientists have had to be given formations of lesser importance.

The first system of lunar coordinates was introduced by T. Mayer in 1775. The first really accurate measures with a heliometer were made by F. W. Bessel in 1839.

The first detailed statement that the Moon has a synchronous rotation was made by G. D. Cassini in 1693, though Galileo may also have realized it. The Moon's rotation is equal to its revolution period (27·3 Earth-days), and so the same face is turned Earthward all the time, though the eccentricity of the lunar orbit means that there are 'libration zones' which are brought alternately into and out of view. From

the Earth, 59 per cent of the Moon's surface can be studied at one time or another. The remaining 41 per cent is inaccessible. There is no mystery about the behaviour; tidal forces over the ages have been responsible. Most other planetary satellites also have synchronous rotation with respect to their primaries.

The first detailed descriptions of lunar rills were given by J. H. Schröter, from 1778, though rills had been observed earlier by Huygens.

The first really good map of the Moon was published in 1837–8 by W. Beer and J. H. Mädler, of Berlin. Though they used a small telescope (Beer's 3·75 in refractor) their map was amazingly good, and remained the best for several decades. They also published a book, *Der Mond*, the first detailed description of the lunar surface.

The first detailed English book about the Moon was published in 1876 by E. Neison (real name, Nevill). Neison included a map, which was basically a revision of that by Beer and Mädler.

The largest lunar map compiled by a visual observer was that by the Welsh selenographer H. P. Wilkins. It was 300 in diameter. The final revision, to one-third this scale, appeared in 1959.

The first photograph of the Moon (a Daguerreotype) was taken on 23 March 1840 by J. W. Draper, using a 5 in reflector. The image was 1 in across, and the exposure time was 20 minutes. **The first good photographic atlas** was published in 1899 by the French astronomers Loewy and Puiseux.

The first report of a 'lunar city' was made in 1822 by the German astronomer F. von P. Gruithuisen, who described a structure with 'dark gigantic ramparts' – which, however, prove to be low and quite haphazard ridges! In 1790 Sir William Herschel had stated his belief in an inhabited Moon, and in 1835 there came the famous 'lunar hoax', when a paper, the New York *Sun*, published some quite imaginary reports of discoveries made by Sir John Herschel at the Cape of Good Hope. The reports were written by a reporter, R. A. Locke, and included descriptions of bat-men and quartz mountains. The first article appeared on 25 August, and the *Sun* admitted the hoax on 16 September.

The first serious suggestion of change on the Moon was made in 1866 by J. Schmidt, from Athens. Schmidt believed that a crater, Linné on the Mare Serenitatis, had been transformed into a pit surrounded by a white patch. The controversy continued for many years, but it is not now believed that any change occurred there.

The first reports of active volcanoes on the Moon were made by Sir William Herschel, in 1783 (May 4) and again in 1787 (April 19 and 20). However, it does not seem that Herschel saw anything more significant than bright areas shining by earthlight. In modern times, lunar events or Transient Lunar Phenomena (TLP) have been reported on many occasions. **The first photographic confirmation** was obtained on 3 November 1958 by N. A. Kozyrev, at the Crimean Astrophysical Observatory, of an event in Alphonsus. On 30 October 1963 an event in the Aristarchus area was observed at the Lowell observatory by J. Greenacre and J. Barr. **The first comprehensive catalogue** of TLP reports was published by NASA, in 1967, by B. Middlehurst and Patrick Moore; Moore published a supplement in 1971. The total number of events reported in these catalogues is 713: though there is no doubt that many of them are due to observational error or misinterpretation, it is generally agreed that some are genuine, and that the Moon is not totally inert.

The most famous 'blue moon' of recent times was that of 26 September 1950. The phenomenon was due to dust in the Earth's atmosphere, raised by vast forest fires in Canada. There was also a blue moon on 27 August 1883, caused by material sent up from the Krakatoa outburst; and green moons were seen in Sweden in 1884 – at Kalmar on 14 February, for 3 minutes, and at Stockholm on 17 January, also for 3 minutes.

The temperatures on the lunar surface were first measured with reasonable accuracy by the fourth Earl of Rosse, from Birr Castle in Ireland. Lord Rosse's papers, from 1869, indicated that at noon the surface temperature approached 100 °C, though later Langley erroneously concluded that the temperature never rose above 0 °C. It is known that the equatorial noon temperature is about 117 °C, falling to −163 °C at minimum.

The first radar echoes from the Moon were obtained in 1946 by Z. Bay, from Hungary. One recent determination of the Moon's distance, by a laser measure (J. Faller, Lick) gave 353 911·218, ±45 m. At the Moon, the laser beam has a diameter of 4 km. The telescope used was the 1·4 m reflector at Catalina (Arizona).

The largest regular 'sea' is the Mare Imbrium, diameter 1300 km. It is 7 km deeper in its centre than at its periphery. Of the other regular maria, Orientale has a diameter of 965 km; Serenitatis, 690 km × 885 km; Crisium, 590 km × 460 km; Humorum, 423 km × 460 km; Humboldtianum, 305 km × 410 km. Mare Crisium is longest in an East-West direction, though from Earth foreshortening makes it appear longest North-South. Of the irregular seas, the Oceanus Procellarum is much larger than the Mare Imbrium. On average the Oceanus Procellarum is only 1 km lower than the mean sphere, whereas the deeper seas, such as Mare Crisium, are 4 km lower. The Mare Orientale (discovered by Wilkins and Moore in 1938) is the only major sea which extends on to the Moon's far side, and is very badly placed for observation from Earth.

The largest crater is Bailly, diameter 295 km (depth 3·96 km). Next, insofar as the Moon's near side is concerned, is Clavius (232 km diameter, 4·91 km deep). Many of the large structures are much shallower – such as Ptolemæus (diameter 148 km, depth 1·2 km). Craters with central structures include Theophilus (101 km diameter, 4·4 km deep); Copernicus (97 km, 3·35 km) and Tycho (84 km, 4·27 km). The great dark formation Plato has a depth of 2·4 km.

The deepest crater is thought to be Newton, with a floor 8·85 km below the crest of its wall; the wall rises 2·25 km above the outer surface.

The brightest crater on the Moon is Aristarchus, which often appears prominently when earthlit. **The dar-**

kest area is the floor of Grimaldi.

The greatest ray-crater is Tycho, whose rays dominate the scene at or near full moon. Second to it is Copernicus, on the Oceanus Procellarum. Other important ray-centres include Kepler, Olbers, Anaxagoras, and Thales.

The most famous plateau on the Moon is Wargentin, which is lava-filled. All other known plateaux are much smaller and less regular.

The most famous rills on the Moon are those in the Mare Vaporum area (Hyginus, Ariadæus) and the Hadley Rill in the Apennine area, visited by the Apollo 15 astronauts; the Hadley Rill is 135 km long, 1 to 2 km wide, and 370 m deep. Other famous rills or rill-systems include those associated with Sirsalis, Bürg, Hesiodus, Triesnecker, Ramsden and Hippalus. Extending from Herodotus, near Aristarchus, is the imposing valley known as Schröter's Valley in honour of its discoverer, J. H. Schröter (in a way rather misleading, since the crater named after Schröter is in a completely different area).

Domes, up to 80 km in diameter, occur in various parts of the Moon – for instance near Arago in the Mare Tranquillitatis, Prinz in the general area of Aristarchus, and on the floor of Capuanus. Many of them have summit pits; their slopes are gentle. Domes were first described in detail by the British selenographer R. Barker in 1932.

The first description of the 'grid system' (families of linear features, aligned in definite directions) was given by the American geologist J. E. Spurr in 1945; Spurr also invented the term. Most astronomers believe it to be of considerable significance in studies of the Moon's past history.

The origin of the lunar craters is still a matter for debate. There are various theories, ranging from the plausible to the eccentric! The main question is whether the craters were produced by internal action – that is to say, vulcanism – or by meteoritic impact. The impact theory was proposed by Gruithuisen in 1824; it was revived by G. K. Gilbert in 1892 and again by R. B. Baldwin in 1949. On the volcanic theory, it is generally assumed that the craters are of the caldera type. No doubt both vulcanism and impact have played a part in the moulding of the lunar surface.

(Among the more erratic theories may be cited those of P. Fauth, who died in 1943 and supported the idea that the Moon is covered with ice; Weisberger, who in 1942 denied the existence of any lunar mountains or craters, and attributed the effects to storms and cyclones in a dense lunar atmosphere; and Ocampo, whose theory appeared in 1951. To Ocampo, the craters were the result of an atomic war between the two races of Moonmen. The fact that some craters have central peaks while others do not shows, of course, that the warring factions used different kinds of bombs!)

The first reports of mascons, or **mass concentrations**, were made by P. Muller and W. L. Sjögren in 1968, from studies of the movements of the probe Orbiter 5. Exceptionally dense areas of material were found to lie below various regular features – Imbrium, Crisium, Smythii, Serenitatis, Humorum, Nectaris, Humboldtianum, Orientale, Æstuum and Grimaldi. On the other hand, the material below some of the other features, such as Copernicus, was of much lower density. (In 1968 B. Middlehurst and Patrick Moore had shown that TLP are often associated with features near the boundaries of the regular seas.)

The first attempted lunar probe was Able 1, launched by the Americans on 17 August 1958. It was not successful.

The first successful probe to the Moon was Luna 1 (2 January 1959). It by-passed the Moon, sending back valuable data.

The first probe to land on the Moon was Luna 2 (13 September 1959).

The first photographs of the Moon's far side were obtained from Luna 3, in October 1959.

The first good close-range pictures of the Moon were obtained from Ranger 7 (31 July 1964). The region, not far from Guericke, has been named the Mare Cognitum, or Known Sea, though the name has not come into general use.

The first successful soft landing

The obligatory footprint on the Moon's surface – Apollo 11, 1969. The footprint is Neil Armstrong's. (NASA)

on the Moon was made by Luna 9, launched on 31 January 1966.

The first lunar satellite was Luna 10, launched on 31 March 1966.

The first manned flight round the Moon was that of Apollo 8, in December 1968.

The first landing on the Moon was that of Apollo 11, on 20 July 1969. Early on 21 July Neil Armstrong stepped out on to the lunar surface, followed shortly afterwards by Edwin Aldrin.

The first unmanned probe to return samples from the Moon was Luna 16, launched on 12 September 1970.

The first 'crawler' on the Moon was Lunokhod 1. It was taken to the Moon by Luna 17, and landed there on 17 November 1970.

The first 'Moon Car' or LRV (Lunar Roving Vehicle) was taken to the Moon in Apollo 15 (30 July 1971). Astronauts Scott and Irwin used it to drive around in the Apennines area.

The first professional geologist to go to the Moon was Harrison ('Jack') Schmitt, in Apollo 17 (11 December 1972). It was Schmitt who found the famous 'orange soil' near the crater which was known unofficially as Shorty. The colour proved to be due to small glassy particles which were very ancient; the age is 3·7 to 3·8 æons. (One æon: a thousand million years.)

The last man on the Moon (so far!) was Eugene Cernan, in Apollo 17. When he re-entered the lunar module, following Dr Schmitt, the first stage in lunar manned exploration was over.

The Apollo results have changed many of our ideas about the Moon. There is a loose upper layer or **regolith**, from 1 to 20 metres deep (on average 4 to 5 metres in the maria, 10 in the highlands); this layer contains many different ingredients. On the highlands, the rock fragments are chiefly anorthosites, with the minerals plagioclase, pyroxene, and/or olivine; on the maria, the rock fragments are basalt, with pyroxene, plagioclase and ilmenite. Below the regolith is a kilometre-thick layer of shattered bedrock. Under this is a layer of more solid rock going down to about 25 km in the Mare regions. Beneath this is a layer of material 30 km thick, with properties similar to the anorthosites and the feldspar-rich gabbros of the highlands. The bottom of this layer is 50 km from the surface. At 50 km another kind of rock appears; there is a sharp boundary. These new rocks are pyroxene and olivine rich, and are dense; they go down to at least 150 km, possibly much more. Beneath this is the core region, which is metallic, and may be 1000 to 1500 km in diameter. A 1-tonne meteorite which hit the Moon on 17 July 1972 indicated, from the moonquake waves, that there is a zone 1000 to 1200 km deep inside the Moon where the rocks are hot enough to be molten. The crust on the far side of the Moon may be thicker than that on the Earth-turned side (75 to 100 km deep).

All the rocks brought home for analysis are igneous or breccias produced by impact processes; the Apollo missions recovered 2196 samples, with a total weight of 381·69 kg, now divided into 35 600 samples. The youngest rocks seem to be about 3·1 thousand million years old; the oldest, 4·7 thousand million – about the same as the age of the Moon itself. There are no sedimentary or metamorphic rocks. The main epoch of lava-flowing ended about 3 æons ago.

In the **lavas**, basalts are dominant. They contain more titanium than terrestrial lavas; over 10 per cent in the Apollo 11 samples (3 to 5 per cent in the Apollo 12 samples, as against 1 to 3 per cent in terrestrial basalts). Small amounts of metallic iron were found. Generally, lunar rocks have one-tenth as much sodium and potassium as terrestrial rocks. A new mineral – an opaque oxide of iron, titanium and magnesium, not unlike ilmenite – has been named armalcolite.

All the highland rocks are igneous, with twice as much calcium and aluminium as the Mare lavas, but less titanium, magnesium and iron. Plagioclase makes up at least half the highland rocks; there is little sodium or potassium, so that the volatiles must have been lost from the Moon at an early stage. The average age of highland rocks is from 4 to 4·2 æons. The so-called 'Genesis Rock' (Apollo 15) is anorthosite, with an age of 4 æons.

Breccias have been collected from every landing site on the Moon. These are complex rocks made up of shattered, crushed and sometimes once-melted pieces of rock. The fragments may be igneous rocks of various types, glasses, or bits of other breccias.

The Apollo 12 sample 12013 (collected by Conrad from the Oceanus Procellarum) is unique. It is about the size of a lemon, and contains 61 per cent of SiO_2, whereas the associated lavas have only 35 to 40 per cent of SiO_2. It also contains 40 times as much potassium, uranium and thorium, making it one of the most radioactive rocks found anywhere on the Moon. It is composed of a dark grey breccia, a light grey breccia and a vein of solidified lava.

(**Pyroxene**, a Ca-Mg-Fe silicate, is the most common mineral in lunar lavas, making up about half of most specimens; it forms yellowish-brown crystals up to a few centimetres in size. **Plagioclase** or feldspar, a Na or Ca-Al silicate, forms elongated white crystals. **Anorthosite** contains over 90 per cent plagioclase; it may even be pure plagioclase. **Olivine**, a Mg-Fe silicate, is made up of pale green crystals a few milli-metres in size, and is not uncommon in the anorthosites. **Ilmenite**, present in the basalts, is an Fe-Ti oxide. It is notable that there are no hydrated materials on the Moon, so that evidently water has never existed there.)

Moonquakes have been recorded regularly; there are about 3000 per year, centred mainly from 600 to 800 km below the surface, though very shallow moonquakes also occur. The impacts from the crashed lunar modules of the Apollo missions showed that the outermost few kilometres of the Moon are made up of cracked and shattered rock, so that signals can echo back and forth for hours; it was even said that after an impact of a module the Moon 'rang like a bell'! Researches by J. Green, B. Middlehurst and Patrick Moore have confirmed that moonquakes are most common at the time of lunar perigee, when the Moon's crust is under maximum strain.

There is virtually no overall magnetic field, though it seems that one did exist in the remote past. But, there are areas of magnetized material, particularly in and near the far-side crater Van de Graaff. All recording Apollo stations on the Moon were shut down on 30 September 1977, partly on the grounds of expense and partly because some of the equipment was showing signs of failure.

There is no Earth-type atmosphere, and the so-called lunar atmosphere is a collisionless gas in which hydrogen, helium, neon and argon have been identified; the density is negligible – about 2×10^5 molecules per cubic centimetre, which corresponds to what we normally call a very high laboratory vacuum. There is no trace of life, either past or present, and the quarantining of Apollo astronauts was dropped after the first three successful missions.

The origin of the Moon remains uncertain, but it is now thought that the Earth-Moon system should be regarded as a double planet rather than as a planet and a satellite (the Moon's orbit is always concave to the Sun.) The age of the Moon is about the same as that of the Earth. The Earthward-turned axis is 2·2 km longer than the rotational axis, so that there is a slight Earthward bulge. In 1968 the **north pole star** of the Moon was Omega Draconis; by 1977 it was 36 Draconis. The **south pole star** is Delta Doradûs.

No minor Earth satellites of natural origin seem to exist. Careful searches for them have been made, as in 1957 by Tombaugh (discoverer of Pluto), but without result. Clouds of loose material in the Moon's orbit, at the Lagrangian points, were reported by K. Kordylewski in 1961, but remain unconfirmed. It therefore seems safe to conclude that the Moon is the Earth's only attendant.

NAMED 'SEAS' ON THE NEAR SIDE OF THE MOON

Sinus Æstuum	The Bay of Heats
Mare Australe	The Southern Sea
Mare Crisium	The Sea of Crises
Palus Epidemiarum	The Marsh of Epidemics
Mare Fœcunditatis	The Sea of Fertility
Mare Frigoris	The Sea of Cold
Mare Humboldtianum	Humboldt's Sea
Mare Humorum	The Sea of Humours
Mare Imbrium	The Sea of Showers
Sinus Iridum	The Bay of Rainbows
Mare Marginis	The Marginal Sea
Sinus Medii	The Central Bay
Lacus Mortis	The Lake of Death
Palus Nebularum	The Marsh of Mists
Mare Nectaris	The Sea of Nectar
Mare Nubium	The Sea of Clouds
Mare Orientale	The Eastern Sea
Oceanus Procellarum	The Ocean of Storms
Palus Putredinis	The Marsh of Decay
Sinus Roris	The Bay of Dews
Mare Serenitatis	The Sea of Serenity
Mare Smythii	Smyth's Sea
Palus Somnii	The Marsh of Sleep
Lacus Somniorum	The Lake of the Dreamers
Mare Spumans	The Foaming Sea
Mare Tranquillitatis	The Sea of Tranquility
Mare Undarum	The Sea of Waves
Mare Vaporum	The Sea of Vapours

These 'seas' are not alike. Some, such as Mare Imbrium, are regular and largely mountain-bordered, whereas others, such as Mare Undarum, are discontinuous, and consist only of patches of dark mare material. Most of the seas are connected; for instance Palus Nebularum (a name omitted from some lists) is part of the Mare Imbrium, while Sinus Iridum is a bay leading out from the Mare Imbrium. Isolated seas are rare; the Mare Crisium is a good example. The Mare Orientale spreads on to the far side of the Moon. The name was given to it by H. P. Wilkins and the present writer, but owing to a policy decision by the International Astronomical Union, reversing lunar east and west, the *Eastern* Sea is now on the *western* limb of the Moon.

LUNAR CRATERS ON THE NEAR SIDE OF THE MOON

The following are named craters. The diameters are in many cases arbitrary, because so many of the craters are irregular in shape, but the values given here are of the right order. c.p. = central peak.

Name	Lat	Long	Diam, km	Notes
Abenezra	21 S	12 E	43	W. of Nectaris. Well-formed. Pair with Azophi.
Abulfeda	14 S	14 E	64	Abenezra area. Pair with Almanon.
Adams	32 S	69 E	64	E. of Vendelinus. Irregular walls.
Agatharchides	20 S	31 W	48	Humorum area. Remnant of c.p. Irregular walls.
Agrippa	4 N	11 E	48	Vaporum area. Regular; c.p. Pair with Godin.
Airy	18 S	6 E	35	Albategnius area: pair with Argelander. Irregular walls.
Albategnius	12 S	4 E	129	Companion to Hipparchus. Terraced, irregular walls; c.p.
Alexander	40 N	14 E	105	N. end of Caucasus. Darkish floor. Low walls.
Alfraganus	6 S	19 E	19	N.W. of Theophilus. Very bright. Minor ray-centre.
Alhazen	18 N	70 E	32	Near border of Crisium.
Aliacensis	31 S	5 E	84	Walter area; pair with Werner. High walls. c.p.
Almanon	17 S	15 E	48	Pair with Abulfeda. Regular.
Alpetragius	16 S	4 W	43	Outside Alphonsus. High, terraced walls. Vast c.p. with summit pit.
Alphonsus	13 S	3 W	129	Ptolemæus group. Low c.p. Rills on floor.
Anaxagoras	75 N	10 W	52	N. polar area; distorts Goldschmidt. Major ray-centre. c.p.
Anaximander	66 N	48 W	87	Pythagoras area; pair with Carpenter. No c.p.
Anaximenes	75 N	45 W	72	Near Philolaus. Rather low walls.
Anděl	10 S	13 E	40	Highlands W. of Theophilus. Low, irregular walls.
Ångström	30 N	42 W	11	N. of Harbingers, in Imbrium.
Ansgarius	14 S	82 E	80	E. of Fœcunditatis. Distinct. Pair with La Peyrouse.
Apianus	27 S	8 E	63	Aliacensis area. High walls.
Apollonius	5 N	61 E	48	Uplands S. of Crisium. Well-formed.
Arago	6 N	21 E	29	On Tranquillitatis. Domes nearby.
Aratus	24 N	5 E	13	In Apennines. Bright, not regular.
Archimedes	30 N	4 W	75	On Imbrium. Very regular. No c.p.
Archytas	59 N	5 E	34	On Frigoris. Bright, distinct; c.p.
Argelander	17 S	6 E	42	Albategnius area; pair with Airy. c.p.
Ariadæus	5 N	17 E	15	Vaporum area. Associated with great rill.
Aristarchus	24 N	48 W	37	On Procellarum. Brightest crater on the Moon. Terraced walls; c.p.
Aristillus	34 N	1 E	56	Archimedes group. Fine c.p.
Aristoteles	50 N	18 E	97	Pair with Eudoxus. High walls.
Arnold	67 N	38 E	81	N.W. of Democritus; N. of Frigoris. Low walls.
Arzachel	18 S	2 W	97	Ptolemæus group. High walls; c.p.
Asclepi	55 S	26 E	32	S. uplands, W. of Hommel. Distinct.
Atlas	47 N	44 E	89	N. of Somniorum; pair with Hercules. High walls; much interior detail.
Autolycus	31 N	1 E	36	Archimedes group. Regular, distinct.
Auwers	15 N	17 E	16	Foothills of Hæmus. Not very bright.
Auzout	10 N	64 E	31	Outside Crisium. Low c.p.
Azophi	22 S	13 E	43	W. of Nectaris; pair with Abenezra. Well-formed
Babbage	58 N	52 W	145	Nr. Pythagoras. Irregular enclosure.
Baco	51 S	19 E	64	Licetus area. High walls; low c.p.
Baillaud	75 N	40 E	80	Uplands N.E. of Meton. Rather low walls.
Bailly	66 S	65 E	294	'Field of ruins'; southern uplands.
Baily	50 N	31 E	19	Frigoris area N. of Bürg.
Ball	36 S	8 W	40	On edge of Deslandres. High walls.
Barocius	45 S	17 E	80	Outside Maurolycus. High but broken walls.
Barrow	73 N	10 E	87	N. of W. C. Bond (N. polar uplands). Low, broken walls.
Bayer	51 S	35 W	52	Outside Schiller. High, terraced walls.
Beaumont	18 S	29 E	48	Bay on Nectaris.
Beer	27 N	9 W	11	On Imbrium. Pair with Feuillé.
Behaim	16 S	71 E	56	E. of Vendelinus. High walls, c. craters.

Name	Lat	Long	Diam, km	Notes
Bellot	12 S	48 E	19	Edge of Fœcunditatis. Bright floor.
Bernouilli	34 N	60 E	40	E. of Geminus. Fairly regular.
Berosus	33 N	70 E	61	E. of Cleomedes; pair with Hahn. Terraced walls.
Berzelius	37 N	51 E	24	Taurus area. Darkish floor; c.p.
Bessel	22 N	18 E	19	On Serenitatis. Associated with long ray.
Bettinus	63 S	45 W	66	Bailly area; one of a line. High walls; c.p.
Bianchini	49 N	34 W	40	In Jura Mts. Rather irregular; c.p.
Biela	55 S	52 E	74	Vlacq area. High walls; c.p.
Billy	14 S	50 W	42	S. edge Procellarum. Very dark floor. Pair with Hansteen.
Biot	23 S	51 E	16	In Fœcunditatis. Very bright.
Birmingham	64 N	10 W	106	N. of Frigoris. Low-walled; irregular.
Birt	22 S	9 W	18	On Nubium; W. of Straight Wall. Profile irregular. Rill to the W.
Blagg	1 N	2 E	5	In Sinus Medii. Quite distinct.
Blancanus	64 S	21 W	92	Near Clavius; pair with Scheiner. High walls.
Blanchinus	25 S	3 E	53	N. of Werner. Uneven walls; rough floor.
Bode	7 N	2 W	18	Outside Æstuum. Bright; minor ray-centre.
Boguslawsky	75 S	45 E	97	Southern uplands. High walls.
Bohnenberger	16 S	40 E	35	Edge of Nectaris. Low walls.
Bond, G. P.	32 N	36 E	23	E. of Posidonius. Fairly regular.
Bond, W. C.	64 N	3 E	160	N. of Frigoris. Old and broken.
Bonpland	8 S	17 W	58	Fra Mauro group (Nubium). Fairly regular.
Borda	25 S	47 E	42	W. of Petavius. Low walls.
Boscovich	10 N	11 E	43	On edge of Vaporum. Low walls; very dark floor.
Bouguer	52 N	36 W	24	In Jura uplands. Very distinct.
Boussingault	70 S	50 E	78	S. uplands. Made up of three rings.
Bouvard	37 S	87 W	129	Schickard area. Central ridge rising to a peak.
Brayley	21 N	37 W	16	On Procellarum. Low c.p.
Breislak	48 S	18 E	32	S.E. of Maurolycus. Fairly regular. No c.p.
Brenner	39 S	39 E	90	Broken. Adjoins Janssen (S. uplands).
Briggs	26 N	69 W	38	On Procellarum (Otto Struve area). Similar to Seleucus.
Brisbane	50 S	65 E	47	Australe area. Fairly regular.
Brown	46 S	18 W	40	N.E. of Longomontanus. Irregular.
Bruce	1 N	0	9	In Sinus Medii. Distinct.
Buch	39 S	18 E	48	Maurolycus area; adjoins Büsching. Regular.
Bullialdus	21 S	22 W	50	On Nubium. Massive, terraced walls; c.p.
Burckhardt	31 N	57 E	48	N. of Cleomedes. Member of complex group.
Bürg	45 N	28 E	48	N. of Mortis. Large c.p. with pit. Associated with major rill-system.
Burnham	14 S	7 E	29	Albategnius area. Low walls.
Büsching	38 S	20 E	58	Adjoins Buch; is less regular.
Byrgius	25 S	65 W	64	W. of Humorum, Byrgius A, on its E. crest, is a ray-centre.
Cabæus	85 S	20 W	140	S. polar uplands. Fairly high walls.
Calippus	39 N	11 E	31	N. end of Caucasus. Irregular.
Campanus	28 S	28 W	38	W. edge of Nubium; pair with Mercator. c.p.
Capella	8 S	36 E	48	Uplands N. of Nectaris. Large c.p. Cut by crater-valley.
Capuanus	34 S	26 W	56	Edge of Epidemiarum. Domes on floor.
Cardanus	13 N	73 W	52	On Procellarum; pair with Krafft. c.p.
Carlini	34 N	24 W	8	On Imbrium. Rather bright.
Carpenter	70 N	50 W	65	N. of Frigoris; adjoins Anaximander.
Carrington	44 N	62 E	25	Messala group.
Casatus	75 S	35 W	104	S. of Clavius; intrudes into Klaproth. High walls.
Cassini	40 N	5 E	58	Edge of Nebularum. Low walls. Contains deep crater, A.
Cassini, J. J.	68 N	16 W	50	Near Philolaus (N. uplands). Irregular ridge-bounded area.
Catharina	18 S	24 E	89	Theophilus group. Rough floor: no c.p.
Cauchy	10 N	39 E	13	On Tranquillitatis. Bright.
Cavalerius	5 N	67 W	64	In Hevel group. Central ridge.
Cavendish	25 S	54 W	52	W. of Humorum. Fairly high walls.
Cayley	4 N	15 E	13	Uplands W. of Tranquillitatis. Very bright.
Celsius	34 S	20 E	45	Rabbi Levi group. Elliptical.
Censorinus	0	32 E	5	Uplands S.E. of Tranquillitatis. Brilliant.
Cepheus	41 N	46 E	48	Somniorum area; pair with Franklin.
Chacornac	30 N	32 E	48	Edge of Serenitatis, near Posidonius. Outline irregular.

Apollo 17; Dr Harrison H. Schmitt, the geologist-astronaut, by a large lunar boulder. The photograph was taken by Commander Eugene Cernan. (NASA)

Name	Lat	Long	Diam, km	Notes
Challis	78 N	9 E	56	N. polar area. Pair with Main.
Chevallier	45 N	52 E	47	Atlas area. Low walls.
Chladni	4 N	1 E	8	Sinus Medii area; abuts on Murchison. Bright.
Cichus	33 S	21 W	32	Just S. of Nubium. Well-formed.
Clairaut	48 S	14 E	64	Maurolycus area. Broken walls.
Clausius	37 S	44 W	24	Schickard area. Bright and distinct.
Clavius	56 S	14 W	232	S. uplands. Massive walls. Curved line of craters on floor. No c.p.
Cleomedes	27 N	55 E	126	Crisium area. Distorted by Tralles.
Cleostratus	60 N	74 W	70	Pythagoras area. Distinct.
Colombo	15 S	46 E	81	In Fœcunditatis. Irregular. Broken by large crater, A.
Condamine	53 N	28 W	48	Edge of Frigoris. Fairly regular.
Condorcet	12 N	70 E	72	Outside Crisium. Regular; no c.p.
Conon	22 N	2 E	21	In Apennine uplands. Fairly distinct.
Cook	18 S	49 E	42	Edge of Fœcunditatis. Darkish floor.

Name	Lat	Long	Diam, km	Notes
Copernicus	10 N	20 W	97	Great ray-crater. Massive walls. Central mountain group.
Crozier	14 S	51 E	24	Fœcunditatis area, S.E. of Colombo, c.p.
Crüger	17 S	67 W	48	S.W. of Procellarum. Very dark floor.
Curtius	77 S	5 E	81	Moretus area. Massive, terraced walls.
Cuvier	50 S	10 E	66	Licetus group. High walls; c.p.
Cyrillus	13 S	24 E	97	Theophilus group. Low c.p.
Cysatus	66 S	7 W	47	Moretus area. High walls.
Daguerre	12 S	34 E	32	In Nectaris. Very low walls.
Damoiseau	5 S	61 W	35	E. of Grimaldi. Very irregular.
Daniell	35 N	31 E	24	Adjoins Posidonius. No c.p.
Darney	15 S	24 W	17	On Nubium (Fra Mauro area). Bright.
D'Arrest	2 N	15 E	30	W. of Sabine-Ritter pair. Low walls.
Darwin	20 S	69 W	130	Grimaldi area. Very irregular; low walls. Contains a large dome.
Da Vinci	9 N	45 E	30	Somnii area. Low walls.
Davy	12 S	8 W	32	E. edge of Nubium. Irregular walls.
Dawes	17 N	26 E	23	Between Serenitatis and Tranquillitatis. Distinct.
Debes	29 N	52 E	48	Outside Cleomedes. Fusion of 2 rings.
Dechen	46 N	67 W	12	On Procellarum (N.W. area). Not bright.
De Gasparis	26 S	50 W	48	W. of Humorum, S. of Mersenius. Fairly regular.
Delambre	2 S	18 E	52	Tranquillitatis area. High walls.
De la Rue	67 N	56 E	160	Adjoins Endymion to N.W. Very low, broken walls.
Delaunay	22 S	3 E	45	Albategnius area, near Faye. Irregular.
De l'Isle	30 N	35 W	22	On Procellarum; pair with Diophantus. c.p.
Delmotte	27 N	60 E	31	N. of Crisium. Not prominent.
Deluc	55 S	3 W	45	Clavius area. Moderate walls.
Dembowski	3 N	7 E	23	E. of Medii. Low walls.
Democritus	62 N	35 E	37	Highlands N. of Frigoris. Very deep.
Demonax	85 S	35 E	121	Boguslawsky area. Fairly regular.
De Morgan	3 N	15 E	8	Uplands W. of Tranquillitatis. Bright.
Descartes	12 S	16 E	48	N.E. of Abulfeda. Low, broken walls.
Deseilligny	21 N	21 E	9	On Serenitatis. Distinct.
Deslandres	32 S	61 W	186	Ruined enclosure W. of Walter.
De Vico	20 S	60 W	24	W. of Gassendi. Deep.
Dionysius	3 N	17 E	19	On edge of Tranquillitatis. Brilliant.
Diophantus	28 N	34 W	18	On Procellarum; pair with De l'Isle. c.p.
Dollond	10 S	14 E	10	W. of Theophilus. Borders a large 'ghost'.
Donati	21 S	5 E	35	S. of Albategnius; pair with Faye. Irregular. c.p.
Doppelmayer	28 S	41 W	68	Bay in Humorum. Remnant of c.p.
Dove	47 S	32 E	18	Janssen area. Low walls.
Draper	18 N	22 W	10	In Imbrium, S. of Pytheas. One of a pair.
Drebbel	41 S	49 W	29	Schickard area. Well-formed.
Drygalski	80 S	55 W	140	Adjoins Legentil (Bailly area).
Dubiago	4 N	70 E	48	Smythii area. Regular.
Dunthorne	30 S	32 W	18	Edge of Epidemiarum. Broad walls.
Egede	49 N	11 E	49	Near Alpine Valley. Low walls. Lozenge-shaped.
Eichstädt	23 S	80 W	52	Orientale area. Regular.
Eimmart	24 N	65 E	42	Near edge of Crisium. Regular.
Einstein	18 N	86 W	160	W. of Procellarum; Otto Struve area. Central crater.
Elger	35 S	30 W	24	Edge of Epidemiarum. Low-walled. Imperfect.
Encke	5 N	37 W	32	On Procellarum; Kepler area.
Endymion	55 N	55 E	117	Humboldtianum area. Darkish floor.
Epigenes	73 N	4 W	52	N. of Frigoris. Broad walls.
Epimenides	41 S	30 W	24	One of a pair E. of Hainzel.
Eratosthenes	15 N	11 W	61	End of Apennines. Very deep. c.p.
Euclides	7 S	29 W	12	Near Riphæan Mts. Lies on bright nimbus.
Eudoxus	44 N	16 E	64	S. of Frigoris. Pair with Aristoteles.
Euler	23 N	29 W	25	On Imbrium. Minor ray-centre.
Fabricius	43 S	42 E	89	Intrudes into Janssen. Rough floor. c.p.
Faraday	42 S	8 E	64	Intrudes into Stöfler. Irregular.
Fauth	6 N	20 W	13	On Procellarum, S. of Copernicus. Double crater (with A).
Faye	21 S	4 E	35	S. of Albategnius; pair with Donati. Irregular. c.p.
Fermat	23 S	20 E	40	Altai area. Distinct.
Fernelius	38 S	5 E	64	N. of Stöfler. Rather irregular.

Crater Copernicus (Cdr H. R. Hatfield)

Craters Aristoteles and Eudoxus (both lower left) with the Alps and Alpine Valley (Patrick Moore)

Name	Lat	Long	Diam, km	Notes
Feuillé	27 N	10 W	13	On Imbrium. Twin with Beer.
Firmicus	7 N	64 E	56	S. of Crisium. Dark floor. No c.p.
Flammarion	3 S	4 W	72	N.W. of Ptolemæus. Irregular enclosure.
Flamsteed	5 S	44 W	19	On Procellarum. Associated with 100-km 'ghost'.
Fontana	16 S	57 W	48	Between Billy and Crüger. c.p.
Fontenelle	63 N	19 W	45	N. edge of Frigoris. Deep, distinct.
Foucault	50 N	40 W	24	Jura Mts. area. Bright and deep.
Fourier	30 S	53 W	58	W. of Humorum. Terraced, with central crater.
Fracastorius	21 S	33 E	97	Bay at S. of Nectaris.
Fra Mauro	6 S	17 W	81	On Nubium; low walls. Trio with Bonpland and Parry.
Franklin	39 N	48 E	55	Regular, S.E. of Atlas.
Franz	16 N	40 E	34	Edge of Somnii. Low-walled.
Fraunhofer	39 S	59 E	48	S. of Furnerius. N.W. wall broken by craters.
Furnerius	36 S	60 E	129	In Petavius chain. Rather broken walls.
Galilaei	10 N	63 W	15	On Procellarum. Inconspicuous.
Galle	56 N	22 E	24	On Frigoris. Distinct.
Gambart	1 N	15 W	26	On Procellarum, 200 km S.S.E. of Copernicus. Regular. Low walls.
Gärtner	60 N	34 E	101	Bay on Frigoris; 'seaward' wall barely traceable.
Gassendi	18 S	40 W	89	Edge of Humorum; 'seaward' wall low. c.p. Many rills on floor.
Gaudibert	11 S	37 E	26	Edge of Nectaris. Low walls.
Gauricus	34 S	12 W	64	Pitatus group. Irregular outline.
Gauss	36 N	80 E	136	Along limb from Humboldtianum. High walls; c.p.
Gay-Lussac	14 N	21 W	24	N. of Copernicus. Irregular enclosure.
Geber	20 S	14 E	40	Between Almanon and Abenezra. Regular.
Geminus	35 N	57 E	90	Crisium area. Broad, terraced walls; central hill.
Gemma Frisius	34 S	14 E	84	N. of Maurolycus. High but broken walls.
Gérard	44 N	75 W	87	W. of Roris. Fairly distinct.
Gioja	North polar		35	Fairly distinct and regular.
Glaisher	13 N	49 E	16	W. border of Crisium. Obscure.
Goclenius	10 S	45 E	52	Edge of Fœcunditatis. Lava-flooded. Low c.p.
Godin	2 N	10 E	43	S. of Vaporum; pair with Agrippa. c.p.
Goldschmidt	75 N	0	109	E. of Anaxagoras (N. of Frigoris). Low walls; broken.
Goodacre	33 N	14 E	48	E. of Aliacensis. Low c.p.
Gould	19 S	17 W	42	'Ghost' in Nubium, E. of Bullialdus.
Grimaldi	6 S	68 W	193	W. of Procellarum. Low, irregular walls. No c.p. Darkest point on the Moon.
Grove	40 N	33 E	26	In Somniorum. Bright and deep.
Gruemberger	68 S	10 W	87	Moretus area. High walls; contains a very deep crater.
Gruithuisen	33 N	40 W	16	On Procellarum. Bright.
Guericke	12 S	14 W	53	In Nubium (Fra Mauro area). Broken walls; irregular.
Gutenburg	8 S	41 E	72	Edge of Fœcunditatis, near Goclenius. Irregular.
Gyldén	5 S	1 E	48	N. of Ptolemæus; floor partly filled with lava. Crater-valley abuts on W.
Hagecius	60 S	46 E	81	Vlacq group. Wall broken by craters.
Hahn	31 N	74 E	56	E. of Cleomedes (Crisium area); pair with Berosus. Regular; c.p.
Haidinger	39 S	25 W	12	S. of Epidemiarum. Not conspicuous.
Hainzel	41 S	34 W	97	N. of Schiller. Compound; 2 coalesced rings.
Hall	34 N	37 E	25	Somniorum area, E. of Posidonius.
Halley	8 S	6 E	35	Hipparchus group; pair with Hind. Regular.
Hanno	57 S	73 E	64	Australe area. Darkish floor.
Hansen	44 S	83 E	36	Crisium area, near Agarum; similar to Alhazen. Regular.
Hansteen	12 S	52 W	52	S. edge of Procellarum; pair with Billy. Regular.
Harding	43 N	70 W	23	Roris area. Low walls.
Harpalus	53 N	43 W	52	Edge of Frigoris. Deep.
Hase	30 S	63 E	77	S. of Petavius. Rather irregular.
Heinsius	39 S	18 W	72	Tycho area. Irregular: 3 craters on its S. wall.
Heis	32 N	32 W	12	On Imbrium. Bright.

OUTLINE MAP of the MOON
Patrick Moore

(Drawn by Patricia A. Cullen)
2nd. Edition 1969

Name	Lat	Long	Diam, km	Notes
Hekatæus	23 S	84 E	180	S.E. of Vendelinus. Irregular walls; c.p.
Helicon	40 N	23 W	29	Iridum area. Pair with Le Verrier. Regular.
Hell	32 S	8 W	31	In W. of Deslandres. Low c.p.
Helmholtz	72 N	78 E	97	S. uplands; Boussingault area. Fairly regular.
Henry, Paul	24 S	57 W	40	W. of Humorum; distinct. Pair with Prosper Henry.
Henry, Prosper	24 S	59 W	45	Pair with Paul Henry.
Heraclitus	49 S	6 E	92	Cuvier-Licetus group, S. of Stöfler. Very irregular.
Hercules	46 N	39 E	72	Companion to Atlas. Bright walls; deep interior crater.
Herigonius	13 S	34 W	16	N.E. of Gassendi. Bright; c.p.
Hermann	1 S	57 W	16	On Procellarum, E. of Lohrmann. Bright.
Herodotus	23 N	50 W	37	Pair with Aristarchus; fairly regular. Associated with great valley.
Herschel	6 S	2 W	45	N. of Ptolemæus. Terraced walls; large c.p.
Herschel, Caroline	34 N	31 W	13	On Imbrium; group with Carlini and De l'Isle. Bright.
Herschel, John	62 N	41 W	145	N. of Frigoris. Ridge-bordered enclosure.
Hesiodus	29 S	16 W	45	Companion to Pitatus (S. of Nubium). A major rill runs S.W. from it.
Hevel	2 N	67 W	122	Grimaldi chain. Convex floor, containing rills. c.p.
Hind	8 S	7 E	26	Hipparchus group; pair with Halley. Regular.
Hippalus	25 S	30 W	61	Bay on Humorum; remnant of c.p. Associated with rills.
Hipparchus	6 S	5 E	145	E. of Ptolemæus; pair with Albategnius. Low, broken walls.
Holden	19 S	63 E	40	S. of Vendelinus. Deep.
Hommel	54 S	33 E	121	S. uplands. 2 large craters on floor.
Hooke	41 N	55 E	43	W. of Messala. Fairly regular.
Horrebow	59 N	41 W	32	Frigoris area, outside J. Herschel. Deep; pair with Robinson.
Horrocks	4 S	6 E	29	Within Hipparchus. Regular.
Hortensius	6 N	28 W	16	On Procellarum. Bright. Domes to N.
Huggins	41 S	2 W	48	Encroaches on Orontius. Irregular.
Humboldt, W.	27 S	81 E	193	E. of Petavius. Irregular. Rills on floor.
Hyginus	8 N	6 E	6	Depression on Vaporum. In great crater-rill.
Hypatia	4 S	23 E	48	S. of Tranquillitatis. Triangular; low walls.
Ideler	49 S	22 E	44	S.E. of Maurolycus. Distinct.
Inghirami	48 S	70 W	97	Schickard area. Regular; c.p.
Isidorus	8 S	33 E	48	Outside Nectaris; pair with Capella. Deep.
Jacobi	57 S	12 E	66	Heraclitus group. Fairly distinct.
Jansen	14 N	29 E	26	In Tranquillitatis. Low walls. Darkish floor.
Janssen	46 S	40 E	170	S. uplands. Great ruin, broken in N. by Fabricius.
Julius Cæsar	9 N	15 E	71	Vaporum area. Low, irregular walls. Very dark floor.
Kaiser	36 S	6 E	48	N. of Stöfler. Well-marked. No c.p.
Kane	63 N	26 E	45	N. of Frigoris. Fairly regular.
Kant	11 S	20 E	30	W. of Theophilus. Large c.p. with summit pit.
Kapteyn	13 S	70 E	35	E. of Langrenus. Not prominent.
Kästner	6 S	80 E	129	Smythii area. Distinct.
Kepler	8 N	38 W	35	On Procellarum. Major ray-centre.
Kies	26 S	23 W	42	On Nubium; Bullialdus area. Very low walls.
Kinau	60 S	15 E	42	Jacobi group. High walls; c.p.
Kirch	39 N	6 W	11	In Imbrium. Bright.
Kircher	67 S	45 W	74	Bettinus chain; Bailly area. Very high walls.
Kirchhoff	30 N	39 E	50	Cleomedes area; larger of 2 craters W. of Newcomb.
Klein	12 N	3 E	43	In Albategnius. Regular; c.p.
König	24 S	25 W	21	In Nubium; Bullialdus area. Deep.
Krieger	29 N	46 W	23	Aristarchus area. Distinct, but walls broken by craterlets.
Krusenstern	26 S	6 E	48	Werner area. Not prominent.
Kunowsky	3 N	32 W	21	On Procellarum, E. of Encke. Low central ridge.
Lacaille	24 S	1 E	53	Werner area. Irregular.
Lacroix	38 S	59 W	38	N. of Schickard. Fairly regular; c.p.
Lade	1 S	10 E	48	Highlands S. of Godin. Low walls.
Lagalla	45 S	23 W	63	Abuts on Wilhelm I. Low-walled, irregular.

Schröter's Valley meanders in from the right in this oblique view of the Moon's Ocean of Storms from the orbiting Apollo-15 spacecraft. The bright crater at upper left is Aristarchus, measuring about 35 kilometres in diameter. At right is the crater Herodotus. South is at the top. The picture was made with the 3-inch (75 mm) mapping camera, one of eight lunar orbit scientific experiments mounted in a bay of the service-propulsion section. Astronaut Al Worden crawled out of the command capsule and retrieved the film during the flight back to Earth.

The Hyginus Rill, from Orbiter. Hyginus itself is the crater at the main bend in the rill. This is one of the most famous rills on the Moon, and a small telescope on Earth will show it, though not of course in great detail.

Name	Lat	Long	Diam, km	Notes
Lagrange	33 S	72 W	165	W. of Humorum. Low walls; rough floor.
Lalande	4 S	8 W	24	Ptolemæus area. Low c.p.
Lambert	26 N	21 W	29	In Imbrium. Bright. Central crater.
Lamé	15 S	65 E	81	On N.E. wall of Vendelinus. Well-formed.
Lamèch	43 N	13 E	95	Eudoxus area. V. low, irregular walls.
Lamont	5 N	23 E	62	On Tranquillitatis; Arago area. Low walls.
Landsberg	0	26 W	42	On Nubium. Massive walls; c.p.
Langrenus	9 S	61 E	137	E. edge of Fœcunditatis. Massive walls; c.p. Petavius chain.
La Peyrouse	10 S	78 E	72	E. of Fœcunditatis; pair with Ansgarius. Distinct.
Lassell	16 S	8 W	23	In Nubium; Alphonsus area. Low walls.
La Voisier	36 N	70 W	71	W. of Procellarum. Well-formed.
Lee	31 S	41 W	42	S. edge of Humorum; damaged by lava.
Legendre	29 S	70 E	74	W. Humboldt area. Central ridge.
Legentil	73 S	80 E	140	S. of Bailly. Distinct.
Lehmann	40 S	56 W	45	Schickard area. Irregular.
Le Monnier	26 N	31 E	55	Bay on Serenitatis; smooth floor.
Lepaute	33 S	34 W	14	Edge of Epidemiarum. Distinct.
Letronne	10 S	43 W	113	Bay at S. edge of Procellarum; N. wall destroyed. Low c.p.
Le Verrier	40 N	20 W	25	Imbrium area; pair with Helicon. Distinct.
Lexell	36 S	4 W	63	Edge of Deslandres; N. wall reduced. Remnant of c.p.
Licetus	47 S	6 E	74	Cuvier-Heraclitus group (Stöfler area). Fairly regular; c.p.
Lichtenberg	32 N	68 W	19	Edge of Procellarum. Minor ray-centre.
Lick	12 N	53 E	35	Edge of Crisium. Incomplete.
Liebig	24 S	48 W	45	W. of Humorum. Moderate walls.
Lilius	54 S	6 E	52	Jacobi group. High walls; c.p.
Lindenau	32 S	25 E	56	Rabbi Levi group. Terraced walls.
Linné	28 N	12 E	11	On Serenitatis. Craterlet surrounded by nimbus.
Lippershey	26 S	10 W	6	On Nubium; Pitatus area. Distinct.
Littrow	22 N	31 E	35	Edge of Serenitatis. Irregular.
Lockyer	46 S	37 E	48	Intrudes into Janssen. Bright walls.
Loewy	23 S	33 W	23	E. edge of Humorum. Distinct.
Lohrmann	1 S	67 W	45	Between Grimaldi and Hevel. c.p.
Lohse	14 S	60 E	32	On N. wall of Vendelinus. Deep; c.p.
Longomontanus	50 S	21 W	145	Clavius area. Complex walls; much floor detail.
Louville	44 N	47 W	35	In Jura Mts. Low walls; darkish floor.
Lubbock	4 S	42 E	13	W. edge of Fœcunditatis. Fairly bright.
Lubiniezky	18 S	24 W	45	In Nubium, Bullialdus area. Low walls.
Luther	33 N	24 E	8	In Serenitatis (N. part). Distinct.
Lyell	14 N	41 E	43	W. edge of Sumnii. Darkish floor.
Maclaurin	2 S	68 E	45	W. of Smythii. Concave floor; uneven walls.
Maclear	11 N	20 E	18	On Tranquillitatis. Darkish floor.
McClure	15 S	50 E	23	In Fœcunditatis (Goclenius area). Bright.
Macrobius	21 N	46 E	68	Crisium area: high walls, compound c.p.
Mädler	11 S	30 E	32	On Nectaris. Irregular.
Maestlin	5 N	41 W	50	On Procellarum, near Encke. Obscure.
Magelhæns	12 S	44 E	40	Edge of Fœcunditatis; pair with A. Darkish floor.
Maginus	50 S	6 W	177	Clavius area; irregular walls. Obscure near Full Moon.
Main	81 N	9 E	48	N. polar area; twin with Challis.
Mairan	42 N	43 W	40	Jura Mts. area. Bright; not perfectly regular.
Mallet	45 S	53 E	50	Rheita Valley area; not conspicuous.
Manilius	15 N	9 E	36	On edge of Vaporum. c.p.; brilliant walls.
Manners	5 N	20 E	16	On Tranquillitatis (Arago area). Bright.
Manzinus	68 S	25 E	90	Boguslawsky area. High, terraced walls.
Maraldi	19 N	35 E	37	In N. of Tranquillitatis. Distinct.
Marco Polo	15 N	2 W	19	Apennines area. Irregular; darkish floor.
Marinus	50 S	75 E	48	Australe area. Distinct, with c.p.
Marius	12 N	51 W	42	In Procellarum; pair with Reiner. Low c.p.
Marth	31 S	29 W	13	In Epidemiarum. Concentric crater.
Maskelyne	2 N	30 E	24	In Tranquillitatis. Low c.p.
Mason	43 N	30 E	31	Bürg area; companion to Plana.
Maupertuis	50 N	27 W	42	In Juras. Irregular mountain enclosure.
Maurolycus	42 S	14 E	109	E. of Stöfler. Rough floor; central peak group.
Maury	37 N	40 E	18	Atlas area. Bright and deep.
Mayer, C.	63 N	17 E	39	N. of Frigoris. Rhomboidal.
Mayer, T.	16 N	29 W	35	In Carpathian Mts. c.p.
Mee	44 S	35 W	114	Abuts on Hainzel. Low, broken walls.

Name	Lat	Long	Diam, km	Notes
Menelaus	16 N	16 E	32	In Hæmus Mts. Brilliant. Interior peak.
Mercator	29 S	26 W	38	Pair with Campanus (Humorum area). Noticeably dark floor.
Mercurius	56 N	65 E	53	Humboldtianum area. Low c.p.
Mersenius	21 S	49 W	72	W. of Humorum. Convex floor. Rills nearby.
Messala	39 N	60 E	128	Humboldtianum area. Oblong; low, broken walls.
Messier	2 S	48 E	13	On Fœcunditatis. Twin with Messier A (=W. H. Pickering). 'Comet' ray to W.
Metius	40 S	44 E	81	Janssen group. Pair with Fabricius. Distinct.
Meton	74 N	25 E	170	N. polar area, near Scoresby. Compound formation; smooth floor.
Milichius	10 N	30 W	13	On Procellarum. Bright. Dome to the W.
Miller	39 S	1 E	48	Orontius group. Fairly distinct.
Mitchell	50 N	20 E	19	Abuts on Aristoteles. Distinct.
Möltke	1 S	24 E	7	N. edge of Tranquillitatis. Distinct.
Monge	19 S	48 E	32	Edge of Fœcunditatis. Rather irregular.
Montanari	46 S	20 W	87	Longomontanus area. Distorted.
Moretus	70 S	8 W	105	Southern uplands. Massive walls; very high c.p.
Mösting	1 S	6 W	26	Medii area. Mösting A, to the S.S.E. is a bright craterlet used as a reference point on older maps.
Müller	7 S	3 E	24	Ptolemæus area. Fairly regular.
Murchison	5 N	0	56	Edge of Medii. Low-walled; irregular.
Mutus	63 S	30 E	81	S. uplands. 2 large craters on floor.
Nasireddin	41 S	0	48	Orontius group. Fairly distinct.
Nasmyth	52 S	53 W	74	Phocylides group. Fairly regular; no c.p.
Naumann	35 N	62 W	10	In N. of Procellarum. Bright walls.
Neander	31 S	40 E	48	Rheita Valley area. Well formed.
Nearch	58 S	39 E	61	Vlacq area. Craterlets on floor.
Neison	68 N	28 E	45	Meton area. Regular; no c.p.
Neper	7 N	83 E	113	Marginis/Smythii area. Deep.
Neumayer	62 S	75 E	80	Boussingault area. Distinct.
Newcomb	30 N	44 E	52	Cleomedes area. S. wall broken by crater.
Newton	78 S	20 W	113	Moretus area. Unusually deep; irregular.
Nicolai	42 S	26 E	43	Janssen area. Regular.
Nicollet	22 S	12 W	16	In Nubium, W. of Birt. Distinct.
Nöggerath	49 S	45 W	35	Schiller area. Low walls.
Nonius	35 S	4 E	32	Stöfler area. Fairly regular.
Œnopides	57 N	65 W	68	Limb area, near Roris. High walls.
Œrsted	43 N	47 E	40	Edge of Somniorum. Rather irregular.
Oken	44 S	78 E	80	Australe area. Darkish floor.
Olbers	7 N	78 W	64	Grimaldi area. Major ray-centre.
Opelt	16 S	18 W	43	'Ghost' in Nubium, E. of Bullialdus.
Oppolzer	2 S	1 W	48	Edge of Medii. Low walls.
Orontius	40 S	4 W	84	N.E. of Tycho; one of a group. Irregular.
Palisa	9 S	7 W	32	Alphonsus area. On edge of larger ring.
Palitzsch	28 S	64 E	97×32	Outside Petavius to the E. Really a crater-chain.
Pallas	5 N	2 W	47	Medii area; adjoins Murchison. c.p.
Palmieri	29 S	48 W	48	Humorum area. Floor darkish.
Parrot	15 S	3 E	64	Albategnius area. Very irregular, compound structure.
Parry	8 S	16 W	42	On Nubium; Fra Mauro group. Fairly regular.
Peirce	18 N	53 E	19	In Crisium. Fairly regular.
Peirescius	46 S	71 E	64	Australe area. Rather irregular.
Pentland	64 S	12 E	72	S. uplands, near Curtius. High walls; c.p.
Petavius	25 S	61 E	170	One of a great chain. High walls; c.p.; interior rill.
Phillips	26 S	78 E	120	W. of W. Humboldt. Central ridge.
Philolaus	75 N	33 W	74	Frigoris area; pair with Anaximenes. Deep regular.
Phocylides	54 S	58 W	97	Schickard group. Much floor-detail.
Piazzi	36 S	68 W	90	Schickard area. Very broken walls.
Piazzi Smyth	42 N	3 W	10	On Imbrium. Bright.
Picard	15 N	55 E	34	Largest crater on Crisium. C. hill.
Piccolomini	30 S	32 E	80	End of Altai Scarp. High walls; c.p.
Pickering	3 S	7 E	16	Hipparchus area. Distinct.
Pictet	43 S	7 W	48	Closely E. of Tycho. Fairly regular.
Pitatus	30 S	14 W	80	S. edge of Nubium. Passes connect it with Hesiodus.
Pitiscus	51 S	31 E	80	S. uplands. Walls narrow in places.

Mare Serenitatis (Cdr H. R. Hatfield)

Near side of the Moon (P. W. Foley)

Name	Lat	Long	Diam, km	Notes
Plana	42 N	28 E	39	Bürg area; pair with Mason. Darkish floor.
Plato	51 N	9 W	97	Edge of Imbrium; Alps area. Regular; smooth, very dark floor.
Playfair	23 S	9 E	43	Abenezra area. Fairly regular.
Plinius	15 N	24 E	48	Between Serenitatis and Tranquillitatis. C. craters.
Plutarch	25 N	75 E	64	N.E. of Crisium. c.p.
Poisson	30 S	11 E	72	Aliacensis area. Very irregular; compound.
Polybius	22 S	26 E	32	Theophilus/Catharina area. Distinct.
Pons	25 S	22 E	32	Altai area. Very thick walls.
Pontanus	28 S	15 E	45	Altai area. Regular; no c.p.
Pontécoulant	69 S	65 E	97	S. uplands. High walls.
Posidonius	32 N	30 E	96	Edge of Serenitatis. Narrow walls. Much floor detail.
Porter	56 S	10 W	40	On wall of Clavius. Very distinct. c.p.
Prinz	26 N	44 W	50	In Procellarum, N.E. of Aristarchus. Incomplete. Domes nearby.
Proclus	16 N	47 E	29	W. of Crisium. Bright; low c.p. Ray-crater.
Proctor	46 S	5 W	43	Maginus area. Fairly regular.
Protagoras	56 N	7 E	19	In Frigoris. Bright, regular.
Ptolemæus	14 S	3 W	148	Trio with Alphonsus and Arzachel. Floor smooth, but with details including a large crater, A.
Puiseux	28 S	39 W	23	On Humorum, near Doppelmayer. Very low walls.

Name	Lat	Long	Diam, km	Notes
Purbach	25 S	2 W	120	Walter group. Rather irregular.
Pythagoras	65 N	65 W	113	N.W. of Iridum. High, massive walls; c.p.
Pytheas	21 N	20 W	19	On Imbrium. Bright; c.p. Minor ray-centre.
Rabbi Levi	35 S	24 E	80	One of a group of 6, S.W. of Piccolomini.
Ramsden	33 S	32 W	24	Edge of Epidemiarum. Rills nearby.
Réaumur	2 S	1 E	45	Medii area. Low walls.
Regiomontanus	28 S	0	129 × 105	Prominent, though distorted between Walter and Purbach. c.p.
Reichenbach	30 S	48 E	48	Rheita area; irregular. Crater-chain to S.E. ('Reichenbach Valley').
Reimarus	46 S	55 E	45	Rheita Valley area. Irregular.
Reiner	7 N	55 W	32	On Procellarum; pair with Marius. Low c.p.
Reinhold	3 N	23 W	48	On Procellarum, S.W. of Copernicus. Pair with Reinhold B, to the N.E.
Repsold	50 N	70 W	140	W. of Roris. One of a group.
Rhæticus	0	5 E	45	Medii area. Low walls.
Rheita	37 S	47 E	68	W. of Furnerius; sharp crests. Associated with great crater-chain ('Rheita Valley').
Riccioli	3 S	75 W	160	Companion to Grimaldi. Broken walls; very dark patches on floor. No c.p.
Riccius	37 S	26 E	80	Rabbi Levi group. Broken walls, rough floor.
Ritchey	11 S	9 E	26	E. of Albategnius. Broken walls.
Ritter	2 N	19 E	31	On Tranquillitatis; pair with Sabine. c.p.
Robinson	59 N	46 W	27	Frigoris area; similar to Horrebow. Distinct.
Rocca	15 S	72 W	97	S. of Grimaldi. Irregular walls.
Rømer	25 N	37 E	37	Taurus area. Very massive c.p. with summit pit.
Rosenberger	55 S	43 E	80	Vlacq group. Darkish floor; c.p.
Ross	12 N	22 E	29	On Tranquillitatis. c.p.
Rosse	18 S	35 E	16	On Nectaris. Bright.
Rost	56 S	34 W	55	Schiller area; pair with Weigel. Fairly regular.
Rothmann	31 S	28 E	42	Altai area. Fairly deep and regular.
Rümker	41 N	58 W	49	On Roris. Semi-ruined plateau.
Rutherfurd	61 S	12 W	40	On wall of Clavius. Distinct; c.p.
Sabine	2 N	20 E	31	On Tranquillitatis; pair with Ritter. c.p.
Sacrobosco	24 S	17 E	84	Altai area. Irregular.
Santbech	21 S	44 E	70	E. of Fracastorius. Darkish floor.
Sasserides	39 S	9 W	97	N. of Tycho. Irregular enclosure.
Saunder	4 S	9 E	43	E. of Hipparchus. Low walls.
Saussure	43 S	4 W	50	N. of Maginus. Interrupts larger ring.
Scheiner	60 S	28 W	113	Clavius area; pair with Blancanus. High walls; interior craterlet.
Schiaparelli	23 N	59 W	29	On Procellarum. Distinct.
Schickard	44 S	54 W	202	Rather low walls; great walled plain.
Schiller	52 S	39 W	180 × 97	Schickard area. Fusion of 2 rings.
Schmidt	1 N	19 E	13	In Tranquillitatis; Sabine/Ritter area. Bright.
Schömberger	76 N	30 E	65	Boguslawsky area. Regular.
Scoresby	77 N	15 E	58	Polar uplands. Deep; prominent. c.p.
Schröter	3 N	7 W	32	Medii area. Low walls.
Schubert	3 N	70 E	74	In Smythii area. Distinct.
Schumacher	42 N	55 E	40	Messala area. Fairly distinct.
Secchi	2 N	43 E	23	In Fœcunditatis. Bright walls; c.p.
Seeliger	2 S	1 E	45	N. of Hipparchus. Irregular.
Segner	59 S	48 W	74	Schiller area; pair with Zucchius. Distinct.
Seleucus	21 N	66 W	45	On Procellarum. Terraced walls; c.p.
Sharp	46 N	40 W	35	In Jura Mts. Deep; small c.p.
Sheepshanks	59 N	17 E	19	N. of Frigoris. Fairly regular.
Short	76 S	5 W	70	Moretus group. Deep; high walls.
Shuckburgh	43 N	53 E	47	Cepheus group. Fairly regular.
Silberschlag	6 N	13 E	13	Near Ariadæus. Bright.
Simpelius	75 S	15 E	80	Moretus area. Deep.
Sinas	9 N	32 E	8	On Tranquillitatis. Not prominent.
Sirsalis	13 S	60 W	32	Just outside Procellarum; twin with Sirsalis A. Associated with great rill.
Snellius	29 S	56 E	80	Furnerius area; pair with Stevinus. High walls; c.p.
Sömmering	0	7 W	27	Medii area. Low walls.
Sosigenes	9 N	18 E	23	Edge of Tranquillitatis. Bright; low c.p.
South	57 N	50 W	98	Frigoris area. Ridge-bounded enclosure.
Spallanzani	46 S	25 E	25	W. of Janssen. Low walls.
Spörer	4 S	2 W	35	N. of Ptolemæus and Herschel. Partly lava-filled.
Stadius	11 N	14 W	70	Ghost ring E. of Copernicus. Pitted.
Steinheil	50 S	48 E	70	S.W. of Janssen. Twin with Watt.

The Straight Wall, showing the immense crater-chain of Walter, Regiomontanus and Purbach (Cdr H. R. Hatfield)

Name	Lat	Long	Diam, km	Notes
Stevinus	33 S	54 E	70	Furnerius area; pair with Snellius. High walls; c.p.
Stiborius	34 S	32 E	44	S. of Piccolomini (Altai area). c.p.
Stöfler	41 S	6 E	145	S.E. of Walter; broken by Faraday. Smooth darkish floor.
Strabo	62 N	55 E	47	Near De la Rue. Minor ray-centre.
Street	46 S	10 W	50	S. of Tycho. Fairly regular; no c.p.
Struve	43 N	65 E	18	Adjoins Messala; lies on dark patch.
Struve, Otto	25 N	75 W	160	—each of 2 ancient rings; W. edge of Procellarum.
Suess	4 N	48 W	15	On Procellarum, W. of Encke. Obscure.
Sulpicius Gallus	20 N	12 E	13	On Serenitatis. Very bright.
Sven Hedin	5 N	75 W	98	W. of Hevel. Broken walls; irregular.
Tacitus	16 S	19 E	40	Catharina area. Polygonal; 2 floor-craters.
Tannerus	56 S	22 E	24	S. uplands, near Mutus, c.p.
Taquet	17 N	19 E	10	Just on Serenitatis, near Menelaus. Bright.
Taruntius	6 N	48 E	60	On Fœcunditatis. Concentric crater, with c.p. Rather low walls.
Taylor	5 S	17 E	40	Delambre area. Rather elliptical.
Tempel	4 N	12 E	8	Uplands W. of Tranquillitatis. Bright.
Thales	59 N	41 E	39	Near Strabo. Major ray-centre.
Theætetus	37 N	6 E	26	On Nebularum. Low c.p.
Thebit	22 S	4 W	60	Edge of Nubium, near Arzachel. W. wall broken by A, which is itself broken by F— a characteristic trio.
Theon Junior	2 S	16 E	16	Near Delambre: pair with Theon Senior. Bright.
Theon Senior	1 S	15 E	17	Pair with Theon Junior. Bright.
Theophilus	12 S	26 E	101	Edge of Nectaris. Very deep; massive walls; c.p. Chain with Cyrillus and Catharina.
Timæus	63 N	1 W	34	Edge of Frigoris. Bright.
Timocharis	17 N	13 W	35	On Imbrium. Central crater. Minor ray-centre.
Timoleon	35 N	75 E	130	Adjoins Gauss. Fairly distinct.
Tisserand	21 N	48 E	33	Crisium area. Regular; no c.p.
Torricelli	5 S	29 E	19	On Nectaris. Irregular, compound structure.
Tralles	28 N	53 E	48	On wall of Cleomedes. Very deep.
Triesnecker	4 N	4 E	23	Vaporum area. Associated with major rill-system.
Trouvelot	49 N	6 E	10	Alpine Valley area. Rather bright.
Turner	2 S	13 W	13	On Nubium, near Gambart. Deep.
Tycho	43 S	11 W	84	Brilliant southern ray-centre—greatest on the Moon. Terraced walls; c.p.
Ukert	8 N	1 E	23	Edge of Vaporum. Rills nearby.
Ulugh Beigh	29 N	85 W	70	W. of Procellarum. High walls; c.p.
Vasco de Gama	15 N	85 W	80	W. of Procellarum. Central ridge.
Vega	45 N	63 E	80	Australe area. Deep.
Vendelinus	16 S	62 E	165	Petavius chain. Irregular and broken.
Vieta	29 S	57 W	52	W. of Humorum. Low c.p.
Vitello	30 S	38 W	38	S. edge of Humorum. Concentric crater.
Vitruvius	18 N	31 E	31	Between Serenitatis and Tranquillitatis. Low but rather bright walls.
Vlacq	53 S	39 E	90	S.W. of Janssen; one of a group of six. Deep, with c.p.
Vogel	15 S	6 E	60×25	S.E. of Albategnius. Chain of 4 craters.
Wallace	10 N	9 W	29	In Imbrium; Apennine area. Imperfect ring; very low walls.
Walter	33 S	1 E	129	Outside Nubium; massive walls, interior peak and craters. Trio with Regiomontanus and Purbach.
Wargentin	50 S	60 W	89	Schickard group. The famous plateau.
Watt	50 S	51 E	72	S.W. of Janssen. Twin with Steinheil.
Webb	1 S	60 E	26	Edge of Fœcunditatis. Darkish floor; c.p. Minor ray-centre.
Weigel	58 S	39 W	55	Schiller area; pair with Rost. Fairly regular.
Weinek	28 S	37 E	30	N.E. of Piccolomini (Altai area). Darkish floor.
Weiss	32 S	20 W	60	Pitatus group. Very irregular.
Werner	28 S	3 E	66	Pair with Aliacensis. Very regular; high walls, c.p.
Whewell	4 N	14 E	8	Uplands W. of Tranquillitatis. Bright.
Wichmann	8 S	38 W	13	On Procellarum. Associated with large ghost-crater.

Crescent Moon. The craters on the terminator, from the top (south), are Furnerius, Petavius, Vendelinus and Langrenus; part of the Mare Crisium, over which the Sun is just rising, is seen below the centre of the terminator.

Name	Lat	Long	Diam, km	Notes
Wilhelm I	43 S	20 W	97	Longomontanus area. Uneven walls.
Wilkins	30 S	20 E	64	Rabbi Levi group. Irregular.
Williams	42 N	37 E	45	Somniorum area. Not prominent.
Wilson	69 S	33 W	74	Bettinus chain, near Bailly. Deep and regular. No c.p.
Wöhler	38 S	31 E	45	Near Rabbi Levi group. Fairly regular.
Wolf	23 S	17 W	32	In S. of Nubium. Irregular; low walls.
Wollaston	31 N	47 W	11	N. of Harbinger Mts. Bright.
Wrottesley	24 S	57 E	55	Petavius group. Twin-peaked c. mountain.
Wurzelbauer	34 S	16 W	80	Pitatus group. Irregular walls; much floor-detail.
Xenophanes	57 N	77 W	108	Limb near Roris. High walls; c.p.
Yerkes	15 N	52 E	25	On W. edge of Crisium. Irregular. Low walls.
Young	42 S	51 E	45	Rheita Valley area. Irregular.
Zach	61 S	5 E	52	E. of Clavius. Fairly regular and deep.
Zagut	32 S	22 E	80	Rabbi Levi group. Irregular.
Zöllner	8 S	19 E	40	N.W. of Theophilus. Elliptical.
Zucchius	61 S	50 W	63	Schiller area; pair with Segner. Distinct.
Zupus	17 S	52 W	26	S. of Billy. Low walls; irregular. Very dark floor.

RECENTLY-NAMED LUNAR CRATERS

Most of the craters in this list are relatively small. The names have however been approved by the International Astronomical Union.

Name	Lat	Long	Former designation
Abbot	6 N	55 W	Apollonius K
Banting	27 N	16 E	Linné E
Bowen	18 N	9 E	Manilius A
Brackett	18 N	24 E	
Cajal	13 N	31 E	Jansen F
Cameron	6 N	46 E	Taruntius C
Carmichael	20 N	40 E	Macrobius A
Clerke	22 N	29 E	Littrow B
Curtis	16 N	57 E	
Daly	5 N	57 E	Apollonius A
Daubrée	16 N	15 E	Menelaus S
Eckert	18 N	58 E	
Franck	23 N	36 E	Rømer K
Freud	26 N	52 W	
Galen	22 N	5 E	Aratus A
Hadley	25 N	3 E	Hadley C
Haldane	1 S	84 E	
Hill	21 N	41 E	Macrobius B
Hornsby	24 N	12 E	Aratus CB
Houtermans	9 S	87 E	
Humason	31 N	57 W	Lichtenberg G
Huxley	20 N	5 W	Wallace B
Joy	25 N	7 E	Hadley A
Kiess	6 S	84 E	
Knox-Shaw	5 N	80 E	Banachiewicz F
Kreiken	9 S	85 E	
Lawrence	8 N	43 E	Taruntius M
Lucian	15 N	37 E	Maraldi B
Nielsen	32 N	52 W	Wollaston C
Peek	3 N	87 E	
Runge	2 S	87 E	
Sarabhai	25 N	21 E	Bessel A
Shapley	10 N	57 W	Picard H
Spurr	26 N	3 W	Archimedes M
Tacchini	5 N	86 E	Neper K
Tebbutt	10 N	54 E	Picard G
Theophrastus	18 N	39 E	Maraldi M
Väisälä	26 N	48 W	Aristarchus A
Very	26 N	25 E	Le Monnier B
Watts	9 N	46 E	Taruntius D
Widmanstätten	6 S	86 E	
Yangel	17 N	5 E	Manilius F
Zinner	27 N	59 W	Schiaparelli B

NAMED CAPES AND PROMONTORIES

	Lat	Long	
Acherusia	17 N	22 E	Serenitatis: E. end of Hæmus Mountains.
Ænarium	19 S	7 W	Nubium: Straight Wall area.
Agarum	14 N	66 E	Crisium, near Condorcet.
Agassiz	42 N	2 E	Imbrium, near Cassini.
Argæus	19 N	29 E	Serenitatis; near Littrow.
Banat	17 N	26 W	Carpathians, near T. Mayer.
Deville	46 N	0	Imbrium: Alpine area, near Agassiz.
Fresnel	29 N	5 E	Serenitatis; end of Apennines.
Heraclides	40 N	34 W	Sinus Iridum.
Kelvin	27 S	33 W	Humorum, near Hippalus.
Laplace	46 N	26 W	Sinus Iridum.

MOUNTAIN RANGES

Alps	N. border of Imbrium.
Altai Scarp	S.W. of Nectaris, from Piccolomini.
Apennines	Bordering Imbrium.
Carpathians	Bordering Imbrium, to the S.
Caucasus	Separating Serenitatis and Nebularum.
Cordillera	Limb range: Grimaldi/Darwin area.
Hæmus	S. border of Serenitatis.
Harbinger	Clumps of hills in Imbrium (Aristarchus area).
Jura	Bordering Iridum.
Percy	N.W. border of Humorum. Not a major range.
Pyrenees	Clumps of hills bordering Nectaris, to the E.
Riphæans	In Nubium. Short range.
Rook	Limb range, associated with Orientale.
Spitzbergen	In Imbrium, N. of Archimedes. Mountain clump.
Stag's-Horn	At S. end of Straight Wall (Nubium, near Thebit).
Straight Range	In Imbrium: Plato area. Very regular.
Taurus	Mountain clumps E. of Serenitatis.

Mare Nubium (Cdr H. R. Hatfield)

Teneriffe	In Imbrium, S. of Plato. Mountain clumps.
Ural	Extension of the Riphæans.

NAMED PEAKS

	Lat	Long	
Ampère	20 N	4 W	In Apennines.
Blanc	45 N	1 E	In Alps.
Bradley	22 N	1 E	In Apennines.
Hadley	27 N	5 E	In Apennines.
Huygens	20 N	3 W	In Apennines.
La Hire	28 N	25 W	In Imbrium (Lambert area).
Pico	46 N	9 W	In Imbrium, S. of Plato (edge of old ring).
Piton	41 N	1 W	In Imbrium, near Agassiz.
Schneckenberg	9 N	9 E	'Spiral mountain', near Hyginus Rill.
Serao	17 N	6 W	In Apennines.
Wolff	17 N	7 W	In Apennines.

NOMENCLATURE OF THE FAR SIDE OF THE MOON

Names allotted to features on the Moon's far side are listed here. Older maps show, on the Moon's near side, the Leibnitz Mountains and the D'Alembert Mountains; these have proved to be not true ranges, and the names have been deleted so that they could be re-allotted. The name of Porter, in honour of the American astronomer Russell W. Porter, was suggested for a far-side crater at latitude 56 °S, longitude 110 °W; but following a suggestion by the present writer, it was officially transferred to a crater inside Clavius, on the near side of the Moon. Early maps of the far side showed a mountain range, recorded by the Russian probe Luna 3 in 1959, and named the Soviet Mountains. The name has been deleted, for the excellent reason that the mountains do not exist. (The feature recorded was nothing more than a bright ray.)

The following craters are actually on the near side of the Moon, but have only been named recently, and are extremely difficult to study from Earth because they are so foreshortened:

Aston	32	N	88 W
Baade	47	S	83 W
Balboa	20	N	84 W
Barnard	32	N	86 E
Boole	65	N	89 W
Bunsen	41	N	85 W
Catalan	46	S	87 W
Dalton	18	N	86 W
Gibbs	19	S	85 E
Graff	43	S	88 W
Gum	40	S	89 E
Hamilton	44	S	83 E
Hartwig	7	S	82 W
Hayn	63	N	85 E
Hubble	22	N	87 E
Jansky	8	N	87 E
Krasnov	31	S	89 W
Liapunov	27	N	88 E
Lyot	48	S	88 W
Nicholson	27	S	85 W
Petrov	61	S	88 E
Pettit	28	S	86 W
Rynin	37	N	86 E
Schlüter	7	S	84 W
Shaler	33	S	88 W
Voskresensky	28	N	88 W
Wright	31	S	88 W

FEATURES ON THE FAR SIDE OF THE MOON

Features on the Moon's far side are named according to recommendations from the International Astronomical Union.

Name	Lat		Long
Abbe	58	S	175 E
Abul Wafa	2	N	117 E
Aitken	17	S	173 E
Al-Biruni	18	N	93 E
Alden	24	S	111 E
Alekhin	68	S	130 W
Alter	19	N	108 W
Amici	10	S	172 W
Amundsen	83	S	103 W
Anders	42	S	144 W
Anderson	16	N	171 E
Antoniadi	69	S	173 W
Apollo	37	S	153 W
Appleton	37	N	158 E
Arrhenius	55	S	91 W
Artamonov	26	N	104 E
Artemev	10	N	145 W
Avicenna	40	N	97 W
Avogadro	64	N	165 E
Babcock	4	N	94 E

Name	Lat		Long	Name	Lat		Long
Backlund	16	S	103 E	Cyrano	20	S	157 E
Baldet	54	S	151 W	Dædalus	6	S	180
Banachiewicz	51	N	135 W	D'Alembert	52	N	164 E
Barbier	24	S	158 E	Danjon	11	S	123 E
Barringer	29	S	151 W	Dante	25	N	180
Bartels	24	N	90 W	Das	14	S	152
Becquerel	41	N	129 E	Davisson	38	S	175 W
Běcvár	2	S	125 E	Dawson	67	S	134 W
Beijerinck	13	S	152 E	Debye	50	N	177 W
Belkovich	60	S	92 E	De Forest	76	S	162 W
Bell	22	N	96 W	Dellinger	7	S	140 E
Bellingshausen	61	S	164 W	Delporte	16	S	121 E
Belopolsky	18	S	128 W	Denning	16	S	143 E
Belyayev	23	N	143 E	De Roy	55	S	99 W
Bergstrand	19	S	176 E	Deutsch	24	N	111 E
Berkner	25	N	105 W	De Vries	20	S	177 W
Berlage	64	S	164 W	Dewar	3	S	166 E
Bhabha	56	S	165 W	Dirichlet	10	N	151 W
Birkeland	30	S	174 E	Donner	31	S	98 E
Birkhoff	59	N	148 W	Doppler	13	S	160 W
Bjerknes	38	S	113 E	Douglass	35	N	122 W
Blazhko	31	N	148 W	Dreyer	10	N	97 E
Bobone	29	N	131 W	Drude	39	S	91 W
Boltzmann	55	S	115 W	Dryden	33	S	157 W
Bolyai	34	S	125 E	Dufay	5	N	170 E
Borman	37	S	142 W	Dugan	65	N	103 E
Bose	54	S	170 W	Dunér	45	N	179 E
Boss	46	N	90 E	Dyson	61	N	121 W
Boyle	54	S	178 E	Dziewulski	21	N	99 E
Bragg	42	N	103 W	Edison	25	N	99 E
Brashear	74	S	172 W	Ehrlich	41	N	172 W
Bredikhin	17	N	158 W	Eijkman	62	S	141 W
Brianchon	77	N	90 W	Einthoven	5	S	110 E
Bridgman	44	N	137 E	Ellerman	26	S	121 W
Brouwer	36	S	125 W	Ellison	55	N	108 W
Brunner	10	S	91 E	Elvey	9	N	101 W
Buffon	41	S	134 W	Emden	63	N	176 W
Buisson	1	S	113 E	Engelhardt	5	N	159 W
Butlerov	9	N	109 W	Eötvös	36	S	134 E
Buys-Ballot	21	N	175 E	Erro	6	N	98 E
Cabannes	61	S	171 W	Esnault-Pelterie	47	N	142 W
Cajori	48	S	168 E	Espin	28	N	109 E
Campbell	45	N	152 E	Evans	10	S	134 W
Cannizzaro	55	N	100 W	Evdokimov	35	N	153 W
Cantor	38	N	118 E	Evershed	36	N	160 W
Carnot	52	N	144 W	Fabry	43	N	100 E
Carver	43	S	127 E	Fechner	58	S	125 E
Cassegrain	52	S	113 E	Fenyi	45	S	105 W
Ceraski	49	S	141 E	Feoktistov	31	N	140 E
Chaffee	39	S	155 W	Fermi	20	S	122 E
Chamberlin	59	S	96 E	Fersman	18	N	126 W
Champollion	37	N	175 E	Firsov	4	N	112 E
Chandler	44	N	171 E	Fitzgerald	27	N	172 W
Chang Heng	19	N	112 E	Fizeau	58	S	133 W
Chant	41	S	110 W	Fleming	15	N	109 E
Chaplygin	6	S	150 E	Focas	34	S	94 W
Chapman	50	N	101 W	Foster	23	N	142 W
Chappell	62	N	150 W	Fowler	43	N	145 W
Charlier	36	N	132 W	Freundlich	25	N	171 E
Chaucer	3	N	140 W	Fridman	13	S	127 W
Chauvenet	11	S	137 E	Froelich	80	N	110 W
Chebyshev	34	S	133 W	Frost	37	N	119 W
Chrétien	33	S	113 E	Gadomski	36	N	147 W
Clark	38	S	119 E	Gagarin	20	S	150 E
Coblentz	38	S	126 E	Galois	16	S	153 W
Cockcroft	30	N	164 W	Gamow	65	N	143 E
Compton	55	N	104 E	Ganswindt	79	S	110 E
Comrie	23	N	113 W	Garavito	17	N	131 E
Comstock	21	N	122 W	Gavrilov	17	N	131 E
Congreve	0		168 W	Geiger	14	S	158 E
Cooper	53	N	176 E	Gerasmović	23	S	124 W
Coriolis	0		172 E	Gernsback	36	S	99 E
Coulomb	54	N	115 W	Ginzel	14	N	97 E
Cremona	68	N	93 E	Giordano Bruno	36	N	103 E
Crocco	47	S	150 E	Glasenapp	2	S	138 E
Crommelin	68	S	147 W	Golitzyn	25	S	105 W
Crookes	11	S	165 W	Golovin	40	N	161 E
Curie	23	S	92 E	Grachev	3	S	108 W

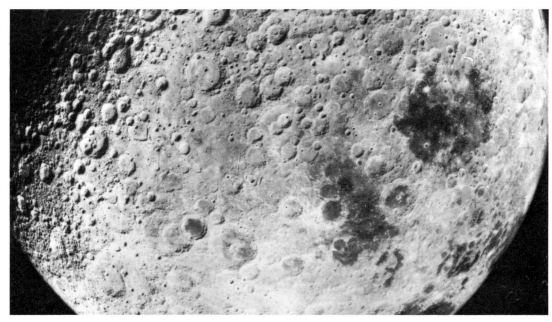

Far side of the Moon (NASA)

Name	Lat	Long	Name	Lat	Long	Name	Lat	Long
Green	4 N	133 E	Joliot	26 N	94 E	Lebedev	48 S	108 E
Gregory	2 N	127 E	Joule	27 N	144 W	Lebedinsky	8 N	165 W
Grigg	13 N	130 W	Jules Verne	36 S	146 E	Leeuwenhoek	30 S	179 W
Grissom	45 S	160 W	Kamerlingh Onnes	15 N	116 W	Leibnitz	38 S	178 E
Grotrian	66 S	128 E	Karpinsky	73 N	166 E	Lemaître	62 S	150 W
Gullstrand	45 N	130 W	Kearons	12 S	113 W	Lenz	3 N	102 W
Guthnick	48 S	94 W	Keeler	10 S	162 E	Leonov	19 N	148 E
Guyot	11 N	117 E	Kékulé	16 N	138 W	Leucippus	29 N	116 W
Hagen	56 S	135 E	Khwolson	14 S	112 E	Levi-Civita	24 S	143 E
Hansky	10 S	97 E	Kibaltchitch	2 N	147 W	Lewis	19 S	114 W
Harriott	33 N	114 E	Kidinnu	36 N	123 E	Ley	43 N	154 E
Hartmann	3 N	135 E	Kimura	57 S	118 E	Lindblad	70 N	99 W
Harvey	19 N	147 W	King	5 N	120 E	Lobachevsky	10 N	113 E
Hatanaka	29 N	122 W	Kirkwood	69 N	157 W	Lodygin	18 S	147 W
Hayford	13 N	176 W	Kleimenov	33 S	141 E	Lomonosov	27 N	98 E
Healy	32 N	111 W	Klute	37 N	142 W	Lorentz	34 N	100 W
Heaviside	10 S	167 E	Koch	43 N	150 E	Love	6 S	129 E
Hedin	68 N	123 E	Kohlschütter	15 N	154 E	Lovelace	82 N	107 W
Helberg	22 N	102 W	Kohlhörster	10 N	115 W	Lovell	39 S	149 W
Henderson	5 N	152 E	Komarov	25 N	153 E	Lowell	13 S	103 W
Hendrix	48 S	161 W	Kondratyuk	15 S	115 E	Lucretius	9 S	121 W
Henyey	13 N	152 W	Konstantinov	20 N	159 E	Lundmark	39 S	152 E
Hertz	13 N	104 E	Kopff	17 S	90 W	Lütke	17 S	123 E
Hertzsprung	0	130 W	Korolev	5 S	157 W	Lyman	65 S	162 E
Hess	54 S	174 E	Kostinsky	14 N	118 E	Mach	18 N	149 W
Heymans	75 N	145 W	Kovalevskaya	31 N	129 W	Maksutov	41 S	169 W
Hilbert	18 S	108 E	Kovalsky	22 S	101 E	Malyi	22 N	105 E
Hippocrates	71 N	146 W	Kramers	53 N	128 W	Mandelstam	4 N	156 E
Hirayama	6 S	93 E	Krasovsky	4 N	176 W	Marci	22 N	169 W
Hoffmeister	15 N	137 E	Krylov	9 S	157 W	Marconi	9 S	145 E
Hogg	34 N	122 E	Kugler	53 S	104 E	Mariotte	29 S	140 W
Hohmann	18 S	94 W	Kulik	42 N	155 W	Maunder	14 S	94 W
Holetschek	28 S	151 E	Kuo Shou Ching	8 N	134 W	Maxwell	30 N	98 E
Houzeau	18 S	124 W	Kurchatov	38 N	142 E	McKeller	16 S	171 W
Hutton	37 N	169 E	Lacchini	41 N	107 W	McLaughlin	47 N	93 W
Ibn Yunis	14 N	91 E	Lamarck	57 S	158 E	McMath	15 N	167 W
Icarus	6 S	173 W	Lamb	43 S	101 E	McNally	22 N	127 W
Idelson	81 S	114 E	Lampland	31 S	131 E	Mees	14 N	96 W
Ingalls	4 S	153 W	Landau	42 N	119 W	Meggers	24 N	123 E
Ingenii, Mare	34 S	163 E	Lane	9 S	132 E	Meitner	11 S	113 E
Innes	28 N	119 E	Langemak	10 S	119 E	Mendel	49 S	110 W
Izsak	23 S	117 E	Langevin	44 N	162 E	Mendeleev	6 N	141 E
Jackson	22 N	163 W	Langmuir	36 S	129 W	Merrill	75 N	116 W
Jeans	53 S	91 W	Larmor	32 N	180	Mesentsev	72 N	129 W
Jenner	42 S	96 E	Lauritsen	27 S	96 E	Meshcerski	12 N	125 E
Joffe	15 S	129 W	Leavitt	46 S	140 W	Metchnikiff	11 S	149 E

Name	Lat	Long
Michelson	6 N	121 W
Milanković	77 N	170 E
Millikan	47 N	121 E
Mills	9 N	156 E
Milne	31 S	113 E
Mineur	25 N	162 W
Minkowski	56 S	145 E
Mitra	18 N	155 W
Möbius	16 N	101 E
Mohorovičić	19 S	165 W
Moiseev	9 N	103 E
Montgolfier	47 N	160 W
Moore	37 N	178 W
Morozov	5 N	127 E
Morse	22 N	175 W
Moseley	23 N	95 W
Moscoviense, Mare	27 N	147 E
Moulton	61 S	97 E
Nagaoka	20 N	154 E
Nassau	25 S	177 E
Nernst	36 N	95 W
Neujmin	27 S	125 E
Niépce	72 N	120 W
Nijland	33 N	134 E
Nikolayev	35 N	151 E
Nishina	45 S	171 W
Nobel	15 N	101 W
Nöther	66 N	114 W
Numerov	71 S	161 W
Nušl	32 N	167 E
Obruchev	39 S	162 E
O'Day	31 S	157 E
Ohm	18 N	114 W
Olcott	20 N	117 E
Omar Khayyám	58 N	102 W
Oppenheimer	35 S	166 W
Oresme	43 S	169 E
Orlov	26 S	175 W
Östwald	11 N	122 E
Paneth	63 N	95 W
Pannekoek	4 S	140 E
Papaleski	10 N	164 E
Paracelsus	23 S	163 E
Paraskevopoulos	50 N	150 E
Parenago	26 N	109 W
Parkhurst	34 S	103 E
Parsons	37 N	171 W
Paschen	14 S	141 E
Pasteur	12 S	105 E
Pauli	45 S	137 E
Pavlov	29 S	142 E
Pawsey	44 N	145 E
Pease	13 N	106 W
Perelman	24 S	106 E
Perepelkin	10 S	128 E
Perkin	47 N	176 W
Perrine	42 N	129 W
Petrie	45 N	108 E
Petropavlovsky	37 N	115 W
Petzval	63 S	113 W
Pirquet	20 S	140 E
Pizzetti	35 S	119 E
Planck	57 S	135 E
Plaskett	82 N	175 E
Plummer	25 S	155 W
Pogson	42 S	111 E
Poincaré	57 S	161 E
Poinsot	79 N	147 W
Polzunov	26 N	115 E
Popov	17 N	99 E
Poynting	17 N	133 W
Prager	4 S	131 E
Prandtl	60 S	141 E
Priestly	57 S	108 E
Purkyne	1 S	95 E
Quételet	43 N	135 W
Racah	14 S	180
Raimond	14 N	159 W
Ramsay	40 S	145 E
Rasumov	39 N	114 W
Rayet	45 N	114 E
Rayleigh	67 S	179 E
Riccó	75 N	177 E
Riedel	49 S	140 W
Riemann	40 N	96 E
Rittenhouse	74 S	107 E
Ritz	15 S	92 E
Roberts	71 N	175 W
Robertson	22 N	105 W
Roche	42 S	135 E
Rowland	57 N	163 W
Rozhdestvensky	86 N	155 W
Rumford	29 S	170 W
Rydberg	47 S	96 W
Safarik	10 N	177 E
Saha	2 S	103 E
Sänger	4 N	102 E
St. John	10 N	150 E
Sanford	32 N	139 W
Sarton	49 N	121 W
Scaliger	27 S	109 E
Schaeberle	26 S	117 E
Schjellerup	69 N	157 E
Schlesinger	47 N	138 W
Schliemann	2 S	155 E
Schneller	42 N	164 W
Schönfeld	45 N	98 W
Schorr	19 S	90 E
Schrödinger	75 S	133 E
Schuster	4 N	147 E
Schwarzschild	71 N	120 E
Seares	74 N	145 E
Sechenov	7 S	143 W
Segers	47 N	128 E
Seidel	33 S	152 E
Seyfert	29 N	114 E
Shajn	33 N	172 E
Sharonov	13 N	173 E
Shatalov	24 N	140 E
Shi Shen	76 N	105 E
Siedentopf	22 N	135 E
Sierpinski	27 S	155 E
Sisakian	41 N	109 E
Sklodowska	18 S	159 E
Slipher	50 N	160 E
Smoluchowski	60 N	96 W
Sniadecki	22 S	169 W
Sommerfeld	65 N	161 W
Spencer Jones	13 N	166 E
Spiru Haret	59 S	176 W
Stark	25 S	134 E
Stebbins	65 N	143 W
Stefan	46 N	109 W
Stein	7 N	179 E
Steklov	37 S	105 W
Steno	33 N	162 E
Sternberg	19 N	117 W
Stetson	40 S	119 W
Stoletov	45 N	155 W
Stoney	56 S	156 W
Størmer	57 N	145 E
Stratton	6 S	165 E
Strömgren	22 S	133 W
Subbotin	29 S	135 E
Sumner	37 N	109 E
Sundman	11 N	91 W
Swann	52 N	112 E
Szilard	34 N	106 E
Teisserenc de Bort	32 N	137 W
ten Bruggencate	9 S	134 E
Terashkova	28 N	147 E
Tesla	38 N	125 E
Thiel	40 N	134 W
Thiessen	75 N	169 W
Thomson	32 S	166 E
Tikhomirov	25 N	162 E
Tikhov	62 N	172 E
Tiling	52 S	132 W
Timiryazev	5 S	147 W
Titius	27 S	101 E
Titov	28 N	150 E
Trümpler	28 N	168 E
Tsander	5 N	149 W
Tsiolkovskii	21 S	129 E
Tsu Chung-chi	17 N	144 E
Tyndall	35 S	117 E
Valier	7 N	174 E
Van de Graaff	27 S	172 E
Van den Bergh	31 N	159 W
Van der Waals	44 S	119 E
Van Gent	16 N	160 E
Van Gu	11 S	139 W
Van Maanen	36 N	127 E
Van Rhijn	52 N	145 E
Van't Hoff	62 N	133 W
Van Wijk	63 S	119 E
Vashakidze	44 N	93 E
Vavilov	1 S	139 W
Vening Meinesz	0	163 E
Ventris	5 S	158 E
Vernadsky	23 N	130 E
Vesalius	3 S	115 E
Vestine	34 N	94 E
Vetchinkin	10 N	131 E
Vilev	6 S	144 E
Volterra	57 N	131 E
Von der Pahlen	25 S	133 W
Von Kármán	45 S	176 E
Von Neumann	40 N	153 E
Von Zeipel	42 N	142 W
Walker	26 S	162 W
Waterman	26 S	128 E
Watson	63 S	124 W
Weber	50 N	124 W
Wegener	45 N	113 W
H. G. Wells	41 N	122 E
Wexler	69 S	90 E
Weyl	16 N	120 W
White	48 S	149 W
Wiechart	84 S	165 E
Wiener	41 N	146 E
Wilsing	22 S	155 W
Winkler	42 N	179 W
Winlock	35 N	160 W
Woltjer	45 N	160 W
Wood	44 N	121 W
Wyld	1 S	98 E
Yablochkov	61 N	127 E
Yamamoto	59 N	161 E
Zeeman	75 S	135 W
Zelinsky	29 S	167 E
Zernike	18 N	168 E
Zhiritsky	25 S	120 E
Zhukovsky	7 N	167 W
Zinger	57 N	176 E
Zsigmondy	59 N	105 W
Planck Rima:	65 S	129 E
	to 54 S	125 E
Schrödinger Rima:	62 S	99 E
	to 71 S	114 E

(The name 'Hedin' commemorates the Swedish scientist Sven Hedin. The name of Sven Hedin, used for a large enclosure near Grimaldi which is visible from Earth, has been deleted from some maps, but is retained here because lunar observers continue to use it.)

ECLIPSES OF THE MOON

Eclipses of the Moon are caused by the Moon's entry into the cone of shadow cast by the Earth. At the mean distance of the Moon, the diameter of the shadow cone is approximately 9170 km; the shadow is 1 367 650 km long on average. Totality may last for up to 1 h 44 m.

Lunar eclipses may be either total or partial. If the Moon misses the main cone, and merely enters the zone of 'partial shadow' or penumbra to either side, there is a slight dimming; but a penumbral eclipse is difficult to detect with the naked eye. Of course, the Moon must pass through the penumbra before entering the main cone or umbra.

During an eclipse the Moon becomes dim, often coppery. The colour during eclipse depends upon conditions in the Earth's atmosphere; thus the eclipse of 19 March 1848 was so 'bright' that lay observers refused to believe that an eclipse was happening at all. On the other hand, it is reliably reported that during the eclipses of 18 May 1761 and 10 June 1816 the Moon became completely invisible to the naked eye. The French astronomer A. Danjon has given an 'eclipse scale' from 0 (dark) to 4 (bright) and has attempted to correlate this with solar activity, though the evidence is still far from conclusive.

Obviously, a lunar eclipse can happen only at full moon: a solar eclipse, at new moon. In the original edition of the famous novel **King Solomon's Mines**, H. Rider Haggard described a full moon, a solar eclipse and another full moon on successive days. When the mistake was pointed out, he altered the second edition, turning the solar eclipse into a lunar one!

LUNAR ECLIPSES, 1923–1987

* = total. For partial eclipses, the percentage maximum phase is given.

1923	March 2	38	1943	August 15	88	1965	June 13/14	18
1923	August 25	17	1945	June 25	60	1967	April 24	*
1924	February 20	*	1945	December 18/19	*	1967	October 18	*
1924	August 14	*	1946	June 14	*	1968	April 13	*
1925	February 8/9	74	1946	December 8	*	1968	October 6	*
1925	August 4	75	1947	June 3	2	1970	February 21	5
1927	June 15	*	1948	April 23	3	1970	August 17	41
1927	December 8	*	1949	April 13	*	1971	February 10	*
1928	June 3	*	1949	October 6/7	*	1971	August 6	*
1928	November 27	*	1950	April 2	*	1972	January 30	*
1930	April 13	11	1950	September 26	*	1972	July 26	55
1930	October 7	3	1952	February 10/11	9	1973	December 10	11
1931	April 2	*	1952	August 5	54	1974	June 4	83
1931	September 26	*	1953	January 29/30	*	1974	November 29	*
1932	March 22	97	1953	July 26	*	1975	May 25	*
1932	September 14	98	1954	January 18/19	*	1975	November 18/19	*
1934	January 30	12	1954	July 15/16	41	1976	May 13	13
1934	July 26	67	1955	November 29	13	1977	April 4	21
1935	January 19	*	1956	May 24	97	1978	March 24	*
1935	July 16	*	1956	November 18	*	1978	September 16	*
1936	January 8	*	1957	May 13/14	*	1979	March 13	89
1936	July 4	27	1957	November 7	*	1979	September 6	*
1937	November 18	15	1958	May 3	2	1981	July 17	58
1938	May 14	*	1959	March 24	27	1982	January 9	*
1938	November 7/8	*	1960	March 13	*	1982	July 6	*
1939	May 3	*	1960	September 5	*	1982	December 30	*
1939	October 28	99	1961	March 2	81	1983	June 25	34
1941	March 13	33	1961	August 26	99	1985	May 4	*
1941	September 5	6	1963	July 6/7	71	1985	October 28	*
1942	March 2/3	*	1963	December 30	*	1986	April 24	*
1942	August 26	*	1964	January 24/5	*	1986	October 17	*
1943	February 20	77	1964	December 19	*	1987	October 1	1

LUNAR ECLIPSES, 1987–2000

For partial eclipses, the percentage maximum phase is given. * = total. P = penumbral eclipse, for which the maximum phase is given in brackets.

		Type	Time of mid-eclipse GMT		Duration of eclipse Totality		Partial	
			h	m	h	m	h	m
1988	March 3	P (101)	16	14	–	–	–	–
1988	August 27	29	11	05	–	–	1	52
1989	February 20	*	15	37	1	18	3	42
1989	August 17	*	03	09	1	36	3	31
1990	February 9	*	19	13	0	42	3	24
1990	August 6	68	14	12	–	–	2	56
1991	January 30	P (88)	06	00	–	–	–	–
1991	June 27	P (31)	03	16	–	–	–	–
1991	July 26	P (25)	18	09	–	–	–	–
1991	December 21	9	10	34	–	–	1	4
1992	June 15	68	04	58	–	–	3	0
1992	December 9	*	23	45	1	14	3	28
1993	June 4	*	13	02	1	36	3	38
1993	November 29	*	06	26	0	56	3	30
1994	May 25	24	03	32	–	–	1	44
1994	November 18	P (88)	06	45	–	–	–	–
1995	April 15	11	12	19	–	–	1	12
1995	October 8	P (83)	16	05	–	–	–	–
1996	April 4	*	00	11	1	26	3	36
1996	September 27	*	02	55	1	10	3	22
1997	March 24	92	04	41	–	–	3	22
1997	September 16	*	18	47	1	2	3	16
1998	March 13	P (71)	04	42	–	–	–	–
1998	August 8	P (12)	02	26	–	–	–	–
1999	January 31	P (100)	16	20	–	–	–	–
1999	July 28	40	11	34	–	–	2	22
2000	January 21	*	04	45	1	16	3	22
2000	July 16	*	13	57	1	0	3	16

Crater Tycho, with its central peak casting a prominent shadow, photographed from Surveyor 7 (NASA)

LUNAR PROBES, 1958–1977

AMERICAN, PRE-RANGER

Name	Launch date	Results
Able 1	17 August 1958	Failed after 77 s, at 20 km (explosion of lower stage of launcher).
Pioneer 1	11 October 1958	Reached 113 000 km.
Pioneer 2	9 November 1958	Failed when third stage failed to ignite.
Pioneer 3	6 December 1958	Reached 106 000 km.
Pioneer 4	3 March 1959	Passed within 60 000 km of the Moon on 5 March.
Able 4	26 November 1959	Failure soon after lift-off.
Able 5A	25 October 1960	Total failure.
Able 5B	15 December 1960	Exploded 70 s after lift-off.

AMERICAN RANGERS

Ranger 1	23 August 1961	Failed to go anywhere near the Moon.
Ranger 2	18 November 1961	Total failure.
Ranger 3	26 January 1962	Missed the Moon by 37 000 km (28 January).
Ranger 4	23 April 1962	Instruments and guidance failure. Probably landed on the night side of the Moon on 26 April.
Ranger 5	18 October 1962	Missed the Moon by over 630 km.
Ranger 6	30 January 1964	Hit the Moon (2 February), but no pictures received.
Ranger 7	28 July 1964	Hit the Moon on 31 July, in the Mare Nubium. Sent back 4308 photographs before impact.
Ranger 8	17 February 1965	Hit the Moon on 20 February, in the Mare Tranquillitatis. Sent back 7137 photographs before impact.
Ranger 9	21 March 1965	Hit the Moon on 24 March; interior of Alphonsus. Sent back 5814 photographs before impact.

AMERICAN SURVEYORS

Surveyor 1	30 May 1966	Landed N. of Flamsteed; returned 11 150 photographs.
Surveyor 2	20 September 1966	Guidance failure. Crash-landed S.E. of Copernicus.
Surveyor 3	17 April 1967	Landed in Oceanus Procellarum, 612 km E. of Surveyor 1, and close to the site of the later Apollo 12 landing, 6315 photographs returned; soil physics studies carried out.
Surveyor 4	14 July 1967	Failure. Crash-landed in Sinus Medii.
Surveyor 5	8 September 1967	Landed in Mare Tranquillitatis, 25 km from the later Apollo 11 site. 18 000 photographs returned; soil physics studied.
Surveyor 6	7 November 1967	Landed in Sinus Medii, 30 000 photographs returned; soil physics, etc.
Surveyor 7	17 January 1968	Landed on N. rim of Tycho. 21 000 photographs returned, plus a great deal of miscellaneous data.

AMERICAN ORBITERS

Orbiter 1	10 August 1966	Successful photographic probe.
Orbiter 2	7 November 1966	Successful photographic probe.
Orbiter 3	25 February 1967	Successful photographic probe.
Orbiter 4	4 May 1967	Successful photographic probe.
Orbiter 5	1 August 1967	Successful photographic probe. Final controlled impact, 31 January 1968.

AMERICAN APOLLO TEST MISSIONS

Number	Launch date	Crew	Duration of flight	
Apollo 7	11 October 1968	Schirra Eisele Cunningham	10 d 20 h	Earth orbiter
Apollo 8	21 December 1968	Borman Lovell Anders	6 d 3 h	Lunar orbiter
Apollo 9	3 March 1969	McDivett Scott Schweickart	10 d 2 h	Testing LM in Earth orbit
Apollo 10	18 May 1969	Stafford Cernan Young	8 d 0 h	Testing LM in Lunar orbit

AMERICAN APOLLO PROBES

Number	Landing date	Crew	Site	EVA duration, hours	Distance covered, km
Apollo 11	20 July 1969	Armstrong Aldrin Collins	Mare Tranquillitatis Lat 00°67′N Long 23°49′E	2·2	—

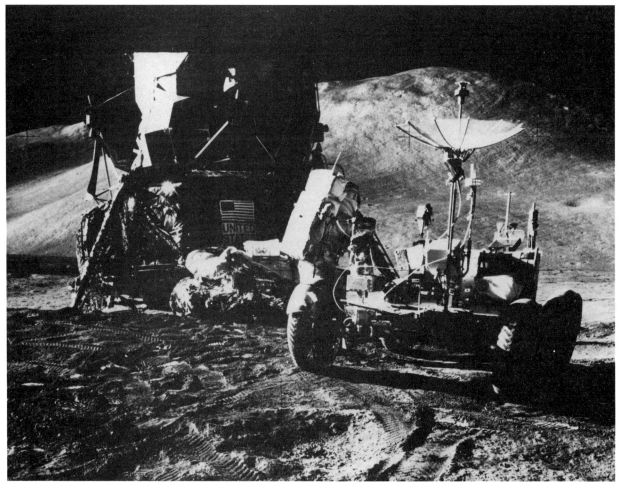

Number	Landing date	Crew	Site	EVA duration, hours	Distance covered, km
Apollo 12	19 November 1969	Conrad Bean Gordon	Oceanus Procellarum Lat 03°12'S Long 23°23'W	7·6	1·4
Apollo 13	11 April 1970 (launch)	Lovell Haise Swigert	No landing; failure on outward trip Splashdown 17 April	–	–
Apollo 14	31 January 1971	Shepard Mitchell Roosa	Fra Mauro Lat 03°40'S Long 17°28'E	9·2	3·4
Apollo 15	30 July 1971	Scott Irwin Worden	Hadley-Apennines Lat 26°06'N Long 03°39'E	18·3	28
Apollo 16	21 April 1972	Young Duke Mattingly	Descartes Lat 08°60'S Long 15°31'E	20·1	26
Apollo 17	11 December 1972	Cernan Schmitt Evans	Taurus-Littrow Lat 20°10'N Long 30°46'E	22	29

Apollo 15; Astronaut James Irwin by the Lunar Rover, in the foothills of the Lunar Apennines. (NASA)

RUSSIAN LUNA AND ZOND PROBES

Name	Launch date	Results
Luna 1	2 January 1959	Passed within 5955 km of the Moon on 4 January.
Luna 2	12 September 1959	Landed on the Moon, 13 September.
Luna 3	4 October 1959	Went round the Moon and photographed the far side. Approached the Moon to 6200 km.
Luna 4	2 April 1963	Missed the Moon by 8529 km; contact lost.
Luna 5	9 May 1965	Crash-landed on the Mare Nubium on 12 May. Unsuccessful soft-lander.

Name	Launch date	Results
Luna 6	8 June 1965	Missed the Moon by 161 000 km (11 June).
Zond 3	18 July 1965	Photographed far side of the Moon; 25 pictures taken. Minimum distance from Moon, 9219 km. Pictures sent back from 2 200 000 km on 27 July.
Luna 7	4 October 1965	Crashed in Oceanus Procellarum. Unsuccessful soft-lander.
Luna 8	3 December 1965	Crashed in Oceanus Procellarum. Unsuccessful soft-lander.
Luna 9	31 January 1966	Successful soft-lander; 100 kg capsule landed on the Moon (Oceanus Procellarum). Photographs obtained.
Luna 10	31 March 1966	First lunar satellite; contact maintained for 2 months (460 orbits). Minimum distance from the Moon, 350 km. Valuable data obtained.
Luna 11	24 August 1966	Lunar satellite. Contact maintained until 1 October. Minimum distance from the Moon, 159 km.
Luna 12	22 October 1966	Lunar satellite. Pictures showed craters down to 15 m. Contact lost on 19 January 1967.
Luna 13	21 December 1966	Soft landing in Oceanus Procellarum. Contact maintained until 27 December. Soil density, etc. studied.
Luna 14	7 April 1968	Lunar satellite. Minimum distance from Moon, 160 km. Valuable data obtained.
Zond 5	14 September 1968	Went round the Moon (minimum distance 1950 km) and returned to Earth on 21 September. Plants, seeds, insects and tortoises carried.
Zond 6	10 November 1968	Went round the Moon (minimum distance 2420 km) and filmed the far side; returned to Earth on 17 November.
Luna 15	13 July 1969	Unsuccessful attempt to return lunar samples. Crash-landed in Mare Crisium, 21 July.
Zond 7	7 August 1969	Circum-lunar flight. Lunar far side photographed from 2000 km; colour pictures of Earth and Moon secured. Returned to Earth.
Luna 16	12 September 1970	Landed in Mare Fœcunditatis (lat 0°41′S, long 56°18′E). Secured 100 g of material, and after 26½ hours lifted off. Returned to Earth; capsule recovered in Kazakhstan on 24 September.
Zond 8	20 October 1970	Circumlunar flight; colour pictures of Moon and Earth. Returned to Earth, 27 October, in Indian Ocean.
Luna 17	10 November 1970	Carried Lunokhod 1 to the Moon; landed 17 November in Mare Imbrium.
Luna 18	2 September 1971	Unsuccessful soft-lander. Contact lost during descent manœuvre to Mare Fœcunditatis.
Luna 19	28 September 1971	Lunar satellite. Contact maintained for over a year and over 4000 lunar orbits; studies of mascons, lunar gravitational field, etc., as well as solar flares.
Luna 20	14 February 1972	Successful sample-recovery probe. Landed 120 km N. of Luna 16's impact point, south of the Mare Crisium. Returned on 25 February.
Luna 21	8 January 1973	Carried Lunokhod 2 to the Moon; landed near Le Monnier, 180 km from Apollo 17's site, on 16 January.
Luna 22	29 May 1974	Successful lunar orbiter.
Luna 23	28 October 1974	Landed in the southern part of Mare Crisium, 6 November. Unsuccessful.
Luna 24	9 August 1976	Landed on 18 August in Mare Crisium, latitude 12°45′N, long 62°12′E. Drilled down to 2 metres, and obtained samples. Lifted off on 19 August, and returned to Earth on 22 August.

RUSSIAN LUNOKHODS

Name	Carrier	Weight, kg	Site	Notes
Lunokhod 1	Luna 17	756	Mare Imbrium	Operated for 11 months after its arrival on 17 November 1970. Area photographed exceeded 80 000 m². Over 200 panoramic pictures and 20 000 photographs returned. Distance travelled 10·5 km.
Lunokhod 2	Luna 21	850	Le Monnier	Operated until mid-May 1973; landed 16 January 1973. 86 panoramic pictures and 80 000 television pictures obtained. Distance travelled, 37 km. On 3 June the Soviet authorities announced that the programme had ended – perhaps prematurely.

MERCURY

The innermost planet is Mercury; it is also the **smallest** of the principal planets, with the exception of Pluto. It was once thought that a still closer-in planet must exist, and it was even named (Vulcan), but it is now known that no planet of appreciable size moves within the orbit of Mercury.

ELONGATIONS OF MERCURY, 1987–2000

Western
1987 Mar. 26, July 25, Nov. 13.
1988 Mar. 8, July 6, Oct. 26.
1989 Feb. 18, June 18, Oct. 10.
1990 Feb. 1, May 31, Sept. 24.
1991 Jan. 14, May 12, Sept. 7, Dec. 27.
1992 Apr. 23, Aug. 21, Dec. 9.
1993 Apr. 5, Aug. 4, Nov. 22.
1994 Mar. 19, July 17, Nov. 6.
1995 Mar. 1, June 29, Oct. 20.
1996 Feb. 11, June 10, Oct. 3.
1997 Jan. 24, May 22, Sept. 16.
1998 Jan. 6, May 4, Aug. 31, Dec. 20.
1999 Apr. 16, Aug. 14, Dec. 2.
2000 Mar. 28, July 27, Nov. 15.
Eastern
1987 Feb. 12, June 7, Oct. 4.
1988 Jan. 26, May 19, Sept. 15.
1989 Jan. 9, May 1, Aug. 29, Dec. 23.
1990 Apr. 13, Aug. 11, Dec. 6.
1991 Mar. 27, July 25, Nov. 19.
1992 Mar. 9, July 6, Oct. 31.
1993 Feb. 21, June 17, Oct. 14.
1994 Feb. 4, May 30, Sept. 26.
1995 Jan. 19, May 12, Sept. 9.
1996 Jan. 2, Apr. 23, Aug. 21, Dec. 15.
1997 Apr. 6, Aug. 4, Nov. 28.
1998 Mar. 20, July 17, Nov. 11.
1999 Mar. 3, June 28, Oct. 24.
2000 Feb. 15, June 9, Oct. 6.

The discovery of Mercury must have been prehistoric. The oldest observation of the planet which has come down to us is dated 15 November 265 BC, when the planet was one lunar diameter away from a line joining the stars Delta and Beta Scorpii. This information has been given by the last great astronomer of Classical times, Ptolemy (*c.* AD 120–180). Plato (*Republic*, X, 14) commented upon the yellowish colour of Mercury, though most naked-eye observers will describe it as being white. Mercury can actually become brighter than any star, but can never be seen against a really dark sky.

The phases of Mercury were first detected by Hevelius, in the first half of the 17th century.

The first prediction of a transit of Mercury across the face of the Sun

---

MERCURY DATA

Mean distance from the Sun:
  57·9 million km = 0·387 a.u.
Maximum distance from the Sun:
  69·7 million km = 0·467 a.u.
Minimum distance from the Sun:
  45·9 million km = 0·306 a.u.
Sidereal period: 87·969 days
Rotation period: 58·6461 days
Mean orbital velocity: 47·87 km/s
Axial inclination: Negligible
Orbital inclination: 7°00′15″·5
Orbital eccentricity: 0·206
Diameter: 4878 km
Apparent diameter from Earth:
  max. 12″·9, min 4″·5
Reciprocal mass, Sun = 1: 6 000 000
Density, water = 1: 5·5
Mass, Earth = 1: 0·055
Volume, Earth = 1: 0·056
Escape velocity: 4·3 km/s
Surface gravity, Earth = 1: 0·38
Mean surface temperature: +350 °C (day), −170 °C (night), (max +467 °C, min −183 °C)
Oblateness: Negligible
Albedo: 0·06
Maximum magnitude: −1·9
Mean diameter of Sun, seen from Mercury: 1°22′40″

---

was made by Kepler, for the transit of 7 November 1631. His prediction enabled Gassendi to observe the transit.

Transits between 1631 and 1988 were on:

1631 Nov. 7	1756 Nov. 7	1878 May 8
1644 Nov. 8	1769 Nov. 8	1881 Nov. 8
1651 Nov. 2	1776 Nov. 2	1891 May 8
1661 May 3	1782 Nov. 12	1894 Nov. 10
1664 Nov. 4	1786 May 3	1907 Nov. 12
1677 Nov. 7	1789 Nov. 5	1914 Nov. 6
1690 Nov. 10	1799 May 7	1924 May 8
1697 Nov. 3	1802 Nov. 9	1927 Nov. 10
1707 May 6	1815 Nov. 12	1937 May 11
1710 Nov. 5	1822 Nov. 5	1940 Nov. 11
1723 Nov. 9	1832 May 5	1953 Nov. 14
1736 Nov. 11	1835 Nov. 7	1957 May 6
1740 May 2	1845 May 8	1960 Nov. 7
1743 Nov. 5	1848 Nov. 9	1970 May 9
1753 May 6	1861 Nov. 12	1973 Nov. 10
	1868 Nov. 5	1986 Nov. 13

The following transits will occur between 1988 and 2000:

1993 Nov. 6, Mid-transit 3h 58m UT.
1999 Nov. 15, Mid-transit 21h 42m UT.

Transits can occur only in May and November. May transits occur with Mercury near aphelion; at November transits Mercury is near perihelion, and November transits are the more frequent in the approximate ratio of 7 to 3. The longest transits (those of May) may last for almost 9 hours.

The first serious telescopic

---

observations of Mercury were made in the late 18th century by Sir William Herschel, who, however, could make out no surface detail. At about the same time Mercury was studied by J. H. Schröter, who recorded some surface patches, and who believed that he had detected high mountains. It seems, however, that these results were illusory.

The first attempted map of Mercury was compiled by G. V. Schiaparelli, from Milan, using 0m.218 and 0m.49 refractors between 1881 and 1889. He recorded various dark markings, and believed the rotation period to be synchronous – that is to say, equal to Mercury's revolution period. This would mean that part of the planet would be in permanent daylight and another part in permanent night, with an intervening 'twilight zone' over which the Sun would rise and set. Schiaparelli was also the first observer to study Mercury in daylight, when both it and the Sun were high above the horizon.

The best pre-Mariner map was compiled by E. M. Antoniadi, using the 0·38 m Meudon refractor. The map was published in 1934, and various features were named. Like Schiaparelli, Antoniadi believed the rotation period to be synchronous, and he also believed in local obscurations, due to material suspended in a thin Mercurian atmosphere. These conclusions are now known to be wrong.

The first (and probably only!) observer to draw 'canal-like' features on Mercury was P. Lowell, at Flagstaff (Arizona) in 1896. These features are completely illusory.

The first disproof of the synchronous rotation theory was obtained in 1962 by W. E. Howard and his colleagues at Michigan, who measured the long-wavelength radiations from Mercury and found that the dark side was much warmer than it would be if it never received any direct sunlight. In 1965 the shorter rotation period was confirmed by radar methods (R. Dyce and G. Pettengill, at Arecibo in Puerto Rico). The true period is 2/3 of the revolution period, and when Mercury is best placed for observation from Earth the same face is always presented to us.

The first (and so far, the only) Mercury Probe, Mariner 10, made three active passes of the planet: in March and September 1974 and March 1975. It was launched on 3 November 1973, and obtained some photographs of the Moon before by-passing Venus (5 February 1974) and making its first rendezvous with Mercury. **The first crater to be identified on Mercury** during the initial encounter was the bright ray-crater now named Kuiper. Closest approach took place on 29 March, and altogether 647 pictures were obtained. The second closest approach was that of 21 September 1974, and the third occurred on 16 March 1975, by which time the equipment was deteriorating. Contact was finally lost on 24 March 1975, though

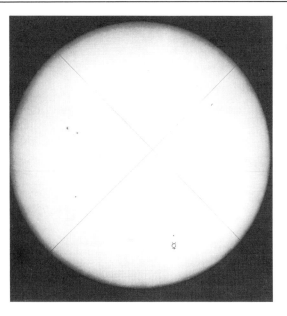

Transit of Mercury across the disk of the Sun on 7 November 1914 (RAS)

Mercury; Mariner 10, 1975. This is a mosaic of over 200 pictures of the south polar region. It shows Mercury as it would appear from a distance of 50 000 km. (NASA)

MERCURY

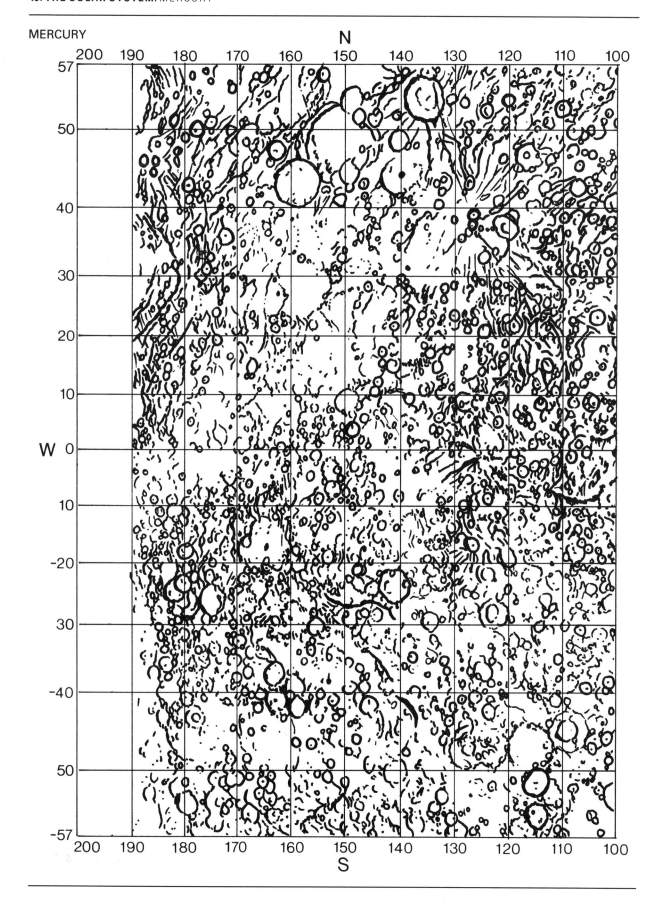

N

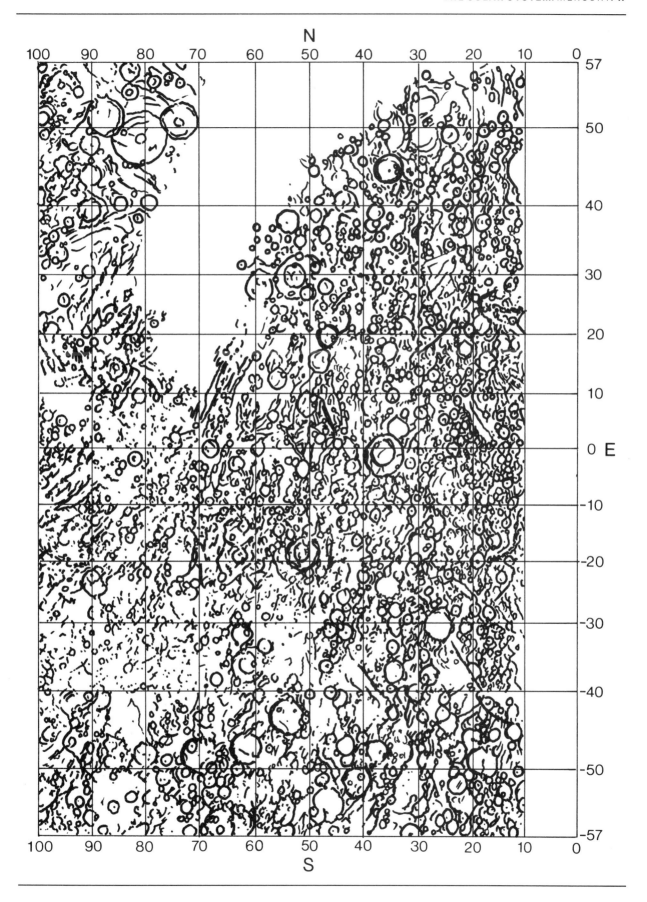

S

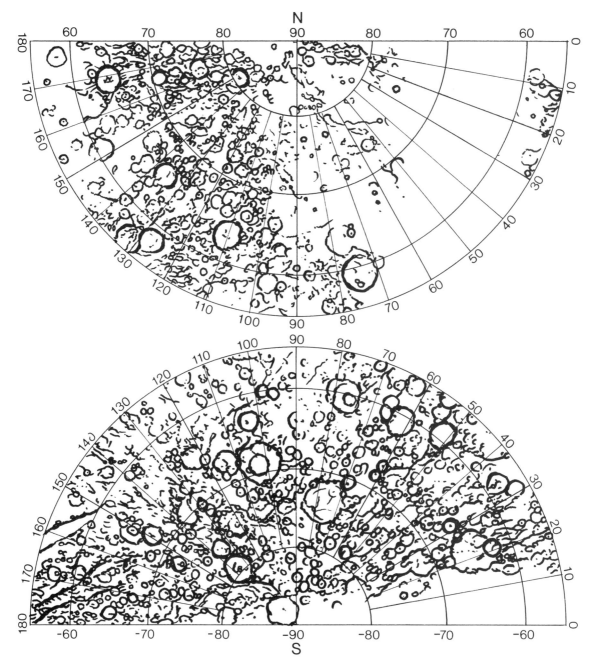

no doubt the probe is still in solar orbit and continues to make periodical approaches to Mercury.

The atmosphere of Mercury was first detected from Mariner 10. The ground pressure is about 10^{-9} millibars, and the main constituent is helium, suggesting that the weak magnetic field traps helium nuclei from the solar wind. Earlier spectroscopic reports of an atmosphere (initially by H. Vogel in 1871) are now known to have been erroneous.

The magnetic field of Mercury was also detected by Mariner 10. The encounter yielded a value of 350 gammas at the surface, or about 1 per cent of the Earth's magnetic field. This means that the magnetic field of Mercury is stronger than that of either Venus, Mars or the Moon. The field is dipolar, with two equal magnetic poles of opposite polarity inclined by 11 degrees to the rotational axis of the planet.

The densest planets are Mercury and the Earth, each with a specific gravity of about 5·5. Mercury has a dense iron core, 3600 km in diameter (somewhat larger than the whole of the Moon) and containing 80 per cent of the planet's mass.

The Mercurian craters were first detected from Mariner 10, as were other features: plains (**planitia**), mountains (**montes**), valleys (**valles**), scarps (**dorsa**) and ridges (**rupes**). The south pole of Mercury lies inside the crater Chao Meng Fu. It has been

Mercury; Mariner 10, 1974. This photomosaic was made to show a region where hills and ridges cut across many craters and intercrater plains. The large, relatively smooth-floored crater near the centre is Petrarch, 160 km in diameter; on the terminator, upper right, Carlo (160 km); lower right, also on the terminator, Pigalle (130 km). (NASA)

agreed that the 20th-degree meridian of Mercury passes through the centre of the 1·5 km crater Hun Kal, 0°·58 south of the Mercurian equator (the name stands for 20 in the language of the Maya, who used a base-20 number system).

The most imposing formation on Mercury is the Caloris Basin (Caloris Planitia), so called because it lies near a 'hot pole', the highest-temperature region of Mercury (there are two 'hot poles'). The Caloris Basin is 1300 km in diameter, and is bounded by a ring of smooth mountain blocks rising 1 to 2 km above the surrounding surface. Unfortunately only half of the Caloris Basin was recorded from Mariner 10; at each encounter the same regions were on view, so that our knowledge of the surface is still very incomplete.

There is an obvious similarity between the surface features of Mercury and those of the Moon, though Mercury lacks the broad lunar-type maria. The distribution of the craters is of the same general type – when a crater breaks into another, it is generally the smaller formation which intrudes into the larger; there are pairs, groups and lines of craters, and some of the formations have central peaks while others do not. Ray-craters also exist. There are also relatively smooth plains, and what are termed intercrater plains, which are level to gently rolling areas with a rough texture caused by large numbers of small craters. There are also characteristic 'lobate scarps' – long, sinuous cliffs which run across the surface for hundreds of kilometres, and are so called because of their rounded and lobed appearance (no such features are found on the Moon).

Mercury is certainly a hostile world in every way. With its excessively thin atmosphere and its extreme temperatures, it is unsuited to any form of life, and there seems little chance of its being visited in the foreseeable future, though no doubt many unmanned probes – both orbiters and landers – will be sent there during the coming decades.

NAMED FORMATIONS ON MERCURY

The naming of Mercurian features is not yet complete, and, of course, less than half the total surface was covered from Mariner 10 – our sole source of information at the time of writing (1988). In general craters have been named after persons who have made great contributions to human culture; plains (**planitia**) from the names of Mercury in various languages (for instance, the Suisei Planitia is the Japanese form); valleys (**valles**) after radar installations; ridges (**rupes**) after famous ships of exploration and discovery, and scarps (**dorsa**) after astronomers who have been particularly associated with the observation of Mercury. There are a few exceptions to these general rules. Caloris Basin is 'the hot basin' because it lies near one of the two hot poles; the late Gerard P. Kuiper is honoured by

having the first-identified crater on Mercury named after him; and as we have seen, Hun Kal is the Mayan word for 'twenty', since, by definition, the 20th meridian on Mercury passes through its centre.

Apart from the Caloris Basin, the largest circular structure is Beethoven (625 km in diameter). There are various ray-craters; those named are Copley, Kuiper, Mena, Tansen, and Snorri.

No doubt the nomenclature of Mercury will be extended in the near future. Meantime, here is the current list, as approved by the International Astronomical Union.

Craters	Latitude (deg)	Longitude (deg)	Diameter (km)
Abu Nuwas	17·5	21	115
Africanus Horton	−50·5	42	120
Ahmad Baba	58·5	127	115
Al-Akhtal	59	97	102
Alencar	−63·5	104	85
Al-Hamadhani	39	89·5	170
Al-Jāhiz	1·5	22	95
Amru Al-Qays	13	176	50
Andal	−47	38·5	90
Aristoxenes	82	11	65
Asvaghosa	11	21	80
Bach	−69	103	225
Balagtas	−22	14	100
Balzac	11	145	65
Bartók	−29	135	80
Bashō	−32	170·5	70
Beethoven	−20	124	625
Belinskij	−76	104	70
Bello	−18·5	120·5	150
Bernini	−79·5	136	145
Bjornson	73	110	90
Boccaccio	−80·5	30	135
Boethius	−0·5	74	130
Botticelli	64	110	120
Brahms	58·5	177	75
Bramante	−46	62	130
Brontë	39	126·5	60
Brueghel	50	108	75
Brunelleschi	−8·5	22·5	140
Burns	54	116	45
Byron	−8	33	100
Callicrates	−65	32	65
Camões	−70·5	70	70
Carducci	−36	90	75
Cervantes	−75	122	200
Cézanne	−8	124	75
Chaikovskij	8	50·5	160
Chao Meng-Fu	−87·5	132	150
Chekov	−35·5	61·5	180
Chiang K'ui	14·5	103	40
Chŏng Ch'ŏl	47	116	120
Chopin	−64·5	124	100
Chu Ta	2·5	106	100
Coleridge	−54·5	66·5	110
Copley	−37·5	85·5	30
Couperin	30	152	75
Darío	−26	10	160
Degas	37·5	127	45
Delacroix	−44·5	129·5	135
Derzhavin	44·5	35·5	145
Despréz	81	92	40
Dickens	−73	153	72
Donne	3	14	90

Craters	Latitude (deg)	Longitude (deg)	Diameter (km)
Dostoevskij	−44·5	177	390
Dowland	−53	180	80
Dürer	22	119·5	190
Dvořák	−9·5	12·5	80
Echegaray	43	19	75
Eitoku	−21·5	157·5	105
Equiano	−39	31	80
Fet	−5	180	24
Flaubert	−14	72	95
Futabatei	−15·5	83·5	55
Gainsborough	−36	183	100
Gauguin	66·5	97	75
Ghiberti	−48	80	100
Giotto	12·5	56	150
Gluck	37·5	18·5	85
Goethe	79·5	44	340
Gogol	−28	147	87
Goya	−6·5	152·5	135
Grieg	51	14	65
Guido d'Arezzo	−38	19	50
Hals	−55	115	100
Handel	4	34	150
Han Kan	−71·5	145	50
Harunobu	15·5	141	100
Hauptmann	−23	180	120
Hawthorne	−51	116	100
Haydn	−26·5	71·5	230
Heine	33	124·5	65
Hesiod	−58	35·5	90
Hiroshige	−13	27	140
Hitomaro	−16	16	105
Holbein	35·5	29	85
Holberg	−66·5	61	66
Homer	−1	36·5	320
Horace	−68·5	52	48
Hugo	39	47·5	190
Hun Kal	−0·5	20	1·5
Ibsen	−24	36	160
Ictinus	−79	165	110
Imhotep	−17·5	37·5	160
Ives	−32·5	112	20
Janáček	56	154	47
Jókai	72·5	136	85
Judah Ha-Levi	11·5	108	85
Kālidāsā	−17·5	180	110
Keats	−69·5	154	110
Kenkō	−21	16·5	90
Khansa	−58·5	52	100
Kōshō	60	138	65
Kuan Han-ch'ing	29	53	155
Kuiper	−11	31·5	60
Kurosawa	−52	23	180
Leopardi	−73	180	69
Lermontov	15·5	48·5	160
Lessing	−29	90	100
Liang K'ai	−39·5	183·5	105
Li Ch'ing-Chao	−77	73	60
Li Po	17·5	35	120
Liszt	−16	168	85
Lu Hsun	0·5	23·5	95
Lysippus	1·5	133	150
Ma Chih-Yuan	−59	77	170
Machaut	−1·5	83	105
Mahler	−19	19	100
Mansart	73·5	120	75
Mansur	47·5	163	75
March	31·5	176	55
Mark Twain	−10·5	138·5	140
Martí	−75·5	164	63
Martial	69	178	45
Matisse	−23·5	90	210
Melville	22	9·5	145
Mena	0·5	125	20
Mendes Pinto	−61	19	170
Michelangelo	−44·5	110	200
Mickiewicz	23·5	102·5	115

Craters	Latitude (deg)	Longitude (deg)	Diameter (km)
Milton	−25·5	175	175
Mistral	5	54	100
Mofolo	−37	29	90
Molière	16	17·5	140
Monet	44	9·5	250
Monteverdi	64	77	130
Mozart	8	190·5	225
Murasaki	−12	31	125
Mussorgskij	33	96·5	115
Myron	71	79·5	30
Nampeyo	−39·5	50·5	40
Nervo	43	179	50
Neumann	−36·5	35	100
Nizāmī	71·5	165	70
Ovid	−69·5	23	40
Petrarch	−30	26·5	160
Phidias	9	150	155
Philoxenus	−8	112	95
Pigalle	−37	10·5	130
Po Chü-I	−6·5	165·5	60
Po Ya	−45·5	21	90
Polygnotus	0	68·5	130
Praxiteles	27	60	175
Proust	20	47	140
Puccini	−64·5	46	110
Purcell	81	148	80
Pushkin	−65	24	200
Rabelais	−59·5	62·5	130
Rajnis	5	96·5	85
Rameau	−54	38	50
Raphael	−19·5	76·5	350
Ravel	−12	38	75
Renoir	−18	52	220
Repin	−19	63	95
Riemenschneider	−52·5	100·5	120
Rilke	−44·5	13·5	70
Rimbaud	−63	148	85
Rodin	22	18·5	240
Rubens	59·5	73·5	180
Rublev	−14·5	157·5	125
Rūdakī	−3·5	51·5	120
Rude	−33	80	75
Rūmī	−24	105	75
Sadī	−77·5	56	60
Saikaku	73	177	80
Sarmiento	−28·5	188·5	115
Sayat-Nova	−27·5	122·5	125
Scarlatti	40·5	99·5	135
Schoenberg	−15·5	136	30
Schubert	−42	54·5	160
Scopas	−81	173	95
Sei	−63·5	88·5	130
Shakespeare	48·5	151	350
Shelley	−47·5	128·5	145
Shevchenko	−53	47	130
Sholem Aleichem	51	86·5	190
Sibelius	−49	145	90
Simonides	−29	45	95
Sinan	16	30	140
Snorri	−8·5	83·5	20
Sophocles	−6·5	146·5	145
Sor Juana	49	24	80
Sōseki	39	38	90
Sōtatsu	−48	19·5	130
Spitteler	−68	62	66
Stravinsky	50·5	73	170
Strindberg	54	136	165
Sullivan	−16	87	135
Sūr Dās	−46·5	94	100
Surikov	−37	125	105
Takanobu	31	108	80
Takayoshi	−37	164	105
Tansen	4·5	72	25
Thākur	−2·5	64	115
Theophanes	−4	143	50
Thoreau	6	133	80

Craters	Latitude (deg)	Longitude (deg)	Diameter (km)
Tintoretto	−47·5	24	60
Titian	−3	42·5	115
Tolstoj	−15	165	400
Ts'ai Wen-chi	23·5	22·5	120
Ts'ao Chan	−13	142	110
Tsurayuki	−62	22·5	80
Tung Yüan	73·5	55	60
Turgenev	66	135	110
Tyagaraja	4	149	100
Unkei	−31	62·5	110
Ustad Isa	−31·5	166	105
Vālmiki	−23·5	141·5	220
Van Dijck	76·5	165	100
Van Eyck	43·5	159	235
Van Gogh	−76	135	95
Velázquez	37	54	120
Verdi	64·5	169	150
Vincente	−56·5	143	85
Vivaldi	14·5	86	210
Vlaminck	28	13	97
Vyāsa	48·5	80	275
Wagner	−67·5	114	135
Wang Meng	9·5	104	170
Wergeland	−37	56·5	35
Whitman	41	111	70
Wren	24·5	36	215
Yakovlev	−40·5	163·5	100
Yeats	9·5	35	90
Yun Sŏn-Do	−72·5	109	61
Zeami	−2·5	148	125
Zola	50·5	178	60

Mountains (Montes)	Latitude (deg)	Longitude (deg)
Caloris	22–40	180

Plains (Planitiae)	Latitude (deg)	Longitude (deg)
Borealis	70	80
Budh	18	148
Caloris	30	195
Odin	25	171
Sobkou	40	130
Suisei	62	150
Tir	3	177

Ridges (Dorsa)	Latitude (deg)	Longitude (deg)
Antoniadi	28	30
Schiaparelli	24	164

Scarps (Rupes)	Latitude (deg)	Longitude (deg)
Adventure	−64	63
Astrolabe	−42	71
Discovery	−53	38
Endeavour	38	31
Fram	−58	94
Gjöa	−65	163
Heemskerck	25	125
Hero	−57	173
Mirni	−37	40
Pourquois-Pas	−58	156
Resolution	−62	52
Santa María	6	20
Victoria	50	32
Vostok	−38	19
Zarya	−42	22
Zeehaen	50	158

Valleys (Valles)	Latitude (deg)	Longitude (deg)
Arecibo	−27	29
Goldstone	−15	32
Haystack	5	46·5
Simeiz	−12·5	65

VENUS

The **brightest planet** is Venus, second in order of distance from the Sun.

PHENOMENA OF VENUS, 1977–2000

E. Elongation	Inferior Conjunction
1977 Jan. 24	1977 Apr. 6
1978 Aug. 29	1978 Nov. 7
1980 Apr. 5	1980 June 15
1981 Nov. 11	1982 Jan. 21
1983 June 16	1983 Aug. 25
1985 Jan. 22	1985 Apr. 3
1986 Aug. 27	1986 Nov. 5
1988 Apr. 3	1988 June 13
1989 Nov. 8	1990 Jan. 10
1991 June 13	1991 Aug. 22
1993 Jan. 19	1993 Apr. 1
1994 Aug. 25	1994 Nov. 2
1996 Apr. 1	1996 June 10
1997 Nov. 6	1998 Jan. 16
1999 June 11	1999 Aug. 20

W. Elongation	Superior Conjunction
1977 June 15	1978 Jan. 22
1979 Jan. 18	1979 Aug. 25
1980 Aug. 24	1981 Apr. 7
1982 Apr. 1	1982 Nov. 4
1983 Nov. 4	1984 June 15
1985 June 13	1986 Jan. 19
1987 Jan. 15	1987 Aug. 23
1988 Aug. 22	1989 Apr. 5
1990 Mar. 30	1990 Nov. 1
1991 Nov. 2	1992 June 13
1993 June 10	1994 Jan. 17
1995 Jan. 13	1995 Aug. 20
1996 Aug. 19	1997 Apr. 2
1998 Mar. 27	1998 Oct. 30
1999 Oct. 30	2000 June 11

The first man to find that Venus is at its brightest when in the crescent stage was Edmond Halley, in 1721. This is because when more of the illuminated hemisphere faces the Earth, Venus is further away from us.

The discovery of Venus must, of course, have been prehistoric. The most ancient observations known to us are Babylonian, and are recorded on the Venus Tablet found by Sir Henry Layard at Konyunjik, now to be seen in the British Museum. Homer (*Iliad*, XXII, 318) refers to Venus as 'the most beautiful star set in the sky'.

Venus is **the only planet referred to by Napoleon Bonaparte!** According to the French astronomer F. Arago, Napoleon was visiting Luxembourg when he saw that the crowd was paying more attention to the sky than to him; it was noon, but Venus was visible, and Napoleon himself saw it. Not surprisingly, his followers referred to it as

---

DATA

Mean distance from the Sun: 108.2 million km = 0·723 a.u.
Maximum distance from the Sun: 109 million km = 0·728 a.u.
Minimum distance from the Sun: 107·4 million km = 0·718 a.u.
Sidereal period: 224·701 days
Rotation period: 243·16 days
Mean orbital velocity: 35·02 km/s
Axial inclination: 178°
Orbital inclination: 3°23'39"·8
Orbital eccentricity: 0·007
Diameter: 12 104 km
Apparent diameter from Earth: max 65"·2, min 9"·5, mean 37"·3
Reciprocal mass, Sun = 1: 408 520
Density, water = 1: 5·25
Mass, Earth = 1: 0·815
Volume, Earth = 1: 0·86
Escape velocity: 10·36 km/s
Surface gravity, Earth = 1: 0·903
Mean surface temperature: cloud-tops −33 °C, surface +480 °C
Oblateness: 0
Albedo: 0·76
Maximum magnitude: −4·4
Mean diameter of Sun, seen from Venus: 44'15"

---

being the star of 'the Conqueror of Italy'.

The first men to refer to shadows cast by Venus were the Greek astronomer Simplicius, in his *Commentary on the Heavens of Aristotle*, and the Roman writer Pliny around AD 60.

The first man to record the phases of Venus telescopically was Galileo, in 1610. This was an important observation, since according to the old Ptolemaic theory, with the Earth in the centre of the planetary system, Venus could never show a full cycle of phases from new to full. The observation therefore strengthened Galileo's faith in the 'Copernican' or Sun-centred system. (The phases had not previously been mentioned specifically, though exceptionally keen-sighted people can see the crescent form with the naked eye.)

The first prediction of a transit of Venus across the disk of the Sun was Johannes Kepler, who in 1627 predicted a transit for 6 December 1631. This transit was not actually observed.

The first observation of a transit of Venus was made by two English amateurs, J. Horrocks and W. Crabtree, on 24 November 1639 (O.S.). (It has been claimed that the Eastern scholar Al-Farabi saw a transit in AD 910, from Kazakhstan. This may well be true.)

The first suggestion of using a transit of Venus to measure the length of the astronomical unit was Edmond Halley. The method was used during the transits of 1761, 1769, 1874 and 1882, but proved to be inaccurate, and is now obsolete. It depended upon timing the exact moment of the beginning of the transit, but when Venus passes on to the Sun it seems to draw a strip of blackness after it, and when this strip disappears the transit is already in progress. The effect, termed the 'Black Drop', is caused by Venus' atmosphere.

The last transit of Venus was that of 1882. The next will be in 2004.

The only occultation of Mars by Venus to be observed was that of 3 October 1590, seen by M. Möstlin at Heidelberg.

The only occultation of Mercury by Venus to be observed was that of 17 May 1737, seen by J. Bevis at Greenwich.

The last occultation of a first-magnitude star by Venus was that of 7 July 1959, when Venus occulted Regulus. The phenomenon was widely observed, and led to a determination of the height of Venus' atmosphere. (The writer of this book observed the occultation from Selsey in Sussex, using a 12 in reflecting telescope.)

The first observation of the Ashen Light was made on 9 January 1643 by G. Riccioli. The Ashen Light is the faint visibility of the night hemisphere of Venus. It may be due to electrical phenomena in the upper atmosphere of Venus, though some astronomers dismiss it as being nothing more than a contrast effect.

The first markings on Venus were reported in 1645 by F. Fontana. However, Fontana was using a small-aperture, long-focus refractor, and there is no doubt that his 'markings on Venus' were illusory.

The first map of Venus was drawn by F. Bianchini in 1727. However, the markings seen by Bianchini, like those reported by Fontana, were certainly illusory. Bianchini believed that he had charted seas and continents on the surface of Venus!

The first observer to report an atmosphere around Venus was the Russian astronomer M. V. Lomonosov,

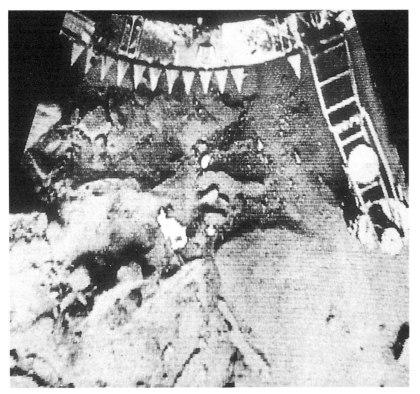

Surface of Venus, from Venera 14, showing also part of the space-craft. (Venera)

during the transit of 1761.

The first really serious observer of Venus telescopically was J. H. Schröter. His series of observations began in 1779, but it was not until 28 February 1788 that he observed markings, which he correctly interpreted as being atmospheric. On 11 December 1798 he believed that he had identified a mountain 43 km high, but this was obviously erroneous. In 1793 he was the first to find that the observed and theoretical phases of Venus do not agree exactly; when Venus is waning, dichotomy (half-phase) is earlier than predicted by a few days, while when Venus is waxing dichotomy is late. The phenomenon is now generally called Schröter's Effect (a term due to the writer of this book!). It is due to effects of Venus' atmosphere.

The first report of a satellite of Venus was made by G. D. Cassini on 18 August 1686. Other reports followed, the last being that of Montbaron, at Auxerre, on 29 March 1764. It is now certain that no satellite exists, and that the observers were deceived by 'telescopic ghosts'.

The first man to estimate the rotation period of Venus was G. D. Cassini in 1666–7; he gave a period of 23h 21m. Many other estimates were made in later years, but there was no general agreement.

The first man to suggest a synchronous rotation period was G. V. Schiaparelli, in 1890. This would mean that the rotation period and the sidereal period of Venus were identical – 224·7 Earth-days – in which case Venus would keep the same hemisphere turned towards the Sun all the time. However, this did not seem to fit the facts, and in 1954 G. P. Kuiper proposed a period of 'a few weeks'.

The first estimate of the rotation period by using spectroscopic methods (the Doppler shift) was made by R. S. Richardson in 1956. Richardson concluded that the rotation was very slow, and retrograde – that is to say, opposite in sense to that of the Earth. This has proved to be correct. The true rotation period is 243 days, so that Venus is the only planet with a rotation period longer than its sidereal period. (The 'solar day' on Venus is equal to 118

Earth-days.) Venus is also exceptional in having a genuinely retrograde rotation, though it is true that the axial tilt of Uranus is slightly more than a right angle (98 degrees). However, during the 1960s, French observers established that the upper clouds have a rotation period of only 4 days; this was confirmed by photographs obtained from the Mariner 10 probe in 1974.

The first substance to be detected spectroscopically in the atmosphere of Venus was carbon dioxide, identified by W. S. Adams and T. Dunham at Mount Wilson in 1932.

The first reliable temperature measurements of the upper clouds of Venus were made in 1923–8 by E. Pettit and S. B. Nicholson, using a thermocouple attached to the 100 in Hooker reflector at Mount Wilson. They gave a value of $-38\,°C$ for the day side and $-33\,°C$ for the dark side, which is in good agreement with modern values.

The first good infra-red and ultra-violet photographs of Venus were taken by F. E. Ross in 1923. The infra-red pictures showed no detail, but vague features were shown in ultra-violet, indicating high-altitude cloud phenomena.

The first suggestion of formaldehyde clouds on Venus was made by R. Wildt in 1937. This is no longer accepted, and it seems more likely that the clouds contain quantities of sulphuric acid. Hydrogen chloride has also been detected; it was found, along with hydrogen fluoride, by W. S. Benedict of the University of Maryland.

The first suggestion that Venus might be entirely water-covered was made by F. L. Whipple and D. H. Menzel in 1954; they believed that the clouds were made up of H_2O. The 'marine theory' was disproved by the Mariner 2 results of 1962. Venus is too hot for liquid water to exist on its surface.

The first radio measurements of Venus at centimetre wavelengths were made by Mayer and his colleagues in the United States in 1958. They indicated a very high temperature for the surface of Venus, and this has been fully confirmed.

The first reliable radar contact with Venus was made in 1961 by the team at the Lincoln Laboratory, United States (the success first claimed in 1958 proved to be premature). Several other groups made radar contact with the planet at about the same time. The measurements led to a better determination of the length of the astronomical unit, and in 1964 the International Astronomical Union officially adopted a revised value of 149 600 000 km.

The first Venus probe was Venera 1 (USSR), launched on 12 February 1961. It was not successful, as contact with it was lost at a distance of 7 500 000 km from Earth.

The first successful Venus probe was Mariner 2 (USA), launched on 26 August 1962, which bypassed Venus at 35 000 km on 14 December 1962, confirming the high surface temperature and the virtual absence of any magnetic field.

The first probe to land on Venus was Venera 3 (USSR), on 1 March 1966. It failed to return any data, because it was crushed by the intense pressure as it parachuted down through the atmosphere of Venus.

The first probe to transmit information during the descent through Venus' atmosphere was Venera 4 (USSR), on 18 October 1967.

The first probe to transmit information from Venus' surface was Venera 7 (USSR), on 15 December 1970. It survived for 23 minutes before being put out of action by the intensely hostile conditions.

The first close-range photographs of the clouds of Venus were obtained from Mariner 10 (USA) on 5 February 1974, as the probe by-passed Venus en route for Mercury.

The first picture to be sent back from the surface of Venus was transmitted by Venera 9 (USSR) on 21 October 1975. Venera 10 landed on 25 October 1975, and also sent back one picture.

Venus is an intensely hostile planet. The first pictures from the surface, sent back by Venera 9 and 10, were obtained under a pressure of about 90 000 milli-bars (90 times as great as that of the Earth's atmosphere at sea-level) and an intolerably high temperature – which had been expected, since the atmosphere consists mainly of carbon dioxide which acts in the manner of a green-house, shutting in the solar heat. The Venera 9 landscape was described as 'a heap of stones', several dozen centimetres in diameter and with sharp edges; the Venera 10 landing-site was smoother, as though it were an older plateau. Wind velocities were sluggish, and the light level was described as similar to that at noon in Moscow on a cloudy winter day.

The first attempts at analysis of the surface materials were made in March 1982 by Veneras 13 and 14, which landed in the general area of Phœbe Regio. Venera 13 dropped a lander which continued to transmit for a record 127 minutes after arrival; the temperature was 457 °C, and the pressure 89 atmospheres. The general colour of the rock was reddish-brown, and it was said that the sky was bright orange.

VENUS PROBES, 1961–1978

The Mariners were American; the other probes, Russian

Name	Launch	Arrival	Closest approach, km	Results
Venera 1	12 February 1961	19 May 1961	100 000	Contact lost at 7 500 000 km from Earth
Mariner 1	22 July 1962	–	–	Total failure
Mariner 2	26 August 1962	14 December 1962	35 000	Fly-by. Data transmitted
Zond 1	2 April 1964	?	?	Contact lost within a few weeks
Venera 2	12 November 1965	27 February 1966	24 000	In solar orbit. No Venus data received
Venera 3	16 November 1965	1 March 1966	Landed	Crushed by Venus' atmosphere. No data received
Venera 4	12 June 1967	18 October 1967	Landed	Data transmitted during 94-minute descent
Mariner 5	14 June 1967	19 October 1967	4000	Fly-by. Data transmitted
Venera 5	5 January 1969	16 May 1969	Landed	Data transmitted during descent
Venera 6	10 January 1969	17 May 1969	Landed	Data transmitted during descent
Venera 7	17 August 1970	15 December 1970	Landed	Lander transmitted for 23 minutes
Venera 8	26 March 1972	22 July 1972	Landed	Lander transmitted for 50 minutes
Mariner 10	3 November 1973	5 February 1974	5800	Pictures of upper clouds, plus other data Went on to Mercury
Venera 9	8 June 1975	21 October 1975	Landed	Transmitted for 53 minutes after arrival. One picture received
Venera 10	14 June 1975	25 October 1975	Landed	Transmitted for 65 minutes after landing. One picture received
Pioneer Venus 1	20 May 1978	4 December 1978	145	In orbit. Sending back information
Pioneer Venus 2	8 August 1978	9 December 1978	Landed	Multi-probe – five probes landed. Data transmitted back
Venera 11	9 September 1978	21 December 1978	Landed	Transmitted for about 60 minutes
Venera 12	14 September 1978	25 December 1978	Landed	Transmitted for about 60 minutes
Venera 13	30 October 1981	1 March 1982	Landed	Landing position lat. 07·5 S., long. 303. Transmitted for 127 minutes after arrival. One picture sent back. Soil analysis
Venera 14	3 November 1981	5 March 1982	Landed	Landing position lat. S.13·25, long. 310·7. Similar programme to Venera 13
Venera 15	2 June 1983	10 October 1983	1000	In 24th orbits: polar, 1000 × 65 000 km Radar mapping of surface
Venera 16	7 June 1983	16 October 1983	1000	Same as Venera 15
Vega 1	15 December 1984	11 June 1985	8890	Balloons dropped into Venus' atmosphere;
Vega 2	20 December 1984	15 June 1985	8030	probes en route to Halley's Comet

Venera 14 came down in a plain, with fewer of the sharp, angular rocks of the Venera 13 site. The temperature was 465 °C, and the pressure 94 atmospheres. In each case, highly alkaline potassium basalts were much in evidence.

According to the space-probe results, the top of Venus' atmosphere lies at 400 km above the surface. The upper clouds lie at 70 km, and at a height of 63 km the temperature is 13 °C, with an outside pressure of 0·5 atmosphere. At an altitude of 50 km above the ground the temperature is 20 °C; below lies a clear layer, and then a layer of denser cloud. Beneath this layer, at 47 km, there is a second clear region. The cloud-deck ends at 30 km above the surface. The wind structure of the atmosphere is remarkable, remembering that the upper clouds have a rotation period of 4 days as against 243 days for the solid surface – in each case, retrograde (east to west) for reasons which are unknown. Winds decrease close to the surface, from about 100 m/s at the cloud-top level to only 50 m/s at 50 km, and only a few metres per second at the surface, though in that thick atmosphere even a slow wind will have tremendous force.

The first accurate maps of the surface were obtained by the radar equipment in the Pioneer probe, from December 1978. About 80 per cent of the surface has been mapped, from latitude 75 °N to 63 °S. There are highlands, lowlands, and a huge rolling plain which covers at least 60 per cent of the surface. Of the two main highland areas, Ishtar Terra in the north has a diameter of 2900 km; the western part, Lakshmi Planum, is a smooth plateau at a height of 3 km. The Maxwell Montes occupy the east end of Ishtar, and rise to 11 km above the mean level of the planet or 8.2 km above the adjoining plateau. Aphrodite Terra, near the equator, is decidedly larger, measuring 9700 × 3200 km, and consists of eastern and western ranges separated by a lower region.

The lowest point on Venus is Diana Chasma, 2 km below the mean radius and 4 km below the adjacent ridges. A third highland region is Beta Regio, containing two shield volcanoes,

Rhea and Theia, each 4 km high and probably active. Another area of probable active vulcanism is the so-called Scorpion's Tail, at the end of Aphrodite. Lightning and thunder are probably almost continuous, and it has been said that Venus approximates to the conventional picture of hell!

Life appears to be out of the question. The atmosphere consists mainly of carbon dioxide, and the clouds contain sulphuric acid drops, which fall as 'rain' though they evaporate before reaching the surface.

If Venus and the Earth are so alike in size and mass, why have they developed so differently? The cause must be Venus' lesser distance from the Sun. It has been suggested that in the early days of the Solar System, when the Sun was much less luminous than it is now, the two planets may have started to evolve along similar lines; but as the Sun became more powerful, the oceans of Venus evaporated, the carbonates were driven out of the rocks, and there was a 'runaway greenhouse' effect which produced the intolerable conditions of today. If this is so, then Venus has had a tragic history; any life which gained a foothold there was ruthlessly destroyed.

FEATURES ON VENUS
Bold numbers indicate map references

Terræ	Lat	Long W
Aphrodite	40S–5N	140–000
Ishtar	52–75N	080–305

Regiones		
Alpha **1**	29–32 S	000
Asteria **2**	18–30 N	228–270
Atla	20 N–05 S	185–210
Bell	25–35 N	045–055
Beta **3**	20–38 N	292–272
Eisila	10–25 N	350–050
Imdr	42 S	211
Metis **4**	72 N	245–255
Mnemosyne	68 N	275–285
Ovda	05 N–10 S	080–108
Phœbe **5**	10–20 N	275–300
Tellus **6**	35 N	080
Tethus **7**	55 N	100
Themis **8**	37–40 S	275–310
Thetis	02–15 S	118–140
Ulfrun	108 N–3 S	220–230

Planitia		
Aino **9**	45 S	090
Atalanta **10**	54 N	162
Guinevere **11**	40 N	310
Helen **12**	55 S	255
Lakshmi Planum **23**	60 N	330
Lavinia **13**	45 S	350
Leda **14**	45 N	065

Planitia	Lat	Long W
Nava	08 S	315
Niobe **15**	138 N–10 S	132–185
Rusa	10 N–10 S	160–185
Sedna **16**	40 N	335

Patera		
Cleopatra	67 N	009
Theodora	23 N	280

Dorsum		
Juno	32–35 S	085–095

Linea		
Antiope	40 S	350
Guor	20 N	000
Hariasa	19 N	015
Hippolyta	42 S	345
Kara	44 S	306
Lampedo	57 N	290
Molpadia	48 S	359
Vihansa	54 N	020

Chasma		
Artemis	30–42 S	121–145

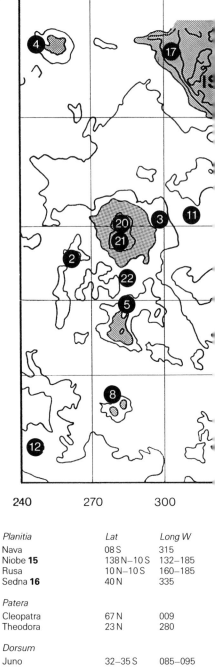

240   270   300

VENUS

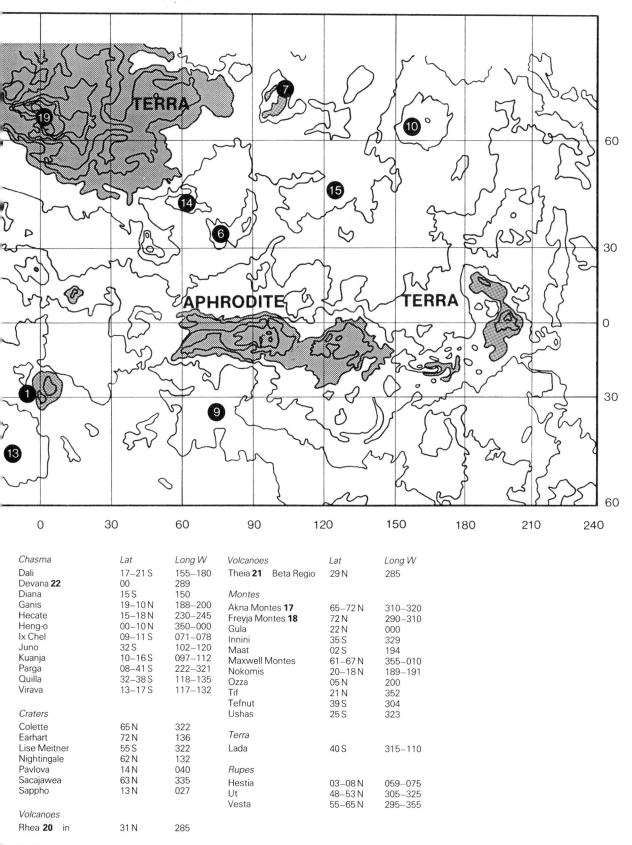

Chasma	Lat	Long W
Dali	17–21 S	155–180
Devana **22**	00	289
Diana	15 S	150
Ganis	19–10 N	188–200
Hecate	15–18 N	230–245
Heng-o	00–10 N	350–000
Ix Chel	09–11 S	071–078
Juno	32 S	102–120
Kuanja	10–16 S	097–112
Parga	08–41 S	222–321
Quilla	32–38 S	118–135
Virava	13–17 S	117–132

Craters		
Colette	65 N	322
Earhart	72 N	136
Lise Meitner	55 S	322
Nightingale	62 N	132
Pavlova	14 N	040
Sacajawea	63 N	335
Sappho	13 N	027

Volcanoes		
Rhea **20** in	31 N	285

Volcanoes	Lat	Long W
Theia **21** Beta Regio	29 N	285

Montes		
Akna Montes **17**	65–72 N	310–320
Freyja Montes **18**	72 N	290–310
Gula	22 N	000
Innini	35 S	329
Maat	02 S	194
Maxwell Montes	61–67 N	355–010
Nokomis	20–18 N	189–191
Ozza	05 N	200
Tif	21 N	352
Tefnut	39 S	304
Ushas	25 S	323

Terra		
Lada	40 S	315–110

Rupes		
Hestia	03–08 N	059–075
Ut	48–53 N	305–325
Vesta	55–65 N	295–355

EARTH

DATA

Mean distance from the Sun: 149·5979 million km (1 a.u.)
Minimum distance from the Sun: 147 million km (0·9833 a.u.)
Maximum distance from the Sun: 152 million km (1·0167 a.u.)
Perihelion (1988): 4 January
Aphelion (1988): 6 July
Equinoxes (1988): 20 March, 09 h 39 m; 22 September, 19 h 29 m
Solstices (1988): 21 June, 03 h 57 m; 21 December, 15 h 28 m
Obliquity of the ecliptic, 1988: 23°·44084
Sidereal period: 365·256 days
Rotation period: 23 h 56 m 04 s
Mean orbital velocity: 29.79 km/s
Orbital inclination: 0 (by definition)
Orbital eccentricity: 0·016719
Equatorial diameter: 12 756 km
Reciprocal mass, Sun = 1: 328 900
Density, water = 1: 5·517
Escape velocity: 11·18 km/s
Oblateness: 0·003
Mean surface temperature: 22 °C
Albedo: 0·36
Mass: $5·976 \times 10^{24}$ kg

The Earth is the largest and most massive of the inner group of planets.

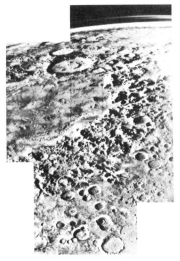

Across Argyre – a mosaic of photographs taken by Mariner 9 *at a range of 18 000 km* (NASA)

MARS

MARS DATA

Mean distance from the Sun: 227·94 million km (1·524 a.u.)
Maximum distance from the Sun: 249·1 million km (1·666 a.u.)
Minimum distance from the Sun: 206·7 million km (1·381 a.u.)
Sidereal period: 686·980 days (= 668·60 sols)
Synodic period: 779·9 days
Rotation period: 24 h 37 m 22·6 s (= 1 sol)
Mean orbital velocity: 24.1 km/s
Axial inclination: 23° 59′
Orbital inclination: 1° 50′ 59″·4
Orbital eccentricity: 0·093
Diameter: 6787 km
Apparent diameter from Earth: max. 25″·7, min 3″·5
Reciprocal mass, Sun = 1: 3 098 700
Density, water = 1: 3·94
Mass, Earth = 1: 0·107
Volume, Earth = 1: 0·150
Escape velocity: 5·03 km/s
Surface gravity, Earth = 1: 0·380
Mean surface temperature: −23 °C
Oblateness: 0·009
Albedo: 0·16
Maximum magnitude: −2·8
Mean diameter of Sun, seen from Mars: 21′

The only red planet is Mars, which comes fourth in order of distance from the Sun, and is the first planet beyond the orbit of the Earth.

OPPOSITION DATES, 1977–2000

Date of Opposition	Closest Approach to Earth	Apparent diameter, "	Magnitude	Constellation
1978 Jan. 22	1978 Jan. 19	14·3	−1·1	Gemini/Cancer
1980 Feb. 25	1980 Feb. 26	13·8	−1·0	Leo
1982 Mar. 31	1982 Apr. 5	14·7	−1·2	Virgo
1984 May 11	1984 May 19	17·5	−1·8	Libra
1986 July 10	1986 July 16	23·1	−2·4	Sagittarius
1988 Sept. 28	1988 Sept. 22	23·7	−2·6	Pisces
1990 Nov. 27	1990 Nov. 20	17·9	−1·7	Taurus
1993 Jan. 7	1993 Jan. 3	14·9	−1·2	Gemini
1995 Feb. 12	1995 Feb. 11	13·8	−1·0	Leo
1997 Mar. 17	1997 Mar. 20	14·2	−1·1	Virgo
1999 Apr. 24	1999 May 1	16·2	−1·5	Virgo

The closest oppositions occur with Mars near its perihelion. During this period the closest approach is that of 1988, with a minimum distance from Earth of 36 300 000 miles or 58 400 000 km. The closest approach during the past half-century has been that of August 1971 (minimum distance 56 200 000 km).

The greatest distance between Mars and the Earth, with Mars at superior conjunction, may amount to 400 000 000 km.

The longest interval between successive oppositions is 810 days; the **shortest interval** is 764 days. The mean synodic period (see Data Table) is 779·9 days.

The least favourable oppositions occur with Mars at aphelion, as in 1980 (least distance from Earth, 101 320 000 km).

The minimum phase of Mars as seen from Earth is 85 per cent. (At opposition, the phase is, of course, virtually 100 per cent.)

Martian Seasons: The seasons are of the same general type as those of the Earth, since the axial tilt is very similar and the Martian day (sol) is not a great deal longer (1 sol = 1·029 days). The lengths of the seasons are as follows:

	Days	Sols
S. spring (N. autumn)	146	142
S. summer (N. winter)	160	156
S. autumn (N. spring)	199	194
S. winter (N. summer)	182	177
	687	669

Southern summer occurs near perihelion. Therefore, climates in the southern hemisphere of Mars show a wider range of temperature than for the north. The effect is much greater than in the case of the Earth, partly because there is no sea on Mars and partly because of the greater eccentricity of the Martian orbit. At perihelion, Mars receives 44 per cent more solar radiation than at aphelion.

The discovery of Mars was unquestionably prehistoric. The planet was recorded in Egypt, China and Assyria. Later, its redness led to its being named in honour of the War-God, Ares (Mars); the study of the Martian surface is still officially known as 'areography'.

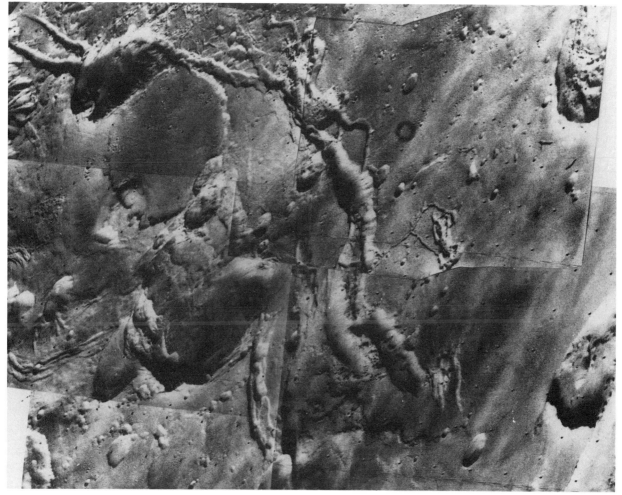

Mars; Viking 1 orbiter, 1976. The region is 2 degrees S. of the equator, and shows fault zones in the Martian crust. (NASA)

The first observation of an occultation of Mars by the Moon was made by Aristotle (384–322 BC). The exact date of the phenomenon is not known.

The first precise observation of the position of Mars dates, according to Ptolemy, back to 17 January 272 BC, when Mars was observed to be close to the star Beta Scorpii.

The best pre-telescopic observations of the movements of Mars were made by the Danish astronomer Tycho Brahe, from his island observatory at Hven between 1576 and 1596. It was these observations which enabled Kepler, in 1609, to publish his first Laws of Planetary Motion, showing that the planets move round the Sun in elliptical orbits.

The first telescopic observations of Mars were made by Galileo in 1610.

No surface details were seen.

The first detection of the phase of Mars was also due to Galileo in 1610, and recorded by him in a letter written to Father Castelli on 30 December of that year.

The first telescopic drawing of Mars was made by F. Fontana, in Naples, in 1636. Mars was shown as spherical, and 'in its centre was a dark cone in the form of a pill'. This feature was, of course, an optical effect. Fontana's second drawing (24 August 1638) was similar.

The first marking to be recorded on Mars was the Syrtis Major, by C. Huygens on 28 November 1659, at 7 p.m. The Syrtis Major is easily recognizable (though exaggerated in size). The sketch has been very useful in confirming the constancy of Mars' rotation period.

The first reasonably accurate estimate of the rotation period of Mars was also due to Huygens. On 1 December 1659 he recorded that the period was about 24 hours.

The first really accurate measurement of the rotation period was made by G. D. Cassini in 1666; his value was 24 h 40 m.

The first record of the polar caps was also due to Cassini in 1666. It has been suggested that a cap was seen by Huygens in 1656, but his surviving drawing is very inconclusive. (However, Huygens undoubtedly saw the south polar cap in 1672.)

The discovery that the polar caps do not coincide with the geographical (or areographical!) poles was made by G. Maraldi, in 1719. In 1704 Maraldi had studied the caps, and had also given a rotation period of 24 h

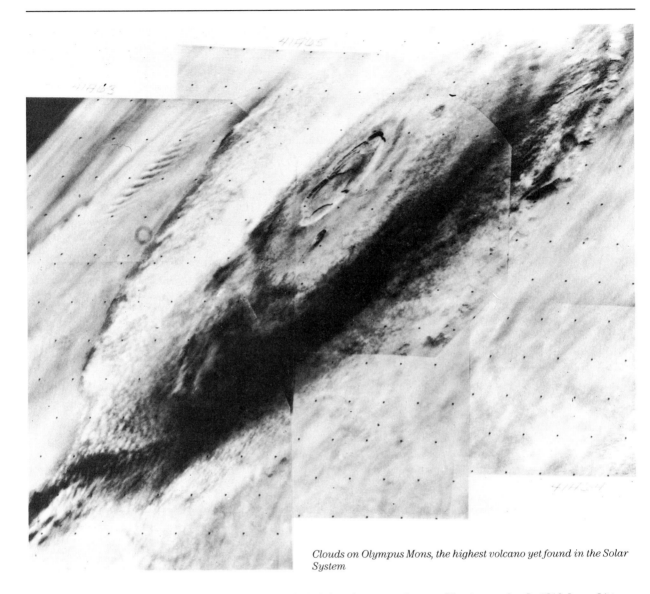

Clouds on Olympus Mons, the highest volcano yet found in the Solar System

39 m, which was very near the truth.

The first known panic due to Mars was that of 1719. Mars was at perihelic opposition, and was so bright that people mistook it for a red comet which might well collide with the Earth.

The first identification of the polar caps with ice and snow was suggested by W. Herschel, from his observations made between 1777 and 1783. Herschel also measured the rotation period, and his observations were later re-worked by W. Beer and J. H. Mädler, yielding a period of 24 h 37 m 23 s·7, which is only one second in error.

The first comment upon the Martian atmosphere was made by Herschel in 1783. From observations of the close approach of a star to Mars, he rightly concluded that the atmosphere could not be very extensive.

The first good determination of the axial inclination of Mars was also due to Herschel at the same period. His value was 28 degrees, which is only 4 degrees too great. The Martian north polar star is at present Deneb (Alpha Cygni), but the inclination varies between 35 degrees and 14 degrees in a period of about 50 000 years, so that precessional effects are important in any long-term consideration of the changing Martian climate.

The first suggestion of signalling to the inhabitants of Mars was made by the great German mathematician K. F. Gauss, about 1802; the plan was to draw vast geometrical patterns in the Siberian tundra. In 1819 J. von Littrow, of Vienna, proposed to use signal fires lit in the Sahara. Later (1874) Charles Cros, in France, put forward a scheme to focus the Sun's heat on to the Martian deserts by means of a huge burning-glass; the glass could be swung around to write messages in the deserts!

The first suggestion of clouds in the atmosphere of Mars was made by the French astronomer H. Flaugergues in 1811.

The first reasonably good charts of Mars were made by W. Beer and J. H. Mädler in 1830–32, from Berlin. They were also the first to report a dark band round the periphery of a shrinking polar cap. This band was later seen by many observers, and in the late 19th century P. Lowell attributed it, wrongly, to moistening of the ground owing to the melting polar ice.

The first report that the southern polar cap has a greater range of size than the northern was made by the French astronomer F. Arago in 1853. (Flaugergues, in 1811, had suggested that this should be so, because the temperature range is more extreme in the south.)

The first observations of 'white' clouds on Mars were made by A. Secchi in 1858.

The first suggestion that the dark areas are vegetation tracts rather than seas was made in 1860 by E. Liais, a French astronomer who spent much of his life in Brazil. This was also the view of G. V. Schiaparelli, who in 1863 pointed out that the dark areas did not show the Sun's reflection as they would be expected to do if they were sheets of water. The vegetation theory was regarded as probable up to the space-probe era.

The first reported detection of water vapour in the Martian atmosphere was by J. Janssen in 1867. Taking his equipment to the top of Mount Etna, to get above the densest and wettest part of the Earth's atmosphere, Janssen compared the spectrum of Mars with that of the airless Moon, and believed that he could see differences. It is now known that these results were spurious.

The first system of Martian nomenclature was that of R. A. Proctor in 1867. The features were named in honour of past and contemporary observers of the planet. The system was superseded in 1877 by that of Schiaparelli.

The first detection of the alleged canal network was due to G. V. Schiaparelli in 1877, when he recorded 40 canals. Streaks had been recorded earlier by various observers, including Beer and Mädler (1830–1840) and the Rev W. R. Dawes (1864), but Schiaparelli's work marked the beginning of the 'canal controversy'. He also produced a new map, and as the zero for Martian longitudes he adopted the feature which had also been selected by Beer and Mädler; it is now called Sinus

Meridiani, or the Meridian Bay. Following the space-probe results from 1965, the zero was fixed as the centre of a small crater in the Sinus Meridiani; this has been named Airy-0, in honour of the 19th-century Astronomer Royal, Sir George Airy (the zero for terrestrial longitudes passes through the Airy transit circle at the old Royal Observatory in Greenwich Park). On this map Schiaparelli introduced a new system of nomenclature, which has only recently been modified in view of the Mariner and Viking results.

The first reports of the twinning or 'gemination' of some canals were made by Schiaparelli in 1879. The first observers to support Schiaparelli's canal network were Perrotin and Thollon, at the Nice Observatory in France, in 1886. Subsequently canals became fashionable, and P. Lowell, who built the observatory which bears his name at Flagstaff and observed Mars between 1896 and 1916, believed the network to be artificial. Only since the close-range photography of Mars by space-probes has it been shown that the canal system is completely illusory.

The first astronomer to use the term 'deserts' for the bright regions of Mars was W. H. Pickering, in 1886. In 1892 Pickering observed features in the dark regions as well as the 'deserts', and this more or less killed the theory that the dark areas were seas.

The first observations of craters on Mars seem to have been made by E. E. Barnard in 1892, using the Lick Observatory 36-in refractor. However, Barnard did not publish his results, for fear of ridicule. Craters were also seen by J. Mellish in 1917, with the 40-in Yerkes refractor, but these observations also remained unpublished. The first prediction of craters on Mars was made by D. L. Cyr in 1944, but it was not until the flight of Mariner 4, in 1965, that the existence of craters was proved.

The first suggestion that the polar caps were made of solid carbon dioxide was due to A. C. Ranyard and Johnstone Stoney in 1898. The theory was more or less disregarded until the early space-probe results, beginning with Mariner 4 in 1965, but was then favoured until the missions of Vikings 1

and 2 in 1976. It is now known that the main caps are of ice, many metres thick, with a thin surface layer of solid carbon dioxide which disappears during Martian summer. However, the caps may not be identical, and the south cap may contain carbon dioxide ice as well as water ice.

The first prize offered for communicating with extraterrestrial beings was the Guzman Prize, announced in Paris on 17 December 1900. The sum involved was 100 000 francs – but Mars was excluded, because it was felt that contact with the Martians would be too easy!

The first good measurements of the surface temperature of Mars were made in 1909, by Nicholson and Pettit at Mount Wilson and by Coblentz and Lampland at Flagstaff. They found that Mars has a mean surface temperature of $-28\,°C$, as against $+15\,°C$ for the Earth.

The first report of the 'Violet Layer' was made by Lampland in 1909. This was supposed to block out the short-wavelength radiations, and prevent them from reaching the Martian surface, except on occasions when the layer cleared away. It could not be seen visually. Following the space-probe results, its existence is now regarded as doubtful.

The first analysis of the Martian atmosphere by using the Doppler method was made by W. S. Adams and T. Dunham in 1933. When Mars is approaching the Earth, the spectral lines should be shifted toward the short-wave end of the band; when Mars is receding, the shift should be to the red. It was hoped that in this way any lines due to gases in the Martian atmosphere could be disentangled from the lines produced by the same gases in the Earth's atmosphere. From their results, Adams and Dunham concluded that the amount of oxygen over Mars was less than 1/10 of 1 per cent of the amount existing in the atmosphere of the Earth.

The first reasonably good determination of the atmospheric pressure on Mars was made by the Russian astronomer N. Barabaschev in 1934. He gave a value of 50 millibars. Later estimates increased this to between 80 and 90 millibars. However, the Mariner and

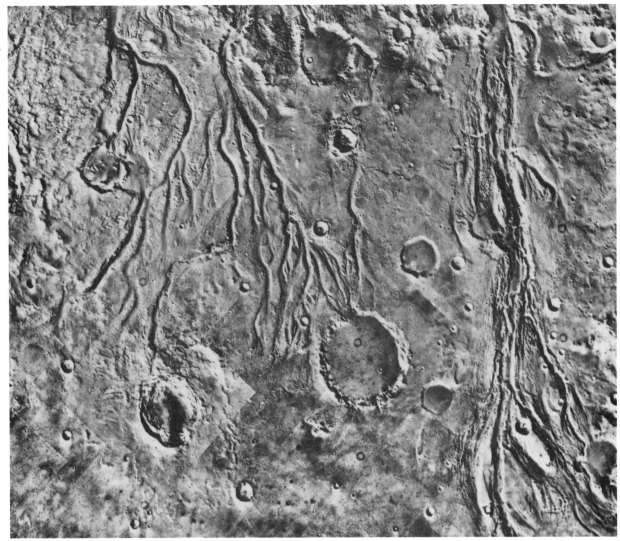

Chryse area of Mars; Viking 1, 1976. The area is centred on lat. 17°N, long. 55°W, near Chryse Planitia. Just to the west of the area are the plains of Lunæ Planum. The channels must certainly indicate past floods. (NASA)

Viking results have shown that the pressure is nowhere as high as 10 millibars.

The first reported spectroscopic detection of carbon dioxide in the Martian atmosphere was due to G. P. Kuiper in 1947. However, it was then thought that the atmosphere must consist chiefly of nitrogen. The Mariner and Viking results show that this is wrong. The atmosphere of Mars is made up of 95 per cent carbon dioxide (CO_2), 2·7 per cent nitrogen (N_2), 0·15 per cent oxygen (O_2), 1·6 per cent argon, and smaller amounts of gases such as krypton and xenon.

The first reported detection of organic matter in the spectra of the Martian dark areas was announced by W. M. Sinton in 1959. The results were later found to be spurious.

The first attempt to send a probe to Mars was made by the Russians in 1962 (Mars-1). The first successful probe was the US Mariner 4, in 1965.

The latest nomenclature for Martian features involves naming the craters in honour of men and women who have been associated with the study of Mars. Other classes of features are as follows:

Catena: crater-chain; a line or chain of craters (e.g. Tithonia Catena).

Cavi: steep-sided hollow (e.g. Angusti Cavi).

Chaos: distinctive area of broken terrain (e.g. Aromatum Chaos).

Chasma: very large linear chain (e.g. Ophir Chasma).

Colles: hills (e.g. Deuteronilus Colles).

Dorsum: ridge; irregular, elongated elevation (e.g. Solis Dorsum).

Fossa: ditch – long, shallow, narrow depression (e.g. Claritas Fossæ).

Labyrinthus: valley complex; network of linear depressions (e.g. Noctis Labyrinthus, the only really major example).

Mensa: mesa. Small plateau or tableland (e.g. Nilosyrtis Mensæ).

Mons: mountain or volcano (e.g. Olympus Mons).

Montes: mountains (e.g. Tharsis Montes).

Patera: saucer-like volcanic structure

A panorama over dunes photographed by Viking Lander 1 (NASA)

with fluted or scalloped edges (e.g. Alba Patera).

Planitia: smooth, low-lying plain (e.g. Hellas Planitia).

Planum: plateau; smooth, high area (e.g. Solis Planum).

Rupes: cliff (e.g. Ogygis Rupes).

Scopulus: irregular, degraded scarp (e.g. Nilokeras Scopulus).

Sulci: intricate network of linear depressions and ridges (e.g. Gigas Sulci).

Terræ: lands, names often given to classical albedo features (e.g. Sirenum Terra).

Tholus: domed hill (e.g. Uranius Tholus).

Vallis: valley. Sinuous channel, often with tributaries (e.g. Valles Marineris).

Vastitas: extensive plain (e.g. the north-polar circumpolar plain, Vastitas Borealis).

There are two major bulges in the Martian lithosphere: those of Tharsis and Elysium. Following the zero contour for Martian datum (taken as the 6.1 millibar at atmospheric pressure level) around the planet, it seems that it describes a great circle inclined to the equator; north of this line the surface is low and lightly cratered, while south of this line it is higher and more cratered. The datum line is inclined to the equator by 30 degrees. The heavily-cratered area is more extensive than the smoother regions; the area ratio is about 2 to 1. The Tharsis bulge, with a mean elevation of 8 km, straddles the boundary between the lower, smoother northern hemisphere and the higher, cratered south.

The greatest Martian dust-storm of the present century may have been that of late 1971, when Mariner 9 reached the vicinity of Mars, but the storm of 1909 was of the same magnitude, and there have been others in 1911, 1924, 1956 and 1973. It seems that major storms are most often seen when Mars is near perihelic opposition – that is to say, at its closest both to the Sun and to the Earth.

The most conspicuous dark area on Mars is the Syrtis Major, formerly known as the Hourglass Sea or Kaiser Sea (or even the Atlantic Canal, though there is no connection with Lowell's canal network). The Syrtis Major is V-shaped, and was drawn by Huygens more than 300 years ago. It was formerly thought to be a vegetation-filled seabed, but is now known to be a plateau. It lies in the equatorial region of Mars; the most prominent dark marking further north is the Acidalia Planitia, formerly known as Mare Acidalium.

The brightest area on Mars, ex-

MARS PROBES, 1962–1988

Name	Date of launch	Date of encounter	Nearest approach (km)	Remarks
Mars 1 (USSR)	1 November 1962	?	190 000?	Contact lost at 106 000 000 km from Earth
Mariner 3 (USA)	5 November 1964	–	–	In solar orbit. Contact lost soon after launch
Mariner 4 (USA)	28 November 1964	14 July 1965	10 000	Successful. Returned 21 pictures, plus miscellaneous data
Zond 2 (USSR)	30 November 1964	August 1965?	?	Contact lost on 2 May 1965
Mariner 6 (USA)	24 February 1969	31 July 1969	3390	Flew over equator of Mars. Returned 76 pictures
Mariner 7 (USA)	27 March 1969	4 August 1969	3500	Flew over southern hemisphere of Mars. Returned 126 pictures
Mariner 8 (USA)	8 May 1971	–	–	Total failure; crashed in the Atlantic immediately after launch
Mars 2 (USSR)	19 May 1971	27 November 1971	Landed	Orbiter achieved Mars orbit: lander crashed on the surface
Mars 3 (USSR)	28 May 1971	2 December 1971	Landed	Orbiter returned data. Lander lost contact 20 seconds after arrival; no useful data received
Mariner 9 (USA)	30 May 1971	13 November 1971	1395	Orbiter. Returned 7329 pictures. Contact finally lost on 27 October 1972
Mars 4 (USSR)	21 July 1973	10 February 1974	?	Failed to orbit; missed Mars
Mars 5 (USSR)	25 July 1973	12 February 1974	?	In Mars orbit
Mars 6 (USSR)	5 August 1973	12 March 1974	Landed	Contact lost at touchdown on Mars
Mars 7 (USSR)	9 August 1973	9 March 1974	?	Failed to orbit; missed Mars
Viking 1 (USA)	20 August 1975	19 June 1976 (in orbit)	Landed	Landed 20 July 1976, lat. 22°·4N, long. 47°·5W (in Chryse)
Viking 2 (USA)	9 September 1975	7 August 1976 (in orbit)	Landed	Landed 3 September 1976 lat. 48°N, long. 226°W (in Utopia). Viking 2 lander is 7420 km north-east of Viking 1 lander

luding the polar caps, may often be Hellas, once thought to be a snow-covered plateau and now known to be a basin. However, Hellas is very variable in brightness.

The highest volcano on Mars is Olympus Mons, formerly known as Nix Olympica (it is visible from Earth as a bright point). It is 25 km high, and has a base measuring 600 km, so that it is the largest known volcano in the Solar System; by contrast, the Hawaiian volcanoes on Earth, which also are shield volcanoes, have a height of only 9 km at most above their bases on the ocean floor. (Heights on Mars are bound to be somewhat arbitrary, since there is no sea-level to act as a standard.)

The largest caldera on a Martian volcano is that of Arsia Mons; the caldera is 110 km in diameter. Other large calderas crown the volcanoes Olympus Mons (85 km), Ascræus Mons (65 km) and Pavonis Mons (45 km); all these volcanoes lie in the Tharsis area. Arsia Mons has a height of 14 km.

The deepest basin on Mars is Hellas, which measures 2200 km by 1800 km and is about 3 km deep (reckoned from the level on Mars where the atmospheric pressure is 6·1 millibars).

The longest valley system on Mars is that of the Valles Marineris, in the Coprates-Tithonius region and not very far from the giant volcanoes of the Tharsis ridge. The system can be traced for a length of 4500 km, with a maximum width of 600 km, and a greatest depth of about 7 km above the rim. There are various other sinuous channels over 1000 km long, giving every impression of having been cut by running water.

The greatest canyon system on Mars is Noctis Labyrinthus (formerly known as Noctis Lacus). The canyons are from 10 to 20 km wide, with slopes of from 10 to 15 degrees. The whole system covers 120 000 km².

The highest atmospheric pressure so far measured on Mars is 8·9 millibars, on the floor of Hellas. The pressure at the top of Olympus Mons is below 3 millibars. When Viking 1 landed in Chryse, the pressure was approximately 7 millibars. A decrease of 0·012 millibar per sol was subsequently measured, due to carbon dioxide

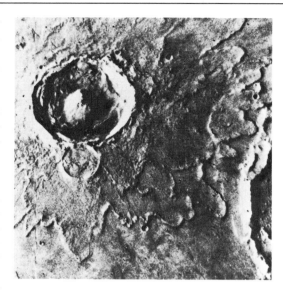

Crater Yuty photographed by Viking Orbiter 1

condensing out of the atmosphere to be deposited on the south polar cap, but this is a seasonal phenomenon, and ceased while Vikings 1 and 2 were still operating.

The maximum temperature of the aeroshell of Viking 1 during descent to Mars was 1500 °C; that of Viking 2 was similar.

The first analysis of the surface material was made from the lander of Viking 1: 44 per cent silicon, 5·5 per cent aluminium, 18·0 per cent iron, 8·0 per cent magnesium, 0·9 per cent titanium, 0·3 per cent potassium. Results from Viking 2 were similar. The general composition seems to be not very different from that of the lunar maria.

The first temperature measurements from the lander of Viking 1 ranged between −86 °C after dawn to a maximum of −31 °C near noon. The results from Viking 2 were similar.

COMPOSITION OF THE MARTIAN ATMOSPHERE AT THE SURFACE

Gas	Proportion
Carbon dioxide, CO_2	95·32%
Nitrogen, N_2	2·7%
Argon, Ar	1·6%
Oxygen, O_2	0·03%
Carbon monoxide, CO	0·07%
Water vapour, H_2O	0·03%
Neon, Ne	2·5 ppm
Krypton, Kr	0·3 ppm
Xenon, Xe	0·08 ppm
Ozone, O_3	0·03 ppm
(ppm = parts per million)	

The minimum summer temperature of the north polar cap was found to be −73 °C; at this temperature

carbon dioxide would evaporate. At this time the temperature of the southern cap was found to be −137 °C.

The first attempts to detect Martian life from the surface were made by Viking 1. There were three main experiments (pyrolitic release, labelled release, and gas exchange) but all were inconclusive; no certain evidence of Martian life has been found. The experiments were repeated from Viking 2, with equally inconclusive results.

The maximum windspeeds on Mars may, it is estimated, reach 400 km/h, but winds in that thin atmosphere will have little force. Windspeeds measured from the Vikings were light, little exceeding 20 km/h.

All our ideas about Mars have been revolutionised by the Mariner and Viking results. The dark areas are no more than albedo features, and are not depressed seabeds; neither are they vegetation-covered. The crust of Mars may be about 200 km deep; below this comes the mantle, and below this again is the core. No magnetic field has been detected. Much of the water on Mars is locked up in the permanent polar caps. If all the water in the atmosphere were condensed, it would cover the Martian surface with a layer only 1/100 mm deep, but the release of all the water in the ice-caps would produce a layer 10 m deep.

The atmospheric pressure is at present too low for liquid water to exist on the surface, though the presence of apparently water-cut channels indicates that the pressure was much greater some tens of thousands of years ago; evidently there are marked climatic changes, due possibly to the fact that when the axial inclination is at a suitable value the volatiles in the caps may be evaporated, leaving no permanent residual ice-caps such as those which exist today. Traces of carbon monoxide, ozone and hydrogen have been detected in the atmosphere; the height of the Martian tropopause is perhaps 120 km, with carbon dioxide or H_2O wave clouds rising to 24 km, water ice clouds to a maximum of 8 km, and dust-storms up to over 40 km (though even during major storms, the tops of the Tharsis volcanoes often protrude above the dusty level). No doubt future probes to Mars will produce further unexpected results!

FEATURES ON MARS

Craters	Lat	Long	Diam km
Achar	+45·7	236·8	6
Adams	+31	197	100
Agassiz	−70	089	110
Airy	+0·5	000	57
Albany	+23·3	049	2
Albi	−41·9	034·7	8
Alexei Tolstoy	−48	234	−
Alga	−24·6	026·5	19
Amsterdam	+23·3	047	2
Annapolis	+23·4	047·8	1
Antoniadi	+22	299	380
Apt	+40·1	009·5	10
Arago	+10	330	150
Arandas	+42·6	015·1	22
Arrhenius	−40	237	120
Aspen	−21·6	022·9	16
Azul	−42·5	042·3	20
Azusa	−05·5	040·5	41
Bacht	+18·9	257·5	7
Bahn	−03·5	043·5	10
Bakhuysen	−23	344	140
Balboa	−03·5	034	20
Baldet	+23	295	190
Balta	−24·1	026·4	15
Baltisk	−42·6	054·5	48
Bamba	−03·5	041·7	21
Bamberg	+39·9	033·1	55
Banff	+17·4	030·5	3
Banh	+19·7	055·5	13
Bar	−25·6	019·3	2
Barabashov	+47	069	130
Barnard	−61	298	122
Basin	+18·1	253·2	16
Batoka	−07·6	036·8	14
Batos	+21·7	029·5	16
Baykonur	+46·7	227·2	4
Becquerel	+22	008	175
Beer	−15	008	80
Belz	+21·8	043·2	10
Bend	−22·6	027·5	2
Bentong	−22·6	018·9	10
Bersaba	−04·4	037·7	36
Bhor	+42	225·5	6
Bianchini	−64	096	50

	Lat	Long	Diam km
Bigbee	−25·0	034·6	18
Bise	+20·5	056·5	9
Bison	−26·6	029·0	15
Bjerknes	−43	189	85
Bled	+21·8	031·4	7
Blitta	−26·3	020·8	11
Blois	+23·9	055·8	11
Bluff	+23·7	250·1	6
Boeddicker	−15	197	110
Bogra	−24·4	028·6	18
Bok	+20·8	031·6	7
Bole	+25·6	054·9	8
Bole	+25·4	053·7	8
Bond	−33	036	120
Bor	+18·1	033·8	4
Bordeaux	+23·4	048·9	2
Boru	−24·6	027·7	11
Bouguer	−19	333	100
Bozkir	−44·5	031·9	77
Brashear	−54	120	115
Bremerhaven	+23·5	048·6	2
Briault	−10	270	90
Bridgetown	+22·2	047·1	2
Bristol	+22·4	047·0	3
Broach	+23·8	056·9	11
Brush	+21·9	248·8	5
Bulhar	+50·6	225·6	19
Burroughs	−72	243	110
Burton	−14	156	120
Buta	−23·5	032·2	12
Butte	−05·1	39·1	12
Byrd	−64	232	120
Byske	−05	031·1	5
Cadiz	+23·4	049·1	1
Cairns	+23·8	047·5	10
Calbe	−25·5	028·7	14
Camiling	−00·8	038·1	21
Camiri	−45·1	041·9	20
Campbell	−54	195	133
Campos	−22·0	027·6	7
Can	+48·4	014·6	6
Canaveral	+47·1	224·0	3
Canberra	+47·5	227·2	3
Cartago	−23·6	017·8	33
Cassini	+24	328	440
Cave	+21·9	035·6	8
Cerulli	+32	338	125
Chamberlin	−66	124	120
Changsong	+23·8	057·3	32
Chapais	−22·6	020·4	33
Charleston	+22·9	047·9	2
Charlier	−69	169	110
Chauk	+23·6	055·9	9
Cheb	−24·5	019·3	8
Chekalin	−24·5	026·6	80
Chimbote	−01·5	039·8	65
Chincoteague	+41·5	236	35
Chinju	−04·6	042·3	65
Chinook	+22·8	055·4	17
Chive	+21·9	56·0	8
Choctaw	−41·5	037·0	20
Chom	+38·7	002·2	6
Chur	+16·7	029·2	4
Circle	−22·4	025·4	11
Clark	−56	134	95
Clogh	+20·8	047·7	10
Cluny	−24·1	027·1	9
Cobalt	−26·1	026·8	10
Coblentz	−55	091	95
Colon	+23·0	047·1	2
Columbus	−29	166	110
Comas Solà	−20	158	130
Conches	−04·2	034·3	18
Concord	+16·6	034·1	20
Copernicus	−50	169	280
Corby	+43·1	222·4	7
Cost	+15·3	256·1	10

	Lat	Long	Diam km
Cray	+44·4	016·2	5
Creel	−06·1	039·0	8
Crewe	−25·2	019·4	3
Crommelin	+05	101	107
Cruls	−43	197	78
Cruz	+38·6	001·7	6
Curie	+29	005	115
Cypress	−47·6	047·0	11
Daet	−07·4	041·9	12
Daly	−66	022	90
Dana	−73	032	90
Dank	+22·2	253·2	7
Darwin	−57	020	173
da Vinci	+02	039	100
Dawes	−09	322	190
Deba	−24·3	017·1	7
Dein	+38·5	002·4	24
Delta	−46·3	039·0	5
Denning	−18	326	150
Dese	−45·8	030·3	12
Dessau	−43·1	053·0	9
Dia-Cau	−0·3	042·8	28
Dingo	−24·0	017·3	13
Dison	−25·4	016·3	20
Dokuchaev	−61	127	−
Douglass	−52	070	90
Dromore	+20·1	049·6	14
du Martheray	−06	266	100
du Toit	−72	046	200
Eads	−28·9	029·8	2
Eagle	+44·0	008·2	12
Echt	−22·2	028·0	2
Edam	−26·6	019·9	20
Eddie	+12	218	100
Eger	−48·7	051·8	12
Eil	+42·0	009·7	4
Eiriksson	−19	174	50
Elath	+46·1	013·7	13
Ely	−23·9	027·1	10
Escalante	000	245	80
Esk	+45·4	007·0	4
Eudoxus	−45	147	90
Evpatoriya	+47·3	225·5	6
Faith	+43·2	011·9	5
Falun	−24·2	024·4	10
Fastov	−25·4	020·2	12
Fesenkov	+22	087	90
Flammarion	+26	312	179
Flat	−25·8	019·4	3
Flaugergues	−17	341	230
Flora	−45	051	15
Focas	+34	347	70
Fontana	−63	073	80
Fournier	−04	287	118
Freedom	+43·6	009·0	12
Funchal	+23·2	049·5	2
Gaan	+38·5	003·2	3
Gagra	−20·9	021·9	14
Gah	−45·1	032·3	2
Gale	−06	222	160
Gali	−44·1	036·9	24
Galilaei	+06	027	133
Galle	−51	031	210
Galu	−22·3	021·5	9
Gandu	−45·8	047·0	5
Gardo	−27·0	024·6	11
Garm	+48·4	009·2	5
Gastre	+24·8	247·6	5
Gatico	−21·2	020·9	18
Gilbert	−68	274	117
Gill	+16	354	85
Glazov	−20·8	026·4	20
Gledhill	−53	274	73
Glide	−08·2	043·3	10
Globe	−24·0	027·1	45
Goba	−23·5	020·9	11
Goff	+23·5	255·2	6

	Lat	Long	Diam km		Lat	Long	Diam km		Lat	Long	Diam km
Gol	+47·4	010·7	7	Kem	−45·3	032·7	2	Maraldi	−62	032	115
Gold	+20·2	031·3	9	Kepler	−47	219	238	Marbach	+17·9	249·2	20
Golden	−22·2	033·3	19	Keul	+46·3	237·7	6	Mariner	−35	164	160
Goldstone	+48·1	225·3	1	Kholm	−07·3	042·1	10	Marth	+13	003	100
Gori	−23·2	028·6	6	Kifri	−46·0	054·1	12	Martz	−34	217	88
Graff	−21	206	160	Kimry	−20·5	016·2	17	Maunder	−50	358	84
Green	−52	008	160	Kin	+20·4	033·4	7	McLaughlin	+22	022	90
Grojec	−21·6	030·6	37	Kingston	+22·4	047·1	2	Mega	−01·5	037·1	13
Groves	−04·1	044·6	10	Kipini	+24·8	032·0	35	Mendel	−59	199	73
Guaymas	+26·2	045·0	20	Kirs	−26·7	019·3	3	Mie	+48	220	100
Guir	−21·8	020·4	12	Kirsanov	+22·4	025·0	15	Mila	−27·5	020·6	10
Gulch	+16·1	251·2	7	Kita	−23·1	017·0	11	Milankovič	+55	147	120
Gusev	−14	184	170	Knobel	−06	226	125	Millochau	−21	275	118
Gwash	+38·9	003·0	5	Kok	+15·7	028·1	6	Misk	−01	035·5	10
Hadley	−19	203	110	Kong	−05·4	038·7	10	Mitchel	−68	284	179
Haldane	−53	231	80	Korolev	+73	196	92	Mohawk	+43·2	005·4	11
Hale	−36	036	150	Kourou	+47·0	227·1	2	Molesworth	−28	211	175
Halley	−49	059	78	Kristofovich	−48	262	—	Moreux	+42	315	135
Ham	−45·0	032·2	1	Kribi	−43·4	043·4	8	Moss	+19·4	250·7	8
Hamaguir	+49	227·3	3	Kuba	−25·6	019·5	25	Muller	−26	232	120
Hamelin	+20·4	032·8	10	Kufra	+40·6	239·7	37	Murgoo	−24·0	022·3	24
Harad	−27·8	027·8	8	Kumara	+43·3	231·4	12	Mut	+22·7	035·7	7
Hartwig	−39	016	100	Kuiper	−57	157	80	Naar	+23·2	042·1	12
Heaviside	−71	095	88	Kunowsky	+57	009	60	Naic	+24·7	252·8	8
Helmholtz	−46	021	110	Kushva	−44·3	035·2	39	Nain	+41·7	233·2	7
Henry	+11	336	170	Lachute	−04·3	039·9	13	Naju	+45·3	237·1	8
Herschel	−14	230	320	Laf	+48·3	005·9	6	Nan	−27·1	019·8	2
Hilo	−44·8	035·5	20	Lagarto	+50·0	008·4	19	Nansen	−50	141	90
Hipparchus	−45	151	90	Lamas	−27·4	020·5	21	Nardo	−27·8	032·7	23
Hit	+47·3	221·5	7	Lambert	−20	335	90	Naukan	+21·4	030·6	6
Holden	−26	034	175	Lamont	−59	114	235	Navan	−26·2	023·2	25
Holmes	−75	292	116	Lampland	−36	079	80	Nazca	−44·0	038·2	13
Honda	−22·7	016·2	9	Land	+48·4	008·8	5	Nema	+20·9	052·1	14
Hooke	−45	044	140	La Paz	+21·2	049·0	1	Nepa	−25·3	019·5	15
Hope	+45·1	010·3	6	Lar	−26·1	028·8	6	New Bern	+21·8	049·2	2
Houston	+48·5	223·9	2	Lassell	−21	063	90	Newcomb	−24	358	250
Hsuanch'eng	+47·0	227·2	2	Lasswitz	−09	222	110	New Haven	+22·3	049·3	2
Huancayo	−03·7	039·9	25	Lau	−74	107	110	Newport	+22·5	049·0	2
Huggins	−49	204	80	Lebu	−20·6	019·4	20	Newton	−40	158	280
Hussey	−54	127	100	Lemgo	−42·9	034·5	11	Nicholson	00	165	100
Hutton	−72	255	90	Le Verrier	−38	343	120	Niesten	−28	302	110
Huxley	−63	259	98	Lexington	+22·1	048·7	5	Nif	+20·2	056·3	7
Huygens	−14	304	495	Liais	−75	253	122	Nitro	−21·5	023·8	28
Ibragimov	−25·5	059	—	Libertad	+23·3	029·4	31	Noma	−25·7	024·0	38
Innsbrück	−06·5	030·1	15	Li Fan	−47	153	100	Nordenskjöld	−53	159	—
Inta	−24·6	024·9	14	Linpu	+18·3	247·0	17	Nune	+17·7	038·6	9
Irbit	−24·6	024·7	13	Lins	+15·9	029·8	6	Nutak	+17·6	030·3	1
Jal	−26·6	028·6	5	Lisboa	+21·5	047·6	1	Ochakov	−42·5	031·6	30
Janssen	+03	322	160	Liu Hsin	−53	172	122	Oglala	−03·2	038·3	15
Jarry-Desloges	−09	276	100	Livny	−27·5	028·9	10	Okhotsk	+23·2	047·4	2
Jeans	−70	206	80	Locana	−03·5	038·3	7	Omura	−25·7	025·0	8
Jen	+40·1	010·6	8	Lockyer	+28	199	80	Oraibi	+17·4	032·4	31
Jezza	−48·8	037·7	6	Lod	+21·2	031·6	7	Ore	+16·9	033·9	7
Jijiga	+25·4	053·9	14	Lohse	−43	016	160	Orinda	+45·6	233·0	9
Jodrell	+47·8	227·6	3	Loja	+41·4	223·8	10	Ostrov	−26·9	027·9	70
Johannesburg	+48·2	226·7	2	Lomonosov	+65	008	135	Ottumwa	+24·9	055·7	55
Joly	−75	042	80	Longa	−20·9	025·7	10	Oudemans	−10	092	120
Jones	−19	020	90	Lorica	−19·9	028·1	68	Pabo	−27·3	022·9	9
Kagoshima	+47·6	224·1	1	Lota	+46·5	011·9	13	Paks	−07·8	042·2	7
Kagul	−24·0	018·9	8	Loto	−22·2	022·3	22	Pasteur	+19	335	125
Kaid	−04·6	044·8	7	Lowell	−52	081	200	Peixe	+20·8	047·6	9
Kaiser	−46	340	205	Luck	+17·4	036·9	7	Perepelkin	+52	065	80
Kaj	−27·4	029·2	2	Luga	−44·6	047·2	42	Peridier	+26	276	100
Kakori	−41·9	029·6	25	Lutsk	+38·5	002·9	5	Pettit	+12	174	90
Kaliningrad	+48·8	224·9	2	Luzin	+26	328·5	115	Philadelphia	+22	048	2
Kampot	−42·1	045·4	9	Lyell	−70	015	120	Phillips	−67	045	175
Kanab	−27·6	018·8	14	Lyot	+50	331	220	Phon	+15·8	257·3	8
Kansk	−20·8	017·1	34	Mädler	−11	357	100	Pickering	−34	133	110
Kantang	−24·8	017·3	48	Madrid	+48·7	224·4	3	Pinglo	−03	037	18
Karpinsk	−46·0	031·8	28	Mafra	−44·4	053·0	12	Playfair	−78	125	40
Karshi	−23·6	019·2	22	Magelhæns	−32	174	110	Plum	−26·4	018·9	3
Kartabo	−41·2	052·3	17	Maggini	+28	350	140	Podor	−44·6	043·0	25
Kashira	−27·5	018·1	67	Maidstone	−41·9	054·1	9	Poona	+24·0	052·3	20
Kasmimov	−25·0	022·8	85	Main	−77	310	115	Polotsk	−20·1	026·1	23
Kaup	+22·5	033·6	3	Manah	−04·7	033·8	9	Port-au-Prince	+21·3	048·2	1
Kaw	+16·7	255·9	10	Manti	−03·7	037·7	15	Porter	−50	114	105
Keeler	−61	152	55	Manzi	−22·3	027·3	7	Porth	+21·5	255·9	10

	Lat	Long	Diam km		Lat	Long	Diam km		Lat	Long	Diam km
Portsmouth	+22·8	049·1	2	Stobs	−05	038·4	10	Windfall	−02·1	043·5	20
Porvoo	−43·7	040·6	9	Stokes	+56	189	70	Wink	−06·6	041·5	8
Priestley	−54	228	120	Ston	+47·2	237·4	7	Wirtz	−49	026	70
Princeton	+21·9	049·1	2	Stoney	−70	138	129	Wislencius	−18	349	150
Proctor	−48	330	160	Sucre	+24	054·5	9	Woomera	+48·4	227·3	3
Ptolemæus	−46	158	160	Suess	−67	179	80	Wright	−59	151	115
Punsk	+20·7	041·2	10	Suf	+16·6	038·2	9	Yakima	+43·3	003·2	12
Pylos	+16·9	030·1	29	Surt	+17	030·6	9	Yala	+17·5	038·5	19
Quénisset	+34	319	130	Sytinskaya	+43	052	–	Yar	+22·5	039·1	5
Quick	+18·5	048·8	8	Taboa	−45·5	034·8	8	Yat	+18·3	029·1	8
Quorn	−05·6	33·8	5	Tak	−26·4	028·5	5	Yegros	−22·5	023·5	14
Rabe	−44	325	108	Talsi	−41·9	049·1	9	Yorktown	+23·1	048·7	8
Radau	+17	005	120	Tara	−44·4	052·7	27	Yoro	+23·0	028·0	8
Rakke	−04·7	043·5	18	Tarakan	−41·6	030·1	37	Yungay	−44·3	044·6	17
Rana	−26·0	021·6	12	Tarata	−03·8	041·4	12	Yuty	+22·4	034·1	18
Raub	+42·6	224·9	7	Tarsus	+23·5	040·2	19	Zir	+18·7	036·6	6
Rauch	+21·6	057·8	8	Taxco	+20·8	040·0	17	Zulanka	−02·3	042·3	47
Rayleigh	−76	240	182	Taza	−44·0	045·1	22	Zuni	+19·3	029·6	25
Redi	−61	267	70	Teisserenc de Bort	+01	315	107				
Renaudot	+42	297	70	Tem	+42·1	009·5	4				

Catena	Lat	Long	Length km
Acheron	+35 to 42	097–103	558
Coprates	−14 to 16	067–058	505
Ganges	−02 to 03	071–067	233
Phlegethon	+35 to 41	095–105	875
Tithonia	−06 to 05	087–080	400
Tractus	+20 to 35	100–104	935

Cavi	Lat	Long	Diam km
Angusti	−73 to 84	055–100	730
Novi	−65 to 70	320–340	535
Sisyphi	−76 to 82	340–000	405
Ultimi	−73 to 76	195–214	320

Chaos	Lat	Long	Diam km
Aram	0 to +05	019–024	275
Aromatum	−01	043·5	–
Arsinoes	−08	021	–
Aureum	−02 to 07	030–024	365
Eos	−16 to 19	043–050	–
Hydaspes	+01 to 04	024–030	350
Hydraotes	−01 to 03	032–040	415
Iani	−05 to 02	020–013	405
Margaritifer	−07 to 13	017–025	430
Pyrrhæ	−11	029	–

Chasma	Lat	Long	Length km
Australe	−80 to 89	284–257	501
Borealis	+85 to 81	050–030	318
Candor	−04 to 08	078–070	400
Capri	−15 to 03	053–031	1275
Coprates	−10 to 16	069–053	975
Echus	−01 to 05	078–032	310
Eos	−16 to 19	043–050	–
Gangis	−06 to 09	055–043	575
Hebes	−02 to 00	079–073	320
Ius	−10 to 14	092–077	890
Juventai	−02 to 06	054–060	250
Melas	−08 to 14	077–069	505
Ophir	−03 to 09	075–064	660
Thyles	−69 to 73	230–235	235
Tithonium	−03 to 07	092–077	880

Colles (hills)	Lat	Long	
Ariadnes	−35	187	
Deuteronilus	+41	338	

Dorsum (Dorsa)	Lat	Long	Diam km
Arena	−11 to 15	291–292	205
Argentea	−76 to 80	015–040	385
Brevia			
Cerberus	−09 to 18	252–258	–
Eumenides	0 to −10	155–158	625
Felis	−20 to 27	063–070	575
Gordii	−01 to 07	142–146	400

	Lat	Long	Diam km
Rengo	−43·9	043·5	12
Reuyl	−10	193	73
Revda	−24·6	028·3	26
Reynolds	−74	150	90
Ribe	+16·6	029·2	11
Richardson	−73	181	80
Rimac	+45·2	223·8	7
Rincon	−08·1	043·1	13
Ritchey	−29	051	73
Romny	−25·7	018·0	5
Rong	+22·7	045·3	9
Ross	−58	108	83
Rossby	−48	192	78
Ruby	−25·6	016·9	25
Rudaux	+38	309	65
Russell	−55	348	140
Rutherford	+19	011	100
Rynok	+44·3	238·2	9
Rypin	−01·3	041·1	18
Salaga	−47·6	051	28
Sandila	−25·9	030·2	8
Sangar	−27·9	024·1	28
San Juan	+23·1	048·1	1
Santa Cruz	+21·5	047·3	2
Santa Fé	+19·5	048·0	20
Sarno	−44·7	054·0	20
Satka	−43·0	036·7	14
Sauk	−45·0	032·3	2
Savannah	+22·3	047·8	1
Say	−28·5	029·5	15
Schaeberle	−24	310	160
Schiaparelli	−03	343	500
Schmidt	−72	079	196
Schoner	+20	309	200
Schröter	−02	304	310
Secchi	−58	258	198
Semeykin	+42	351	–
Seminole	−24·5	018·9	21
Sfax	−07·8	043·5	7
Shambe	−20·7	030·5	29
Sharonov	+27	059	115
Shawnee	+22·7	031·5	16
Sibu	−23·3	019·6	31
Sigli	−20·5	030·6	31
Singa	−22·8	017·2	11
Sitka	−04·3	039·4	11
Sklodowska	+34	003	120
Slipher	−48	084	90
Smith	−66	103	70
Sogel	+21·7	055·1	29
Sokol	−42·8	040·5	20
Soochow	+16·8	028·9	30
South	−77	339	105
Spallanzani	−58	273	68
Spry	−03·8	038·6	6
Spur	+22·2	052·3	7
Steno	−68	115	105

	Lat	Long	Diam km
Terby	−28	286	170
Thule	−23·6	025·5	14
Tikhov	−51	254	115
Tile	+17·8	028·6	8
Timaru	−25·6	022·2	17
Timbuktu	−5·6	037·6	60
Timoshenko	+42	064	–
Tiwi	−27·9	024·6	17
Tombe	−42·8	044·4	12
Tono	−45·2	052·2	10
Torso	−44·7	051·0	14
Troika	+17·1	255·0	13
Trouvelot	+16	013	150
Troy	+23·4	052·6	10
Trud	+17·7	030·8	5
Trümpler	−62	151	72
Tsau	+49·8	238·9	6
Tsukuba	+48·9	225·9	2
Tura	−27·0	021·8	14
Turbi	−40·9	051·2	25
Tuskegee	−02·9	036·2	62
Tycho Brahe	−50	214	90
Tyndall	+40	190	83
Tsau			
Umatac	+42·7	222·8	12
Vaals	−04·0	033·1	8
Valga	−44·6	036·3	16
Valverde	+20·3	055·8	35
Vato	−43·9	053·2	10
Vaux	+18·1	072·8	–
Very	−50	177	140
Viana	+19·5	255·3	29
Vils	+39·2	011·7	6
Vinogradsky	−56	217	150
Vishniac	−77	276	78
Vivero	+49·2	241·1	27
Vögel	−37	013	115
Volgorad	+48·4	224·8	2
Volsk	+23·2	051·3	8
von Kármán	−64	059	80
Voo	−27·3	019·8	2
Wabash	+21·5	033·7	42
Wahoo	+23·5	033·7	63
Wallace	−53	249	150
Wallops	+46·9	227·2	2
Warra	+20·8	037·5	11
Waspam	+20·7	056·6	40
Wassamu	+25·5	053·0	17
Wau	−45·2	042·4	3
Weert	+19·8	051·4	9
Wegener	−65	004	70
Weinbaum	−66	245	82
Wells	−60	238	80
Wer	+45·6	036·2	3
Wicklow	−02·0	040·7	21
Wien	−11	220	120
Williams	−18	164	120
Wilmington	+21·9	047·5	1

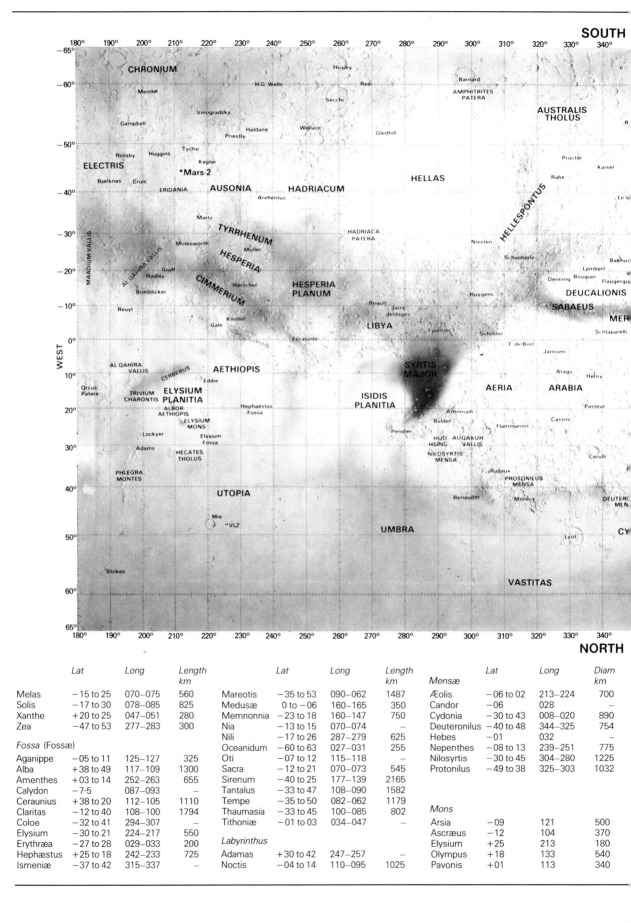

SOUTH

WEST

NORTH

	Lat	Long	Length km		Lat	Long	Length km		Lat	Long	Diam km
Melas	−15 to 25	070–075	560	Mareotis	−35 to 53	090–062	1487	**Mensæ**			
Solis	−17 to 30	078–085	825	Medusæ	0 to −06	160–165	350	Æolis	−06 to 02	213–224	700
Xanthe	+20 to 25	047–051	280	Memnonnia	−23 to 18	160–147	750	Candor	−06	028	–
Zea	−47 to 53	277–283	300	Nia	−13 to 15	070–074	–	Cydonia	−30 to 43	008–020	890
				Nili	−17 to 26	287–279	625	Deuteronilus	−40 to 48	344–325	754
Fossa (Fossæ)				Oceanidum	−60 to 63	027–031	255	Hebes	−01	032	–
Aganippe	−05 to 11	125–127	325	Oti	−07 to 12	115–118	–	Nepenthes	−08 to 13	239–251	775
Alba	+38 to 49	117–109	1300	Sacra	−12 to 21	070–073	545	Nilosyrtis	−30 to 45	304–280	1225
Amenthes	+03 to 14	252–263	655	Sirenum	−40 to 25	177–139	2165	Protonilus	−49 to 38	325–303	1032
Calydon	−7·5	087–093	–	Tantalus	−33 to 47	108–090	1582				
Ceraunius	+38 to 20	112–105	1110	Tempe	−35 to 50	082–062	1179				
Claritas	−12 to 40	108–100	1794	Thaumasia	−33 to 45	100–085	802				
Coloe	−32 to 41	294–307	–	Tithoniæ	−01 to 03	034–047	–	**Mons**			
Elysium	−30 to 21	224–217	550					Arsia	−09	121	500
Erythræa	−27 to 28	029–033	200	**Labyrinthus**				Ascræus	−12	104	370
Hephæstus	+25 to 18	242–233	725	Adamas	+30 to 42	247–257	–	Elysium	+25	213	180
Ismeniæ	−37 to 42	315–337	–	Noctis	−04 to 14	110–095	1025	Olympus	+18	133	540
								Pavonis	+01	113	340

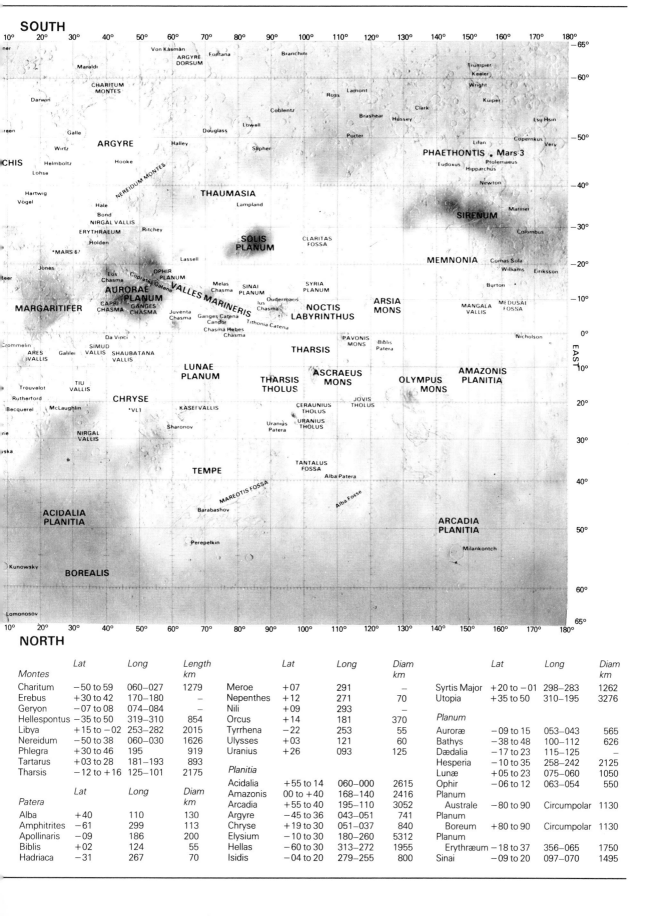

Montes	Lat	Long	Length km		Lat	Long	Diam km			Lat	Long	Diam km
Charitum	−50 to 59	060–027	1279	Meroe	+07	291	–		Syrtis Major	+20 to −01	298–283	1262
Erebus	+30 to 42	170–180	–	Nepenthes	+12	271	70		Utopia	+35 to 50	310–195	3276
Geryon	−07 to 08	074–084	–	Nili	+09	293	–					
Hellespontus	−35 to 50	319–310	854	Orcus	+14	181	370		Planum			
Libya	+15 to −02	253–282	2015	Tyrrhena	−22	253	55		Auroræ	−09 to 15	053–043	565
Nereidum	−50 to 38	060–030	1626	Ulysses	+03	121	60		Bathys	−38 to 48	100–112	626
Phlegra	+30 to 46	195	919	Uranius	+26	093	125		Dædalia	−17 to 23	115–125	–
Tartarus	+03 to 28	181–193	893						Hesperia	−10 to 35	258–242	2125
Tharsis	−12 to +16	125–101	2175	Planitia					Lunæ	+05 to 23	075–060	1050
				Acidalia	+55 to 14	060–000	2615		Ophir	−06 to 12	063–054	550
	Lat	Long	Diam km	Amazonis	00 to +40	168–140	2416		Planum			
Patera				Arcadia	+55 to 40	195–110	3052		Australe	−80 to 90	Circumpolar	1130
Alba	+40	110	130	Argyre	−45 to 36	043–051	741		Planum			
Amphitrites	−61	299	113	Chryse	+19 to 30	051–037	840		Boreum	+80 to 90	Circumpolar	1130
Apollinaris	−09	186	200	Elysium	−10 to 30	180–260	5312		Planum			
Biblis	+02	124	55	Hellas	−60 to 30	313–272	1955		Erythræum	−18 to 37	356–065	1750
Hadriaca	−31	267	70	Isidis	−04 to 20	279–255	800		Sinai	−09 to 20	097–070	1495

	Lat	Long	Length km
Solis	−20 to 30	098–088	1000
Syria	−10 to 20	112–097	900
Rupes			
Amenthes	00	248–252	–
Argyre	−61 to 65	065–073	347
Arena	+11 to 16	288–291	375
Bosporus	−39 to 47	053–060	525
Morpheos	−36 to 39	231–240	–
Ogygis	−32 to 35	053–055	185
Olympus	−15 to 24	129–013	2010
Phison	+25 to 28	308–310	160
Ultima	−65 to 73	198–204	495
Scopulus			
Eridana	−47 to 60	208–235	1344
Frigoris	+78 to 81	045–323	948
Hyperboreus	+80 to 85	060–160	1059
Hypernotius	−75 to 81	244–320	1918
Nilokeras	+30 to 42	051–063	1231
Œnotria	−15 to 10	270–292	–
Promethei	−75 to 83	190–320	2508
Sulci			
Cyane	+23 to 28	126–130	300
Gigas	+06 to 14	125–136	750
Lycus	+18 to 34	130–148	1004
Sulci Gordii	+16 to 20	125–127	250

	Lat	Long	Diam km
Tholus			
Albor	+19	210	115
Australis	−59	323	40
Ceraunius	+24	097	135
Charitum	−55	041	30
Hecates	+32	210	152
Iaxartes	+72	015	53
Jovis	+18	117	50
Kison	+73	358	50

	Lat	Long	Diam km
Ortygia	+70	008	100
Tharsis	+14	091	150
Uranius	+26	098	65

	Lat	Long	Length km
Vallis (Valles)			
Al-Qahira	−19 to 14	200–194	420
Arda	−20 to 22	030–034	156
Ares	+12 to 02	023–017	915
Arnus	− +8	290	–
Auqakuh	+30 to 27	300–297	195
Axius	−53 to 58	285–295	435
Bahram	+20 to 22	055–057	135
Brazes	−2·5 to 6	192–197	–
Clota	−25 to 27	020–022	115
Dao	−23 to 40	265–275	430
Drava	−17 to 23	115–125	–
Drinus	−22 to 24	165–195	–
Evros	−12 to 13	343–349	–
Granicus	−27 to 30	225–231	–
Harmakhis	−33 to 40	265–275	698
Himera	−21 to 23	022–023	–
Hrad	+38 to 45	225–240	825
Huo Hsing	+34 to 28	299–292	662
Indus	−17 to 21	317–322	–
Kasei	+27 to 18	075–056	1090
Ladon	−02 to 25	027–030	280
Licus	−01 to 05	233–234	–
Locras	+07 to 11	311–313	265
Loire	−17·5 to 22	014–017	–
Louros	+08 to 09	033–041	–
Ma'adim	−28 to 16	184–181	955
Maja	+16 to 21	049–057	398
Mamers	+32 to 39	338–345	628
Mangala	−04 to 09	150–152	272
Marineris	+01 to −18	024–113	5272
Maumee	+18 to 20	054–056	130

	Lat	Long	Length km
Mawrth	+14 to 20	019–027	500
Naktong	000	330–321	–
Nanedi	00 to +09	047–051	570
Nirgal	−27 to 30	037–047	665
Oltis	−24 to 26	021–022	105
Osuga	−15 to 17	038–040	–
Parana	−23 to 26	009–012	240
Ravi	−02 to +02	039–043	365
Reull	+37 to 43	248–265	1065
Rubicon	+43 to 37	115–117	230
Scamander	−14 to 18	330–331	–
Shalbatana	00 to +15	040–045	995
Simud	−03 to 14	035–042	690
Tagus	−07	245	35
Tiu	+03 to 14	030–035	680
Uzboi	−27 to 37	035–037	290
Vedra	+19 to 20	054–056	122
Warrego	−42 to 44	091–095	160

	Lat	Long	Diam km
Vastitas			
Vastitas Borealis	+55 to 70	Circumpolar	9999
Terræ			
Aonia	−25 to 65	060–120	2917
Arabia	00 to +43	024–280	5625
Cimmeria	−45	210	–
Margaritifer	+02 to −27	012–045	1924
Meridiani	−05	000	–
Noachis	−15 to 83	040–300	1025
Promethei	−30 to 65	240–300	2967
Sabæa	−01	325	–
Sirenum	−50	150	–
Tempe	+24 to 54	050–093	1628
Tyrrhena	−10	280	–
Xanthe	+19 to −13	015–065	2797

THE SATELLITES OF MARS

The first mention of possible Martian satellites was fictional – by Jonathan Swift, in Gulliver's *Voyage to Laputa* (1727). Swift described two satellites, one of which had a revolution period shorter than the rotation period of its primary; but there was no scientific basis at that time for such an object. Two satellites were also described in another novel, Voltaire's *Micromégas* (1750). The reasoning was, apparently, that since the Earth had one satellite and Jupiter was known to have four, Mars could not possibly manage with less than two!

The first systematic search for Martian satellites was made by W. Herschel in 1783. The result was negative. H. D'Arrest at Copenhagen, in 1862 and 1864, was similarly unsuccessful.

The first satellite to be discovered was Deimos, by Asaph Hall at Washington on 10 August 1877. Hall discovered Phobos on 16 August 1877. The names Phobos and Deimos were

DATA	Phobos	Deimos
Mean distance from centre of Mars: km	9270	23 400
astronomical units	0·000 062 5	0·000 157 0
Mean angular distance from Mars, at mean opposition:	24″·6	1′01″·8
Mean sidereal period, days:	0·3189	1·2624
Mean synodic period:	7 h 39 m 26 s·6	1 d 6 h 21 m 15 s·7
Orbital inclination to Martian equator:	1°·1	1°·8
Orbital eccentricity:	0·0210	0·0028
Diameter, kilometres:	20 × 23 × 28	10 × 12 × 16
Escape velocity, km/s:	0·016	0·008
Magnitude at mean opposition:	11·6	12·8
Maximum apparent magnitude as seen from Mars:	−3·9	−0·1
Apparent diameter as seen from Mars:		
maximum	12′·3	2′
minimum	8′	1′·7

suggested to Hall by Mr Madan of Eton.

The first colour estimates of the satellites were made by E. M. Antoniadi, who in 1930 reported that Phobos was white and Deimos bluish. These results were certainly spurious, even though Antoniadi was using the 83 cm Meudon refractor. It is now known that the satellites are dust-covered, with very low albedoes.

The first proof that Phobos is irregular in shape was obtained in 1969, when Mariner 7 photographed the shadow of Phobos cast on to Mars

and revealed that it was elliptical. The first accurate size measurements were obtained by Mariner 9 in 1971–2. Both satellites are irregular, so that diameter values are bound to be somewhat arbitrary.

The first detection of craters on the satellites was also due to Mariner 9, in 1971–2.

The largest crater on Phobos is Stickney, 5 km in diameter. (Stickney was the maiden name of Asaph Hall's wife.) The principal craters on Deimos have been named Swift and Voltaire.

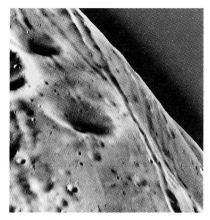

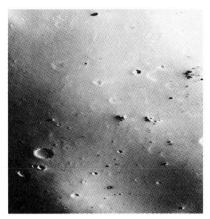

Phobos; Viking 1 orbiter, 20 February 1977. (NASA)

Deimos; Viking 2 orbiter, showing resolution down to 300 metres. The area covered measures 1·2 × 1·5 km. (NASA)

Map of Phobos

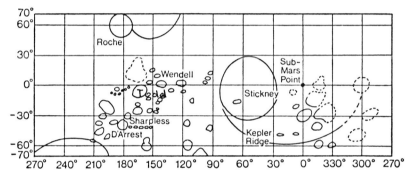

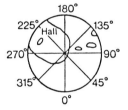

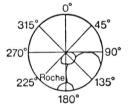

South Polar region

North Polar region

Map of Deimos

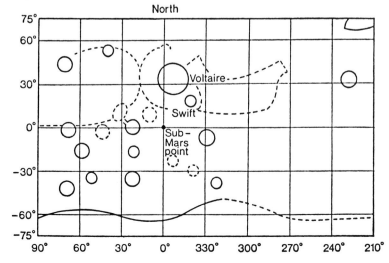

The Martian satellites are very dark (albedo *c.* 5 per cent) and dense (probably about $2·9\,g/cm^2$), and their material may be similar to that of carbonaceous chondrites. Each is covered with a regolith a few mm thick; that of Deimos is thicker than Phobos', so that its surface is more subdued. Phobos also shows strange parallel grooves inclined by 30° to the equator, around 100–200 m wide, and 10–20 m deep. It may well be that they are captured asteroids rather than bona-fide satellites, though there are admittedly difficulties in the way of this hypothesis. Phobos would be invisible to a Martian observer above latitude 69 degrees; the limiting latitude for Deimos would be 82 degrees. To a Martian observer, Phobos would be less than half the apparent diameter of the Moon as seen from Earth, and would give only about as much light as Venus does to us; from Phobos, Mars would subtend an angle of 42 degrees! The apparent diameter of Deimos as seen from Mars would be only about twice the maximum apparent diameter of Venus as seen from Earth, and with the naked eye the phases would be almost imperceptible. Deimos would remain above the Martian horizon for $2\frac{1}{2}$ sols consecutively; Phobos would cross the sky in only $4\frac{1}{2}$ hours, moving from west to east, and the interval between successive risings would be only just over 11 hours. This curious behaviour is due to the fact that Phobos has a revolution period less than the rotation period of its primary. Total solar eclipses could never occur; Phobos would transit the Sun 1300 times a year, taking 19 seconds to cross the disk, while Deimos would show an average of 130 transits, each taking 1 minute 48 seconds. To a Martian observer, eclipses of the satellites by the shadow of Mars would be very frequent.

The rotation periods of the satellites are synchronous, though there would be marked 'libration' effects similar to those of the Moon.

CRATERS ON PHOBOS D'Arrest, Hall, Roche, Sharpless, Stickney, Todd, Wendell

(There is also one named ridge: Kepler Dorsum.)

CRATERS ON DEIMOS Swift, Voltaire

THE MINOR PLANETS

The Minor Planets are also known as asteroids or planetoids. Approximately 3000 have had their orbits computed, and it is estimated that the total membership of the swarm may exceed 40 000.

THE FIRST KNOWN MINOR PLANETS

In the following table:

q = perihelion distance, in astronomical units.
P = revolution period, in years.
i = orbital inclination in degrees.
D = diameter in kilometres.
a = aphelion distance, in astronomical units.

e = orbital eccentricity.
M = opposition magnitude.
A = albedo.

(Asteroid diameters have recently been re-measured. Asteroids whose diameter remains very uncertain are indicated by an asterisk.)

No.	Name	Discoverer	q	a	P	e	i	M	D	A
1	Ceres	Piazzi, 1 January 1801	2·55	2·99	4·61	0·079	10·60	7·4	940	0·054
2	Pallas	Olbers, 28 March 1802	2·11	3·42	4·607	0·237	34·85	8·0	538 (mean)	0·074
3	Juno	Harding, 1 September 1804	1·98	3·35	4·358	0·257	13·00	8·7	288	0·151
4	Vesta	Olbers, 29 March 1807	2·15	2·57	3·630	0·089	7·14	6·5	555	0·229
5	Astræa	Hencke, 8 December 1845	2·10	3·06	4·139	0·187	5·34	9·8	117	0·140
6	Hebe	Hencke, 1 July 1847	1·93	2·92	3·778	0·203	14·77	8·3	195	0·164
7	Iris	Hind, 13 August 1847	1·84	2·94	3·686	0·230	5·50	7·8	209	0·154
8	Flora	Hind, 18 October 1847	1·86	2·55	3·267	0·156	5·89	8·7	151	0·144
9	Metis	Graham, 26 April 1848	2·09	2·68	3·684	0·123	5·58	9·1	151	0·139
10	Hygeia	De Gasparis, 12 April 1849	2·84	3·46	5·593	0·100	3·81	10·2	450	0·041
11	Parthenope	De Gasparis, 11 May 1850	2·20	2·70	3·840	0·102	4·62	9·9	150	0·126
12	Victoria	Hind, 13 September 1850	1·82	2·85	3·568	0·218	8·37	9·9	126	0·114
13	Egeria	De Gasparis, 2 November 1850	2·36	2·80	4·135	0·085	16·53	10·8	224	0·041
14	Irene	Hind, 19 May 1851	2·16	3·01	4·163	0·164	9·13	9·6	158	0·162
15	Eunomia	De Gasparis, 29 July 1851	2·15	3·14	4·300	0·188	11·73	8·5	272	0·155
16	Psyche	De Gasparis, 17 March 1851	2·53	3·32	5·000	0·135	3·09	9·9	250	0·093
17	Thetis	Luther, 17 April 1852	2·13	3·52	3·880	0·138	5·59	10·7	109	0·103
18	Melpomene	Hind, 24 June 1852	1·80	2·80	3·480	0·218	10·14	8·9	150	0·144
19	Fortuna	Hind, 22 August 1852	2·06	2·83	3·816	0·158	1·56	10·1	215	0·032
20	Massalia	De Gasparis, 19 September 1852	2·06	2·76	3·737	0·145	0·70	9·2	131	0·164
21	Lutetia	Goldschmidt, 15 November 1852	2·04	2·83	3·800	0·162	3·08	10·5	115	0·093
22	Calliope	Hind, 16 November 1852	2·61	3·21	4·962	0·104	13·72	10·6	177	0·130
23	Thalia	Hind, 15 December 1852	2·01	3·24	4·252	0·236	10·17	10·1	111	0·164
24	Themis	De Gasparis, 5 April 1853	2·76	3·52	5·559	0·121	0·77	11·8	234	0·030
25	Phocæa	Chacornac, 6 April 1853	1·79	3·01	3·720	0·256	21·57	10·3	72	0·184
26	Proserpina	Luther, 5 May 1853	2·42	2·89	4·328	0·089	3·56	11·3	80*	?
27	Euterpe	Hind, 8 November 1853	1·94	2·75	3·600	0·172	1·59	9·9	108	0·147
28	Bellona	Luther, 1 March 1854	2·35	3·20	4·625	0·153	9·41	10·8	126	0·132
29	Amphitrite	Marth, 1 March 1854	2·37	2·72	4·083	0·073	6·09	10·0	255	0·140
30	Urania	Hind, 22 July 1854	2·07	2·67	3·637	0·127	2·10	10·6	91	0·144
31	Euphrosyne	Ferguson, 1 September 1854	2·45	3·86	5·606	0·223	26·29	11·0	370	0·030
32	Pomona	Goldschmidt, 26 October 1854	2·38	2·80	4·163	0·082	5·50	11·3	93	0·140
33	Polyhymnia	Chacornac, 28 October 1854	1·89	3·84	4·842	0·340	1·90	11·0	45*	?
34	Circe	Chacornac, 6 April 1855	2·40	3·00	4·404	0·108	5·51	12·3	111	0·039
35	Leucothea	Luther, 19 April 1855	2·32	3·67	5·177	0·225	8·04	12·4	45*	?
36	Atalantë	Goldschmidt, 5 October 1855	1·92	3·58	4·557	0·301	18·45	11·6	118	0·024
37	Fides	Luther, 5 October 1855	2·18	3·10	4·296	0·175	3·08	10·7	95	0·186
38	Leda	Chacornac, 12 January 1856	2·32	3·17	4·535	0·155	6·97	12·2	45*	?
39	Lætitia	Chacornac, 8 February 1856	2·46	3·08	4·608	0·112	10·37	10·3	163	0·169
40	Harmonia	Goldschmidt, 31 March 1856	2·16	2·37	3·413	0·047	4·26	10·5	100	0·123
41	Daphne	Goldschmidt, 22 May 1856	2·02	3·51	4·591	0·270	15·87	10·3	204	0·056
42	Isis	Pogson, 23 May 1856	1·89	2·99	3·812	0·226	8·53	10·2	97	0·125
43	Ariadne	Pogson, 15 April 1857	1·83	2·57	3·270	0·168	3·47	10·2	85	0·113
44	Nysa	Goldschmidt, 27 May 1857	2·05	2·79	3·769	0·151	3·71	9·7	82	0·377
45	Eugenia	Goldschmidt, 27 June 1857	2·50	2·94	4·489	0·081	6·60	11·5	226	0·030
46	Hestia	Pogson, 16 August 1857	2·09	3·00	4·011	0·170	2·32	11·1	133	0·028
47	Aglaia	Luther, 15 September 1857	2·49	3·27	4·881	0·135	4·99	12·1	158	0·027
48	Doris	Goldschmidt, 19 September 1857	2·93	3·30	5·496	0·060	6·55	11·9	250*	0·03?
49	Pales	Goldschmidt, 19 September 1857	2·40	3·79	5·455	0·224	3·16	11·4	70*	?
50	Virginia	Ferguson, 4 October 1857	1·89	3·41	4·313	0·287	2·82	11·6	35*	?
51	Nemausa	Laurent, 22 January 1858	2·21	2·52	3·639	0·066	9·95	10·8	151	0·050
52	Europa	Goldschmidt, 4 February 1858	2·75	3·43	5·447	0·111	7·46	11·1	289	0·035
53	Calypso	Luther, 4 April 1858	2·08	3·16	4·232	0·206	5·15	11·7	45*	?
54	Alexandra	Goldschmidt, 10 September 1858	2·17	3·25	4·458	0·200	11·82	10·9	180	0·030
55	Pandora	Searle, 10 September 1858	2·37	3·15	4·585	0·142	7·20	11·6	65*	?
56	Melete	Goldschmidt, 9 September 1859	1·99	3·21	4·187	0·235	8·07	11·5	146	0·026
57	Mnemosyne	Luther, 22 September 1859	2·84	3·48	5·611	0·100	15·17	12·2	109	0·140
58	Concordia	Luther, 24 March 1860	2·58	2·82	4·435	0·045	5·05	13·1	110	0·030
59	Elpis	Chacornac, 12 September 1860	2·38	3·03	4·471	0·117	8·63	11·4	75*	?
60	Echo	Ferguson, 14 September 1860	1·95	2·83	3·703	0·183	3·59	11·5	51	0·154
61	Danaë	Goldschmidt, 9 September 1860	2·51	3·47	5·168	0·161	18·19	11·7	70*	?
62	Erato	Förster, 14 September 1860	2·60	3·67	5·550	0·169	2·22	13·0	45*	?
63	Ausonia	De Gasparis, 10 February 1861	2·09	2·70	3·706	0·127	5·78	10·2	91	0·128
64	Angelina	Tempel, 4 March 1861	2·35	3·02	4·391	0·124	1·31	11·3	56	0·342

No.	Name	Discoverer	q	a	P	e	i	M	D	A
65	Cybele	Tempel, 8 March 1861	3·01	3·83	6·333	0·121	3·54	11·8	309	0·022
66	Maia	Tuttle, 9 April 1861	2·19	3·10	4·306	0·172	3·05	12·5	85	0·029
67	Asia	Pogson, 17 April 1861	1·97	2·87	3·768	0·184	6·01	11·6	58	0·157
68	Leto	Luther, 29 April 1861	2·27	3·30	4·645	0·184	7·97	10·8	126	0·126
69	Hesperia	Schiaparelli, 29 April 1861	2·49	3·47	5·142	0·166	8·55	11·3	90*	?
70	Panopæa	Goldschmidt, 5 May 1861	2·14	3·09	4·227	0·181	11·62	11·5	151	0·039
71	Niobe	Luther, 13 August 1861	2·28	3·23	4·575	0·173	23·27	11·3	115	0·140
72	Feronia	Safford, 29 May 1861	1·99	2·54	3·411	0·120	5·41	12·0	96	0·032
73	Clytia	Tuttle, 7 April 1862	2·55	2·78	4·352	0·048	2·38	13·4	35*	?
74	Galatea	Tempel, 29 August 1862	2·12	3·44	4·632	0·238	4·01	12·0	40*	?
75	Eurydice	Peters, 22 September 1862	1·85	3·49	4·367	0·306	4·99	11·1	40*	?
76	Freia	D'Arrest, 21 October 1862	2·73	4·03	6·213	0·193	2·08	12·3	65*	?
77	Frigga	Peters, 12 November 1862	2·32	3·02	4·362	0·131	2·43	12·1	67	0·113
78	Diana	Luther, 15 March 1863	2·07	3·17	4·238	0·209	8·70	10·9	60*	?
79	Eurynome	Watson, 14 September 1863	1·97	2·92	3·822	0·194	4·62	10·8	76	0·137
80	Sappho	Pogson, 3 May 1864	1·84	2·76	3·481	0·200	8·65	12·8	83	0·113
81	Terpsichore	Tempel, 30 September 1864	2·25	3·46	4·820	0·212	7·89	12·1	45*	?
82	Alcmene	Luther, 27 November 1864	2·15	3·37	4·593	0·220	2·84	11·5	65	0·138
83	Beatrix	De Gasparis, 26 April 1865	2·22	2·64	3·790	0·085	5·00	12·0	123	0·030
84	Clio	Luther, 25 August 1865	1·80	2·92	3·631	0·236	9·34	11·2	90	0·037
85	Io	Peters, 19 September 1865	2·15	3·16	4·325	0·192	11·95	11·2	147	0·042
86	Semele	Tietjen, 4 January 1866	2·42	3·78	5·454	0·220	4·81	12·6	45*	?
87	Sylvia	Pogson, 16 May 1866	3·14	3·82	6·500	0·099	10·85	12·6	90*	?
88	Thisbe	Peters, 15 June 1866	2·32	3·22	4·605	0·162	5·22	10·7	210	0·045
89	Julia	Stéphan, 6 August 1866	2·09	3·01	4·080	0·180	16·08	10·3	155	0·086
90	Antiope	Luther, 1 October 1866	2·59	3·68	5·558	0·174	2·24	12·4	55*	?
91	Ægina	Stéphan, 4 November 1866	2·31	2·87	4·168	0·107	2·11	12·2	104	0·031
92	Undina	Peters, 7 July 1867	2·97	3·43	5·724	0·072	9·93	12·0	250*	0·03?
93	Minerva	Watson, 24 August 1867	2·37	3·14	4·571	0·141	8·57	11·5	168	0·039
94	Aurora	Watson, 6 September 1867	2·83	3·48	5·600	0·102	8·03	12·5	188	0·029
95	Arethusa	Luther, 23 November 1867	2·61	3·53	5·377	0·149	13·00	12·1	230	0·019
96	Ægle	Coggia, 17 February 1868	2·65	3·46	5·340	0·134	16·06	12·4	60*	?
97	Clotho	Tempel, 17 February 1868	1·98	3·36	4·360	0·257	11·79	10·5	95	0·121
98	Ianthe	Peters, 18 April 1868	2·19	3·20	4·417	0·187	15·60	12·6	35*	?
99	Dike	Borrelly, 28 May 1868	2·14	3·19	4·348	0·197	13·88	13·5	25*	?
100	Hecate	Watson, 11 July 1868	2·57	3·61	5·424	0·168	6·42	12·1	60*	?

SOME FURTHER MINOR PLANET DATA

The following asteroids are of note in some way or other – either for orbital peculiarities, physical characteristics, or because they are the senior members of 'families' – or even because of their unusual names! Diameters are given only in cases where reasonably reliable measurements have been made.

No.	Name	Discoverer	q	a	P	e	i	M	D
103	Hera	Watson, 7 September 1868	2·48	2·92	4·439	0·081	5·42	11·5	96
107	Camilla	Pogson, 17 November 1868	3·25	3·73	6·519	0·070	9·92	12·7	211
110	Lydia	Borrelly, 19 April 1870	2·51	2·95	4·517	0·080	6·00	11·6	
132	Æthra	Watson, 13 June 1873	1·61	3·61	4·222	0·383	25·16	11·9	
137	Meliboea	Palisa, 21 April 1874	2·47	3·78	5·524	0·210	13·44	12·1	150
153	Hilda	Palisa, 2 November 1875	3·37	4·59	7·926	0·153	7·85	13·3	
155	Scylla	Palisa, 8 November 1875	2·00	3·51	4·582	0·273	11·45	13·7	
174	Phædra	Watson, 2 September 1877	2·44	3·28	4·830	0·146	12·15	12·6	
175	Andromache	Watson, 1 October 1877	2·55	3·89	5·770	0·208	3·22	13·1	
194	Procne	Peters, 21 March 1879	1·99	3·24	4·232	0·239	18·47	10·6	191
215	Œnone	Knorre, 7 April 1880	2·67	2·86	4·600	0·033	1·70	14·1	
216	Cleopatra	Palisa, 10 April 1880	2·09	3·50	4·668	0·252	13·09	9·9	
232	Russia	Palisa, 31 January 1883	2·11	2·99	4·078	0·173	6·08	13·6	
258	Tyche	Luther, 4 May 1886	2·08	3·15	4·227	0·206	14·28	11·2	
279	Thule	Palisa, 25 October 1888	4·15	4·42	8·864	0·032	2·34	15·4	
298	Baptistina	Charlois, 9 September 1890	2·05	2·48	3·406	0·096	6·29	14·3	
300	Geraldina	Charlois, 3 October 1890	3·15	3·27	5·750	0·020	0·76	14·8	
311	Claudia	Charlois, 11 June 1891	2·89	2·91	4·934	0·003	3·24	14·9	
323	Brucia	Max Wolf, 20 December 1891	1·67	3·10	3·676	0·301	24·20	12·9	
324	Bamberga	Palisa, 25 February 1892	1·78	3·59	4·403	0·336	11·16	9·2	246
344	Desiderata	Charlois, 15 November 1892	1·79	3·40	4·184	0·310	18·49	10·7	
349	Dembowska	Charlois, 9 December 1892	2·66	3·19	5·001	0·090	8·27	10·6	144
360	Carlova	Charlois, 11 March 1893	2·46	3·54	5·197	0·179	11·68	12·7	
386	Siegena	Max Wolf, 1 March 1894	2·40	3·39	4·927	0·170	20·27	11·3	191
399	Persephone	Max Wolf, 23 February 1895	2·82	3·28	5·327	0·074	13·14	14·2	
403	Cyane	Charlois, 18 May 1895	2·55	3·09	4·724	0·096	9·13	13·4	
408	Fama	Max Wolf, 13 October 1895	2·70	3·64	5·631	0·150	9·05	14·2	
434	Hungaria	Max Wolf, 11 September 1898	1·80	2·09	2·711	0·073	22·50	13·5	11
451	Patientia	Charlois, 4 December 1899	2·82	3·30	5·354	0·077	15·23	11·9	276
466	Tisiphone	Max Wolf, 17 January 1901	3·16	3·59	6·201	0·063	19·05	13·5	
471	Papagena	Max Wolf, 7 June 1901	2·21	3·57	4·906	0·024	14·95	9·7	143
475	Ocllo	Stewart, 13 August 1901	1·61	3·60	4·180	0·380	18·80	13·3	
477	Italia	Carnera, 23 August 1901	1·96	2·87	3·754	0·188	5·30	12·7	
499	Venusia	Max Wolf, 24 December 1902	3·08	4·84	7·891	0·222	2·08	14·3	

No.	Name	Discoverer	q	a	P	e	i	M	D
505	Cava	Frost, 21 August 1902	2·03	3·34	4·403	0·244	9·80	11·8	
511	Davida	Dugan, 30 May 1903	2·66	3·72	5·700	0·166	15·72	10·5	323
694	Ekard	Metcalf, 7 November 1909	1·80	3·54	4·363	0·324	15·79	11·5	101
699	Hela	Hellfrich, 5 June 1901	1·55	3·68	4·225	0·408	15·30	13·7	
704	Interamnia	Cerulli, 2 October 1910	2·58	3·53	5·345	0·155	17·30	11·0	350
724	Hapag	Palisa, 21 October 1911	1·82	3·07	3·836	0·260	11·77	15·9	
747	Winchester	Metcalf, 7 March 1913	1·96	4·02	5·181	0·344	18·14	10·7	205
944	Hidalgo	Baade, 31 October 1920	2·00	9·61	14·041	0·657	42·49	14·9	15
985	Rosina	Reinmuth, 4 October 1922	1·66	2·94	3·489	0·277	4·07	14·6	
1009	Sirene	Reinmuth, 31 October 1923	1·44	3·82	4·260	0·454	15·75	15·9	
1011	Laodamia	Reinmuth, 5 January 1924	1·56	3·23	3·703	0·350	5·47	13·4	7
1019	Strackea	Reinmuth, 3 March 1924	1·77	2·05	2·624	0·072	27·00	15·6	
1038	Tuckia	Max Wolf, 24 November 1924	3·00	4·87	7·808	0·238	9·25	15·6	
1108	Demeter	Reinmuth, 31 May 1929	1·81	3·05	3·784	0·257	24·90	14·2	
1178	Irmela	Max Wolf, 13 March 1931	2·18	3·18	4·385	0·185	6·92	15·1	20
1180	Rita	Reinmuth, 9 April 1931	3·29	4·67	7·943	0·173	7·21	14·7	
1345	Potomac	Metcalf, 4 February 1908	3·27	4·70	7·948	0·179	11·35	15·2	
1355	Magœba	Jackson, 30 April 1935	1·76	1·94	2·523	0·049	22·68	14·8	
1373	Cincinnati	Hubble, 30 August 1935	2·31	4·51	6·300	0·321	38·90	17·6	
1383	Limburgia	Van Gent, 9 September 1934	2·49	3·66	5·394	0·191	0·01	15·8	
1453	Fennia	Väisälä, 8 March 1938	1·84	1·95	2·613	0·028	23·68	15·6	
1486	Marilyn	Delporte, 23 August 1938	1·92	2·47	3·261	0·124	0·08	15·9	
1578	Kirkwood	Cameron, 10 January 1951	3·08	4·84	7·876	0·223	0·82	15·8	
1625	The NORC	Arend, 1 September 1953	2·39	3·95	5·634	0·247	15·48	14·1	
1796	Riga	Chernykh, 16 May 1966	3·14	3·56	6·130	0·062	22·62	15·8	

THE TROJANS

These asteroids move in virtually the same orbit as Jupiter, but are about 60 degrees from Jupiter, so that there is no fear of collision. They are named after the characters of the Trojan War. Of the first fifteen Trojans, ten lie east of Jupiter; the western group includes Patroclus, Priamus, Æneas, Anchises and Troilus – it is unfortunate that the Greeks and the Trojans are mixed up! Many Trojans are now known, but, of course, their great distance makes them extremely faint.

THE BRIGHTEST TROJANS

No.	Name	Discoverer	q	a	P	e	i	D	M	A
588	Achilles	Max Wolf, 22 February 1906	4·44	5·98	11·896	0·148	10·32		15·3	?
617	Patroclus	Kopff, 17 October 1906	4·48	5·94	11·881	0·140	22·01	147	15·2	0·037
624	Hector	Kopff, 10 February 1907	4·99	5·25	11·589	0·024	18·28	190	15·2	0·038
659	Nestor	Max Wolf, 23 March 1908	4·68	5·84	12·058	0·110	4·51		15·8	
884	Priamus	Max Wolf, 22 September 1917	4·56	5·81	11·809	0·120	8·90		16·1	
911	Agamemnon	Reinmuth, 19 March 1919	4·81	5·50	11·700	0·067	21·90	158	15·1	
1143	Odysseus	Reinmuth, 28 January 1930	4·73	5·70	11·898	0·093	3·15		15·6	
1172	Æneas	Reinmuth, 17 October 1930	4·64	5·70	11·740	0·102	16·71	130	15·7	0·044
1173	Anchises	Reinmuth, 17 October 1930	4·44	5·89	11·743	0·140	6·95	92	16·2	0·034
1208	Troilus	Reinmuth, 31 December 1931	4·64	5·65	11·741	0·093	33·70		16·0	
1404	Ajax	Reinmuth, 17 August 1936	4·62	5·80	11·888	0·112	18·09		16·6	
1437	Diomedes	Reinmuth, 3 August 1937	4·85	5·32	11·460	0·046	20·61	191	15·7	
1583	Antilochus	Arend, 20 September 1950	4·99	5·56	12·120	0·054	28·30		16·3	
1647	Menelaus	Nicholson, 23 June 1957	5·08	5·37	11·935	0·028	5·64		18·0	
1749	Telamon	Reinmuth, 23 September 1949	4·68	5·85	12·088	0·111	6·07		17·5	

CLOSE-APPROACH MINOR PLANETS

		No.	Name	Discoverer	Distance from Sun, a.u.		Period, years	Eccentricity	Inclination	Notes
					Perihelion	Aphelion				
1898	DQ	433	Eros	Witt (Berlin)	1·133	1·783	1·76	0·223	10·83	Close approaches 1931, 1975
1911	MT	719	Albert	Palisa (Vienna)	1·190	2·258	4·42	0·540	10·82	Lost
1918	DB	887	Alinda	Wolf (Heidelberg)	1·148	3·883	3·99	0·544	9·07	Observed in 1969–73
1924	TB	1036	Ganymed	Baade (Bergedorf)	1·220	2·646	4·33	0·542	26·30	Secure
1929	SH	1627	Ivar	Hertzsprung (Johannesburg)	1·125	2·604	2·55	0·397	8·43	Keeps pace with Earth!
1932	EA	1221	Amor	Delporte (Uccle)	1·083	2·758	2·66	0·436	11·95	Seen every 8 years
1932	HA	1862	Apollo	Reinmuth (Heidelberg)	0·647	2·293	1·78	0·566	6·36	Seen in 1973. Close approaches 1980, 1982
1936	CA	2101	Adonis	Delporte (Uccle)	0·441	3·229	2·56	0·764	1·37	Recovered in 1977. No close approach before 2000
1937	UB	–	Hermes	Reinmuth (Heidelberg)	0·617	2·662	2·10	0·624	6·21	Lost
1948	EA	1863	Antinoüs	Wirtanen (Lick)	0·891	3·629	3·40	0·605	18·45	Recovered in 1972
1948	IA	1685	Toro	Wirtanen (Lick)	0·771	1·964	1·60	0·436	9·37	Secure
1949	MA	1566	Icarus	Baade (Palomar)	0·187	1·969	1·12	0·827	22·95	Last close approach (0·15 a.u.) in 1987
1950	DA	1980	Tezcatlipoca	Wirtanen (Lick)	0·839	2·528	2·18	0·501	12·15	Lost
1950	KA	1580	Betulia	Johnson (Johannesburg)	1·119	3·271	3·25	0·490	52·04	Seen in 1976. No close approach before 2000

		No.	Name	Discoverer	Distance from Sun, a.u. Peri-helion	Ap-helion	Period, years	Eccen-tricity	Inclina-tion	Notes
1950	LA	1980	–	Wilson and Wallenquist (Palomar)	1·049	2·334	2·24	0·365	26·84	Recovered in 1975
1951	RA	1620	Geographos	Wilson and Wallenquist (Palomar)	0·827	1·661	1·39	0·335	13·33	Recovered 1969
1953	EA	1915	Quetzalcoatl	Wilson (Palomar)	1·052	3·989	4·50	0·582	20·55	Secure
1953	RA	1916	Boreas	Arend (Uccle)	1·250					Recovered in 1974
1960	UA	2061	Anza	Giclas (Flagstaff)	1·049	3·481	3·41	0·537	3·70	Recovered in 1977
1963	UA	2059	Baboquivari	Goethe (Link)	1·240					Recovered in 1976
1968	AA	1917	Cuyo	Cesco and Samuel (Barreal)	1·063	3·234	3·15	0·505	24·01	Recovered 1970. Next close approach (0·12 a.u.) 1989
1971	FA	1864	Dædalus	Gehrels (Palomar)	0·563	2·359	1·77	0·615	22·14	Secure
1971	UA	1865	Cerberus	Kohoutek (Hamburg)	0·576	1·584	1·12	0·467	16·09	Secure. Orbit wholly within that of Mars
1972	RA	–	–	Klemola (Lick)	1·124	3·467	3·48	0·510	8·80	Probably recoverable
1972	RB	–	–	Gehrels (Palomar)	1·101	3·234	3·19	0·492	5·24	Not recoverable until 1988
1972	XA	1866	Sisyphus	Wild (Berne)	0·871	2·916	2·61	0·540	41·12	Secure
1973	EA	1981	Midas	Kowal (Palomar)	0·623	2·930	2·37	0·650	39·85	Recovered in 1975. Close approaches, 1987–1992
1973	EC	1943	Anteros	Gibson (El Leoncite)	1·064	1·798	1·71	0·256	8·70	Recovered in 1975
1973	NA	–	–	Helin (Palomar)	0·880	4·014	3·83	0·641	68·06	No close approach before 1992
1974	MA	–	–	Kowal (Palomar)	0·422	3·083	2·53	0·760	37·65	Probably recoverable
1975	YA	2102	Tantalus	Kowal (Palomar)	0·905	1·675	1·47	0·298	64·01	Probably recoverable
1976	AA	2062	Aten	Helin (Palomar)	0·790	1·143	0·95	0·183	18·93	Discovered 7 January 1976. First asteroid known to have a period less than 1 year.
1976	UA	2340	Hathor	Kowal, Sebok and Helin (Palomar)	0·464	1·221	0·76	0·450	5·85	Very faint. Shortest known period. Close approach in 1983.

	Discoverer	Distance from Sun, a.u.: Peri-helion	Aphe-lion	Period, years	Eccen-tricity	Incli-nation	Notes
1976 WA	Schuster (La Silla)	0·827	3·878	3·73	0·656	24·26	No close approach before 1991.
1977 HA	Helin (Palomar)	0·795	2·398	2·02	0·502	22·98	Probably recoverable.
1977 HB	Kowal (Palomar)	0·701	1·453	1·12	0·349	9·41	Close approach in 1985. Makes 3 approaches in successive years, then unobservable for 6 years.
1977 RA	Wild (Berne)	1·24					
1977 VA	Helin (Palomar)	1·13					
1978 DA	Schuster (La Silla)	1·025	2·471	3·89	0·585	15·62	Orbit similar to Alinda and Quetzalcoatl.

Eros on 21 January 1975 (P. B. Doherty)

| | Discoverer | Distance from Sun, a.u.: | | Period, years | Eccen-tricity | Incli-nation | Notes |
		Peri-helion	Aphe-lion				
1978 CA	Giclas (Lowell)	0·883	1·124	1·19	0·214	26·07	
1978 PA	Schuster (La Silla)						
1978 RA (2100)	Ra-Shalom Helin (Palomar)	0·469	0·832	0·8	0·436		Identical with 1975 T.B.
1982 DV	Helin (Palomar)						

Other possible minor planets of this type are 1959 LM (Hoffmeister, Boyden, Bloemfontein) and 1947 XC (Giclas, Flagstaff), which were insufficiently observed and are probably lost.

Close approach asteroids are of three types:
1. Amor type, whose orbits do not actually cross that of the Earth,
2. Apollo type, whose orbits cross that of the Earth,
3. Aten type, whose orbits lie normally inside that of the Earth.

Only three Atens are known so far – Aten itself, Ra-Shalom and Hathor – but new Amor and Apollo type asteroids are being found yearly.

All the 'close approach' asteroids are small. Ganymed, with a diameter of about 35 km, is probably the largest of them. Eros is irregular in shape (it has been likened to a cosmical sausage) with a mean diameter of 23 km; Betulia is perhaps 8 km in diameter, Toro 3, Alinda 4 and Icarus possibly no more than 1 km. It is therefore difficult to keep track of them. Apollo (diameter about 3 km) was 'lost' for many years before its recovery in 1973; it passed within 0·01 astronomical units of Venus in 1950 and within 0·07 astronomical units of Venus in 1968; it also by-passed Mars at 0·05 units in 1946. It approached the Earth to within 0·06 units in November 1980 and again in May 1982.

Toro was within 0·14 units of the Earth on 8 August 1972. It makes periodical fairly close approaches, and there were popular reports that it ranked as a minor Earth satellite. This is, of course, incorrect; it is an ordinary asteroid in orbit round the Sun, though admittedly its mean distance from the Sun is very nearly the same as that of the Earth. (It is not, however, moving in the Earth's orbit, since its path is considerably more eccentric than ours.)

RECORD CLOSE APPROACHES

Name	Date	Minimum distance, a.u.
Hermes	1937 October 30	0·006
Hathor	1976 October 20	0·008
Adonis	1936 February 7	0·015
1977 HA	1977 April 1	0·03
Tantalus	1975 December 26	0·04
Icarus	1967 June 14	0·04
Quetzalcoatl	1953 March 4	0·05
1950 DA	1950 March 12	0·06
Geographos	1969 August 27	0·06
Anza	1960 October 7	0·06
Apollo	1932 May 25	0·07
1973 NA	1973 July 2	0·08

Asteroids are divided into various types according to their surface characteristics, as derived from their spectral affinities. The main classes are:

C (Carbonaceous). These are the most numerous, increasing in number from 10 per cent at a distance of 2·2 astronomical units from the Sun out to 80 per cent at 3 astronomical units. They are of low albedo – often below 5 per cent (darker than coal) – and have flat, featureless spectra similar to those of carbonaceous chondrites.

S (Silicaceous). These are most numerous in the inner part of the main zone, making up 60 per cent of the total at 2·2 astronomical units, but only 15 per cent at 3 astronomical units. Their albedoes range from 15 to 25 per cent, and they seem to resemble the metal-bearing meteorites known as chondrites: generally they are reddish.

M (Metallic). Moderate albedoes, and may be the metal-rich cores of large 'parent asteroids' which have been broken up by collision. Asteroid 16 (Psyche) is almost pure nickel-iron alloy in composition.

E (Enstatite). Relatively rare; high albedo (sometimes over 40 per cent). They may resemble some types of chondrites, in which enstatite ($MgSiO_3$) is a major constituent.

There are also dark reddish asteroids (RD) and some which must be listed as unclassified (U). Overall, 75 per cent of asteroids seem to be of type C, and 15 of type S. Perhaps the most unusual spectrum is that of 2201 Oljato, which has strong ultra-violet reflectance.

In 1987 D. P. Cruikshank and R. H. Brown, using the NASA Infra-Red Telescope at the high-altitude Mauna Kea Observatory in Hawaii, reported organic compounds in the spectrum of asteroid 130, Electra. The compounds seemed to resemble those found in meteorites of the carbonaceous chondrite type. The results showed a 3·4 micrometre absorption band in Electra's spectrum. Other asteroids similar to Electra show bound water in their spectra at 3 micrometres, which is consistent with the bound water in meteorites. These observations indicate widespread distribution of organic matter throughout the Solar System.

The largest asteroid is Ceres, which has one-third the total mass of all the asteroids combined. In 1977 L. Lebofsky detected an absorption band at 3 microns, due to H_2O embedded in the mineral structure of the asteroid. It is a spherical body, grey in colour, and of type C. Second in size is Vesta, whose surface seems to be coated with formerly molten igneous rock: the spectrum shows indications of pyroxene. Colour changes show that the surface is not uniform. Pallas, formerly believed to be larger than Vesta, is triaxial, measuring $559 \times 525 \times 532$ km; its spectrum is not unlike that of Ceres.

Asteroid diameters are not easy to measure. One method depends upon infra-red studies. The apparent brightness of a body in sunlight will depend partly on its size and partly on its albedo; everyone knows that in sunlight a white car will feel cooler than a black one! The amount of infra-red emission gives a clue to the albedo, and hence to the diameter. Another method has been pioneered by G. Taylor; this involves the occultations of stars by asteroids. The first good results were obtained with 6 Hebe, whose diameter was found to be 195 km with an uncertainty of only 6 km either way. These results are of great value, since if the masses are known they lead to estimates of the densities of the asteroids – and this information is needed for studies of the evolution of the Solar System. The asteroids known to have diameters of 250 km or over are:

1	Ceres	940
4	Vesta	555
2	Pallas	559×525×532
10	Hygeia	450
31	Euphrosyne	370
704	Interamnia	350
511	Davida	323
65	Cybele	309
3	Juno	288×230
52	Europa	289
451	Patientia	276
15	Eunomia	272
16	Psyche	250
48	Doris	250
92	Undina	250

Masses are derived from mutual perturbations. The values for the largest asteroids are: Ceres, 1.17×10^{24} gr; Vesta, 2.75×10^{23} gr; Pallas, 2.26×10^{23} gr.

The smallest known asteroid is probably Hathor, with an estimated diameter of half a kilometre; but this is very uncertain, and no doubt many still smaller asteroids exist. Masses of some small asteroids have been given as 5.0×10^{12} kg (Icarus), 5.0×10^{15} kg (Eros) and 5.0×10^{10} kg (Adonis), but with great uncertainty.

The brightest asteroid is Vesta, which is the only member of the swarm ever visible with the naked eye. There are only 10 asteroids which can exceed the ninth magnitude, and one of these (Eros) does so only during its rare close approaches, as in 1931 and 1975. The list is as follows (mean opposition magnitudes):

4	Vesta	6·4
1	Ceres	7·3
2	Pallas	7·5
7	Iris	7·8
3	Juno	8·1
6	Hebe	8·3
433	Eros	8·3 (maximum at close approach)
15	Eunomia	8·5
8	Flora	8·7
18	Melpomene	8·9

The most reflective asteroid so far measured is 44 Nysa, with an albedo of 0·377. Others are 64 Angelina (0·342) and 434 Hungaria (0·300).

The darkest asteroid so far measured is 95 Arethusa, with an albedo of 0·019 (blacker than a blackboard); it is darker than the previous record-holder, 342 Bamberga (0·032).

The shortest known rotation periods (based upon variations in magnitude) are for 1566 Icarus (2 h 16 m) and 321 Florentina (2 h 52 m) (magnitude range 10·6 to 11·4). **The**

slowest known spinner is 280 Glauke (1500 h). Some other derived rotation periods, in hours, are:

Ceres	9·08
Pallas	10·00
Juno	7·22
Vesta	10·68
Hebe	7·28
Iris	7·12
Hygeia	10·00
Eunomia	6·08
Psyche	4·30
Eros	5·27
Davida	5·17
Geographos	5·23

The greatest magnitude range due to rotation is for 1864 Daedalus, with a range of 0·9 magnitude and a period of 8 h 34 m.

Asteroids to be detected by radar include the 'earth-grazers' Eros, Icarus, Toro and Quetzalcoatl. Excellent results were obtained from Eros during its 1975 approach; the surface was found to be rough on a scale of centimetres. The longest and shortest diameters are 36 and 15 km. In 1981 the Arecibo radio telescope was used to contact asteroids by radar: Psyche, Klotho, Apollo and Quetzalcoatl. At that time Apollo was 12 800 000 km from Earth and Psyche 40 000 000 km. Psyche is **the largest known iron asteroid**, with an estimated mass of 50 000 million million tonnes.

The first systematic search for a planet between the orbits of Mars and Jupiter was initiated in 1800, by six astronomers meeting at Lilienthal – at Johann Schröter's observatory. They based their search on Bode's Law (actually first described by Titius, but popularized by Bode). This so-called Law may be summed up as follows:

Take the numbers 0, 3, 6, 12, 24, 48, 96, 192 and 384, each of which (apart from the first) is double its predecessor. Add 4 to each. Taking the Earth's distance from the Sun as 10, the remaining figures give the mean distances of the planets with considerable accuracy out to as far as Saturn, the outermost planet known when Bode popularized the relationship in 1772. Uranus, discovered in 1781, fits in well. Neptune and Pluto do not, but of course these planets were not discovered until considerably later.

Planet	Distance by Bode's Law	Actual distance
Mercury	4	3·9
Venus	7	7·2
Earth	10	10
Mars	16	15·2
–	28	–
Jupiter	52	52·0
Saturn	100	95·4
Uranus	196	191·8
Neptune	–	300·7
Pluto	388	394·6

Whether the Law is anything more than a coincidence is not known, but it was enough for the 'Celestial Police' to hunt for the missing planet corresponding to Bode's number 28. Schröter became president of the association; the secretary was the Hungarian astronomer Baron Franz Xavier von Zach. They very logically concentrated their searches in the region of the ecliptic.

The first asteroid to be discovered was Ceres, by G. Piazzi at Palermo on 1 January 1801. (He was working on a star catalogue, and was not a member of the 'Police', though he joined later.) Ceres was found to have a distance of 27·7 on Bode's scale. Four asteroids were known by the time that the 'Police' disbanded in 1815; the fifth, Astræa, was not found until 1845 as a result of painstaking solo searches by a German amateur, Hencke.

The first asteroid to be discovered photographically was 323 Brucia, on 20 December 1891 by Max Wolf.

The astronomer to have discovered the greatest number of asteroids to which permanent numbers have been given is K. Reinmuth, with 246 discoveries; Max Wolf has 232, and J. Palisa 121. By 1900, 452 asteroids were known.

Only three numbered asteroids have been lost (that is to say, they cannot now be traced): 719 Albert, 724 Hapag and 878 Mildred. Asteroid 330 Adalberta never existed at all; it was claimed by Max Wolf in 1892, but was photographed only twice, and it was then found that the images were of two separate stars.

The first asteroid known to come within the orbit of Mars was 433 Eros, discovered by Witt at Berlin in 1898. Its variability was detected by von Oppolzer in 1900; and in 1931 Finsen and van den Bos, at Johannesburg, saw its elongated shape. Eros can come to within 23 000 000 km of the Earth, and at the approach of 1931 was intensively studied in an attempt to improve our knowledge of the length of the astronomical unit – though the final value (derived by H. Spencer Jones) is now known to have been rather too great. The last close approach occurred in 1975.

The closest approach of an asteroid was that of Hermes, on 28 October 1937, when it approached the Earth to 800 000 km. It was then of the eighth magnitude, moving 5 degrees per hour – so that it crossed the sky in 9 days. (It is not included in the list of the brightest asteroids, since it is so exceptional in every way.)

The first asteroid found to approach the Sun within the orbit of Mercury is 1566 Icarus, discovered by W. Baade in 1949. It has a period of 409 days, and can approach the Sun to a distance of 28 000 000 km, so that its surface must then be at a temperature of approximately 500 °C. The only other asteroid to approach the Sun to within the orbit of Mercury is 3200 Phæthon, discovered in 1983 by J. Davis and S. Green from data supplied by the IRAS satellite. Phæthon is about 5 km in diameter, and its distance from the Sun ranges between 21 000 000 and 390 000 000 km. It seems to be darkish, with a rotation period of about 4 hours. Phæthon's orbit is not unlike that of the Geminid meteor stream, but the surface composition does not suggest that it may be a dead comet.

The first known asteroid to have a revolution period of less than a year is 2062 Aten, discovered by Mrs Helin in January 1976. The period is 346·8 days. It, 2100 Ra-Shalom and Hathor are the only known asteroids with their aphelion distances less than 1 astronomical unit.

The asteroid with the smallest orbit is Ra-Shalom, discovered by Mrs Helin on 25 October 1976. The period is 283 days. The distance from the Sun ranges between 69 000 000 and 126 000 000 km, so that the orbit lies well within that of the Earth.

The greatest perihelion distance for a numbered asteroid (excluding Chiron) is that of 588 Achilles; 5·98

Eros, close to Procyon, on 9 February 1975 (P. B. Doherty)

astronomical units.

The greatest aphelion distance for a named asteroid (again excluding Chiron) is that of 944 Hidalgo (9·61 astronomical units, slightly greater than the mean distance between the Sun and Saturn). Hidalgo also has the **largest orbit**, and the **longest revolution period** (over 14 years). The magnitude ranges between 10 and 19. This range is greater than for any other asteroid. The greatest aphelion distance for a 'normal' asteroid is that of 1038 Tuckia (4·87 a.u.).

Another exceptional asteroid is 1973 SC2, which was identified in 1978 on plates which had been taken by T. Gehrels with the Palomar 48-in Schmidt telescope in 1973. (These were the same plates which resulted in the identification of 34 new Trojans.) 1973 SC2 has an orbit intermediate between those of 944 Hidalgo and 2060 Chiron. The revolution period is 20 years, the orbital eccentricity is 0·55, and the aphelion distance is 11·4 a.u., well outside the orbit of Saturn. Unfortunately, 1973 SC2 has been lost, as it was photographed on only four nights (29 Sept.–5 Oct. 1973) and its orbit is only approximately known. Its rediscovery will be largely a matter of luck! 1981 VA is another tiny asteroid whose orbit takes it out almost as far as Saturn.

The greatest orbital eccentricity is for 1566 Icarus (0·83). Then follow Adonis (0·76) and Hidalgo (0·66).

The least orbital eccentricity is for 311 Claudia (0·0031). Then follow

903 Nealley (0·0039) and 1300 Marcelle (0·0050).

The greatest orbital inclination is for 1973 NA (66·8 degrees). The greatest inclination for a named asteroid is for 1580 Betulia (52·0 degrees).

The least orbital inclination is for 1383 Limburgia (0·014 degrees). Then follows 1486 Marilyn (0·084 degrees).

Gaps in the asteroid zone were predicted by D. Kirkwood in 1857, and confirmed in 1866; they are due to the perturbing influence of Jupiter. **The existence of asteroid 'families'** was demonstrated in 1917 by K. Hirayama. For example, the Hilda asteroids have periods of approximately 2/3 that of Jupiter. Families include:

Family	Distance from Sun, astronomical units
Hungaria	1·9
Flora	2·2 (complex region)
Budrosa	2·9 (6 members; 338 Budrosa is metallic)
Coronis	2·9 (S-type asteroids; largest 208 Lachrimosa, 45 km)
Eos	3·0 (types intermediate between C and S)
Hilda	4·0
Thule	4·3 (isolated, but brightest of a family?)

The Phocæa group, at 2·4 a.u., is not a true family.

The first asteroid known to have a satellite is 532 Herculina. On 7 June 1978 Herculina occulted a star (SAO 120 774) and a second occultation showed the existence of a 50-km satellite. Herculina itself is 217 km in diameter (±3 km). The satellite is assumed to be 3 mag fainter than Herculina, so that its magnitude is about 13;

the angular separation is 0″·9. The observers were Baron, A'Hearn, McMahon and Horne (USA). On 11 December 1978 a satellite of 18 Melpomene was reported by R. M. Williamson, using a 91 cm reflector; confirmation was obtained from the US Naval Observatory and from Silver Spring, Maryland. Another possibility concerns the large Trojan asteroid Hektor, though here the effects may be due to its irregular shape rather than an associated satellite. On 24 January 1975 Eros occulted Kappa Geminorum. The fact that the Pioneers and Voyagers passed unharmed through the asteroid zone en route for Jupiter may indicate that there are fewer very small asteroids than had been anticipated.

The first asteroids were given mythological names (Ceres, for instance, is the patron of Sicily, and it was from Palermo in Sicily that Piazzi made the discovery). Later, the mythological names began to become exhausted; one of the first names of more 'modern' type was that of 125 Liberatrix, which seems to have been given either in honour of Joan of Arc or else of Adolphe Thiers, first President of the French Republic. Asteroids may be named after countries (136 Austria, 1125 China, 1279 Uganda, 1197 Rhodesia), people (1123 Shapleya, in honour of Dr. Harlow Shapley; 1462 Zamenhof, after the inventor of Esperanto; 1486 Marilyn, after the daughter of Paul Herget, director of Cincinnati Observatory). I feel very honoured that Asteroid 2602 has been named after myself!

The first masculine name was that of 433 Eros.

The only asteroid to be named after an electric calculator is 1625 The NORC – the Naval Ordnance Research Calculator at Dahlgren, Virginia. Plants are represented (978 Petunia, 973 Aralia); universities (694 Ekard – 'Drake' spelled backwards; the orbit was computed by students at Drake University), musical plays (1047 Geisha), foods (518 Halawe – the discoverer, R. S. Dugan, was particularly fond of the Arabian sweet *halawe*), social clubs (274 Philagoria – a club in Vienna), and even shipping lines (724 Hapag – Hamburg Amerika Linie).

The only name to be auctioned was that of 250 Bettina, whose discoverer, Palisa, offered to sell his right of naming for £50. The offer was accepted by Baron Albert von Rothschild, who named the asteroid after his wife. (The Rothschilds are well represented; 719 is named Albert, and 977 Philippa after Philip von Rothschild.) 443 Photographica commemorates the new method of asteroid discovery, and 1581 Abanderada – someone who carries a banner – is said to be associated with Eva Perón, wife of the former ruler of Argentina! Finally, 1000 is, appropriately enough, named Piazzia.

CHIRON

In 1977 C. Kowal, using the Schmidt telescope at Palomar, discovered an extraordinary object which has been named Chiron. It has been given an asteroidal designation (1977 UB) and a number (2060) but it is far from being an ordinary asteroid, since its path lies mainly between the orbits of Saturn and Uranus! At perihelion it comes to within 1278 million km of the Sun, and about one-sixth of its orbit lies within that of Saturn; but at aphelion it recedes to 2827 million km greater than the minimum distance between the Sun and Uranus. The orbital inclination of approximately 7 degrees prevents any encounter with Uranus at the present epoch, but it is worth noting that in 1664 BC Chiron approached Saturn to within about 16 000 000 km. This is not so very much greater than the distance between Saturn and its outermost satellite, Phœbe.

The orbital elements of Chiron are as follows:
Orbital period: 50·68 years.
Perihelion distance: 8·5 a.u. (1 278 000 000 km).
Aphelion distance: 18·5 a.u. (2 827 000 000 km).
Orbital inclination: 6·923 degrees.
Orbital eccentricity: 0·3786.

The last two perihelia occurred in 1895 and 1945; the next will be in 1996. At discovery the magnitude was 18, but at perihelion it will rise to about 15. The diameter may be as much as 300 km, though this is very uncertain. G. Taylor has found that no occultations of catalogue stars are due in the near future, but eventually an occultation must occur, and this could be the best way of measuring the diameter of Chiron. It seems to have a fairly low albedo with a dusty or rocky surface.

Kowal's discovery was made on 1 November 1977, from plates exposed on 18 and 19 October 1977. It has since been traced on photographs taken as far back as 1895.

The nature of Chiron remains problematical. Suggestions that it could be a huge comet have been discounted. It could therefore be: (*a*) an exceptional asteroid, (*b*) a planetesimal – a surviving remnant of the 'building-blocks' which condensed to form the outer planets, (*c*) the brightest member of a trans-Saturnian asteroid group, or even (*d*) an escaped satellite of Saturn. For the moment it seems best to treat it as asteroidal in nature, but Kowal summed the situation up neatly when he commented that Chiron was – well, 'just Chiron'!

JUPITER

The largest and most massive planet is Jupiter, which is fifth in order of distance from the Sun.

Jupiter is well placed for observation for several months in every year, and the opposition magnitude has a range of only about 0 m·5. Generally speaking it 'moves' about one constellation per year; thus the 1979 opposition occurred in Cancer and the 1980 opposition in Leo.

On average Jupiter is the brightest of the planets apart from Venus, though for relatively brief periods during perihelic oppositions Mars may outshine it. Note that in the period between 1977 and 2000, the only years in which Jupiter does not come to opposition are 1978 and 1990. This is, of course, because oppositions occurred in the last part of December in 1977 and 1989.

The shortest 'day' in the Solar System is that of Jupiter, insofar as the principal planets are concerned (some of the asteroids rotate much more quickly). It is not possible to give an overall value for the axial rotation period, because Jupiter does not rotate in the way that a solid body would do. The equatorial zone has a shorter rotation than the rest of the planet. Conventionally, System I refers to the region between the north edge of the South Equatorial Belt and the south edge of the North Equatorial Belt; the mean rotation period is 9 h 50 m 30 s, though individual features may have periods which differ perceptibly from this. System II, comprising the rest of the planet, has a period of 9 h 55 m 41 s, though again different features have their own periods; that of the Great Red Spot is approximately 9 h 55 m 37 s·6. In addition there is System III, which relates not to optical features but the bursts of decametre radio radiation; the period seems to be 9 h 55 m 29 s·4.

There are no true 'seasons' on Jupiter, because the axial inclination to the perpendicular of the orbital plane is only just over 3 degrees – less than for any other planet.

The latitudes of the various belts and

DATA

Mean distance from the Sun: 778·34 million km (5·203 a.u.)
Maximum distance from the Sun: 815·7 million km (5·455 a.u.)
Minimum distance from the Sun: 740·9 million km (4·951 a.u.)
Sidereal period: 11·86 years = 4332·59 days
Synodic period: 398·9 days
Rotation period: System I (equatorial) 9 h 50 m 30 s; System II (rest of planet) 9 h 55 m 41 s; System III (radio methods) 9 h 55 m 29 s
Mean orbital velocity: 13·06 km/s
Axial inclination: 3°04'
Orbital inclination: 1°18'15"·8
Orbital eccentricity: 0·048
Diameter: equatorial 142 200 km; polar 134 700 km
Apparent diameter from Earth: max 50"·1, min 30"·4
Reciprocal mass, Sun = 1: 1047·4
Density, water = 1: 1·33
Mass, Earth = 1: 317·89
Volume, Earth = 1: 1318·7
Escape velocity: 60·22 km/s
Surface gravity, Earth = 1: 2·64
Mean surface temperature: −150 °C
Oblateness: 0·06
Albedo: 0·43
Maximum magnitude: −2·6
Mean diameter of Sun, seen from Jupiter: 6'09"

OPPOSITIONS OF JUPITER, 1977–2000

Year and date	Diameter seconds of arc	Magnitude
1977 Dec. 23	47·4	−2·3
1979 Jan. 24	45·8	−2·2
1980 Feb. 24	44·7	−2·1
1981 Mar. 26	44·2	−2·0
1982 Apr. 25	44·4	−2·0
1983 May 27	45·5	−2·1
1984 June 29	46·8	−2·2
1985 Aug. 4	48·5	−2·3
1986 Sept. 10	49·6	−2·4
1987 Oct. 18	49·8	−2·5
1988 Nov. 23	48·7	−2·4
1989 Dec. 27	47·2	−2·3
1991 Jan. 28	45·7	−2·1
1992 Feb. 28	44·6	−2·0
1993 Mar. 30	44·2	−2·0
1994 Apr. 30	44·5	−2·0
1995 June 1	45·6	−2·1
1996 July 4	47·0	−2·2
1997 Aug. 9	48·6	−2·4
1998 Sept. 16	49·7	−2·5
1999 Oct. 23	49·8	−2·5
2000 Nov. 28	48·5	−2·4

zones are also variable, over a limited range. The following values are therefore approximate only:

N.N. Temperate Belt:	lat.	+36°
N. Temperate Belt:		+27°
N. Equatorial Belt:	N. edge	+20°
	S. edge	+ 7°
S. Equatorial Belt:	N. edge	− 6°
	S. edge	−21°
S. Temperate Belt:		−29°
S.S. Temperate Belt:		−44°

The first man to find that System I rotates faster than System II was G. D. Cassini, in 1690.

The most famous marking on Jupiter is the Great Red Spot. It may be identical with a feature recorded by Robert Hooke in 1664, and it was certainly seen by Cassini in 1665, but it became famous in 1878, when it was described as very prominent and brick-red. Subsequently it has shown variations in both intensity and colour, and at times (as in 1976–7) it has been invisible, but it always returns after a few years, so that during telescopic times it has been to all intents and purposes a permanent feature – unlike any other of the spots. At its greatest extent it may be 40 000 km long and 14 000 km wide, so that its surface area is then greater than that of the Earth. On 5 September 1831 H. Schwabe found that it encroached into the region of the southern part of the South Equatorial Belt, making up what is now always termed the Red Spot Hollow.

Though the latitude of the Red Spot varies little, it drifts about in longitude. Over the past century the total longitude drift amounts to approximately 1200 degrees! The latitude is generally very close to −22 degrees.

It was once thought that the Red Spot might be a solid or semi-solid body floating in Jupiter's outer gas, in which case it would be expected to disappear if its level sank for any reason (possibly a decrease in the density of the outer gas). In 1963 R. Hide suggested that it might be the top of a 'Taylor Column', a sort of standing wave above a mountain or depression below the gaseous layer. However, the Pioneer and Voyager results have shown that ideas of this kind are untenable. The Red Spot is a phenomenon of Jovian meteorology. Its longevity may well be due to its exceptional size, but there are signs that it is decreasing in size, and it may not be permanent. It must be added that several smaller red spots have been seen occasionally, but have not lasted for long.

The longest-observed 'disturbance' on Jupiter was the South Tropical Disturbance, discovered by P. Molesworth on 28 February 1901 and last indicated by many observers dur-

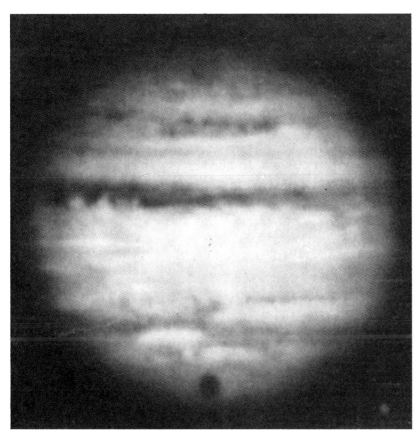

Jupiter, showing the Great Red Spot; 200 inch reflector. (Palomar)

ing the apparition of 1939–40. It took the form of a shaded zone between white spots. The rotation period of the Disturbance was shorter than that of the Red Spot, so that periodically the Disturbance caught the Spot up and passed it; the two were at the same latitude, and the interactions were of great interest. Nine conjunctions were observed, and possibly the beginning of a tenth in 1939–40, though by then the Disturbance had practically vanished. It has not reappeared, and there is no real reason to suggest that it will do so in the future. Its average rotation period was 9 h 55 m 27 s·6.

The most prominent belt on Jupiter is generally the North Equatorial, while the South Equatorial Belt is often double and becomes obscure at times. However, during the apparition of 1988 the South Equatorial Belt was fully equal to the North Equatorial, and this seems also to have been the case at times before 1900, though the records are far from complete. Other belts are also subject to marked variations in intensity.

The first proof that Jupiter is not a miniature Sun was given by H. Jeffreys in his classic papers published in 1923 and 1924. Jeffreys proposed a model in which Jupiter would have a rocky core, a mantle composed of solid water and carbon dioxide, and a very deep, tenuous atmosphere.

The first identification of methane and ammonia in Jupiter's atmosphere was due to R. Wildt in 1932; it followed that much of the planet must be composed of hydrogen. In 1934 Wildt proposed a model giving Jupiter a rocky core 60 000 km in diameter, overlaid by an ice shell 27 000 km thick above which lay the hydrogen-rich atmosphere. (This was certainly more plausible than a strange theory proposed by E. Schoenberg in 1943. Schoenberg believed Jupiter to have a solid surface, with volcanic rifts along parallels of latitude; heated gases rising from these rifts would produce the belts!)

The first 'hydrogen models' were proposed independently in 1951 by W. Ramsey in England and W. DeMarcus in America. On Ramsey's theory, the 120 000 km diameter core was composed of hydrogen, so compressed that it assumed the characteristics of a metal. This core was overlaid by an 8000 km deep layer of ordinary solid hydrogen, above which came the atmosphere.

The first suggestion that Jupiter could be mainly liquid seems to have been made by G. W. Hough as long ago as 1771. Hough, incidentally, believed the Red Spot to be a kind of floating island.

The latest models of Jupiter are based mainly upon work by J. D. Anderson and W. B. Hubbard in the United States. There is a relatively small, rocky core made up of iron and silicates, and at a temperature of about 30 000 °C. Around this is a thick shell of liquid metallic hydrogen. At about 46 000 km from the centre of the planet there is a sudden transition from liquid metallic hydrogen to liquid molecular hydrogen; in the transition region the temperature is assumed to be about 11 000 °C, with a pressure about three million times that of the Earth's air at sea-level. Above the liquid molecular hydrogen comes the gaseous atmosphere, which is about 1000 km deep, and is made up of 82 per cent hydrogen, 17 per cent helium and 1 per cent of other elements. It contains water droplets, ice crystals, ammonia crystals and ammonium hydrosulphide crystals.

Gases warmed by the internal heat of the planet rise into the upper atmosphere, and cool, forming clouds of ammonia crystals floating in gaseous hydrogen. The clouds form the bright zones on Jupiter, which are both colder and higher than the dark belts. The colours of the various features have yet to be satisfactorily explained, but are certainly due to characteristics of Jovian chemistry.

Jupiter radiates 1·7 times more energy than it would do if it depended only upon radiation received from the Sun. On the liquid-planet model, this excess heat must be nothing more than the remnant of the heat generated when Jupiter was first formed. It had been suggested that Jupiter might be slowly contracting, and that this contraction was the cause of the excess radiation emitted; but this is not now generally favoured.

The first detection of radio radiation from Jupiter was due to B. F. Burke and F. L. Franklin, in the United States, in 1955. (It must be admitted that the discovery was accidental!) The radio emissions are concentrated in wavelengths of tens of metres (decametric emissions) and tenths of metres (decimetric). Efforts to correlate the radio burst with visual features, such as the Red Spot, were not successful, but it seems that the decametric radiation is affected by the orbital position of Io, the innermost of the large satellites of Jupiter. From these early results it was surmised – correctly – that Jupiter has an extremely powerful magnetic field.

The first probe to Jupiter was Pioneer 10. It was launched from Cape Canaveral on 2 March 1972, and by-passed Jupiter on 3 December 1973 at a distance of 131 400 km. Studies of the Jovian magnetosphere and atmosphere were carried out, and more than 300 pictures returned. Pioneer 10 was the first probe to leave the Solar System; it will never return. In case it should ever be found by some alien civilization, it carries a plaque to give a clue as to the planet of its origin – though whether any other beings would be able to decipher the message on the plaque has been questioned!

The second Jupiter probe was Pioneer 11, launched on 5 April 1973 and which by-passed Jupiter at 46 400 km on 2 December 1974. It also studied Jupiter's magnetosphere and atmosphere, and sent back excellent pictures; it was then put into a path en route for a rendezvous with Saturn, after which it too will leave the Solar System permanently.

The third Jupiter probe, Voyager 1, was actually launched a few days later than its twin Voyager 2, but travelled in a more economical path. (To confuse matters still further, initial faults detected in the first space-craft caused a switch in numbers, so that the original Voyager 1 became Voyager 2, and vice versa!) The Voyager was much more elaborate than either of the Pioneers, and the results obtained were of far higher quality. In addition to Jupiter itself, Voyager 1 also obtained good images of Io, Ganymede and Callisto.

The fourth Jupiter probe, Voyager 2, followed much the same programme, but was sent close to Europa, the only Galilean satellite not satisfactorily observed by its predecessor.

The Voyagers confirmed the Pioneer data concerning Jupiter's magnetic field, and also confirmed the association between the radiation zones and the active satellite Io. It also seems that at times some of the trapped particles are ejected as bursts of cosmic rays. Jupiter and Io are connected by a flux tube of electrons and ions carrying a current of five million ampères; material from Io produces an ionized plasma ring or torus round Jupiter in Io's orbit.

The Red Spot was found to be in rotation; it spins anti-clockwise, with a period of 12 days at its outer edge and 9 days somewhere inside. Material rises in its centre, and spirals outward toward the edge. The centre is some 8 km above the surrounding clouds, and the Spot is colder than the adjacent regions. Adjacent spots may be 'sucked in', though during the Voyager 2 encounter there seemed to be some sort of barrier east of the Red Spot which had earlier been absent. The colour of the Red Spot may be due to phosphorus, produced by the action of sunlight upon phosphine sent up from the planet's interior.

The Jovian atmosphere was found to consist of 89 per cent hydrogen and 11 per cent helium, with smaller amounts of other substances. Hydrogen compounds include ammonia, water, and ammonium hydrosulphide. The upper clouds are ammonia cirrus; these show up as the bright zones, while the dark belts seem to be due to a lower layer of ammonium hydrosulphide cloud. At a greater depth there may be crystals of water ice. The Anderson-Hubbard picture of the internal structure of Jupiter was confirmed.

Auroræ on Jupiter were first recorded by Voyager 1 during its passage around the night side of the planet. It may be, however, that the main cause of Jovian auroræ is not the solar wind, as with Earth, but material from the volcanoes of Io. Lightning was detected, and is probably very violent; Jovian 'noise' was recorded, due to high-energy particles in motion, so that at some wavelengths Jupiter is a noisy world as well as a turbulent one. The findings were confirmed by Voyager 2.

The ring system of Jupiter was also discovered from Voyager 1. There is in fact only one major ring, extending from between 122 000 and 129 000 km from the planet's centre, so that they

JUPITER PROBES 1972–83

Name	Launch date	Encounter date	Nearest approach, km	Remarks
Pioneer 10	2 Mar 1972	3 Dec 1973	131 400	Complete success; images and data. Now on its way out of the Solar System. Will probably be tracked into the 1990s.
Pioneer 11	5 Apr 1973	2 Dec 1974	46 400	Passed quickly over the Jovian equator to avoid radiation damage. Complemented the results of Pioneer 10. Went on to a rendezvous with Saturn (1 Sep 1979). Now on its way out of the Solar System. Contact probable until into the 1990s.
Voyager 1	5 Sep 1977	5 Mar 1979	350 000	More detailed information about Jupiter and the Galilean satellites Io, Ganymede and Callisto; volcanoes on Io discovered. Went on to rendezvous with Saturn, 12 Nov 1980. Now on its way out of the Solar System. Images obtained of several satellites, including Titan.
Voyager 2	20 Aug 1977	9 Jul 1979	714 000	Complemented Voyager 1, also obtaining close-range data from satellites not well covered by Voyager 1 (Iapetus, Hyperion, Tethys, Enceladus). Encountered Uranus (1986) and will go on to Neptune (1989), after which it too will leave the Solar System.

reach to over 50 000 km above the cloud-tops. There are two main sections, a 5000-km wide segment and a brighter, outer 800-km segment. The thickness of the rings is probably not much more than 1 km and the tiny particles which make up the system orbit Jupiter in from 5 to 7 hours. The smallness of the ring particles is shown by the fact that the rings are best seen when 'backlit' – that is to say, when the space-craft's camera is behind them with respect to the Sun, so that the sunlight is reaching the camera via the ring. It has been suggested that the particles may slowly spiral downward to the planet; if so, they must be continually replenished, and once more the active Io may be involved. Clearly, Jupiter's ring is very different from the glorious ring system of Saturn, and it may not even be a permanent feature. There would have been no hope of discovering it except by using space-probes. Note also that the two small inner satellites, Metis and Adrastea, move in the same region, and certainly have profound effects upon the rings in spite of their small size and mass.

The next Jupiter probe, Galileo, is scheduled for launching in 1989. It will consist of an orbiter and also an entry vehicle, which should send back fascinating information as it plunges to destruction in the Jovian clouds. Despite the cutback in NASA funds, it is very much to be hoped that Galileo will not be cancelled.

The magnetic field of Jupiter was found to be stronger than for any other planet. Except near the planet, the field is dipolar, like that of the Earth, but the direction of the field is opposite to ours – so that if a terrestrial compass were taken to Jupiter, it would point south instead of north. The axis of the dipole field is inclined to the axis of rotation by 10·8 degrees. Inside about three Jupiter radii the field is so complex that it cannot be regarded as truly dipolar. The main dipole field has a strength of about 4 gauss (as against 0·3 to 0·8 gauss at the surface of the Earth). Pioneer 10 crossed the shock wave (i.e. the boundary between the Jovian magnetosphere and the general interplanetary field) on 26 November 1973 – the first probe to do so. Energetic particles were de-

tected as far out as 300 Jovian radii from the planet, and during its passage through the radiation belts it received 200 000 rads from electrons and 50 000 rads from energetic protons; the lethal dose for a man is a mere 500 rads. The Jovian radiation belts – that is to say, the regions in which protons and high-energy electrons are trapped by the magnetic field – are about 10 000 times more intense than the Van Allen belts associated with the Earth. The electric-

al instruments on Pioneer 10 were 95 per cent saturated, and if the probe had approached much closer to Jupiter the mission would have failed. Pioneer 11 was put into a different trajectory, passing quickly over the equatorial zone so as to avoid the worst of the radiation. The Jovian magnetosphere is very extensive, and it has even been found that the magnetic 'tail' extends so far that when suitably placed, Saturn may lie within it.

SATELLITES OF JUPITER

No.	Name	Discoverer	Mean distance from Jupiter, km	Mean angular distance from Jupiter, at mean opposition distance ' "
XVI	Metis	Synott 1980	127 960	
XIV	Adrastea	Jewitt and Danielson 1979	128 980	
V	Amalthea	Barnard 1892	181 300	0 59·4
XV	Thebe	Synott 1980	221 900	
I	Io	Galileo	421 600	2 18·4
II	Europa	and	670 900	3 40·1
III	Ganymede	Marius	1 070 000	5 51·2
IV	Callisto	1610	1 880 000	10 17·6
XIII	Leda	Kowal 1974	11 094 000	60 45
VI	Himalia	Perrine 1904	11 480 000	62 45
X	Lysithea	Nicholson 1938	11 720 000	64 05
VII	Elara	Perrine 1905	11 737 000	64 10
XII	Ananke	Nicholson 1951	21 200 000	116
XI	Carme	Nicholson 1938	22 600 000	123
VIII	Pasiphaë	Melotte 1908	23 500 000	129
IX	Sinope	Nicholson 1914	23 700 000	130

	Sidereal period days	Mean synodic period d h m s	Apparent diameter as seen from Jupiter ' "
Metis	0·295		
Adrastea	0·298		
Amalthea	0·498	0 11 57 27·6	7 24
Thebe	0·675		
Io	1·769	1 18 28 35·9	35 40
Europa	3·551	3 13 17 53·7	17 30
Ganymede	7·155	7 03 59 35·9	18 06
Callisto	16·689	16 18 05 06·9	9 30
Leda	238·7	254	0 0·15
Himalia	250·6	266	0 8·2
Lysithea	259·2	276	0 0·03
Elara	259·7	276	0 1·4
Ananke	631	551	0 0·2
Carme	692	597	0 0·2
Pasiphaë	735	635	0 0·2
Sinope	758	645	0 0·2

	Orbital inclination degrees	Orbital eccentricity	Diameter km	Density water=1	Escape velocity km/s
Metis	0	0	40	3?	0·02?
Adrastea	0	0	24 × 20 × 16	3?	0·01?
Amalthea	0·45	0·003	270 × 170 × 150	3?	0·16?
Thebe	0·9	0·013	110 × 100 × 90	3?	0·8?
Io	0·04	0·004	3632	3·55	2·56
Europa	0·47	0·009	3126	3·04	2·10
Ganymede	0·21	0·002	5276	1·93	2·78
Callisto	0·51	0·007	4820	1·81	2·43
Leda	26·1	0·148	15	3?	0·005?
Himalia	27·6	0·158	185	3?	0·1?
Lysithea	29·0	0·107	35	3?	0·01?
Elara	24·8	0·207	75	3?	0·05?
Ananke	147	0·17	30	3?	0·01?
Carme	164	0·21	40	3?	0·02?
Pasiphaë	145	0·38	50	3?	0·02?
Sinope	153	0·28	35	3?	0·01?

	Reciprocal mass Jupiter = 1	Albedo	Magnitude at mean opposition distance
Metis		?	17·4
Adrastea		?	18·9
Amalthea		0·05	14·1
Thebe		?	15·5
Io	21 300	0·63	5·0
Europa	39 000	0·64	5·3
Ganymede	12 700	0·43	4·6
Callisto	17 800	0·17	5·6
Leda		?	20·2
Himalia		?	14·8
Lysithea		?	18·4
Elara		?	16·7
Ananke		?	18·9
Carme		?	18·0
Pasiphaë		?	17·7
Sinope		?	18·3

The four large satellites are known as the Galileans, though they were observed at about the same time independently by Simon Marius. There is, however, considerable evidence that a naked-eye observation of one of the satellites (probably Ganymede, or even two satellites close together) was made in China by Gan De as long ago as 364 BC. The four outermost satellites (Ananke, Carme, Pasiphaë and Sinope) have retrograde orbits. All the satellites beyond Callisto are so far from Jupiter, and so subject to solar perturbation, that their orbits are not even approximately circular, and orbital elements are subject to change. Indeed, Pasiphaë was 'lost' after its discovery in 1908, found again in 1922, lost once more until 1938, and again between 1941–55.

The last satellite to be discovered by visual means was Amalthea, by E. E. Barnard in 1892; all subsequent discoveries have been photographic or by space-probe.

The first attempted maps of the surfaces of the Galilean satellites were due to A. Dollfus and his colleagues at the Pic du Midi Observatory, in 1961. Some features were recorded, but it is hardly to be expected that the results could have been reliable.

Since the apparent diameter of the Sun as seen from Jupiter is less than 6 minutes of arc, Amalthea and all the Galileans can produce total eclipses. To a Jovian observer, Io would appear much the largest of the satellites.

The satellites beyond Callisto may well be in the nature of captured asteroids, and do not in any way resemble the massive Galileans. Their diameters are very uncertain; only Himalia can be

as much as 100 km in diameter. It is believed that they are darkish bodies, but at present it must be admitted that very little is known about them.

Metis and Adrastea, the two satellites closest to Jupiter, are less than 60 000 km above the cloud-tops, and are associated with the form of Jupiter's ring. They seem to have rather low albedoes, in which case they will be rocky rather than icy, but we have no further information.

Amalthea, on the other hand, has been known since 1892; it was found by E. E. Barnard during a deliberate search with the great Lick refractor. Barnard commented that it was 'much more difficult to see than the satellites of Mars'. Voyager 1 approached it to within 420 100 km and Voyager 2 to 558 270 km, and both took pictures. The gravitational pull of Jupiter has drawn Amalthea out into an irregular shape, and, as expected, the longest axis (270 km) points toward Jupiter. In view of its irregular shape, Amalthea may be fairly dense, and probably the surface is covered with a thick regolith.

The surface is very red, due probably to sulphur contamination from Io. The albedo is generally low, but the bright, rather greenish patches have a reflectivity of up to 20 per cent, and may be due to recently-exposed material which contains a lesser quantity of sulphur-rich glass. The two main craters are bowl-shaped. Pan is 90 km in diameter and well over 8 km deep. Gaea, 75 km in diameter, has a depth of between 10 and 20 km. If the latter figure is correct, then the slope angle of the wall must be 30°, therefore making them the highest known scarps in the Solar System. (It would be interesting to watch a piece of material fall from the crest to the floor; the descent time would be about 10 minutes!) Both Pan and Gaea are deeper, relatively, than craters of similar size on the Moon. Between them from long 0 to 60 deg W. is a complex region of troughs and ridges, tens of kilometres long and sometimes at least 20 km wide. The two bright patches, Ida and Lyctos, are about 15 km in diameter, and are presumably mountains.

Thebe was also found on Voyager pictures during an examination to con-

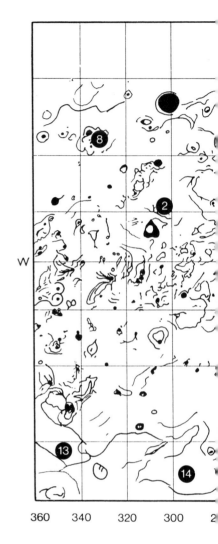

360   340   320   300   2

firm the existence of Adrastea. Thebe moves between the orbits of Amalthea and Io. All we know about it is that its revolution period is 16 h 16 m, and that with a diameter of about 75 km it is decidedly larger than either Metis or Adrastea.

IO

Io is slightly larger than our Moon, and is the densest of the four Galilean satellites. Before the space missions it was tacitly assumed to be a rocky, cratered world, but in the event nothing could have been further from the truth. Io is bright orange; it is covered with sulphur, and there are active volcanoes.

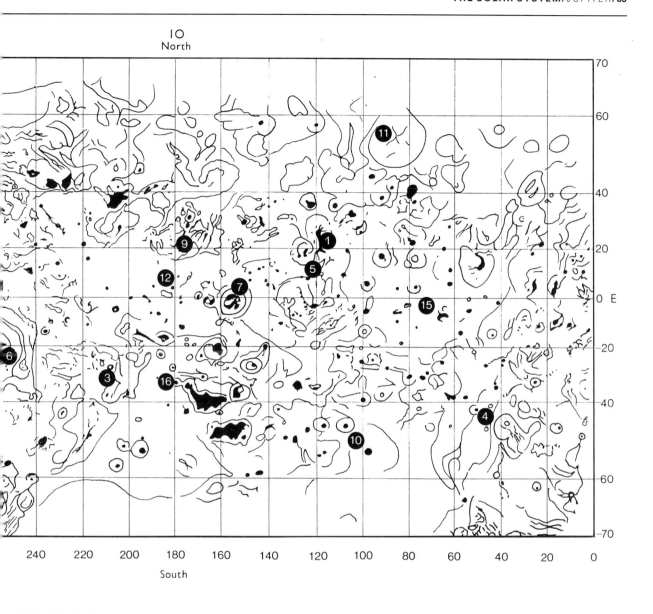

IO
North

South

240 220 200 180 160 140 120 100 80 60 40 20 0

FEATURES ON IO
Bold numbers indicate map references

Volcanoes	Lat	Long W
Amirani (Plume 5) **1**	27 N	119
Loki (Plume 2) **2**	19 N	305
Marduk (Plume 7) **3**	28 S	210
Masubi (Plume 8) **4**	45 S	053
Maui (Plume 6) **5**	19 N	122
Pele (Plume 1) **6**	19 S	257
Prometheus (Plume 3) **7**	03 S	153
Surt **8**	46 N	336
Volund (Plume 4) **9**	22 N	177

Pateræ	Lat	Long W
Amaterasu	38 N	307
Asha	09 N	226
Atar	30 N	279
Aten	48 S	311
Babbar	40 S	272
Bochica	61 S	022
Creidne	52 S	345
Culann	20 S	150
Dædelus	19 N	175

	Lat	Long
Dazhbog	54 N	302
Emakong	00	110
Fuchi	28 N	328
Galai	11 S	289
Gibil	15 S	295
Heno	57 S	312
Hephæstus	02 N	290
Hiruko	65 S	331
Horus	10 S	340
Huo Shen	15 S	329
Inti	68 S	349
Kane	48 S	015
Kibero	12 S	306
Loki	13 N	310
Maasaw	40 S	341
Mafuike	15 S	261
Malik	34 S	128
Manua	35 N	322
Masaya	22 S	350
Maui	20 N	125
Mbali	32 S	007
Mihr	16 S	306
Nina	40 S	165

	Lat	Long
Nusku	63 S	007
Nyambe	00	345
Ra	08 S	325
Reiden	24 S	236
Ruwa	00	002
Sengen	33 S	304
Shakuru	23 N	267
Shamash	36 S	152
Shoshu	19·5 S	324·5
Svarog	48 S	267
Talos	26 S	356
Taw	33·5 S	000·5
Tohil	28 S	157
Ülgen	41 S	288
Uta	35 S	027
Vahagn	27 S	359
Viracocha	62 S	284

Other Features	Lat	Long W
Apis Tholus	11 S	349
Bactria Regio **10**	45 S	125
Chalybes Regio **11**	55 N	085
Colchis Regio **12**	10 N	170

	Lat	Long W
Dodona Planum **13**	60 S	350
Hæmus Mons	70 S	050
Inachus Tholus	29 S	354
Lerna Regio **14**	65 S	300
Mazda Catena	08 S	315
Media Regio **15**	00	070
Mycenæ Regio **16**	35 S	170
Nemea Planum	80 S	270
Silpium Mons	62 S	282
Tarsus Regio	30 S	055

The first indication of vulcanism was found on 9 March 1979 by Linda Morabito, a member of the Voyager team who was looking for a faint star, AGK-10021, to check on Io's position. The volcanic plume rose to 280 km. Other volcanoes were then found; altogether eight were observed during the Voyager 1 mission. There are also many volcanic craters, generally looking like black spots a few tens of kilometres in diameter and frequently surrounded by irregular haloes nearly as black as the central spot. There are extensive lava-flows, and the entire surface is in a state of activity. Voyager 1 temperature scans found some 'hot spots', notably a region near the volcano Loki, where the temperature was 17 °C – contrasting with the surrounding surface temperature, which was −146 °C. The dark feature may well have been a lava lake of molten rock or molten sulphur; the melting point of sulphur is 112 °C. Subsequently it has been found that the volcanic vents are very hot indeed, with temperatures of well over 500 °C.

The heights of the volcanic plumes during the Voyager 1 flyby were very considerable; 280 km for Pele, 100 for Loki, 120 for Marduk, and between 70 and 95 km for the rest. Oddly enough Pele, the most violent of the volcanoes, had become inactive by the time of the Voyager 2 pass a few months later, though there is no reason to suppose that it has become extinct. On the other hand Loki and Prometheus were more active than before, with the Loki plume exceeding 200 km in height.

Why is Io so active? According to one theory, the crust may be a 'sea' of sulphur and sulphur dioxide about 4 km deep, with only the uppermost kilometre solid. Heat escapes from the interior in the form of lava, erupting beneath the sulphur ocean, and the re-

sult is a violent outpouring of a mixture of sulphur, sulphur dioxide gas, and sulphur dioxide 'snow'. It has been proposed that the interior is constantly flexed by tidal effects produced by Jupiter; the orbit of Io is slightly eccentric (due mainly to the perturbations produced by the other Galileans) and so the tidal effects vary.

Io has an atmosphere, but its density is so low that it corresponds to what we normally call a vacuum. As has been noted, material from it produces a torus round Jupiter centred on Io's orbit. It is indeed a strange, colourful place; but since it moves within Jupiter's radiation zones, it must be just about the most lethal world in the entire Solar System.

EUROPA

Europa was not closely studied by Voyager 1, which by-passed it at 732 230 km, so that our main information comes from Voyager 2, where the minimum distance was only just over 204 000 km. The whole scene is utterly different from that on Io. The overall hue of the disk is white, with no volcanoes, and virtually no craters; only three have been identified, none more than about 20 km in diameter. There is little doubt that the surface is made up of ice. The observed features are unlike any previously found anywhere, and new terms have had to be introduced: 'linea' (a dark or bright elongated marking, either straight or curved); 'flexus' (a very low curved ridge with a scalloped pattern) and 'macula' (a dark, sometimes irregularly-shaped patch).

Europa is a smooth world, with an almost complete absence of vertical relief. It has been said that the satellite is as smooth as a billiard-ball. It has also been likened to a cracked eggshell, and that the dark lines look as though they have been drawn on a white ball with a felt-tipped pen.

We know little about the interior, but there may well be an ice crust perhaps 100 km deep, overlying either an ocean of water or else a region of 'slushy ice' below which is a silicate core.

Europa is a map-maker's nightmare; one region looks very much like another. If craters ever existed, they have been obliterated, perhaps by

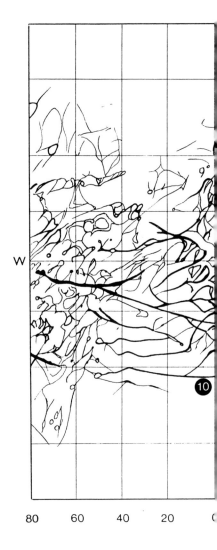

water seeping out from below and then freezing. Yet unless the surface is very young, some craters would be expected. We also have to ask why Europa is so completely unaffected by the forces which have moulded the surface of Io; after all, Europa is less than twice Io's distance from Jupiter. In its way, Europa is as puzzling a world as we have yet found in the Solar System.

FEATURES ON EUROPA
Bold numbers indicate map references

Maculæ	Lat	Long W
Thera Macula **16**	45 S	178
Thrace Macula **17**	44 S	169
Tyre Macula **18**	34 N	144

EUROPA
North

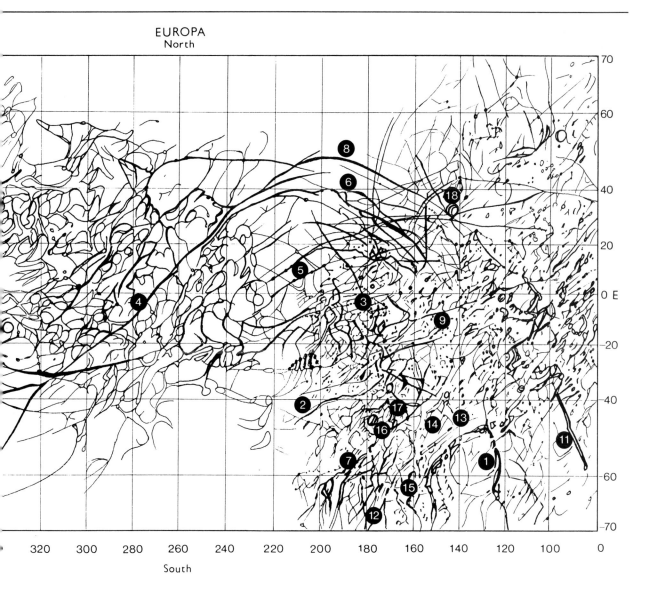

GANYMEDE

Ganymede is the largest satellite in the Solar System, its only possible rival being Triton, the senior attendant of Neptune. Its overall density is less than twice that of water, and it seems to be composed of ice and rock in about equal proportions. Both Voyagers obtained good views of it, Voyager 1 from 112 030 km and Voyager 2 from only 59 530 km, so that 80 per cent of the surface has been mapped down to a resolution of 5 km or less.

The surface is certainly icy, and gives every impression of being ancient. There are dark areas, of which the largest, Galileo Regio, has a diameter of 4000 km, so that it is nearly as large as the continental United States. There are also brighter, presumably younger regions distinguished by 'sulci', i.e. grooves or furrows, with ridges which may rise to a kilometre or so. The two types of terrain intermingle, so that the overall picture is highly complex. There are many craters, some of which are ray-centres.

Probably Ganymede has an icy crust less than 100 km deep, below which is a convecting mantle of water or slushy ice 400–800 km deep; there is presumably a silicate core. The surface could once have been active enough to produce what may be termed tectonic blocks – icy versions of the Earth's tectonic plates – but Ganymede today is completely inert.

An ingenious attempt to detect an atmosphere was made from Voyager 1. Ganymede occulted a star, and the event was carefully monitored to see if any trace of atmosphere could be found. The results were negative, and any atmosphere must have a density of less than one hundred thousand millionth of the Earth's air.

FEATURES ON GANYMEDE
Bold numbers indicate map references

Regiones	Lat	Long W
Barnard **10**	22 N	010
Galileo **11**	35 N	145
Marius **12**	10 S	200
Nicholson **13**	20 S	000
Perrine **14**	40 N	030

Sulci	Lat	Long W
Anshar **15**	15 N	200
Apsu **16**	40 N	230
Aquarius **17**	50 N	010
Dardanus **18**	20 S	013
Harpagia **19**	00	317
Kishar **20**	15 S	220
Mashu **21**	22 N	200
Mysia **22**	10 N	340
Nun **23**	50 N	320
Philus **24**	37 N	215
Phrygia **25**	20 N	005
Sicyon **26**	44 N	003
Tiamat **27**	03 S	210
Uruk **28**	00	157

Craters	Lat	Long W
Achelous **1**	66 N	004
Adad	62 N	352
Adapa	83 N	022
Ammura	36 N	337
Anu	68 N	332
Asshur	56 N	325
Aya	67 N	303
Ba'al	29 N	326
Daniel	04 N	021
Diment	29 N	346
Enkidu	17 S	322
Enlil	52 N	301
Eshmun **2**	22 S	187
Etana	78 N	310
Gilgamesh **3**	58 S	124
Gula	68 N	001
Hathor	70 S	265
Isis **4**	64 S	197
Keret	22 N	034
Khumbam	15 S	332
Kishar	78 N	330
Melkart **5**	13 S	182
Mor	35 N	323
Nabu	36 S	002
Namtar	49 S	343
Nigitsu	48 S	308
Nut **6**	61 S	268
Osiris **7**	39 S	161
Ruti	15 N	304
Sapas	59 N	031
Sebek **8**	65 N	348
Sin	56 N	349
Tanit	59 N	032
Teshub	02 N	016
Tros **9**	20 N	028
Zaqar	60 N	031

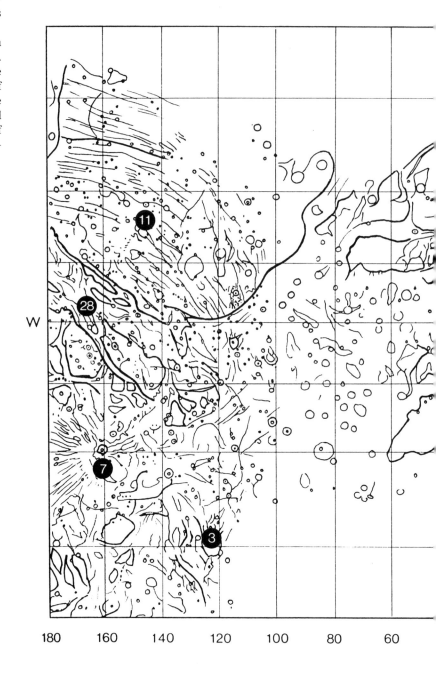

W

| 180 | 160 | 140 | 120 | 100 | 80 | 60 |

CALLISTO

Callisto is slightly smaller, slightly less massive and slightly less dense than Ganymede, and it also has a lower albedo or reflecting power, so that as seen from Earth it is actually the faintest of the four Galileans. Voyager 1 passed it at 123 950 km and Voyager 2 at 212 510 km, so that the surface has not been mapped as accurately as Ganymede's.

The crust has often been described as being made up of 'dirty ice', and may go down to a depth of 200 to 300 km; below this there may be a 1000 km deep mantle of water or soft ice overlying the silicate core. The surface temperatures

GANYMEDE
North

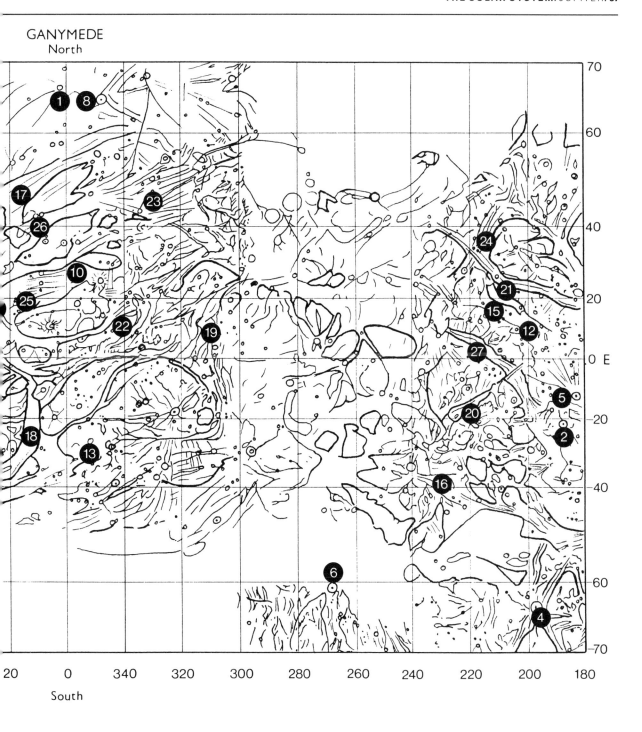

are very low; about −118 °C during the daytime, −193 °C just before dawn (remembering that Callisto has a long 'day'; like all the Galileans its rotation is synchronous, so that its rotation period of 16·7 Earth-days is the same as the time which it takes to orbit Jupiter).

Callisto is the most cratered world known to us, though in general the cra-

ters are fairly shallow. The geology seems to be much more straightforward than that of Ganymede; apparently there has been virtually no activity since a very early stage in the history of the satellite, so that we may well be looking at the oldest planetary surface in the Solar System. There are two huge ringed basins, Valhalla and Asgard.

Valhalla has a brighter circular region 600 km in diameter, and is surrounded by concentric rings, the largest of which has a diameter of 3000 km; Asgard is similar, though rather smaller. It is surely appropriate to name the ancient features of Callisto after the dwelling-places and the gods of ancient Norse mythology!

CALLISTO
North

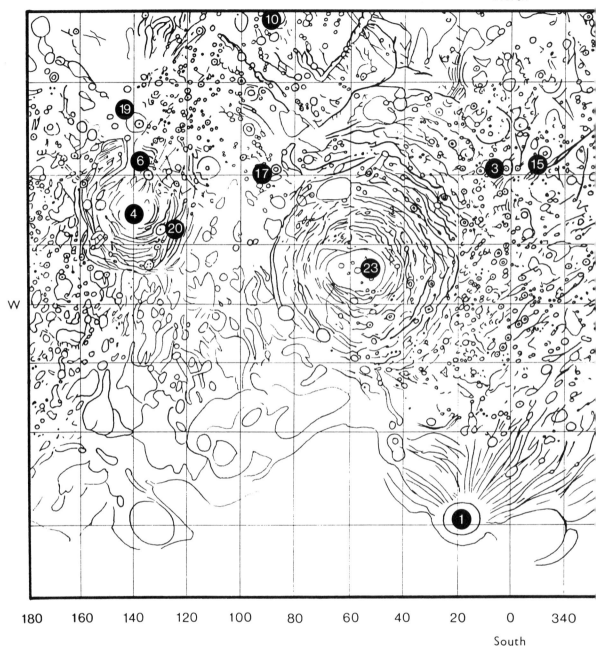

180 160 140 120 100 80 60 40 20 0 340

South

The lack of past tectonic activity may indicate that Callisto's slight inferiority in size and mass compared with Ganymede meant that the core temperature never rose sufficiently for major movements to begin.

Dead though Callisto is, it may well be the first world in the Jovian system to be visited by future astronauts, because it lies outside the main radiation zones of Jupiter itself.

THE OUTER SATELLITES

Of the eight known satellites moving round Jupiter beyond the orbit of Callisto, only Himalia is of any size; its diameter is thought to be about 170 km, while none of the others attain 100 km. They fall into two main groups. Leda, Himalia, Lysithea and Elara move round Jupiter between 11 and 12 mil-

lion km, while Ananke, Carme, Pasiphaë and Sinope move between 20 and 24 million km. The outer four have retrograde orbits, which strengthens the suggestion that these small satellites may be captured asteroids – particularly since they seem to have fairly low albedoes, in which case they are rocky rather than icy. Neither of the Voyagers went close enough to obtain surface pictures of any of the eight, and there-

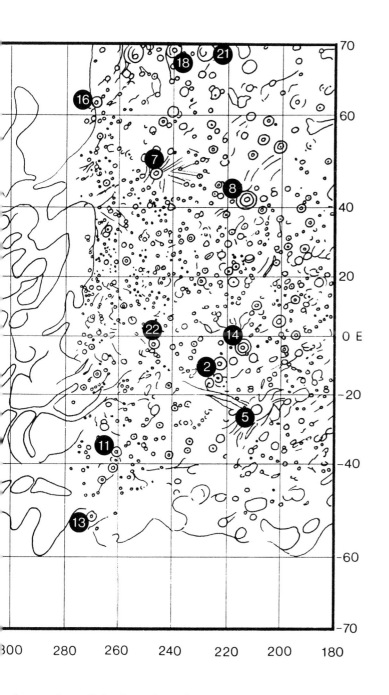

	Lat	Long W
Bavorr	48 N	023
Beli	61 N	079
Bragi	77 N	069
Brami	26 N	018
Bran **5**	25 S	207
Buga	22 N	326
Buri	43 S	044
Burr **6**	40 N	136
Dag	56 N	074
Danr	61 N	075
Dia	73 N	056
Dryops	77 N	029
Durinn	66 N	087
Egdir	31 N	035
Erlik	66 N	358
Fadir	56 N	015
Fili	65 N	349
Finnr	14 N	014
Freki	82 N	010
Frodi	69 N	136
Fulla	74 N	102
Geri	66 N	353
Gipul Catena	65 N	055
Gisl	56 N	035
Gloi **7**	48 N	246
Goll	58 N	323
Gondul	59 N	115
Grimr **8**	43 N	214
Gunnr	64 N	100
Gymir	61 N	055
Habrok	77 N	129
Haki **9**	26 N	315
Har	06 N	357
Hepti	64 N	027
Hodr **10**	69 N	087
Hoenir	36 N	261
Hogni	15 S	005
Igaluk **12**	05 N	315
Ivarr	06 S	322
Jumo	62 N	015
Kari	47 N	103
Lodurr **13**	52 S	270
Loni **14**	04 S	215
Losy	68 N	329
Mera	63 N	073
Mimir	30 N	054
Mitsina	57 N	097
Modi	67 N	115
Nama	57 N	336
Nar	04 S	045
Nerivik	22 S	055
Nidi	66 N	093
Nori **15**	46 N	347
Nuada **16**	62 N	269
Oski	56 N	266
Ottar	60 N	100
Pekko	17 N	006
Reginn **17**	42 N	088
Rigr **18**	69 N	240
Sarakka	08 S	053
Sequinek	55 N	027
Sholmo	52 N	018
Sigyn	33 N	027
Skoll	57 N	317
Skuld	06 N	037
Sudri **19**	53 N	137
Sumbur	69 N	332
Tindr	05 S	355
Tornarsuk **20**	25 N	130
Tyn **21**	68 N	229
Valfodr **22**	03 S	246
Valhalla **23**	10 N	055
Vali	09 N	327
Vestri	42 N	054
Vitr	23 S	347
Ymir	51 N	097

fore we know little about them. One, Sinope, was discovered by accident. On 21 July 1914 S. B. Nicholson set out to photograph Pasiphaë, which had been discovered six years earlier; when he developed the plate, he found not only Pasiphaë but also the newcomer. No doubt there are more faint asteroidal-type satellites awaiting discovery.

FEATURES ON CALLISTO

Bold numbers indicate map references

	Lat	Long W
Adal	77 N	079
Adlinda **1**	58 S	020
Ägröi	42 N	012
Akycha	74 N	325
Alfr **2**	09 S	222
Ali	57 N	058
Anarr **3**	43 N	003
Aningan	51 N	011
Asgard **4**	30 N	140
Askr	53 N	327
Balkr	27 N	012

SATURN

The most remote planet known in ancient times was Saturn, which is sixth in order of distance from the Sun.

OPPOSITION DATES, 1977–2000:

Date	Magnitude
1977 Feb. 2	0·0
1978 Feb. 16	+0·3
1979 Mar. 1	+0·5
1980 Mar. 13	+0·8
1981 Mar. 27	+0·7
1982 Apr. 8	+0·5
1983 Apr. 21	+0·4
1984 May 3	+0·3
1985 May 15	+0·2
1986 May 27	+0·2
1987 June 9	+0·2
1988 June 20	+0·2
1989 July 2	+0·2
1990 July 14	+0·3
1991 July 26	+0·3
1992 Aug. 7	+0·4
1993 Aug. 19	+0·5
1994 Sept. 1	+0·7
1995 Sept. 14	+0·8
1996 Sept. 26	+0·7
1997 Oct. 10	+0·4
1998 Oct. 23	+0·2
1999 Nov. 6	0·0
2000 Nov. 19	−0·1

The opposition magnitude is affected both by Saturn's varying distance and by the angle of presentation of the rings. The last edgewise presentation was in 1980. The next will be in 1995. The intervals between successive edgewise presentations are 13 years 9 months and 15 years 9 months. During the shorter interval, the south pole is sunward, the southern ring-face is seen, and Saturn passes through perihelion.

RING DIAMETERS

Ring A: outer 272 300 km; inner 239 600 km.
Ring B: outer 234 200 km; inner 181 100 km.
Ring C: inner 149 300 km.

The first observations of Saturn must date back before recorded history, since the planet is a bright naked-eye object (at its most brilliant it surpasses any of the stars apart from Sirius and Canopus).

The first recorded observations of Saturn seem to have been those made in Mesopotamia in the mid-7th century BC. About 650 BC there is a record that Saturn 'entered the Moon', which is presumably a reference to an occultation of the planet.

The first observation of Saturn

DATA

Mean distance from the Sun:
1427·0 million km (9·539 a.u.)
Maximum distance from the Sun:
1507 million km (10·069 a.u.)
Minimum distance from the Sun:
1347 million km (9·008 a.u.)
Sidereal period: 10 759·20 days = 29·46 years
Mean synodic period: 378·1 days
Rotation period (equatorial):
10 hours 39·4 minutes
Mean orbital velocity: 9·6 km/s
Axial inclination: 26°44′
Orbital inclination: 2°29′21″·6
Orbital eccentricity: 0·056
Diameter: equatorial 119 300 km; polar 107 700 km
Apparent diameter from Earth: max. 20″·9, min 15″·0
Reciprocal mass, Sun = 1: 3498·5
Density, water = 1: 0·71
Mass, Earth = 1: 95·17
Volume, Earth = 1: 744
Escape velocity: 32·26 km/s
Surface gravity, Earth = 1: 1·16
Mean surface temperature: −180 °C
Oblateness: 0·1
Albedo: 0·61
Opposition magnitude: max −0·3, min +0·8
Mean diameter of Sun, seen from Saturn: 3′22″

---

by Copernicus was made on 26 April 1514, when Saturn lay in a line with the stars 'in the forehead of Scorpio'. Copernicus made three other recorded observations of Saturn: on 5 May 1514, 13 July 1520 and 10 October 1527.

The first observation of Saturn by Tycho Brahe was made on 18 August 1563, when Saturn was in conjunction with Jupiter.

The first telescopic observation of Saturn was made in July 1610 by Galileo, using a magnification of 32 on his largest telescope. He recorded that 'the planet Saturn is not one alone, but is composed of three, which almost touch one another and never move nor change with respect to one another. They are arranged in a line parallel to the Zodiac, and the middle one is about three times the size of the lateral ones.' Galileo's telescope was inadequate for him to see the ring-system in its true guise, and subsequently he lost sight of the 'companions', as the rings were edgewise-on in December 1612; later he recovered them.

The earliest-dated drawing of Saturn to be published was made by Pierre Gassendi on 19 June 1633, though it, together with other drawings

---

made by Gassendi, was not actually published until 1658.

The first true explanation of the nature of the ring system was given by Christiaan Huygens in 1656, from his observations made in 1655. His explanation was challenged, and was not finally accepted by all astronomers until 1665.

The first indications of the Crêpe Ring (Ring C) may have been given by Campani in 1664, though they were not interpreted as being such. Campani also appears to have recorded a belt on the planet's disk.

The discovery of the Cassini Division, separating Rings A and B, was made by G. D. Cassini in 1675 (claims that W. Ball had observed the Division ten years earlier have been discounted). Cassini also recorded the South Equatorial Belt on the planet's disk.

The first suggestion that the rings are not solid was made by J. Cassini in 1705. Theoretical confirmation was provided by James Clerk Maxwell in 1875, who showed that no solid ring could exist; it would be disrupted by the gravitational pull of Saturn.

The discovery of the solar flattening of Saturn was made by W. Herschel in 1789. Herschel gave the ratio of the equatorial to the polar diameter as 11:10, which is approximately correct.

The first well-defined spots to be detected on Saturn were observed by J. H. Schröter and his assistant, K. Harding, at Schröter's observatory at Lilienthal, near Bremen, in 1796.

The discovery of Encke's Division in Ring A was made by J. F. Encke at Berlin on 28 May 1837.

The discovery of the Crêpe Ring, Ring C, was made by W. Bond at Harvard in November 1850, using the 15 in Merz refractor there. Independent confirmation was provided by Lassell and Dawes in England.

The transparency of Ring C was discovered in 1852, independently by Lassell and by C. Jacob at Madras.

The first explanation of the divisions in the ring system was given by D. Kirkwood in 1866. Kirkwood attributed the Divisions to the perturbing effects of Saturn's inner satellites.

Saturn: Voyager 2, 21 July 1981, at a distance of 33 900 000 km from the planet. The two satellites are Rhea (the brighter) and Dione. (NASA)

The discovery of dark absorption bands in the spectrum of Saturn was made by A. Secchi in 1863.

The first brilliant white spot to be recorded on Saturn was discovered by A. Hall on 8 December 1876.

The first observation of the globe of Saturn seen through the Cassini Division was made in 1883 by A. Hall and C. A. Young.

The first spectroscopic proof of the meteoritic nature of the rings was given by J. E. Keeler in 1895. Keeler showed that the rings did not rotate as a solid mass would do; the inner sections had the fastest rotation. This was proved by their Doppler shifts.

The first report of a dusky ring exterior to Ring A was made by G. Fournier, using an 11 in refractor, at Mont Revard (France) in 1907. (There are in fact three exterior rings, detect-able only by space-probes.)

The most famous observation of the passage of a star behind the rings was made by M. A. Ainslie and J. Knight, independently, on 9 February 1917. The dimming of the star as it passed behind the rings and divisions provided valuable information about the transparencies of the ring sections. The star concerned was B. D. +21°1714. Ainslie used a 9 in refractor at Blackheath; Knight a 5 in refractor at Rye in Sussex.

The first theoretical demonstration that Saturn has a cold surface, and is therefore not a 'miniature sun', was given by H. Jeffreys in 1923.

The spectroscopic detection of methane and ammonia in Saturn's atmosphere was made by T. Dunham at Mount Wilson, California, in 1932. Bands had been seen by V. M. Slipher in 1909, but had not previously been identified.

The most prominent white spot on Saturn to be seen during the present century was discovered by W. T. Hay on 3 August 1933. Hay was using a 6 in refractor, and is the only famous comedian to have discovered a spot on Saturn – he is best remembered as Will Hay! The spot remained identifiable until 13 September 1933. W. H. Wright considered it to have been of an 'eruptive nature'.

The first modern-type model of Saturn was described by R. Wildt in 1938. Wildt believed that there was a rocky core, overlaid by a thick layer of ice and then by the gaseous atmosphere. The model has now been rejected, as has an alternative model proposed in 1951 by W. Ramsey, who considered that hydrogen was the main constituent but that near the core it acquired metallic characteristics.

The first serious observation of Saturn with the Hale reflector at Palomar was made by G. P. Kuiper in 1954. Kuiper concluded that the Cassini Division was the only genuine gap, Encke's Division being merely a 'ripple' and the other reported minor divisions illusory.

The first observations of infrared thermal emission from Saturn were made in 1969, from equipment carried in an aircraft.

The first radar reflections from the rings were observed in 1972. It was concluded that the ring particles were icy, with diameters of between 4 and 30 cm.

Much of our present knowledge of Saturn has been drawn from the spaceprobes, of which there have so far been three. The first was Pioneer 11. Its prime target had been Jupiter; but the success of its predecessor, Pioneer 10, and the fact that there was adequate power reserve meant that it could be swung back across the Solar System to a rendezvous with Saturn in 1979. The results were preliminary only, but involved several important discoveries. Perhaps the most significant fact was that Pioneer was not destroyed by a ring particle collision; at that time nobody knew what conditions near Saturn would be like, and estimates of the probe's chances of survival ranged from 1 per cent to 99 per cent! In fact there was a near-encounter with a hitherto-unknown small satellite, but Pioneer emerged unscathed.

Details of the three probes are as follows:

Name	Launch date	Encounter date	Nearest approach km	Remarks
Pioneer 11	5 Apr 1973	1 Sep 1979	20 880	Preliminary results.
Voyager 1	5 Sep 1977	12 Nov 1980	124 200	Complete success. Good. images also of Titan, Dione, Rhea and Mimas.
Voyager 2	20 Aug 1977	25 Aug 1981	101 300	Success apart from one failure after rendezvous. Good images of Iapetus, Hyperion, Tethys and Enceladus.

At this planning stage, Titan was regarded as almost as important a target as Saturn itself. Voyager 1 was programmed to encounter it. Had Voyager 1 failed, then Voyager 2 would have been re-programmed to bypass Titan – but in this case it would have been unable to go on to rendezvous with Uranus and Neptune. There was therefore tremendous relief when Voyager 1 succeeded, leaving its follower free to carry out the full programme.

The latest models of Saturn, supported by the Voyager results, indicate that despite its smaller mass and lower density Saturn bears a strong resemblance to Jupiter. The outer clouds contain more hydrogen but less helium than with Jupiter (the helium content for Saturn's clouds is 6 per cent, as against 11 per cent for Jupiter's). Below the clouds comes liquid hydrogen, at first molecular and then, below a depth of 30 000 km, metallic; the rocky core seems to be little larger than the Earth, though three times as massive. The core temperature is of the order of 15 000 degrees.

Like Jupiter, Saturn sends out more energy than it would do if it depended entirely upon what it receives from the Sun, but the cause is different. Saturn must now have lost all its heat produced during its formation stage, whereas the more massive Jupiter has not had time to do so. It is likely that the excess energy of Saturn is gravitational, produced by helium droplets as they fall through the lighter hydrogen toward the centre of the planet. This also explains why Saturn has less high-altitude helium than with Jupiter. The rotation period was found to be 10 h 39·4 minutes, somewhat longer than the previously-accepted value.

The magnetic field is 1000 times stronger than that of the Earth, though 20 times weaker than Jupiter's. The magnetic axis is within one degree of the rotational axis, in which respect Saturn differs from both Jupiter and the Earth. The magnetic field is stronger at the north pole than at the south (0·69 gauss, against 0·53 gauss), and the centre of the field is displaced some 2400 km northward along the planet's axis. The magnetosphere is decidedly variable; its average boundary lies near the orbit of Titan, so that Titan is sometimes inside the magnetosphere and sometimes outside it. Radiation zones are present, though they cannot compare with those of Jupiter. The numbers of electrons fall off quickly at the outer edge of the main ring system (i.e. the outer edge of Ring A), because the electrons are absorbed by the ring particles; the region between Ring A and the globe is the most radiation-free area in the Solar System. Auroræ were recorded by the Voyagers, but are ten times weaker than those of Jupiter.

Saturn, like Jupiter, has a violently disturbed cloud surface. Features over 1000 km in diameter are not common, and even the largest ovals are no more than half the size of Jupiter's Great Red Spot. A red spot on Saturn was detected from Voyager 1 in August 1980, and was still present in September 1981; it too may owe its colour to phosphorus, but it is not likely to be so long-lived as the Great Red Spot on Jupiter. Smaller reddish and brownish spots were also found on Saturn. The lower temperature as compared with Jupiter means that ammonia crystals form at higher levels, covering the planet with 'haze' and giving it a somewhat bland appearance.

One surprise was that the wind patterns are different from those of Jupiter. The wind zones do not follow the light and dark bands, but are symmetrical with the equator; between latitudes 35°N and 35°S there is an eastward flow or jet-stream with winds up to 1500 km/h, four times faster than any winds found on Jupiter. One prominent 'ribbon' at latitude 47°N was taken to be a wave pattern in a particularly unstable jet-stream.

Seasonal effects occur. The Sun crossed into Saturn's northern hemisphere in 1980, but there is a definite lag effect, and during the Voyager missions the northern hemisphere was still the colder of the two; the temperature difference between the north and south poles amounted to 10 degrees.

THE RINGS

Despite the fascinating information about the globe, it is fair to say that the main emphasis was upon the ring-system. As the results came into Mission Control at Pasadena, California, there was a feeling not only of excitement, but also incredulity. The rings were so completely different from anything which had been envisaged. Instead of being more or less homogeneous, they proved to be immensely complex, made up of thousands of small ringlets and narrow gaps; it was even said that the rings had more grooves than any gramophone record. Also, the system turned out to be much more extensive than expected. The so-called Ring D is not a true ring, and is unobservable from Earth, but there were three rings outside the main system (one of which, Ring F, had been found by Pioneer 11). Even the Cassini Division contained ringlets, and, most remarkable of all, some of the rings were eccentric instead of perfectly circular. Details, with the satellites involved in the system, are as follows:

Feature	Distance from centre of Saturn km	Period hours
Cloud-tops	60 330	10·66
Inner edge of Ring D	67 000	4·91
Inner edge of Ring C	73 200	5·61
Inner edge of Ring B	92 200	7·93
Outer edge of Ring B	117 500	11·41
Middle of Cassini Division	119 000	11·75
Inner edge of Ring A	121 000	11·92
Encke Division	133 500	13·82
Outer edge of Ring A	135 200	14·14
Atlas	137 670	14·61
Prometheus	139 353	14·71
Ring F	140 600	14·94
Pandora	141 700	15·07
Epimetheus	151 422	16·65
Janus	151 472	16·68
Ring G	170 000	19·90
Mimas	185 600	22·60
Middle of Ring E	230 000	31·3
Enceladus	238 100	32·88

The system is very varied. The D 'ring' has no sharp inner edge, and the particles may go down close to the cloud-tops. Ring C, the Crêpe Ring, is 19 000 km broad, and has various gaps, including two with widths of 200 and 300 km respectively; the outer gap contains a narrow, dense and slightly eccentric bright ring, 90 km wide when furthest from Saturn but only 35 km

wide when at its closest. The C-ring particles seem, on average, to be about 2 m in diameter.

The main ring is, of course, Ring B, where the particles range in size from 10 cm to about a metre. The particles are redder than those of the C Ring and D region, and are made up mainly of water ice, as had already been suspected from Earth-based radar work. The ring temperatures ranged between −180 °C in sunlight down to −200° in shadow. It was also found that there is a rarefied cloud of neutral hydrogen extending to 60 000 km above and below the ring-plane. Strange radial spokes were seen in the B ring, and were decidedly puzzling, because logically no such features should form; the difference in rotation period between the inner and outer edges of the ring is well over three hours – and yet the spokes persisted for hours after emerging from the shadow of the globe; when they were distorted and broken up, new spokes emerged from the shadow to replace them. Presumably they are due to particles of a certain definite size elevated away from the ring-plane by magnetic or electrostatic forces. They are confined wholly to Ring B. Another point finally cleared up concerned the thickness of the ring. The maximum cannot be more than 150 m, and is more probably as little as 100 m.

The 4000 km Cassini Division is not empty, but contains many rings a few hundred kilometres wide; the particles are less red than those of Ring B, and more closely resemble those of Ring C. At least one of the rings in the Cassini Division is markedly eccentric.

Ring A is made up of particles ranging from fine 'dust' to larger blocks around 10 m in diameter. The Encke Division in Ring A is 270 km wide, and contains ringlets which are both discontinuous and 'kinky'. (The Encke Division has been referred to as the Keeler Gap, but it has now very properly been decided that the old, familiar name should be retained; three other narrow gaps in the ring system have been named Huygens, Maxwell and Keeler.) Atlas moves close to the outer edge of Ring A, and is responsible for the sharp border to the ring.

Outside the main system we come to

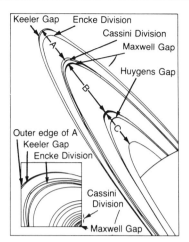

Nomenclature of Saturn's rings.

Ring F. It is faint, but of special interest because it seems to be made up of several rings which are intertwined; the effect was startling in the Voyager 1 pictures, but less so from Voyager 2. Ring F is stabilized by two small satellites, now named Prometheus and Pandora. The satellite slightly closer-in to Saturn will be moving faster than the ring particles, and will tend to speed a particle up if it strays away from the ring region; the outer satellite, moving more slowly, will drag any errant ring particles back and return them to the central region. For obvious reasons, these two tiny bodies are commonly known as the 'shepherd' satellites. It was thought for a while that the numerous divisions in the main rings could be similarly explained, and Voyager 2 was re-programmed to make a careful search for any satellites moving in Rings A and B, but none was found.

Beyond Ring F and its shepherds come the two co-orbital satellites, Epimetheus and Janus. Next there is the very tenuous Ring G. Still further out comes the familiar satellite Mimas. The final ring, E, is nearly 90 000 km wide, but very tenuous indeed; its brightest part is slightly closer-in than the orbit of Enceladus, the second of the well-known satellites, and it has even been suggested that material ejected from Enceladus may be concerned in the formation of the ring. Enceladus has an icy surface, and there can no longer be any doubt that the ring-particles are made up simply of water ice.

An ingenious experiment was carried out from Voyager 2. As seen from the space-craft, the third-magnitude star δ Scorpii was occulted by the rings. It had been expected that there would be regular 'winks', as the starlight would shine through any gaps but would be blocked by ring material. In fact thousands of 'blips' were recorded, but it seems that there is very little empty space anywhere in the main ring system.

What can be the cause of this amazing complexity? The Cassini Division had been explained as due to Mimas; a particle moving in the Division will have a revolution period exactly half that of Mimas, and cumulative perturbations will drive it away from the 'forbidden zone', leaving the Division swept clear. Yet the Voyagers have shown that this explanation is clearly inadequate, and it now seems more likely that we are dealing with a density wave effect. A satellite such as Mimas can alter the orbit of a ring particle and make it elliptical. This causes 'bunching' in various areas; a spiral density wave will be created, and particles in it will collide, moving inward toward the planet and leaving a gap just outside the resonance orbit. This theory, due to Peter Goldreich and Scott Tremaine, does at least sound plausible, but it would be wrong to claim that we really understand the mechanics of the ring system.

The data for the minor satellites are extremely uncertain, and different authorities give different values. As the Sun's apparent diameter as seen from Saturn is only 3'22", total eclipses may be produced on Saturn by Mimas, Enceladus, Tethys, Dione, Rhea and Titan.

At the opposition of 1966, when the rings were edgewise-on to the Earth, A. Dollfus reported the detection of a new satellite, closer-in than Mimas, the quondam nearest satellite. It was also glimpsed by several others (including myself) and was even given a name: Janus. However, it is now clear that there is no satellite moving in the orbit worked out for Dollfus' satellite, and the name Janus has been given to the satellite known formerly as 1980 S–1. There seems little doubt that the object seen in 1966 was either the present Janus or else its co-orbiter, Epimetheus.

Phœbe has retrograde motion, and may possibly be a captured asteroid, particularly as preliminary results show that its surface is darkish and reddish – quite unlike the familiar icy satellites.

THE SATELLITES OBSERVED BY THE VOYAGERS

Apart from Phœbe, all the satellites were reasonably well observed from the Voyager. Minimum distances were as follows:

Satellite	Min. distance; km:	
	Voyager 1	Voyager 2
Atlas	219 000	287 170
Prometheus	300 000	246 590
Pandora	270 000	107 000
Epimetheus	121 000	147 010
Janus	297 000	222 760
Mimas	88 440	309 990
Enceladus	202 040	87 140
Tethys	415 760	93 000
Calypso	237 332	284 396
Telesto	432 295	153 518
Dione	161 520	502 250
Helene	230 000	318 200
Rhea	73 980	645 280
Titan	6 490	665 960
Hyperion	880 440	470 840
Iapetus	2 470 000	909 070
Phœbe	13 537 000	1 473 000

SATELLITES OF SATURN

	Discoverer	Mean distance from Saturn km	Mean angular distance from Saturn, at mean opposition distance ' "
Atlas	From Voyager	137 670	
Prometheus	photographs	139 350	
Pandora	1980	141 700	
Janus	Fountain and	151 470	
Epimetheus	Larson 1978	151 420	
Mimas	W. Herschel 1789	185 540	0 30·0
Enceladus	W. Herschel 1789	238 040	0 38·4
Tethys	G. D. Cassini 1684	294 670	0 47·6
Telesto	Group led by	294 670	0 47·6
Calypso	B. Smith 1980	294 670	0 47·6
Dione	G. D. Cassini 1684	377 420	1 01·1
Helene	Lacques and Lecacheaux 1980	377 420	1 01·1
Rhea	G. D. Cassini 1672	527 040	1 25·1
Titan	C. Huygens 1655	1 221 860	3 17·3
Hyperion	W. Bond 1848	1 481 100	3 59·4
Iapetus	G. D. Cassini 1671	3 561 300	9 35
Phœbe	W. H. Pickering 1898	12 954 000	34 51

	Sidereal period days	Mean Synodic period d h m s	Apparent diameter seen from Saturn ' "
Atlas	0·602		
Prometheus	0·613		
Pandora	0·629		
Janus	0·695		
Epimetheus	0·694		
Mimas	0·942	0 22 37 12.4	10 54
Enceladus	1·370	1 8 53 22	10 36
Tethys	1·888	1 21 18 55	17 36
Telesto	1·888		
Calypso	1·888		
Dione	2·737	2 17 42	12 24
Helene	2·737		
Rhea	4·518	4 12 18	10 42
Titan	15·945	15 23 16	17 10
Hyperion	21·277	21 7 39	0 43
Iapetus	79·331	79 22 05	1 48
Phœbe	550·4	523 13	0 3·2

	Orbital Inclination degrees	Orbital Eccentricity	Diameter km	Density water = 1	Escape velocity km/s
Atlas	0·3	0·002	38 × 30 × 26		
Prometheus	0·0	0·004	140 × 100 × 74		
Pandora	0·1	0·004	110 × 84 × 66		
Janus	0·1	0·007	220 × 180 × 160		
Epimetheus	0·3	0·009	140 × 114 × 100		
Mimas	1·52	0·020	392	1·4	0·1
Enceladus	0·02	0·004	500	1·2	0·2
Tethys	1·86	0·000	1060	1·1	0·4
Telesto	2	0·0	24 × 22 × 20		
Calypso	2	0·0	30 × 24 × 16		
Dione	0·02	0·002	1120	1·4	0·9
Helene	0·2	0·005	36 × 34 × 28		
Rhea	0·35	0·001	1530	1·3	0·6
Titan	0·33	0·029	5140	1·9	2·47
Hyperion	0·43	0·104	410 × 260 × 220	3?	0·2
Iapetus	7·52	0·028	1460	1·2	0·67
Phœbe	175	0·163	220	3?	0·1

	Reciprocal Mass Saturn = 1	Albedo	Magnitude at mean opposition distance
Atlas		0·4?	18·1
Prometheus		0·6?	16·5
Pandora		0·6?	16·3
Janus		0·5?	14·5
Epimetheus		0·5?	15·5
Mimas	15 000 000	0·7	12·9
Enceladus	7 000 000	1·0	11·8
Tethys	910 000	0·8	10·3
Telesto		0·8?	19·0
Calypso		0·6?	18·5
Dione	490 000	0·6	10·4
Helene		0·5?	18·5
Rhea	250 000	0·6	9·7
Titan	4150	0·2	8·4
Hyperion	5 000 000	0·2	14·2
Iapetus	300 000	0·5 to 0·05	10–12
Phœbe		0·1	16·5

Atlas is the 'A-ring shepherd'; Prometheus and Pandora are the 'F-ring shepherds'. Reports of a small satellite co-orbital with Mimas, and a second Dione co-orbital, are unconfirmed. In 1904 W. H. Pickering reported a satellite between the orbits of Titan and Hyperion, and it was even given a name – Themis – but seems not to exist.

Little is known of the physical nature of the first three satellites, except that they are irregular in shape. However, Epimetheus and Janus are of interest because they move in almost identical orbits, and there seems little doubt that they represent two parts of a former larger satellite which broke up. Their behaviour reminds one of a game of cosmic musical chairs. At present (1988) Epimetheus is slightly closer to Saturn, and is moving faster than Janus, so that it will eventually catch its companion up. As they approach each other there will be mutual interactions, so that Epimetheus will be slowed down and Janus speeded up. The end result is that the satellites will exchange orbits – something which happens every four years or so. Obviously they cannot collide, as otherwise they would not continue to exist as separate bodies, but they may come within a few kilometres of each other.

A few craters have been found and named. There are two named craters on Epimetheus (Hilairea and Pollux) and four on Janus (Castor, Idas, Lynceus and Phoibe).

MIMAS

Mimas, first of the satellites known before the Space Age, is only slightly denser than water, and may be made up mainly of ice all the way through to its centre, though there must be some rock as well. The surface is dominated by a huge crater, named Herschel, which has a diameter of 130 km – one-third that of the satellite itself – with walls which rise 5 km above the floor, the lowest part of which is 10 km deep. Herschel has a massive central mountain, with a base measuring 30×20 km; it rises to 6 km above the floor. There are many other craters (such as Modred and Bors) but few as much as 50 km across. Grooves (chasma) are also much in evidence, and some are more or less parallel, indicating that the crust has been subjected to considerable strain. If Herschel were formed by a gigantic impact there would have been a serious risk of breaking up the entire satellite, and its origin may be internal rather than external, though opinions differ. Incidentally, it was originally to be named 'Arthur', and the other main features of Mimas have been named after characters in the Arthurian legend.

A very small satellite has been suspected moving in virtually the same orbit as Mimas, but nothing definite is known about it.

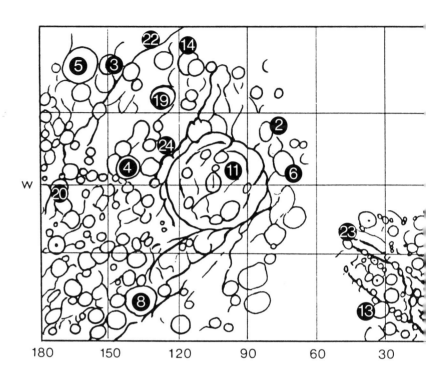

FEATURES ON MIMAS

Bold numbers indicate map references

Craters	Lat	Long W
Accolon	68 S	166
Arthur **1**	35 S	190
Balin **2**	22 N	082
Ban **3**	47 N	149
Bedivere **4**	10 N	145
Bors **5**	45 N	165
Dynas **6**	08 N	075
Elaine	44 N	102
Gaheris **7**	46 S	187
Galahad **8**	47 S	135
Gareth **9**	44 N	280
Gawain	60 S	254
Gwynevere **10**	12 S	312
W. Herschel **11**	00	104
Igraine **12**	40 S	225
Iseult **13**	48 S	035
Kay **14**	46 N	116
Lamerok	65 S	283
Launcelot **15**	10 S	317
Lot	30 S	227
Mark	28 S	297
Merlin **16**	38 S	215
Modred **17**	05 N	213
Morgan **18**	25 N	240
Palomides	04 N	157
Pellinore **19**	35 N	128

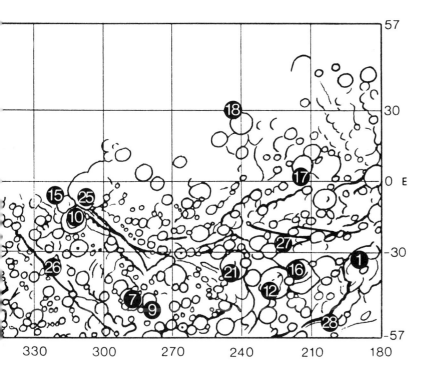

	Lat	Long W
Percivale **20**	01 S	171
Tristram	58 S	026
Uther **21**	35 S	244
Chasma	*Lat*	*Long W*
Avalon **22**	20–57 N	160–120
Camelot **23**	15–60 S	000–045
Oeta **24**	10–35 N	130–105
Ossa **25**	10–30 S	305–280
Pangea **26**	25–55 S	340–290
Pelion **27**	20–25 S	235–200
Tintagel **28**	43–60 S	235–190

ENCELADUS

Enceladus, less than twice the size of Mimas, is completely different. Instead of one huge crater and many others of fair size, we find several completely different types of terrain. Craters exist in many areas, but give the impression of being young and fairly sharp, while there is an extensive plain which is almost crater-free and is instead dominated by long grooves. Like Mimas, Enceladus is not much denser than water, and its albedo is virtually 100 per cent, making it the most reflective body in the Solar System.

It is suggested that despite its small size, Enceladus may be an active world. We have to explain the paucity of craters over wide areas, and it may be that the interior is 'flexed' by the tidal forces of Saturn and of the outer, much more massive satellite Dione, whose revolution period is twice that of Enceladus. If

FEATURES ON ENCELADUS
Bold numbers indicate map references

Craters	*Lat*	*Long W*
Ahmad	58 N	304
Aladdin	63 N	017
Ali Baba **1**	55 N	011
Dalilah **2**	53 N	244
Duban	58 N	176
Dunyazad **3**	34 N	200
Gharib	81 N	245
Julnar **4**	54 N	340
Musa	73 N	003
Peri-Banu	63 N	315
Salih	06 S	000
Samad	61 N	353
Shahrazad **5**	49 N	200
Shahryar **6**	58 N	222
Sindbad	66 N	210
Fossæ	*Lat*	*Long W*
Bassorah **7**	40–50 N	023–345
Daryabar **8**	05–10 N	020–335
Isbanir **9**	10 S–20 N	000–350
Planitia	*Lat*	*Long W*
Diyar **10**	00	250
Sarandib **11**	05 N	300
Sulci	*Lat*	*Long W*
Harran **12**	5 S–35 N	270–210
Samarkand **13**	10 S–75 N	300–340

so, then there may be periods when soft ice or even water wells up over the surface, obliterating features there. The surface formations have been given names from the Arabian Nights legend.

Another interesting fact is that Enceladus is exceptionally cold even by Saturnian standards. Equipment on Voyager 2 gave an average daytime temperature of −200 °C.

TETHYS

Most of our detailed knowledge of Tethys comes from Voyager 2 (Voyager 1 did not make a close approach), and it is unfortunate that some eagerly-awaited images of the satellite were lost when Voyager 2 developed a fault after it had started on its journey from Saturn to Uranus. Tethys seems to be made up of almost pure ice. There is one huge crater, Odysseus, with a diameter of 400 km – larger than Mimas! It is not very deep, and is presumably old, but it is certainly puzzling, since it may well be that an impact violent enough to produce such a formation would have shattered Tethys completely; we must admit that for such a situation all impact theories are suspect. On the other hand there can be nothing in the way of vulcanism on a globe like Tethys. There are many other craters, such as Penelope and Telemachus (the names given are from the Trojan War).

However, the main feature is Ithaca Chasma, a tremendous trench 2000 km in length, running from near the north pole across the equator and along to the south polar region. Nothing similar is known in the Solar System. It was presumably caused when the water inside Tethys froze, expanding as it did so. It extends three-quarters of the way round Tethys; its average width is 100 km, and it is 4 to 5 km deep, with a rim rising to half a kilometre above the outer terrain.

The orbit of Tethys is shared by two small satellites, Calypso and Telesto. They are irregular in shape, with longest diameters of less than 40 km. We have as yet no positive information about their surface features.

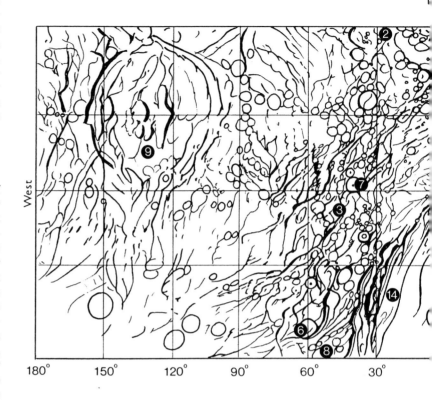

FEATURES ON TETHYS
Bold numbers indicate map references

Craters	Lat	Long W
Ajax **1**	30 S	285
Anticleia **2**	55 N	038
Antinoüs	62 S	275
Arete	04 S	300
Circe **3**	08 S	049
Elpenor **4**	54 N	268
Eumæus	27 N	047
Eurycleia **5**	56 N	247
Lærtes **6**	50 S	060
Melanthius	62 S	204
Mentor **7**	03 N	039
Nausicaa	80 N	352
Nestor **8**	57 S	058
Odysseus **9**	30 N	130
Penelope **10**	10 S	252
Phemius **11**	12 N	290
Polyphemus **12**	05 S	285
Teiresias	62 N	005
Telemachus **13**	50 N	338

Chasma	Lat	Long W
Ithaca **14**	60 S–35 N	030–340

DIONE

Dione is only a little larger than Tethys, but it is much more massive, and may be the densest of all Saturn's satellites apart from Titan. There are suspicions that it may have an effect upon Saturn's radio emission, since there seems to be

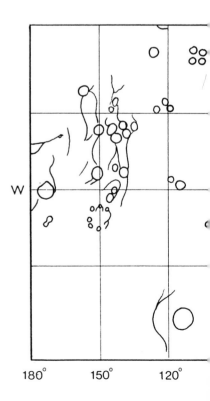

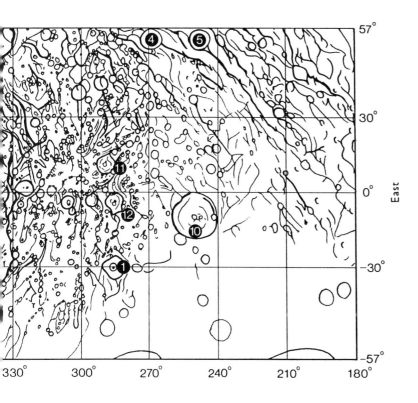

FEATURES ON DIONE
Bold numbers indicate map references

Craters	Lat	Long W
Adrastus	64 S	040
Æneas **1**	26 N	047
Amata **2**	07 N	287
Anchises **3**	35 S	063
Antenor **4**	06 S	008
Butes	68 N	050
Caieta **5**	25 S	080
Cassandra **6**	42 S	245
Catillus **7**	01 S	275
Coras **8**	03 N	268
Creusa	48 N	078
Dido **9**	22 S	015
Halys	60 S	045
Ilia **10**	03 N	344
Italus **11**	20 S	076
Latagus	16 N	026
Lausus **12**	38 N	023
Magus	20 N	024
Massicus **13**	36 S	052
Palinurus	05 S	062
Remus **14**	10 S	030
Ripheus	56 S	029
Romulus **16**	08 S	024
Sabinus **17**	44 S	190
Turnus **18**	21 N	342

Chasma	Lat	Long W
Larissa **19**	20–48 N	015–065
Latium **20**	03–45 N	064–075
Palatine	055–73 S	075–320
Tibur **21**	48–80 N	060–080

Linea	Lat	Long W
Carthage **22**	20–10 N	337–310
Padua **23**	05 N–40 S	245–190
Palatine **24**	10–55 S	285–320

DIONE
North

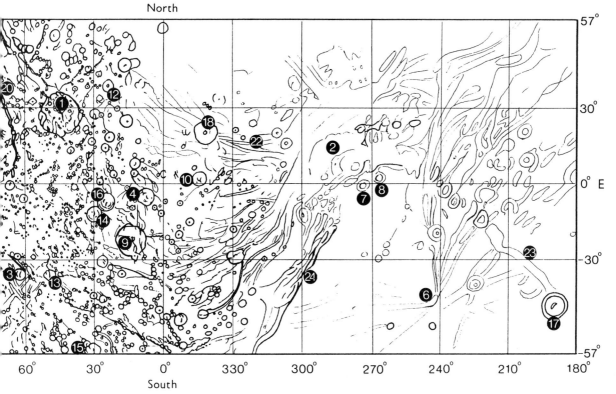

a cycle of 2·7 days – which is also the revolution period of Dione. Moreover, as has been noted, Dione may play a rôle in the flexing of the interior of Enceladus, whose revolution period is almost exactly half that of Dione.

The surface is not uniform. The trailing hemisphere is relatively dark, with an albedo of 0·3, while the brightest features on the leading hemisphere have albedo 0·6. The most prominent feature is Amata, 240 km in diameter, which may be either a crater or a basin; its precise nature is uncertain, but it is associated with a system of bright wispy features which extend over the trailing hemisphere, and are accompanied by narrow linear troughs and ridges. It is believed that these wispy features have been produced by bright, new ice which has seeped out from the interior, so that Dione shows considerably more past activity than Tethys. There are some large, well-marked craters with central peaks, of which Æneas is a good example (the names of the features are, in the main, associated with Roman history). Valleys exist, though some of them, notably Latium Chasma, appear to be essentially craterlet-chains.

Dione has one small co-orbital satellite, Helene, and a second has been suspected.

RHEA

With Rhea, considerably larger than Tethys or Dione, we come to yet another heavily-cratered world, with features named after characters in Far Eastern legends. The most prominent crater is Izanagi, but there are few large craters anywhere on the surface; moreover, the craters tend to be irregular. As with Dione, the trailing hemisphere is darkish, with wispy features reminiscent of those on Dione although much less prominent; the leading hemisphere is brighter, with one suspected ray-centre. It has been found that there are two separate types of cratered area, and it has been suggested that the first cratering period – whether internal or impact-produced – gave rise to some large craters; there was then a period of resurfacing, presumably by material welling up to the surface, followed by a second cratering era in which the structures were smaller. Rhea seems to be made up of ice and rock in about equal amounts. It has certainly been inactive for a very long time indeed, and parts of the surface look not unlike the heavily-cratered surface of Callisto in Jupiter's system.

An interesting observation was made on 8 April 1921 by six English observers independently: A. E. Levin, P. W. Hepburn, L. J. Comrie, E. A. L. Attkins, F. Burnerd and C. J. Spencer. Rhea was eclipsed by the shadow of Titan, and according to the first three observers Rhea vanished completely for over half an hour.

FEATURES ON RHEA
Bold numbers indicate map references

Craters	Lat	Long W
Aananin	39 N	330
Adjua	46 N	126
Aguana	70 N	065
Ameta	59 N	014
Arunaka	14 S	021
Atum	45 S	000
Bulagat **1**	35 S	014
Bumba	70 N	040
Burkhan	69 N	288
Con	24 S	010
Djuli **2**	26 S	046
Ellyay	78 N	095
Faro **3**	52 N	121
Haik **4**	34 S	026
Heller **5**	09 N	310
Haoso	09 N	008
Iraca	45 N	120
Izanagi **6**	49 S	298
Izanami **7**	46 S	310
Jumo	56 N	065
Karora	07 N	016
Khado	45 N	349
Kiho **8**	10 S	354
Kumpara	11 N	321
Leza **9**	19 S	304
Lowa	45 N	009
Malunga	74 N	049
Manoid	33 N	002
Melo **10**	51 S	006
Mubai	61 N	011
Num	23 N	093
Ormazd	62 N	052
Pan Ku	72 N	115
Pedn	48 N	340
Qat **11**	23 S	347
Sholmo	13 N	340
Taaroa	14 N	099
Thunapa **12**	51 N	015
Tika	25 N	087
Tore	00	335
Torom	68 S	345
Uku	85 N	115
Whanin	74 N	121
Wuraka	28 N	357
Xamba **13**	04 N	347
Xu	61 N	070
Yu-ti **14**	55 N	085

Chasma	Lat	Long W
Kun Lun **15**	37–50 N	275–300
Pu Chou **16**	10–35 N	085–115

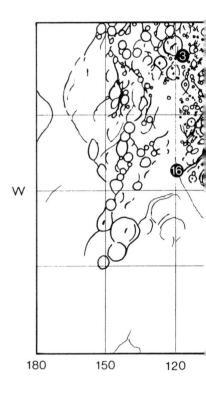

W

180   150   120

TITAN

The first suggestion of an atmosphere round Titan was made in 1903 by the Spanish astronomer J. Comas Solá, who noted that the satellite appeared brighter at its centre than near the limb. The existence of an atmosphere was proved spectroscopically in 1944 by G. P. Kuiper, but it was tacitly assumed that the main constituent would be methane, and it was thought that the atmospheric density must be low; after all, the escape velocity of Titan is a mere 2·5 km/s.

As Voyager 1 approached Saturn's system, astronomers were still divided as to whether any surface details would be seen on Titan. In the event, they were not. Titan is permanently shrouded by its orange clouds. The northern hemisphere was the darker of the two, and there were vague indications of 'banding', but that was all.

However, important measurements were made. Titan's atmosphere proves to be mainly nitrogen, with a ground pressure 1·6 times that of the Earth's air at sea-level. There is a little helium

RHEA
North

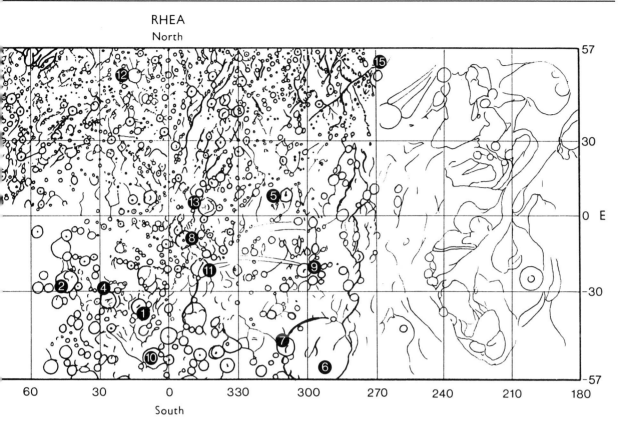

and some methane, and there are many organic compounds, giving the clouds their characteristic orange hue. Titan's globe may consist of about 55 per cent rock and the rest ice, and according to one picture there is a rocky core surrounded by a mantle of liquid water with some dissolved ammonia and methane – above which comes the icy crust. The surface temperature is given as −180 °C, which means that there is no Venus-type 'greenhouse effect', and there is no magnetic field.

As to surface conditions, we can only speculate. The temperature is near the triple point of methane, i.e. the temperature at which methane can exist as a liquid, solid or gas; there may be cliffs of solid ethane or methane, oceans or rivers of liquid methane, and a constant rain of organics from the orange clouds. Alternatively, C. Sagan and S. Dermott have suggested that the surface may be covered with a methane ocean at least 350 m deep. All the ingredients for life exist, though it seems that the very low temperature has probably prevented life from appearing there.

Unfortunately, we cannot hope to find out much more until a probe is sent to orbit Titan and map the surface by radar, as Pioneer has done for Venus. This will certainly be done in the future. Meanwhile, Titan remains very much of an enigma, totally unlike any other planet or satellite in the Solar System.

HYPERION

Hyperion moves between the orbits of Titan and Iapetus. (In 1904 W. H. Pickering reported another satellite in this region, and it was even given a name – Themis – but it has never been confirmed, and in all probability Pickering mistook a faint star for a satellite.) Hyperion is irregular in shape, with a longest diameter of 400 km and a shortest diameter of 240 km. This is strange, because an object as large as this ought to be regular in form; and to make matters still stranger, the longest axis does not point directly at Saturn, as dynamically it ought to do. It has been suggested that Hyperion is part of a larger satellite which broke up – but if so, then

where is the rest of it? The rotation period is 'chaotic', and it has been said that Hyperion is 'tumbling through space'. The 1983 rotation period was about 13 days.

Hyperion is less reflective than the inner icy satellites, but it too is probably composed largely of ice; there may be a 'dirty' layer covering wide areas. Several craters are seen, with diameters of up to 120 km, and there is one long ridge or scarp, extending for 300 km, which has been named Bond-Lassell in honour of the discoverers of Hyperion in 1848. Bond actually found the satellite first, but Lassell's confirmation shortly afterwards was independent.

FEATURES ON HYPERION

Craters	Lat N	Long W
Bahloo	036	196
Helios	071	132
Jarilo	061	183
Meri	003	151

Dorsum	Lat	Long W
Bond-Lassell	012–084	171

IAPETUS
North

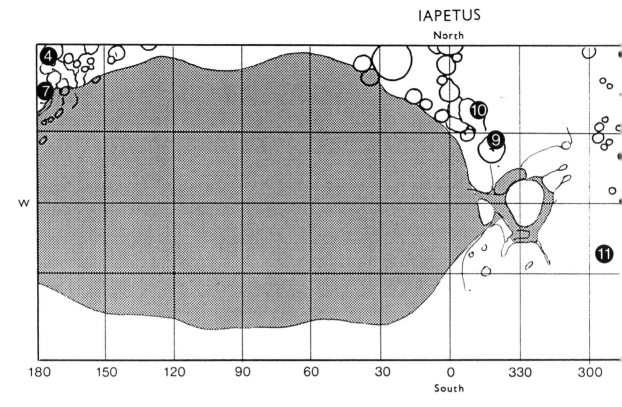

180 150 120 90 60 30 0 330 300

South

IAPETUS

Iapetus was the second of Saturn's satellites to be discovered, by G. D. Cassini in 1671. Cassini soon found that the magnitude is very variable. When west of Saturn in the sky, Iapetus is bright enough to be seen with a small telescope; when east, it becomes much fainter. The magnitude range is from about 9 to below 12. Assuming that the rotation is synchronous, this would mean that one side of Iapetus must be bright and the other side dark.

The Voyager results confirmed this view, though neither probe made a close fly-by, and the surface of Iapetus is less well known than those of the other icy satellites. The leading hemisphere is as black as a blackboard, with an albedo of 0·04 to 0·05; the trailing hemisphere is conventionally bright, with an albedo of 0·5. The line of demarcation is not abrupt; there is a 200–300 km 'transition zone'.

In view of the low density of Iapetus, it seems certain that the satellite itself is bright and icy, so that the dark material represents a deposit of some kind of material whose nature is unknown. There have been suggestions that the

FEATURES ON IAPETUS
Bold numbers indicate map references

Craters	Lat	Long W
Almeric	53 N	274
Baligant **1**	15 N	225
Basan **2**	30 N	197
Berenger	59 N	220
Besgun	72 N	296
Charlemagne **3**	54 N	266
Geboin **4**	56 N	175
Godefroy	78 N	253
Grandoyne **5**	18 N	215
Hamon **6**	10 N	271
Lorant	64 N	165
Marsilion **7**	41 N	177
Milon	75 N	270
Ogier **8**	42 N	274
Oliver	61 N	203
Othon **9**	24 N	344
Roland	78 N	030
Turpin **10**	43 N	000

Regio	Lat	Long W
Cassini **11**	48 S–55 N	210–340

Terra	Lat	Long W
Roncevaux **12**	30 S–90 N	300–130

material has been wafted on to Iapetus from the outermost satellite, Phœbe, but this seems highly unlikely, partly because Phœbe is so small and distant and partly because some of the craters on the bright trailing hemisphere of Iapetus are dark-floored. Presumably, then, the dark material has welled out from the interior of Iapetus. We know

nothing about its depth. Carl Sagan has suggested that it may be thick, and made up of organics; others believe it to be no more than a millimetre or two deep. Obviously we know little about the dark areas, but the bright regions contain the usual craters, named from the Charlemagne period.

It is worth noting that future space-travellers will see Saturn well from Iapetus, because the rings will not be edgewise-on. The orbital inclination is nearly 15°, and Saturn will indeed be a glorious object in the Iapetan sky. All the inner satellites move practically in the plane of the rings.

PHŒBE

The outermost satellite, Phœbe, differs from all the rest. It moves in a retrograde orbit, and its rotation period is not synchronous; Phœbe takes over 550 days to complete one journey round Saturn, but only 9·4 hours to spin round. Apparently it is a darkish body, and may be of the same type as the carbonaceous asteroids. Indeed, there is every chance that Phœbe itself is a captured asteroid rather than a bona-fide

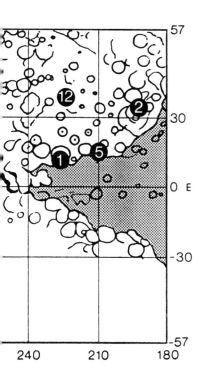

satellite. Investigations carried out by D. Cruikshank at the high-altitude Mauna Kea Observatory have shown that the dark material of Phœbe differs from that on the leading hemisphere of Iapetus, but its nature is unknown, and since neither Voyager made a close fly-by of Phœbe our knowledge of its surface is still very incomplete. There are suggestions that Phœbe may be of the same type as the exceptional Chiron. As we have noted, Chiron approached Saturn in the year 1664 BC to a distance of 16 million km – only about three million km beyond the orbit of Phœbe.

Saturn's satellite Dione, showing the large craters Dido and Æneas. Voyager picture. (NASA)

URANUS

The first planet to be discovered in modern (telescopic) times was Uranus, which comes seventh in order of distance from the Sun.

OPPOSITION DATES, 1977–2000

1977 Apr. 30
1978 May 5
1979 May 10
1980 May 14
1981 May 19
1982 May 24
1983 May 29
1984 June 1
1985 June 6
1986 June 11
1987 June 16
1988 June 20
1989 June 24
1990 June 29
1991 July 4
1992 July 7
1993 July 12
1994 July 17
1995 July 21
1996 July 25
1997 July 29
1998 Aug. 3
1999 Aug. 7
2000 Aug. 11

During this period Uranus passes through several constellations:

1977–1981, Uranus is in Libra
1982–1984, Uranus is in Scorpio
1985–1988, Uranus is in Ophiuchus
1989–1995, Uranus is in Sagittarius
1996–2000, Uranus is in Capricornus

In June 1989 Uranus will reach its greatest southerly declination ($-23°·7$). (Greatest northerly declination had been reached in March 1950.)

The most recent aphelion passage was that of 1 April 1925. Uranus was at its maximum distance from the Earth (21·09 a.u.) on 13 March of that year. Aphelion will next be reached on 27 February 2009.

The most recent perihelion passage was that of 20 May 1966. Uranus was at its closest to the Earth (17·29 a.u.) on 9 March of that year. Perihelion will next be reached on 13 August 2050.

The unique axial inclination of Uranus (98°) means that the rotation is technically retrograde, though not generally reckoned as such; all the satellites move virtually in the equatorial plane of Uranus, and move in the same sense as the rotation of the

---

DATA

Mean distance from the Sun: 2869·6 million km (19·181 a.u.)
Maximum distance from the Sun: 3004 million km (20·088 a.u.)
Minimum distance from the Sun: 2735 million km (18·275 a.u.)
Sidereal period: 84·01 years = 30 684·9 days
Synodic period: 369·7 days
Rotation period: 17·24 hours
Mean orbital velocity: 6·80 km/s
Orbital eccentricity: 0·047
Diameter: 51 200 km
Apparent diameter from Earth: max 3″·7, min 3″·1
Reciprocal mass, Sun = 1: 22 800
Density, water = 1: 1·27
Mass, Earth = 1: 14·6
Volume, Earth = 1: 67
Escape velocity: 22·5 km/s
Surface gravity, Earth = 1: 1·17
Mean surface temperature: −214 °C
Oblateness: 0·24
Albedo: 0·35
Maximum magnitude: +5·6
Mean diameter of Sun, seen from Uranus: 1′41″

---

planet. From Earth, the equator of Uranus is regularly presented (as in 1923 and 1966); at other times a pole is presented (the south pole in 1901 and 1985, the north pole in 1946 and 2030).

The first record of Uranus seems to have been an observation by the Astronomer Royal, the Rev. John Flamsteed, on 23 December 1690, when the planet was in Taurus; Flamsteed took it for a star (34 Tauri). Altogether, 22 pre-discovery observations have been listed, as follows:

Flamsteed 1690, 1712, four times in 1715.
J. Bradley, 1748 and 1750.
P. Le Monnier, twice in 1750, 1764, twice in 1768, six times in 1769, 1771.
T. Mayer, 1756.

It is interesting that Le Monnier failed to identify Uranus because of its movement. He observed it eight times in four weeks (27 December 1768–23 January 1769) without realizing that it was anything but a star. He has often been ridiculed for this, but it seems that when he made his observations Uranus was near its stationary point – it is hardly surprising that Le Monnier failed to identify it.

The discovery of Uranus was made on 13 March 1781 by William Herschel, using a 6·2 in reflector of 7 ft focal length and a magnification of 227. Herschel recognized that the object (in Gemini) was not a star, and he believed

it to be a comet. His communication to the Royal Society was indeed entitled 'An Account of a Comet'.

The first recognition of the new body as a planet seems to have been due, independently but at about the same time (May 1781) by the French amateur astronomer J. de Saron – who was guillotined in 1794 during the Revolution – and by the Finnish mathematician Anders Lexell. Lexell calculated an orbit, finding that the distance of the planet from the Sun was 19 a.u. – only slightly too small. He gave a period of between 82 and 83 years, and stated that the apparent diameter was between 3″ and 5″, making the planet larger than any other apart from Jupiter and Saturn.

The first proposal to name the planet 'Uranus' was made in 1781 by the German astronomer J. E. Bode. Other names were proposed – for instance 'Hypercronius' (J. Bernouilli, 1781) and 'the Georgian Planet' (by Herschel himself in 1782, in honour of his patron, King George III). Others called it simply 'Herschel'. Until 1850 the *Nautical Almanac* continued to call it the Georgian Planet, but in that year the famous mathematician John Couch Adams suggested changing over to 'Uranus'. This was done, and the name became universally accepted.

The first attempt to measure the apparent diameter of Uranus was made by Herschel in 1781. His value (4″·18) was rather too great. In 1788 he gave the diameter as 34 217 miles (55 067 km), with a mass 17·7 times that of the Earth; these values were also slightly too high. In 1792–4 Herschel made an attempt to measure the polar flattening, and from his results concluded that Uranus must have a rapid rotation.

The first observation of an occultation of Uranus by the Moon was made on 6 August 1824 by Captain (later Rear-Admiral) Sir John Ross, using a power of 500 on a reflector of 25 ft (7·26 m) focal length. On 4 October 1832 Thomas Henderson observed an occultation of Uranus by the Moon from the Cape of Good Hope.

The first estimate of the axial rotation period was made by J. Houzeau, in France in 1856. Houzeau

Dust in the rings of Uranus; Voyager 2, January 1986. When taking the picture, Voyager was in the shadow of Uranus, at a distance of 236 000 km; resolution about 33 km. Lanes of fine dust particles are shown. Exposure 96 seconds, so that there is a noticeable 'smear' as well as streaks due to trailed stars. (NASA)

were obtained in 1970, from a height of 24 km; the balloon was *Stratoscope II*. The resolution was 0″·15, ten times better than can be obtained by a telescope on the ground.

On Uranus, sunlight would be relatively strong, ranging between 1068 and 1334 times that of full moonlight on Earth. Saturn would be fairly bright when well placed (every $45\frac{1}{2}$ years); Neptune would just be visible with the naked eye when near opposition; Jupiter would have an apparent magnitude of 1·7, but would remain inconveniently close to the Sun in the Uranian sky.

MOVEMENTS OF URANUS, 1987–91

Date		R.A.			South Dec.		
		h	m	s	°	′	″
1987	Jan 5	17	33	07	23	24	46
	Apr 5	17	45	42	23	32	17
	Jul 4	17	33	58	23	27	47
	Oct 2	17	29	57	23	25	22
1988	Jan 10	17	52	06	23	35	57
	Apr 9	18	04	32	23	37	46
	Jul 8	17	52	46	23	38	50
	Oct 6	17	48	50	23	37	59
1989	Jan 4	18	08	26	23	38	54
	Apr 4	18	23	14	23	34	55
	Jul 3	18	13	10	23	40	46
	Oct 1	18	06	39	23	42	12
1990	Jan 9	18	27	13	23	34	00
	Apr 9	18	41	47	23	23	47
	Jul 8	18	31	42	23	34	11
	Oct 6	18	25	18	23	38	12
1991	Jan 4	18	43	14	23	23	24
	Apr 4	18	59	47	23	04	50
	Jul 3	18	51	49	23	17	19
	Oct 1	18	43	10	23	26	42

THE SATELLITES OF URANUS

The first (and so far, the only) space-probe to by-pass Uranus is Voyager 2, which made its closest approach (80 000 km) on 24 January 1986. Much information was obtained, including the discovery of ten new satellites.

It now seems that Uranus contains more heavy elements than Jupiter or Saturn, and is intermediate in type between the hydrogen- and helium-rich Jupiter and Saturn on the one hand, and the rocky, metal- and oxygen-rich inner planets on the other. According to the so-called 'three-layer' model, there is a rocky core surrounded by a liquid ocean which is in turn overlaid by the atmosphere; on the 'two-layer' model, which from the Voyager results is the more probable, there is a dense atmosphere in which gases are mixed with

stated that the rotation period must be between $7\frac{1}{4}$ hours and $12\frac{1}{2}$ hours.

The first reports of markings on the disk of Uranus were made by J. Buffham on 25 January 1870; the telescope used was a 9 in refractor, magnification 320. Two bright round spots were recorded, and on 19 March a light streak was described. It seems rather improbable that these markings were genuine features, though one cannot be sure.

The first observations of dark lines in the spectrum of Uranus were made by A. Secchi in 1869, from Italy. Further lines were detected by the English amateur W. Huggins in 1871, and in 1889 Huggins obtained the first photographs of these lines.

The first really good tables of the motion of Uranus were compiled in America by Simon Newcomb, in 1875.

The first spectroscopic confirmation of the retrograde rotation was obtained by the French astronomer H. Deslandres, in 1902. Final confirmation

was obtained by P. Lowell and V. M. Slipher, at the Flagstaff Observatory in Arizona, in 1911. For many years the rotation period was thought to be 10 h 49 m, but is now known to be 17·24 hours.

The first identification of methane in the atmosphere of Uranus was made by R. Mecke, at Heidelberg, in 1933 (it had been suggested, on theoretical grounds, by R. Wildt in 1932). Confirmation was obtained by V. M. Slipher and A. Adel, at Flagstaff, in 1934.

The first widely-accepted model of Uranus was that of R. Wildt, who proposed in 1934 that the planet must have a rocky core, overlaid by a thick layer of ice which is in turn overlaid by the atmosphere. In 1951 W. R. Ramsey, of Manchester University, proposed an alternative model, according to which Uranus was made up largely of methane, ammonia and water.

The first photographs of Uranus from a balloon-borne telescope

'ices', i.e. substances which would be frozen at the low temperatures at the surface. These ices are mainly water, ammonia and methane. It seems that the bulk of the atmosphere is made up of water, and that water, ammonia and methane condense in that order to form thick, icy cloud-layers. Methane freezes at the lowest temperature, and so forms the top layer, above which comes the predominately hydrogen atmosphere containing appreciable quantities of helium and neon.

Features at different latitudes have rotation periods from 14 to 17 hours, so that we are observing winds blowing in an east–west direction; cloud bands were also seen concentric with the south pole, which was at this time turned toward the Sun and the Earth. The equator was in twilight, and the north pole in darkness, but the temperature at the poles and the equator was found to be about the same.

Uranus seems to have little internal heat compared with the other giants. Only about 30 per cent of the heat radiated by the planet comes from the interior, as against 70 per cent for Jupiter and Saturn (and 0·01 per cent for the Earth).

The magnetic field proved to be surprising inasmuch as the magnetic axis is displaced by almost 60 degrees to the rotational axis. It has been suggested, though without proof, that Uranus may be experiencing a 'magnetic reversal', or that the dynamo region is closer to the surface than it is on the other planets. The Uranian magnetosphere extends to 590 000 km on the day side and 6 000 000 km on the night side. Ultra-violet observations showed strong emissions on the day side, producing what has been called the electroglow. It may be caused by electrons exciting the hydrogen molecules in the upper atmosphere, though it is not known how these electrons get their energy. Neither did Voyager 2 throw any light on the problem of why the rotational axis of Uranus is tilted at so great an angle.

The rings of Uranus were discovered on 10 March 1977. It had been predicted that the planet would occult the star SAO 158687, magnitude 8·9, and the occultation would give a good

Ariel; Voyager 2, 24 January 1986. Voyager was then 130 000 km from Ariel; resolution about 2·4 km. (NASA)

opportunity for measuring the diameter of Uranus. Calculations made by Gordon Taylor at the Royal Greenwich Observatory indicated that the occultation would be visible only from a restricted area in the southern hemisphere, and observations were made by J. Elliot, T. Dunham and D. Mink, flying at 12·5 km above the southern Indian Ocean in the Kuiper Airborne Observatory (KAO), which is in fact a modified C–141 aircraft carrying a 36 in reflecting telescope. Close watches were also being kept at ground-based observatories, including those in South Africa. Thirty-five minutes before occultation the KAO observers saw the star 'wink' five times, so that it was apparently being temporarily obscured by material in the vicinity of Uranus. The occultation by Uranus began at 20·52 h UT, and

lasted for 25 minutes. After emersion there were more winks, and it was later found that these were symmetrical with the first set, indicating a system of rings. The post-emersion winks were also recorded from Cape Town by J. Churms.

Subsequent occultations confirmed the existence of the ring system. The first Earth-based picture of them was obtained in 1978 by G. Neugebauer and his colleagues at Palomar, using the Hale reflector, and in 1984 D. A. Allen, at Siding Spring in Australia, obtained clear pictures of them in infra-red.

The rings were surveyed in detail by Voyager 2; details are given in the accompanying table. The outermost ring, the E-ring, is not symmetrical, and is narrowest when closest to Uranus; the newly-discovered satellites Cordelia and Ophelia are 'shepherds' to it. There is little dust in the main rings, but Voyager 2 took a final picture, when on its outward journey from Uranus when the planet hid the Sun, showing 200 very diffuse, nearly transparent bands of microscopic dust surrounding the main rings. The rings seem to be made up of particles a few metres in diameter, with not many centimetre- and millimetre-sized objects; they are very dark – as dark as coal. It has been suggested, though without proof, that they may be relatively young, and perhaps not even permanent features of the Uranian system.

RING-SYSTEM OF URANUS

Ring	Distance from Uranus km	Eccentricity	Inclination	Width km
1986 U2R	37 000–39 500	0	0	2500
6	41 850	1	63	1–3
5	42 240	1·9	52	2–3
4	42 580	1·1	32	2
α	44 730	0·8	14	8–11
β	45 670	0·4	5	7–11
η	47 180	0	2	2
γ	47 630	0	11	1–4
δ	48 310	0	4	3–9
1986 U1R	50 040	0	0	1–2
ε	51 160	7·9	1	22–93

	Reciprocal Mass, Uranus = 1	Albedo	Magnitude at mean opposition distance
Miranda	1 000 000		16·5
Ariel	67 000	0·40	14·4
Umbriel	67 000	0·19	15·3
Titania	20 000	0·28	14·0
Oberon	30 000	0·24	14·2

As the diameter of the Sun as seen from Uranus is below 2′, all the satellites known before the Voyager 2 pass could produce total eclipses. To an observer on Uranus, Ariel would appear much the largest of the satellites. As the satellites move round Uranus in the same sense as the planet rotates, their movements are technically retrograde, though not generally regarded as such.

The first satellites to be discovered were Oberon and Titania, by William Herschel in 1787. Both were seen on 11 January, though Herschel delayed making any announcement until he was certain of their nature. In 1797 Herschel announced the discovery of four more satellites, but three of these are certainly non-existent, and although the fourth may have been Umbriel there is considerable uncertainty. Ariel and Umbriel were discovered on 24 October 1851 by the English amateur W. Lassell, with his 24 in reflector (previous observations by Lassell in 1847 had been inconclusive). In 1894, 1897 and 1899 W. H. Pickering had searched unsuccessfully for new satellites.

The last satellite to be discovered visually was Miranda, on 16 February 1948, by G. P. Kuiper, with the 82 in reflector at the McDonald Observatory in Texas. The ten inner satellites were discovered during the Voyager 2 pass in 1985–6.

The names of the first four satellites were suggested by Sir John Herschel – two Shakespearean; one (Ariel) from both Shakespeare and Pope's *Rape of the Lock*, and Umbriel from *Rape of the Lock*. The name 'Miranda' was suggested by Kuiper, and the Shakespearean names of the ten inner satellites were adopted by the International Astronomical Union in 1987.

The nine innermost satellites are very small (less than 100 km in diameter). Cordelia and Ophelia act as

THE SATELLITES OF URANUS

	Discoverer	Mean distance from Uranus km	Angular distance from Uranus, at mean opposition distance ″
Cordelia	⎫	49 471	
Ophelia	⎪	53 796	
Bianca	⎪	59 173	
Cressida	⎪ From	61 777	
Desdemona	⎬ Voyager 2 1986	62 676	
Juliet	⎪	64 352	
Portia	⎪	66 085	
Rosalind	⎪	69 941	
Belinda	⎭	75 258	
Puck	Voyager 2 1985	86 000	
Miranda	Kuiper 1948	129 400	9·9
Ariel	Lassell 1851	191 000	14·5
Umbriel	Lassell 1851	266 300	20·3
Titania	W. Herschel 1787	435 000	33·2
Oberon	W. Herschel 1787	583 500	44·5

	Sidereal Period, days	Mean Synodic Period d h m s	Apparent Diameter as seen from Uranus ′ ″
Cordelia	0·330		
Ophelia	0·372		
Bianca	0·433		
Cressida	0·463		
Desdemona	0·475		
Juliet	0·493		
Portia	0·513		
Rosalind	0·558		
Belinda	0·622		
Puck	0·762		
Miranda	1·414	1 9 55 31	17 54
Ariel	2·520	2 12 29 39·0	30 54
Umbriel	4·144	4 3 28 25·8	14 12
Titania	8·706	8 17 00 1·2	15 00
Oberon	13·463	13 11 15 36·5	9 48

	Orbital Inclination, degrees	Orbital Eccentricity	Diameter km	Density water = 1	Escape Velocity km/s
Cordelia			24		
Ophelia			32		
Bianca			48		
Cressida			72		
Desdemona	Very Low	Very Low	48		Very Low
Juliet			72		
Portia			88	Low?	
Rosalind			48		
Belinda			48		
Puck			145		
Miranda	0·0	0·017	484	1·3	0·5?
Ariel	0·0	0·003	1160	1·6	1·2?
Umbriel	0·0	0·004	1190	1·4	1·2?
Titania	0·0	0·002	1610	1·6	1·6?
Oberon	0·0	0·001	1550	1·5	1·5?

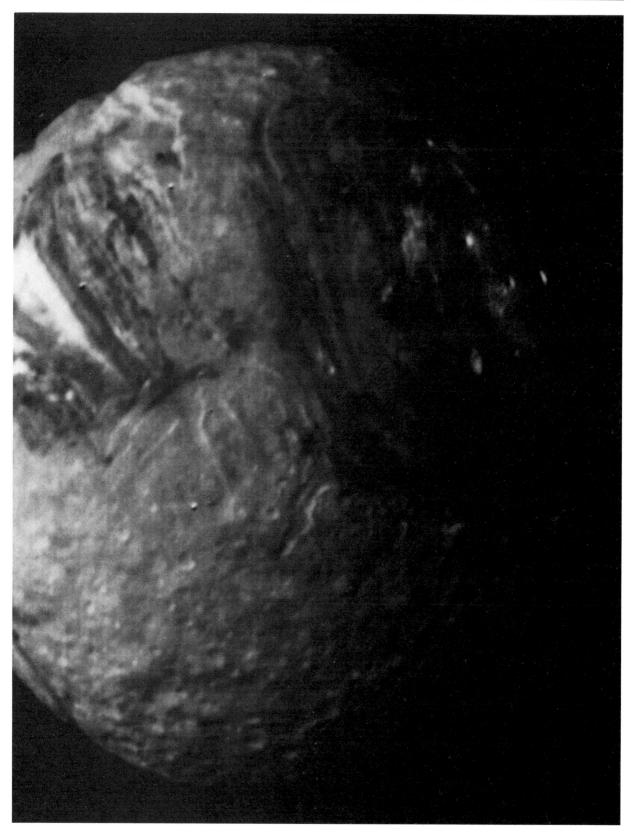

Miranda; Voyager 2, 24 January 1986, from 150 000 km; resolution 2·7 km. Note the different types of terrain! The main dark structure is Arden Corona. (NASA)

MIRANDA

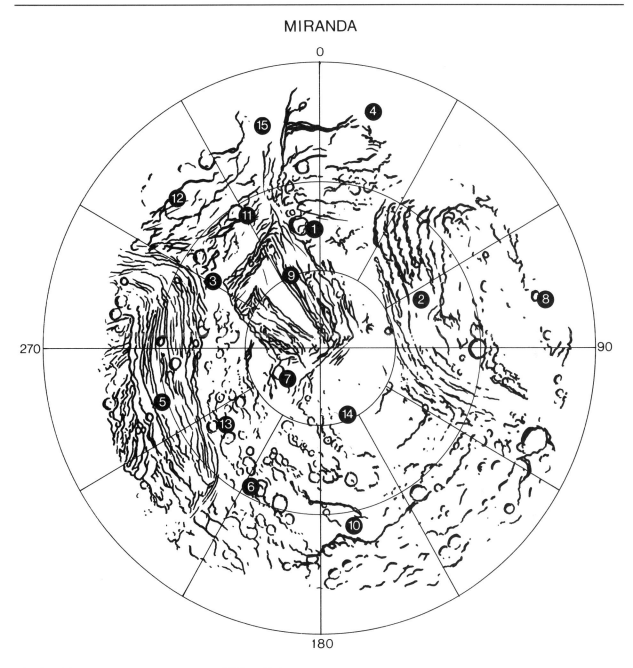

'shepherds' to the ring. Nothing definite is known about the physical characteristics of these satellites.

PUCK

Puck was the first of the inner satellites to be discovered from Voyager 2 (on 30 December 1985). On 24 January 1986 a single image was obtained of it, from a range of 500 000 km. The resolution is of the order of 10 km. Puck is roughly spherical, with a low albedo (7 per cent). Three craters were recorded: Bogle, Lob and Butz.

MIRANDA

As with the other satellites, only one half of Miranda was available for study from Voyager 2; the northern hemisphere was in darkness. Miranda was imaged from close range, and has proved to have an incredibly varied landscape. There are several distinct types of terrain: old, cratered plains, brighter areas with cliffs and scarps, and 'ovoids', large trapezoidal-shaped regions 200 to 300 km across (Arden Corona, Inverness Corona and Elsinore Corona). The

FEATURES ON MIRANDA

	Lat.S	Long.E
Alonso	47	345
Arden Corona	10–60	30–120
Argier Rupes	40–50	310–340
Dunsinane Regio	20–75	345–65
Elsinore Corona	10–42	215–305
Ferdinand	36	208
Francisco	70	246
Gonzalo	13	75
Inverness Corona	38–90	0–360
Mantua Regio	10–90	75–300
Prospero	35	323
Silicia Regio	10–50	295–340
Stephano	36	239
Trinculo	67	168
Verona Rupes	10–40	340–350

ARIEL

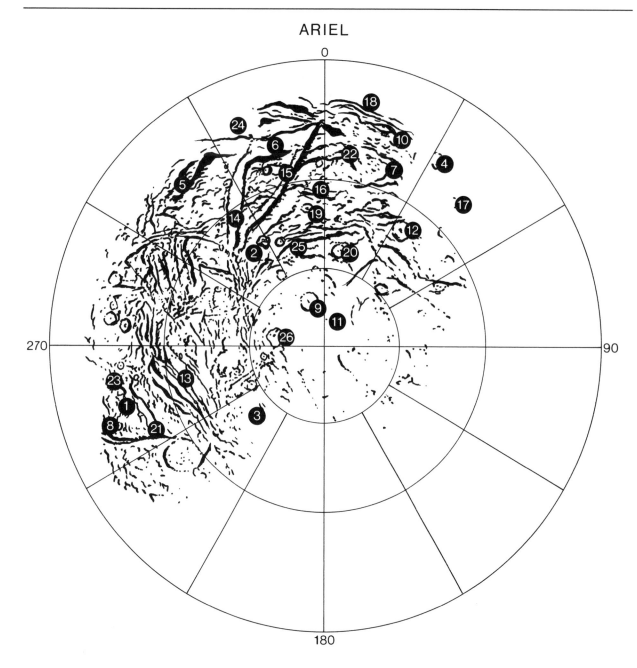

cliffs, presumably of ice, may be as much as 20 km high. There are few really large craters.

ARIEL

Ariel was imaged from 130 000 km, giving a resolution down to 2·4 km. There are plenty of craters, in terrain transected by fault scarps and graben, but the dominant features are broad, branching, smooth-floored valleys such as Korrican Chasma and Kewpie Chasma, and although Ariel is now totally inert it has clearly experienced tremendous tectonic activity earlier in its history. Its surface certainly appears to be younger than those of Umbriel, Titania or Oberon.

FEATURES ON ARIEL

	Lat.S	Long.E
Abans	16	251
Agape	47	336
Ataksak	53	225
Befanak	17	32
Berylune	23	328
Brownie Chasma	5–21	325–357
Deive	23	23
Djadek	12	251
Domovoy	72	339

	Lat.S	Long.E
Finvara	16	19
Gwyn	78	23
Huon	39	33
Kachina Chasma	24–40	210–280
Kewpie Chasma	15–42	307–335
Korrigan Chasma	25–46	328–353
Kra Chasma	32–36	355–2
Laica	22	44
Leprachaun Vallis	5–15	350–25
Mab	39	353
Melusine	53	9
Onagh	22	244
Pixie Chasma	18–25	350–20
Rima	18	260
Sprite Vallis	12–17	332–355
Sylph Chasma	45–50	328–15
Yangoor	68	260

UMBRIEL

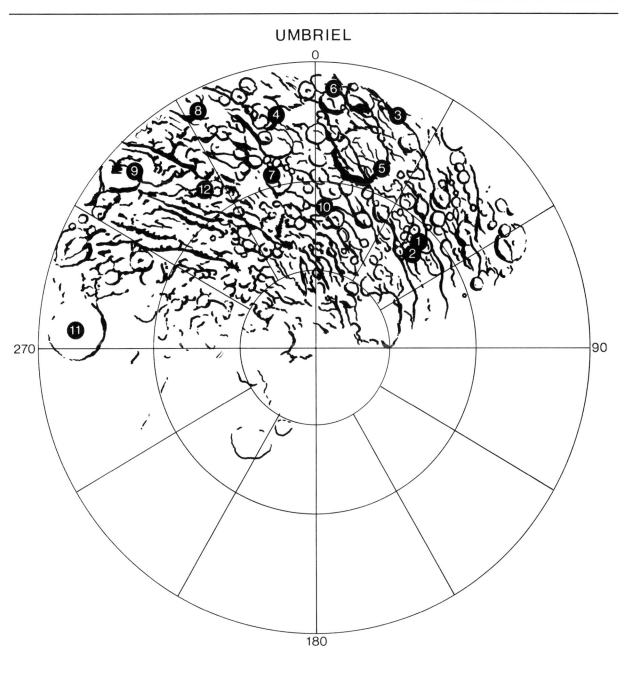

UMBRIEL

Umbriel has a much darker, more sub-dued surface than Ariel, and there are no ray-craters. The most prominent crater, Skynd, is 110 km in diameter, with a bright central peak. Wunda, diameter 140 km, lies near Umbriel's equator at around longitude 270°; its nature is not known, but it is the most reflective feature shown, and may be a crater.

FEATURES ON UMBRIEL

	Lat.S	Long.E
Alberich	31	43
Fin	36	41
Gob	9	26
Kanaloa	11	351
Malingee	22	13
Peri	9	6
Setibos	31	350
Skynd	1 (N)	335
Vuver	2	311
Wokolo	33	7
Wunda	6	274
Zlyden	24	330

TITANIA

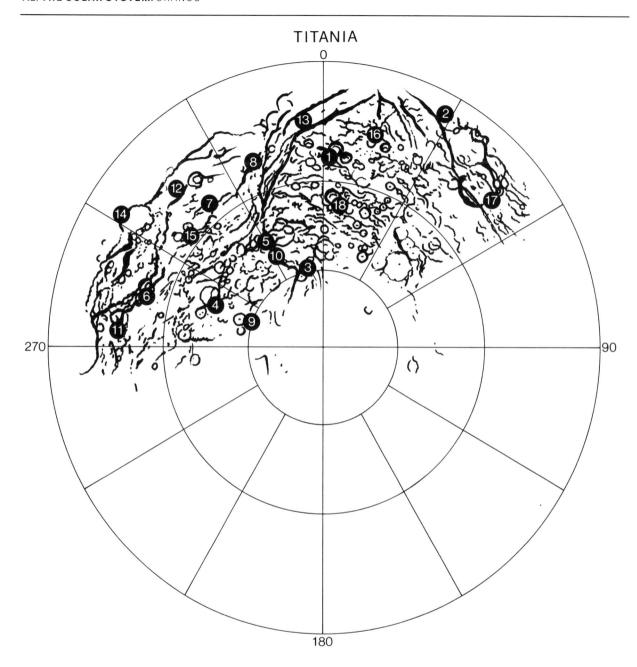

TITANIA

Like Ariel, though to a lesser extent, Titania has clearly been the site of past tectonic activity. There are many craters, together with ice-cliffs; there are prominent fault valleys, such as Messina Chasmata, up to 1500 km in length. The 200 km crater Ursula is cut by a younger fault valley over 100 km wide. The largest crater recorded, Gertrude, is well over 200 km in diameter.

FEATURES ON TITANIA

	Lat.S	Long.E
Adriana	20	4
Belmont Chasma	4–25	25–35
Bona	55	351
Calpurnia	42	292
Elinor	44	334
Gertrude	15	288
Imogen	25	318
Iras	19	339
Jessica	55	286
Katherine	51	333
Lucetta	9	277
Marina	15	316
Messina Chasma	8–28	325–5
Mopsa	11	302
Phrynia	24	309

	Lat.S	Long.E
Rousillon Rupes	7–25	17–38
Ursula	13	44
Valeria	34	4

OBERON

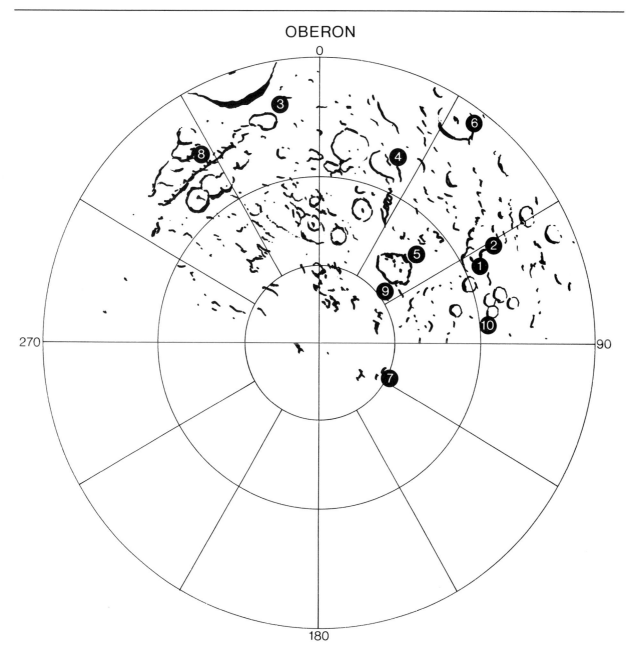

OBERON

Oberon shows less obvious signs of past tectonic activity than Titania, and differs also inasmuch as some of the major craters such as Hamlet, Othello and Falstaff have dark material inside them; this may be a mixture of ice and carbonaceous material erupted from the interior. Oberon was imaged at a minimum distance of 660 000 km, with a resolution down to 12 km. One interesting feature is the high mountain, 6 km in altitude, seen on the limb near Macbeth.

FEATURES ON OBERON

	Lat.S	Long.E
Anthony	28	65
Caesar	27	61
Coriolanus	11	345
Falstaff	22	19
Hamlet	46	45
Lear	5	31
Macbeth	59	112
Mommur Chasma	16–20	240–343
Othello	65	44
Romeo	28	88

NEPTUNE

The first planet to be discovered by mathematical calculation was Neptune, which comes eighth in order of distance from the Sun. Because of its great distance from the Sun and from the Earth, Neptune reaches opposition only about two days later every year.

The first recorded observation of Neptune seems to have been that of the French astronomer J. Lalande, who recorded it as a star on 8 and 10 May 1795.

The first suggestion of an unknown planet beyond Uranus was made by the Rev. T. J. Hussey, Rector of Hayes, in a letter written to the Astronomer Royal (Airy) on 17 November 1834. Uranus had been observed to depart from its predicted position, and in 1832 Airy had himself pointed out that the best available tables of Uranus, compiled by Alexis Bouvard, were in error by half a minute of arc. Various theories were later advanced to explain this (an unknown satellite of Uranus, braking due to 'cosmic fluid', and even a cometary collision!) but Hussey's suggestion was supported in 1835 by J. E. B. Valz, and subsequently by others.

The first plan to search for the trans-Uranian planet was announced by F. W. Bessel in 1840. Bessel hoped to collaborate with his pupil F. W. Flemming, but Flemming died soon after; Bessel himself became ill, and was never able to follow the problem up (he died in 1846).

The first estimated position of the new planet was calculated by John Couch Adams, of Cambridge, in 1845. Adams had made up his mind to tackle the problem while still graduating, in 1841. On 21 October 1845 Adams sought an interview with Airy, who had given him little encouragement, but was unsuccessful. No search was instigated. Meanwhile, U. J. J. Le Verrier, in France, had begun working on the problem (1845), and reached similar conclusions. Le Verrier's memoir reached Airy in December 1845, and finally, on 9 July 1846, Airy requested Challis, professor of astronomy at Cambridge, to begin searching for a new

DATA

Mean distance from the Sun: 4496·7 million km (30·058 a.u.)
Maximum distance from the Sun: 4537 million km (30·316 a.u.)
Minimum distance from the Sun: 4456 million km (29·800 a.u.)
Sidereal period: 164·8 years = 60 190·3 days
Synodic period: 367·5 days
Rotation period: 17 h 52 m
Mean orbital velocity: 5·43 km/s
Axial inclination: 28° 48′
Orbital inclination: 1° 45′ 19″·8
Orbital eccentricity: 0·009
Diameter: 49 500 km
Apparent diameter from Earth: max 2″·2, min 2″·0
Reciprocal mass, Sun = 1: 19 300
Density, water = 1: 1·77
Mass, Earth = 1: 17·2
Volume, Earth = 1: 57
Escape velocity: 23·9 km/s
Surface gravity, Earth = 1: 1·2
Mean surface temperature: −220 °C
Oblateness: 0·02
Albedo: 0·35
Maximum magnitude: +7·7
Mean diameter of Sun, seen from Neptune: 1′04″

planet on the basis of Adams' calculations. Challis used the 11·75 in 'Northumberland' refractor, but had no proper maps of the area, and his searches were laborious – and, it must be admitted, carried out with a strange lack of energy.

The first identification of Neptune was made by J. Galle and H. D'Arrest, at Berlin, on 23 September 1846, on the basis of Le Verrier's calculations. It was found that Le Verrier's predicted position was in error by only 55 minutes of arc. The telescope used by Galle and D'Arrest was a 9 in refractor, which was powerful enough to show the planet as a tiny disk. Subsequently, Challis found that he had seen the planet twice – the second occasion being on 12 August – but had failed to identify it. Following the announcement from Berlin the planet was widely observed; in England J. R. Hind, using a 7 in refractor, saw it on 30 September.

The first suggested name for the new planet was 'Janus', by Galle. Le Verrier suggested 'Neptune', but then changed his mind and proposed to have the planet named after himself; Challis proposed 'Oceanus', but before long the name 'Neptune' became universally accepted.

The first announcement of

Adams' independent work (almost as accurate as Le Verrier's) was made by Sir John Herschel on 3 October 1846. The announcement caused deep resentment in France, and led to acrimonious disputes in which, however, neither Adams nor Le Verrier took much part; they remained on cordial terms for the rest of their lives. Nowadays they are recognized as co-discoverers of Neptune.

(As an aside, I have always wondered why Adams did not make a search personally! He was confident of the rough position of the planet, and it was by no means faint. He had finished his calculations long before Le Verrier, and he would have been quite capable of identifying the planet with modest optical aid if he had tried to do so!)

A ring of Neptune was suspected by the English amateur W. Lassell on 14 October 1846. Later observations showed that the Neptunian ring, like Herschel's reported ring round Uranus, does not exist. Lassell did however discover Triton, the senior satellite, shortly after the identification of Neptune itself.

Neptune is a twin of Uranus – but it is a non-identical twin. In particular, it is appreciably denser and more massive. Unlike Uranus, it appears to have an internal heat-source, so that despite its greater distance from the Sun it has virtually the same effective temperature. Interior heat indicates convective activity of some kind, and the aqueous mantle, compressed sufficiently, may well conduct electricity well enough to produce a dynamo – in which case Neptune, alone of the planets, may have a magnetic field which is generated outside its core. We can only hope that Voyager 2 will still be working well enough in 1989 to clear the matter up.

Neptune's atmosphere also differs from that of Uranus inasmuch as it contains a variable haze, made up either of aerosols, ice crystals or some unknown substance. There are times when about half of the planet is covered by thin atmospheric haze, which dissipates and re-forms over short periods – a few days or a few weeks. In 1982 M. Belton, L. Wallace and S. Howard at Kitt Peak Observatory stated that there are zonal winds with velocity differences of over

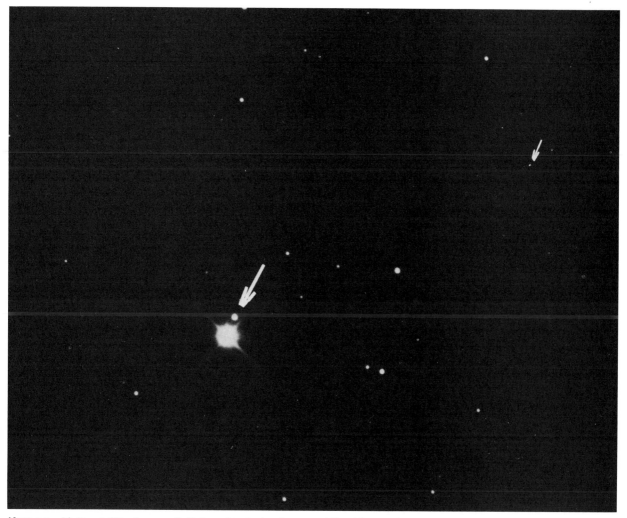

Neptune, with its satellites (arrowed); Triton, close to the planet, and Nereid, much further away. (Science Photo Library)

100 m/s between different latitudes. The rotation period is now believed to be 17 h 52 m.

The diameter of Neptune has been measured and re-measured. On 7 April 1968 the planet occulted a star, and this enabled G. Taylor, at Herstmonceux, to announce a revised diameter value of 50 940 equatorial and 49 920 km polar – slightly larger than the previously-accepted value, but still decidedly smaller than the diameter of Uranus.

THE SATELLITES OF NEPTUNE

	Triton	Nereid
Discoverer	Lassell, 1846	Kuiper, 1949
Mean distance from Neptune:	353 000 km	5 560 000 km
	0·0024 a.u.	0·0312 a.u.
Mean angular distance from Neptune, at mean opposition distance:	16″·9	4′ 23″·9
Apparent diameter, as seen from surface of Neptune	1° 01′	average 19″
Mean sidereal period, days	5·877	359·881
Mean synodic period	5 d 21 h 3 m 29 s·8	362 d 1 h
Orbital inclination	159°·9	27°·2
Orbital eccentricity	0·000	0·749
Diameter, km	6000?	500?
Reciprocal mass, Neptune=1	750	?
Density, water=1	5	?
Escape velocity, km/s	4·9	0·3
Magnitude	13·5	19

As the apparent diameter of the Sun as seen from Neptune is only about 1′, Triton can produce a total eclipse. Triton is the only large satellite in the Solar System to have retrograde motion; it was discovered by Lassell on 10 October 1846. There is probably a tenuous atmosphere, and it has even been suggested that the surface may have oceans of liquid nitrogen, but at present (1988) our knowledge of Triton is very slight. Even the diameter is uncertain, and may be almost anything between 3600 and 6000 km. At any rate, Triton is certainly larger than the Moon, and much larger than Pluto. H. L. Alden has given its mass as 0·022 that of the Earth, or 1·8 times as massive as the Moon.

Nereid, discovered by Kuiper on 1 May 1949, has the most eccentric orbit of any natural satellite in the Solar System. Its distance from Neptune ranges between 140 000 km and 9 500 000 km, and it has been said that the orbit is more like that of a comet than a satellite!

In 1982 a third satellite of Neptune was suspected, again by the occultation method. Observations were made by a team from the Lunar and Planetary Laboratory in Arizona, when Neptune made a close approach to a faint star. Two telescopes, 6 km apart, simultaneously recorded an 8-second 'blink' of the star which could have been due to a Neptunian satellite about 100 km in diameter, moving at a distance of around 76 000 km from the centre of the planet – closer-in than either Triton or Nereid. The fact that the 'blink' was observed by two telescopes some distance apart seems to eliminate instrumental or atmospheric effect, though one cannot altogether rule out the chance that it was due to an asteroid. As yet the third satellite remains unconfirmed, but it may well exist. Neither do we know whether there is a ring-system. None has yet (1988) been found, though there are suspicions of incomplete ring 'arcs'.

MOVEMENTS OF NEPTUNE 1984–90

Date		R.A.			South Dec.		
		h	m	s	°	′	″
1984	Jan 1	17	57	11	22	16	43
	May 10	18	04	36	22	13	35
	Sep 7	17	54	12	22	15	39
1985	Jan 5	18	07	04	22	19	11
	May 5	18	14	37	22	14	36
	Sep 2	18	03	47	22	18	43
1986	Jan 10	18	16	58	22	19	23
	May 10	18	23	55	22	13	29
	Sep 7	18	13	13	22	20	03
1987	Jan 5	18	25	15	22	18	18
	May 5	18	33	52	22	10	03
	Sep 2	18	22	53	22	18	57
1988	Jan 10	18	35	08	22	14	21
	May 9	18	43	10	22	04	46
	Sep 6	18	32	17	22	16	13
1989	Jan 4	18	43	23	22	09	48
	May 4	18	53	00	21	57	01
	Sep 1	18	42	01	22	10	43
1990	Jan 9	18	53	14	22	01	43
	May 9	19	02	18	21	47	38
	Aug 7	18	53	25	22	00	20

Voyager 2, which made successive rendezvous with Jupiter, Saturn and Uranus, and is due to by-pass Neptune in 1989.

PLUTO

The most recently discovered planet is Pluto, whose mean distance from the Sun is much greater than for any other planet. As yet our knowledge of it is very far from complete.

Pluto comes to opposition during April at the present epoch; the 1988 opposition occurred on 1 May, and the planet is in Virgo. It is, of course, a very slow mover! The next perihelion is due in 1989, and the next aphelion will occur in 2113.

The magnitude of Pluto is often quoted as 14·9. This, however, is the mean opposition magnitude, and Pluto is at present approaching perihelion. Estimates by T. J. C. A. Moseley and by the present writer make the magnitude exactly 14, so that Pluto is visible in a telescope of 25 cm aperture or more.

The first search for a trans-Neptunian planet seems to have been made in 1877 by David Peck Todd, of the US Naval Observatory. From perturbations of Uranus, he predicted a planet at a distance of 52 astronomical units from the Sun, with a diameter of 80 000 km. He conducted a visual search, using the 66-cm USNO reflector with powers of 400 and 600, hoping to detect an object showing a sensible disk. He continued the hunt for 30 clear, moonless nights between 3 November 1877 and 5 March 1878, but with negative results. A second investigation was conducted by the French astronomer Camille Flammarion in 1879. Flammarion based his suggestion upon the fact that several comets appeared to have their aphelia at approximately the same distance, well beyond the orbit of Neptune. Further comments were made in 1900 by G. Forbes, who assumed a planet larger than Jupiter, moving at a distance of about 100 a.u. in a period of 1000 years; in 1902 by T. Grigull of Münster, whose hypothetical planet was the size of Uranus, moving at 50 a.u. in a period of 360 years (Grigull even gave it a name – 'Hades'); by Gaillot in France – who believed that two planets existed – and by T. J. J. See in America, who increased this number to three.

The first systematic photo-

DATA

Mean distance from the Sun: 5·900 million km (39·44 a.u.)
Maximum distance from the Sun: 7375 million km (49·19 a.u.)
Minimum distance from the Sun: 4425 million km (29·58 a.u.)
Sidereal period: 247·7 years = 90 465 days
Synodic period: 366·7 days
Rotation period: 6 days 9 hours 17 minutes
Axial inclination: 118°
Mean orbital velocity: 4·7 km/s
Orbital inclination: 17°·2
Orbital eccentricity: 0·248
Diameter: 2290 km
Apparent diameter from Earth: below 0·25″
Reciprocal mass, Sun = 1: below 4 000 000
Mean surface temperature: about −230 °C
Albedo: about 0·4
Opposition magnitude at perihelion: 14
Mean diameter of Sun, as seen from Neptune: 49″

graphic and visual search for the planet was undertaken by Percival Lowell, at Flagstaff, from 1905 to 1907. The results were negative. Lowell's 'Planet X' was believed to have a period of 282 years and a mass 7 times that of the Earth, with a rather eccentric orbit (0·202), and the date of next perihelion was given as 1991. A second search at Flagstaff, carried out by C. O. Lampland in 1914, was equally fruitless, and was given up on Lowell's death in 1916. Lowell's final calculations about his planet were published in 1915.

The second prediction was made by W. H. Pickering in 1909; his planet

had a mass twice that of the Earth and a period of 373·5 years. His method – unlike Lowell's – was essentially graphical, but the conclusions were much the same. On the basis of his results, Milton Humason, at Mount Wilson Observatory, undertook a photographic search in 1919, but again the results were negative.

The discovery of Pluto was made at the Flagstaff Observatory by Clyde Tombaugh. Using a 13 in refractor, Tombaugh began work in 1929, again using photographic methods. Pluto was detected upon plates taken on 23 and 29 January, but the announcement was made on 13 March – 149 years after the discovery of Uranus and 78 years after Lowell's birth.

Earlier searches had failed because Pluto was fainter than had been predicted. However, Humason's failure in 1919 was sheer bad luck. When his plates were re-examined, it was found that Pluto had been recorded twice; but once the image fell upon a flaw in the plate, and on the second occasion Pluto was masked by an inconveniently-placed bright star. Moreover, the orbital inclination of 17 degrees was unexpectedly high. At perihelion, Pluto comes 56 000 000 km within the orbit of Neptune, but at the present epoch the distance between the two planets cannot be less than 384 000 000 km, so that there is no fear of collision.

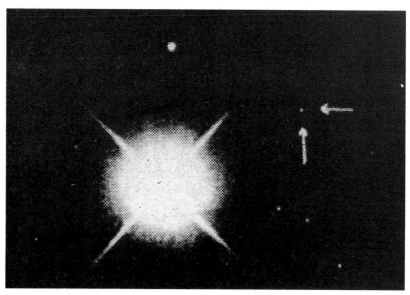

Pluto; Clyde Tombaugh's discovery plate, 1930. Pluto is arrowed. The bright over-exposed star is Delta Geminorum.

Early diameter estimates indicated that Pluto is small by planetary standards – so small, indeed, that it could produce no measurable perturbations on Uranus or Neptune; yet it was by these alleged perturbations that the planet had been tracked down. In 1936 A. C. D. Crommelin, of Greenwich, suggested the theory of specular reflection. If Pluto were highly reflective, the bright image of the Sun might falsify the diameter estimates, so that Pluto could be much larger – and hence more massive – than it seemed.

The first suggestion that Pluto might be an escaped satellite of Neptune was made by R. A. Lyttleton, of Cambridge, in 1936; it was supported by G. P. Kuiper in 1956. It is true that Pluto is much smaller than Triton, Neptune's large satellite, and Triton is also the only large satellite with a retrograde orbit, so that the idea of great disturbances in the remote past was logical enough. However, the theory has been weakened by the discovery of Pluto's satellite, Charon.

The first measurement of Pluto's diameter was made by Kuiper, using the 82 in reflector at the McDonald Observatory in Texas, in 1949. The diameter was given as 10 200 km, with a mass 8/10 of that of the Earth. However, in 1950 Kuiper and Humason made new measurements with the Palomar 200 in reflector, and the diameter of Pluto was reduced to 5800 km – smaller than Mars, and possibly smaller than Triton. **The first partial occultation of a star by Pluto** was observed in 1965. From this, it appeared that the diameter could not be more than 5800 km. Even this is now known to be a marked over-estimate.

The first measurement of the rotation period of Pluto was made in 1955 by M. Walker and R. Hardie. They found that the brightness of the planet varied by 20 per cent in 6·39 days, and this was presumably due to axial rotation. There are thus over 14 100 Plutonian 'days' in each Plutonian 'year'!

The first report of a satellite of Pluto was made on 22 June 1978 by James W. Christy at the US Naval Observatory at Flagstaff, Arizona. Pluto had been repeatedly photographed with the 1·55 metre reflector with a

view to predicting possible occultations of stars by the planet. Pictures taken in April and May showed that Pluto's image was elongated. Plates taken earlier (1965, 1970, 1971) were then checked, and the same effects were noted. Intensive studies were carried out at Flagstaff, and confirmation was obtained on 6 July 1978 by J. Graham, using the 401 cm reflector at the Cerro Tololo Observatory in Chile. Subsequently, the existence of the satellite as a separate body was confirmed by D. Bonneau and P. Foy, using the 3·6 metre Canada–France–Hawaii Telescope on Mauna Kea. Using the technique of speckle interferometry, they recorded the two bodies separately. The satellite has been named Charon.

As soon as Charon had been identified, it became possible to make a better evaluation of the size and mass of Pluto itself. It now seems that Pluto has a diameter of 2250 km, and Charon 1200 km, so that both are smaller than the Moon. The distance between the two is no more than 19 000 km, and the revolution period of Charon is the same as the rotation period of Pluto – 6·3867 days – so that the two are 'locked' in a way which is, so far as we know, unique in the Solar System.

The atmosphere of Pluto was discovered in 1980; methane ice had been detected spectroscopically in 1976. The methane atmosphere (density about one-millionth that of the Earth's air) may envelop both Pluto and Charon when Pluto is near perihelion, and more methane ice is evaporated from the surface.

Infra-red observations of Pluto were made in 1983 by equipment on the IRAS satellite. From these, M. Sykes, R. Cutri, L. Lebofsky and R. Binzel infer that methane frost does not cover the whole surface; there is a warmer band at the planet's equator, bright in infra-red but dark at visible wavelengths. It is believed that this band is largely methane frost free, and varies in size over a Plutonian year, with a maximum spread down to latitude 45 degrees. The thickest atmosphere that Pluto's icecaps could support is equivalent to a 9-metre column of the Earth's atmosphere.

Mutual eclipses of Pluto and Charon occur only for a few years every 248 years, when the Earth passes through the plane of the system; fortunately the mid and late 1980s provided one such opportunity. In 1987 it was found by M. Buie, D. P. Cruikshank, L. Lebofsky and E. F. Tedesco that the surface compositions of Pluto and Charon are not alike; this was possible because the spectrum of Pluto could be observed during a total eclipse of Charon, and once this had been studied it could be subtracted from the spectrum of Pluto and Charon combined. Surprisingly, Charon shows indications of water ice, with no evidence of methane or ammonia frost. If both bodies were of the same origin, Charon's original surface methane has presumably sublimated and escaped, while the lack of any sign of water frost on Pluto indicates that any water frost there has been covered with methane frost. Presumably the differences are due to the

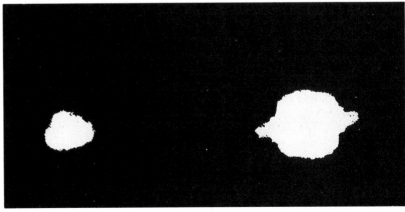

Pluto and Charon; CCD photograph from Mauna Kea, Hawaii. The 'bars' through the disks are optical effects, but Charon (left) is clearly separate from Pluto (right).

The 33-cm refractor (officially the Lawrence Lowell Telescope) at Flagstaff, used by Clyde Tombaugh in the Pluto search. It has now been transferred to a new site at nearby Anderson Mesa. (Patrick Moore)

lower albedo and lower mass of Charon (Pluto contributes about 80 per cent of the total light of the system).

The nature of Pluto is still a matter for debate. Its mass is so small (even combined with that of Charon) that it could not possibly perturb the motions of giant planets such as Uranus and Neptune to any measurable extent, so that it was not the planet for which Lowell was searching. Either the reasonably correct prediction, and Tombaugh's discovery, was sheer luck, or else an extra planet does exist, presumably at a considerably greater distance. This was suggested by Pickering after Pluto had been found, and also in 1950 by K. H. Schütte of Munich on the basis of cometary aphelia. In 1962 A. G. W. Cameron wrote that 'it is difficult to escape the conclusion that there must be a tremendous mass of small solid material on the outskirts of the Solar System'. In 1972 K. A. Brady (USA) investigated the motion of Halley's Comet, and suggested that there were perturbations due to a planet moving round the Sun in a retrograde orbit at about 8000 million km in a period of 512 years; its mass would be greater than that of Saturn. Brady gave a position for his 'planet' in Cassiopeia; unsuccessful searches were made, but it now seems that Brady's calculations were unsound.

It does seem that there are slight, unexplained irregularities in the movements of Uranus and Neptune which were more evident a century ago than they are now. In 1987 J. Anderson, of NASA, proposed that the planet responsible moves in a very eccentric and highly inclined orbit; the mean distance from the Sun was given as around 80 astronomical units (12 000 million km), with a period of from 700 to 1000 years and a mass five times that of the Earth. If this planet exists, and is at present within what may be called 'reasonable range', it could possibly be tracked down by observations of the movements of the probes now on their way out of the Solar System (Pioneers 10 and 11, and Voyagers 1 and 2), though the planet would have to be in just the right place at just the right time.

So far as Pluto and Charon are concerned, there may be grounds for regarding the system as a double asteroid. Unfortunately none of the current deep-space probes will go anywhere near the pair, so that in the foreseeable future our knowledge must remain very incomplete.

MOVEMENTS OF PLUTO, 1984–90

Date		R.A.			Dec.		
		h	m	s			
1984	Jan 1	14	21	21	03	36	32 N
	Jul 9	14	12	47	04	40	03
1985	Jan 15	14	31	29	02	40	40
	Jul 4	14	22	09	03	43	35
1986	Jan 10	14	40	13	01	40	42
	Jul 9	14	31	17	02	41	47
1987	Jan 5	14	48	51	00	41	17
	Jul 4	14	40	45	01	43	54
1988	Jan 10	14	58	19	00	17	19 S
	Jul 8	14	49	52	00	41	36 N
1989	Jan 4	15	07	48	01	16	24 S
	Jul 3	14	59	25	00	17	14
1990	Jan 9	15	16	24	02	15	06
	Jul 8	15	08	32	01	19	35

COMETS

Comets are members of the Solar System, but are made up of small particles together with extremely tenuous gas, so that they are very different from the planets and are of slight mass. Generally their orbits are highly eccentric.

For all practical purposes, comets may be divided into two classes: (1) Periodical comets, which are seen regularly and which move around the Sun in periods ranging from 3·3 years to over 150 years; and (2) Non-periodical comets, which cannot be predicted because their periods are too long. (There are also comets which may well travel in parabolic orbits, in which case they will never return to the neighbourhood of the Sun; but it is almost impossible to distinguish between a section of a parabola and of an extremely eccentric ellipse.) **The first telescopic observation of a comet** was made in 1618 by J. Cysat.

For many centuries comets were regarded as atmospheric phenomena – though Anaxagoras, around 500 BC, attributed them to clusters of faint stars. The first man to prove that they were extraterrestrial was Tycho Brahe, who found that the comet of 1577 showed no diurnal parallax, and must therefore be at least six times as remote as the Moon (actually, of course, it was much further away than that). There have been occasional panics due to comets – as in France in 1773, when a mathematical paper by J. J. de Lalande was misinterpreted, and people believed that a comet would collide with the Earth on 20 or 21 May. (Seats in Paradise were sold by members of the clergy at high prices.) Another alarm occurred in 1832, when it was suggested – wrongly – that there might be a near-encounter between the Earth and the now-lost Biela's Comet. In 1910, when Halley's Comet was in view, a manufacturer in America made a large sum of money by selling anti-comet pills, and in 1970 some Arabs mistook Bennett's Comet for an Israeli war weapon. Since 1950 there have appeared several books by an eccentric psychiatrist, Dr Immanuel Velikovsky,

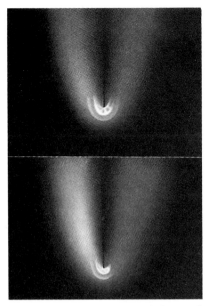

Drawing of the head of Donati's Comet of 1858.

who confused planets with comets, and believed Venus to have been a comet only a few thousands of years ago!

The brightest periodical comet is Halley's, with a period of 76 years. Most comets are named in honour of their discoverers, but this is not the case for Halley's Comet, which has been recorded regularly since well before the time of Christ. In 1682 it was observed by Edmond Halley (the actual discovery was made on 15 August of that year by G. Dorffel), and subsequently Halley decided that it must be identical with comets previously seen in 1607 and in 1531. He predicted a return for 1758. On Christmas Night 1758 the comet was duly found, by a German amateur, J. Palitzsch, and it passed perihelion in 1759. This was **the first predicted cometary return**, and it was appropriate that the comet concerned should be known as Halley's.

The first definitely recorded return of Halley's Comet is that of 240 BC, though it seems that a comet observed by the Chinese in 1059 BC may well have been Halley's. Since 240 BC the comet has been seen at every return:

240 BC	141	530	912	1301	1682
164 BC	218	607	989	1378	1759
87 BC	295	684	1066	1456	1835
12 BC	374	760	1145	1531	1910
AD 66	451	837	1222	1607	1986

The brightest return was that of 837, when the comet passed within 0·04 astronomical units of the Earth (11 April). The magnitude was about −3·5, and the tail had a length of 93 degrees. The return of 1986 was the worst for 2000 years, though the comet was visible with the naked eye. The next return, that of 2061, will also be unfavourable.

The only comet to have been referred to by a Roman emperor was Halley's. Comets were once nicknamed 'hairy stars' and were regarded as evil. (In AD 79 Vespasian commented that a comet seen in that year – not Halley's – 'menaced rather the King of the Parthians; for he is hairy, while I am bald' – however, Vespasian died in the same year.) Another Roman emperor, Macrinus, died at the time of Halley's return in 218.

The first surviving drawing of Halley's Comet refers to the return of 684. It comes from the *Nürnberg Chronicle*.

The only comet recorded in the Bayeux Tapestry is Halley's. The comet was seen in spring 1066, as the Normans were preparing to invade England, and in the Tapestry it appears clearly, with King Harold toppling on his throne and the Saxon courtiers showing signs of great alarm. (Some scholars consider that the Tapestry was woven by William I's wife.)

The only comet to have been officially referred to by a Pope is again Halley's. This was at the 1456 return, when Pope Calixtus III regarded it as an agent of the Devil. (Some scholars doubt this story, but it may well be true!)

The first observer to see Halley's Comet in 1835 was Dumouchel, from Rome; the last to see it in 1836 (May) was Sir John Herschel, from the Cape. The comet was recovered on 12 September 1909 by Max Wolf in Germany, and remained under observation until June 1911, when its distance from the Sun was more than 800 000 000 km. Aphelion was reached in 1948.

The last perihelion occurred on 9 February 1986. The comet was recovered on 16 October 1982 by a team of astronomers at Palomar (Jewitt, Schneider and Dressler) who used the Hale reflector to detect the comet as a tiny blur of magnitude 24·3 – making it one of the faintest objects ever recorded. It was a mere 8 seconds of arc away from its predicted position. The discovery was confirmed shortly afterwards at Kitt Peak. At the time of its recovery, the comet was still moving between the orbits of Saturn and Uranus.

The first attempt to observe a cometary transit across the face of the Sun was made on 18–19 May 1910, with Halley's Comet. The American astronomer F. Ellerman went to Hawaii to observe under the best possible conditions, but could see no trace of the comet.

The first comet known to hit the Sun was 1979 XI (Howard-Kooman-Michels), on 31 August of that year. Several other cases have since been established, and it may well be that cometary suicides are much more common than has been previously supposed.

The first short-period comet to be predicted was Encke's, again now named in honour of the mathematician who computed the orbit. The comet was first seen on 17 January 1786 by P. Méchain, from France. It was again recorded on 7 November 1795 by Caroline Herschel; the next returns to be seen were those of 1805 (discovered by Thulis at Marseilles, 19 October) and 1818 (Jean Pons, 26 November). Encke, at Berlin, decided that these comets must be identical, and predicted a return for 1822. He was correct, and since then the comet has been seen at every return except that of 1944 (when it was badly placed, and most astronomers were otherwise engaged).

The comet with the shortest known period is also Encke's: 3·3 years. (Comet Wilson-Harrington, discovered in 1949, had a calculated period of only 2·3 years, but has never been seen again.) The period of Encke's comet has shortened slightly since its discovery; the decrease amounted to over a day between 1822 and 1858. Evaporation processes taking place inside the comet are responsible – not, as Encke himself supposed, some resisting medium in the inner part of the Solar System.

The comet which has presented most observed returns is Encke's.

The 1987 return was the 54th to be seen. Encke's Comet is therefore far ahead of its nearest rivals: Halley (28 returns, last perihelion passage 1986) and Pons-Winnecke (18 returns, last perihelion passage 1976). Modern instruments are now capable of recording Encke's Comet when it is near aphelion.

The following data relate to comets with periods of between 3·3 and 20 years, and which have been observed at more than one return.

Greatest orbital eccentricity: Encke (0·85)
Smallest orbital eccentricity: Schwassmann-Wachmann 1 (0·11)
Greatest orbital inclination: Giacobini-Zinner (31°·7)
Smallest orbital inclination: Du Toit-Neujmin-Delporte (2°·4)
Greatest perihelion distance: Schwassmann-Wachmann 1 (5·45 a.u.)
Smallest perihelion distance: Encke (0·34 a.u.)
Greatest aphelion distance: Neujmin 1 (12·16 a.u.)
Smallest aphelion distance: Arend (4·00 a.u.)

A few comets, with orbits of exceptionally low eccentricity, are visible throughout their orbits. The first-discovered of these was Schwassmann-Wachmann 1, first seen in 1925. Its orbit lies entirely between those of Jupiter and Saturn. Normally it is a very faint object, but it can show sudden, unpredictable outbursts which bring it within the range of small telescopes. A recent occasion was in the autumn of 1976, when the magnitude rose to above 12. Other comets which may be kept in view throughout their orbits are Gunn and Smirnova-Chernykh. Oterma's Comet used to have a very low eccentricity and a period of 7·9 years, but a close approach to Jupiter in 1963 altered the period to 19·3 years.

The most numerous comet 'family' is Jupiter's. The mean distance of Jupiter from the Sun is 5·2 a.u., and a glance at the table on page 125 shows that many short-period comets have their aphelia at about this distance. 'Comet families' of Saturn and Uranus are very questionable. However, the Neptune 'family' is rather more definite. The mean distance of Neptune from the Sun is approximately 30 a.u., which corresponds to the aphelia of at least four comets, including Halley's. It must however be added that some authorities are highly sceptical about the

significance of these alleged comet families, and attempts to predict the existence of a still more remote planet by means of cometary aphelia are very speculative.

The comet with the longest confirmed period is Herschel-Rigollet (1788 and 1939; 156 years). This is the only comet which has been seen at more than one return and whose period is longer than that of Halley's Comet.

The most famous 'lost' periodical comet is Biela's. At the 1846 return it separated into two; the twins returned in 1852, but have never been seen since, though the comet's orbit is associated with a meteor shower. Westphal's Comet was seen in 1852 and again in 1913, but it evidently failed to survive the perihelion passage of 1913, and did not appear on schedule in 1976. Brorsen's Comet was seen at five returns between 1846 and 1879, but has not appeared since, and has evidently disintegrated. However, one must beware of jumping to conclusions. Comet Di Vico-Swift was 'lost' for 38 years after 1897, but was recovered in 1965; Holmes' 'Comet' 'went missing' for 58 years prior to its recovery in 1964. Both these latter successes were due to the mathematical work of the British astronomer B. Marsden.

The closest cometary approach to the Earth ever recorded was that of Lexell's Comet of 1770 (discovered by C. Messier on 14 June of that year; A. Lexell of St. Petersburg computed the orbit). The distance was reduced to 1·2 million km, and the comet was visible with the naked eye. It has never been seen again, since an approach to Jupiter in 1779 has altered the orbit, and the comet now comes nowhere near the Earth.

The closest known cometary approach to Jupiter was that of Comet Brooks 2, in March 1886 (the approach was not, of course, actually observed). Evidently the planet passed within the orbit of Io, the closest to Jupiter of the four large satellites. The comet suffered partial disruption, and at the 1889 return was seen to be attended by four minor companions which were classified as 'splinters', but which did not last for long. The comet has returned regularly since then, the last appearance being that of 1973.

The largest comet ever recorded seems to have been that of 1811; the diameter of the coma was approximately 2 000 000 km, larger than the Sun. The maximum length of the tail was over 160 000 000 km.

The longest tail ever recorded was that of the Great Comet of 1843: 330 000 000 km – considerably greater than the distance between the Sun and the orbit of Mars.

The brightest comet of modern times was probably that of 1843. According to the famous astronomer Sir Thomas Maclear, it was much more brilliant than the comet of 1811; and Maclear saw both.

The most beautiful comet of modern times is said to have been Donati's of 1858, with its wonderfully curved main tail and its two shorter ones. It was discovered by G. Donati, from Florence, on 2 June 1858, and was last seen on 4 March 1859. During October 1838 the length of the tail was as much as 80 000 000 km. The period is unknown; 2000 years has been suggested, but with no certainty.

The most-tailed comet on record was that of 1744, discovered on 9 December 1743 by Klinkenberg in Holland (and independently on 13 December by De Chéseaux in Switzerland; rather unfairly, it is usually called De Chéseaux' Comet). There were at least six bright, broad tails, but records are sparse, as the comet remained brilliant for only a few nights in March 1744.

The comet with the longest computed period is Delavan's Comet of 1914: 24 000 000 years. This is naturally very arbitrary, and all we can really say is that the period is very long indeed.

The only comet to be associated with port wine was that of 1811. In that year the vintage in Portugal was unusually good; for years afterwards 'Comet Wine' appeared in the price-lists of wine-merchants! It would be interesting to know whether any of it remains to be drunk.

The first time that the Earth was known to pass through the tail of a comet was in late June 1861. The comet concerned – discovered by J. Tebbutt from Australia – was brilliant, but the nucleus never came within 1·8 mil-

Brooks' Comet on 21 October 1911 (Lick Observatory)

lion km of the Earth. No effects of the comet's tail were noticed with certainty.

The first comet to be discovered during a total eclipse of the Sun was Tewfik's Comet of 1882; the eclipse was photographed from Egypt, and the comet was named after the then ruler of Egypt. It was not seen again, so that this is our only record of it. The bright comet of 1948 was also discovered during a total solar eclipse.

The first known attempt to photograph a comet was made by the British pioneer W. de la Rue, with the Great Comet of 1861. However, he was unsuccessful.

The first really good comet photograph was taken in 1882 at the Cape by Sir David Gill. The comet was very brilliant, and was well shown. Many stars were also shown, and it was this picture which made Gill appreciate the endless potentialities of stellar photography.

The first cometary spectrum was studied in 1864 by G. Donati. The comet concerned was reasonably bright, and Donati saw that the spectrum was not merely a reflected solar spectrum; there were lines which could be due only to materials in the comet itself.

The brightest comet so far seen during the 20th century is probably the so-called Daylight Comet of 1910, discovered by some diamond miners in the Transvaal on 12 January. It was certainly brighter than Halley's, which was on view later in the year; and people who claim to remember Halley's Comet are warned that the comet they saw was probably the Daylight! The orbit was virtually parabolic.

The greenest comet of modern times was Jurlov-Achmarov-Hassell (1939). The green colour was evident even in a small telescope.

The most famous 'spiked' comet was Arend-Roland, which was easily visible with the naked eye in April 1957.

There was an apparent tail directed sunward – though the appearance was actually due to thinly-spread matter in the comet's orbit, catching the sunlight at a suitable angle. (The Skylab astronauts also reported a 'spike' with Kohoutek's Comet of 1973.)

The first discovery of a hydrogen cloud associated with a comet was made in 1969, when Comet Tago-Sato-Kosaka was studied from the space-vehicle OAO2 (Orbiting Astronomical Observatory 2) and was found to be surrounded by a hydrogen cloud approximately 1·6 million km in diameter. Similar clouds were associated later with Bennett's Comet of 1970 and Kohoutek's of 1973. Kohoutek's Comet was also found to contain large quantities of the ionized water molecule, H_2O^+.

The greatest cometary disappointment of modern times was Kohoutek's Comet of 1973, which was found on 7 March by L. Kohoutek at Hamburg when it was still almost 700 000 000 km from the Sun. Few comets are detectable as far away as this, and the comet was expected to become a magnificent object in the winter of 1973–4, but it failed to come up to expectations even though it was visible with the naked eye. It was, however, scientifically interesting, and was carefully studied by the astronauts then aboard the US space-station Skylab (Carr, Gibson and Pogue). Perhaps the comet will do better when it next returns to the Sun, in approximately 75 000 years' time!

The last brilliant comet (to the spring of 1988) was West's, first recorded on a photograph taken at La Silla, in Chile, on 10 August 1975. During March 1976 it was a prominent naked-eye object with a major tail, but after perihelion, and as it receded, it showed signs of disintegration. The estimated period is 300 000 years.

The most distant comet ever observed was Bowell's Comet of 1982, which was followed out to a distance of more than 1 500 000 000 km – so that it was then between the orbits of Saturn and Uranus. It was under observation for more than four years, which is a record for a non-periodical comet. Had it come closer it would have been brilliant; it had an extensive coma and a peculiar narrow tail.

The greatest perihelion distance for any comet is that of Schuster's Comet of 1976, discovered by H. Schuster from Chile on 25 February. The perihelion distance was just over one million km, not far from midway between the orbits of Jupiter and Saturn. (The previous record-holder had been Van den Bergh's Comet of 1974: 900 000 000 km.)

The record for comet discoveries is held by J. L. Pons, who discovered a total of 37. C. Messier discovered 13. **The record year** for discoveries of new comets or recoveries of previously known ones was 1987 (33 comets).

The most successful 20th-century comet-hunter is the Australian amateur Bill Bradfield, with 13 discoveries to his credit (as of 1988).

The first comet to be encountered by a space-probe was Giacobini-Zinner, in September 1985. The probe was ICE (International Cometary Explorer), formerly known as ISEE (International Sun-Earth Explorer) which had been launched in 1978 mainly to study the solar wind. On 10 June 1982 a complicated series of manœuvres began, involving five passes of the Moon and leading to a rendezvous with the comet – though since ICE carried no camera, there were no pictures returned.

The first comet to be studied from close range by a number of space-probes was Halley's, in 1986. There were five probes in all: two Japanese (Suisei and Sakigake), two Russian (Vega 1 and Vega 2), and one European (Giotto). Of these only Giotto was scheduled to pass into the comet's inner coma, and image the nucleus, but prior information sent back from the earlier vehicles was of immense value. Vega 1 made its closest approach on 6 March 1986 and Vega 2 on 9 March; Giotto passed within 605 km of the comet's nucleus on the night of 13–14 March.

It was generally believed that the 'dirty ice-ball' theory proposed by F. L. Whipple in 1950 was correct. On this theory a comet's nucleus was composed of rocky fragments held together with ices such as frozen methane, ammonia, carbon dioxide and water; when the comet approached perihelion these ices would start to evaporate, producing a coma and tail (or tails). Tails were of two kinds: ion and dust. Both would point more or less away from the Sun, as the ions would be repelled by sunlight and the dust by the solar wind, and when receding from the Sun a comet would move tail-first – though in general an ion tail would be straight and a dust-tail curved. (Smaller comets, of course, would never develop tails of either type.)

All the probes to Halley's Comet were successful. The camera on Giotto (the HMC, or Halley Multicolour Camera) functioned until 14 seconds before the closest approach, when the space-craft was made to gyrate by the impact of a 'dust' particle about the size of a grain of rice, and communications were temporarily interrupted; in fact the camera did not produce any further results, and the closest image was obtained at 1703 km from the nucleus. The nucleus itself was found to be $15 \times 8 \times 8$ km in dimension, with a total volume of over 500 cubic km, and a mass of from 50 000 million to 100 000 million tons and a density of 0·1 to 0·2 grammes per cubic centimetre. (It would take 60 000 million comets of this mass to equal the mass of the Earth.) To the surprise of most people, the nucleus proved to be very dark, with an albedo from 2 to 4 per cent.

Water ice appeared to be the main constituent of the nucleus (84 per cent) followed by formaldehyde and carbon dioxide (each around 3 per cent) and smaller amounts of other volatiles, including nitrogen and carbon monoxide. The shape of the terminator showed that the central region was smoother than the ends; a bright patch 1·5 km in diameter was taken to be a hill, and there were features which appeared to be craters, around 1 km across. Dust-jets were active, though only from a small area of the nucleus on the sunward side. The sunward side was found to have a temperature of 47 degrees C, far higher than expected, and from this it was inferred that the icy nucleus was covered with a layer of warm, dark dust. The nucleus is eroded at around 1 cm per day near perihelion, and at each return the comet must lose about 300 000 000 tons of material. The rotation period was found to be 53 hours with respect to the long axis of the nucleus, with a 7·3-day rotational period around this axis; the nucleus is in fact 'precessing' rather in the manner of a toppling gyroscope.

At its latest return, Halley's Comet was first recorded, from Palomar, on 16 October 1982. It will probably be followed until at least 1990, and perhaps for much longer. It will next reach aphelion in the year 2024.

The origin of comets is still a matter for debate, but at least we may be sure that they are bona-fide members of the Solar System. According to the Dutch astronomer J. H. Oort, there is a reservoir or 'cloud' of comets orbiting the Sun at a distance of a light-year or so, and when a comet is perturbed for some reason it may enter an orbit which swings it sunward, with the inevitable risk of being perturbed by a planet (usually Jupiter) and forced into a smaller orbit. It seems that short-period comets have limited lifetimes, since they lose some material by evaporation at every perihelion passage so that there must be, presumably, some source of replenishment to maintain the numbers of short-period comets.

It has been suggested that Apollo and Aten type asteroids are nothing more nor less than the nuclei of dead comets, and it is certainly true that two periodical comets, Arend-Rigaux and Neujmin I, now appear stellar. In 1982 a theory was proposed by two astronomers of the Royal Observatory Edinburgh, V. Clube and W. Napier, that the Apollo-type asteroid Hephaistos (diameter 10 km), discovered in 1978, was a comet in near-historical times, and was brilliant enough to give rise to some of the old myths about comets and disasters. They also maintain that every time the Sun passes through a spiral arm of the Galaxy, the Oort cloud is replenished, so that for a definite period after the Sun emerges from the arm brilliant comets and, later, collisions from Apollo or Aten type asteroids are more frequent than at other times. They also believe that such an impact occurred some 65 000 000 years ago, and altered conditions on Earth so markedly that

dinosaurs became extinct. The theory is admittedly highly speculative!

COMET NOMENCLATURE

Normally, a comet is named after its discoverer (e.g. Comet Lovas of 1974), after independent discoverers (Arend-Roland of 1957) or after the mathematicians who computed the orbits (Lexell, 1770). A comet is given a letter to correspond with its discovery; thus the first comet to be discovered in 1987 was 1987a, the next 1987b and so on. This is a provisional designation, and a permanent designation depends upon the time of perihelion; thus the first comet to reach perihelion in 1987 was 1987 I, the next 1987 II, and so on. These two systems may disagree with respect to the year; thus Comet 1990m may well become 1991 III!

The periodical comets are identified by P/ (e.g. P/Halley, P/Encke). They may be named according to their discoverers at different returns (P/Pons-Winnecke, P/Di Vico-Swift). Occasionally a name is altered; thus the comet formerly called Pons-Coggia-Winnecke Forbes is now known as P/Crommelin, since the identity of the four observed comets was established by A. C. D. Crommelin of Greenwich Observatory.

It must be remembered that these elements alter considerably for each revolution. The values given here apply to the beginning of 1988. Du Toit 1 (Comet 1944 III) was reported again in 1974, but the observations cannot be regarded as definite.

Orbits with inclinations over 90° indicate retrograde motion. There are two comets in the list with retrograde orbits, Tempel-Tuttle and Halley.

SOME PERIODICAL COMETS WHICH HAVE BEEN OBSERVED AT MORE THAN ONE RETURN

Comet	Period years	Distance from Sun, a.u. Perihelion	Aphelion	Eccentricity	Inclination
Encke	3·3	0·34	4·09	0·85	12·0
Grigg-Skjellerup	5·1	1·00	4·94	0·66	21·1
Tempel 2	5·3	1·36	4·68	0·55	12·5
Honda-Mrkós-Pajdusáková	5·3	0·58	5·49	0·58	13·1
Neujmin 2	5·4	1·34	4·84	0·57	10·6
Tempel 1	5·5	1·50	4·73	0·52	10·5
Tuttle-Giacobini-Kresák	5·6	1·15	5·13	0·63	13·6
Tempel-Swift	5·7	1·15	5·22	0·64	5·4
Wirtanen	5·9	1·26	5·16	0·61	12·3
D'Arrest	6·2	1·17	5·61	0·66	16·7
Du Toit-Neujmin-Delporte	6·3	1·68	5·15	0·51	2·9
Di Vico-Swift	6·3	1·62	5·21	0·52	3·6
Pons-Winnecke	6·3	1·25	5·61	0·64	22·3
Forbes	6·4	1·53	5·36	0·56	4·6
Kopff	6·4	1·57	5·34	0·55	4·7
Schwassmann-Wachmann 2	6·5	2·14	4·83	0·39	3·7
Giacobini-Zinner	6·5	0·99	5·98	0·71	31·7
Churyumov-Gerasimenko	6·6	1·30	3·51	0·63	7·12
Wolf-Harrington	6·6	1·62	5·38	0·54	18·4
Tsuchinshan 1	6·6	1·49	5·57	0·58	10·5
Perrine-Mrkós	6·7	1·27	5·85	0·64	17·8
Reinmuth 2	6·7	1·94	5·19	0·45	7·0
Borrelly	6·8	1·32	5·84	0·63	30·2
Johnson	6·8	2·20	4·96	0·39	13·9
Tsuchinshan 2	6·8	1·78	5·40	0·51	6·7
Harrington	6·8	1·58	5·60	0·56	8·7
Gunn	6·8	2·45	4·74	0·32	10·4
Arend-Rigaux	6·8	1·44	5·76	0·60	17·8
Brooks 2	6·9	1·84	5·39	0·49	5·6
Finlay	6·9	1·10	6·19	0·70	3·6
Taylor	7·0	1·95	3·64	0·47	20·6
Holmes	7·0	2·16	5·20	0·41	19·2
Daniel	7·1	1·66	5·72	0·55	20·1
Harrington-Abell	7·2	1·77	5·68	0·52	16·8
Shajn-Schaldach	7·3	2·23	5·28	0·41	6·2
Faye	7·4	1·62	5·98	0·58	9·1
Ashbrook-Jackson	7·4	2·29	5·33	0·40	12·5
Whipple	7·5	2·48	5·16	0·35	10·2
Reinmuth 1	7·6	2·00	5·76	0·49	8·3
Arend	7·9	1·84	4·00	0·54	20·0
Oterma	7·9	3·39	4·53	0·14	4·0
Schaumasse	8·2	1·20	6·92	0·70	12·0
Jackson-Neujmin	8·4	1·43	6·83	0·65	14·1
Wolf	8·4	2·52	5·78	0·40	27·3
Smirnova-Chernykh	8·5	3·60	4·20	0·15	6·6
Comas Solá	8·6	1·77	6·59	0·58	13·4
Kwerns-Kwee	9·0	2·23	6·43	0·49	9·0
Denning-Fujikawa	9·0	0·78	4·33	0·82	8·9
Swift-Gehrels	9·3	1·36	4·40	0·69	9·3
Neujmin 3	10·6	1·98	7·66	0·59	3·9
Gale	11·0	1·18	8·70	0·76	11·7
Klemola	11·0	1·76	4·94	0·64	10·6
Väisälä 1	11·3	1·87	8·19	0·63	11·5
Slaughter-Burnham	11·6	2·54	7·72	0·50	8·2
Van Biesbroeck	12·4	2·40	5·35	0·55	6·62
Wild	13·3	1·98	9·24	0·65	19·9
Tuttle	13·8	1·02	10·46	0·82	54·4
Du Toit 1	15·0	1·29	10·85	0·79	18·7
Schwassmann-Wachmann 1	15·0	5·45	6·73	0·11	9·7
Neujmin 1	17·9	1·54	12·16	0·78	15·0
Crommelin	27·9	0·74	17·65	0·92	28·9
Tempel-Tuttle	32·9	0·98	19·56	0·90	162·7
Stephan-Oterma	38·8	1·60	21·34	0·86	17·9
Olbers	69·5	1·18	32·62	0·93	44·6
Pons-Brooks	71·0	0·77	33·51	0·96	74·2
Brorsen-Metcalf	71·9	0·49	34·11	0·97	19·2
Halley	76·1	0·59	35·33	0·97	162·2
Herschel-Rigollet	156·0	0·75	56·94	0·97	64·2

Halley's Comet, 1910, when it was much brighter than it was in 1986: 12 and 15 May, as seen from Honolulu. (Palomar Obs.)

LOST PERIODICAL COMETS

Comet	Period, years	Distance from Sun, a.u. Perihelion	Aphelion	Eccentricity	Inclination	Number of returns	Last return
Brorsen	5·5	0·59	5·61	0·81	29·4	5	1879
Biela	6·6	0·86	6·19	0·76	12·6	6	1852
Westphal	61·9	1·25	30·03	0·92	40·9	2	1913

It seems certain that these comets have disintegrated.

Periodical comets seen at only one return

There are various reasons for 'losing' comets. A few, listed separately, seem definitely to have disintegrated. In other cases the orbit has been so violently perturbed that we have lost contact; the classic case is that of Lexell's Comet of 1770. The list also includes several short-period comets which have been discovered recently, and which will no doubt return on schedule; these are identified by an asterisk.

Comet	Period in years	Last seen
Wilson-Harrington	2·3	1949
Helfenzrieder	4·5	1766
Blanpain	5·1	1819
Du Toit 2	5·3	1945
La Hire	5·4	1678
Barnard 1	5·4	1884
Schwassmann-Wachmann 3	5·4	1930
Grischow	5·4	1743
Haneda-Campos*	5·4	1978
Machholz	5·4	1986
Clark*	5·6	1973
Brooks 1	5·6	1886
Lexell	5·6	1770
Kulin	5·6	1939
Pigott	5·9	1783
West-Kohoutek-Ikemura*	6·1	1975
Singer-Brewster	6·1	1986
Wild 2*	6·2	1978
Kohoutek*	6·2	1975
Hartley 2	6·2	1985
Spitaler	6·4	1890
Harrington-Wilson	6·4	1951
Russell 4	6·4	1984
Wiseman-Skiff	6·5	1987
Barnard 3	6·6	1892
Lovas 2	6·6	1986
Giacobini	6·7	1896
Schorr	6·7	1918
Russell 3	6·8	1983
Longmore*	7·1	1974
Russell 2	7·1	1980
Swift 2	7·2	1895
Tritton*	7·3	1978
Takmizawa	7·3	1984
Shoemaker 1	7·3	1984
Kowal-Mrkos	7·3	1984
Denning 2	7·4	1894
Metcalf	7·9	1906
Gehrels 2*	7·9	1973
Gehrels 3*	8·4	1975
Shoemaker 2	8·6	1984
Swift 1	8·9	1889
Lovas	9·1	1980
Boethin*	11·0	1975
IRAS	13·3	1983
Peters	13·4	1846
Gehrels 1*	14·6	1973

Comet	Period in years	Last seen
Kowal*	15·1	1977
Bowell-Skiff	15·2	1983
Perrine	16·4	1916
Shoemaker 3	17·0	1985
Hartley-IRAS	20·7	1983
Pons-Gambart	63·9	1827
Ross	64·6	1883
Dubiago	67·0	1921
Di Vico	75·7	1846
Väisälä 2	85·5	1942
Swift-Tuttle	119·6	1862
Barnard 2	128·3	1889
Mellish	145·3	1917

Brilliant Comets seen since 1700

1744 De Chéseaux' six-tailed comet (actually discovered by Klinkenberg).

1811 Great Comet: discovered by H. Flaugergues.
1843 Great Comet, superior to that of 1811.
1858 Donati's Comet, with its curved tails—possibly the most beautiful comet ever seen.
1861 Brilliant comet, discovered by Tebbutt.
1862 Bright comet, discovered by Swift.
1874 Coggia's Comet; bright naked-eye object.
1882 Great Comet, photographed by Gill from South Africa.
1901 Bright southern comet, discovered by Paysandu.
1910 Daylight Comet; brightest of the century (so far).
1927 Skjellerup's Comet. Bright for a brief period.
1947 Bright southern comet; brief visibility.
1965 Comet Ikeya-Seki; brilliant as seen from the southern hemisphere for a brief period (also well seen from the USA).
1970 Bennett's Comet; fairly bright.
1976 West's Comet; bright naked-eye object during early mornings in March.

To this list must be added Halley's Comet at its returns of 1759, 1835 and 1910.

Kohoutek's Comet of 1973–4; this picture was taken at the Catalina Observatory on 15 January 1974, with the 42-cm Schmidt telescope, 10 minute exposure. (by R. B. Minton. NASA)

METEORS

Meteors may be regarded as the debris of the Solar System. They are small and variable, and cannot reach ground level intact. Sporadic meteors may appear from any direction at any moment, but many meteors are members of definite showers.

THE PRINCIPAL METEOR SHOWERS

The showers listed in the table occur annually, though not all of them are consistent: the Perseids are the most reliable, whereas the Leonids produce major showers only occasionally (the last being in 1966). The ZHR or zenithal hourly rate is the probable hourly rate of meteors for an observer who has the radiant at his zenith, and is observing under ideal conditions. In practice, of course, these conditions are never fulfilled, so that the ZHR is the theoretical maximum only.

The only meteor shower named after a discarded constellation is that of the Quadrantids. The old constellation, Quadrans Muralis, has been long since rejected.

The maximum velocity of meteors entering the atmosphere is around 72 km/s. Most meteors vaporize above altitudes of 80 km. Micrometeorites, about 0·1 mm in diameter, are too small to cause luminous effects.

The total number of meteors entering the atmosphere daily has been estimated at 75 000 000 for meteors of magnitude 5 or brighter. An observer under ideal conditions could normally be expected to see about 10 naked-eye meteors per hour (except during a shower, when the number would naturally be higher).

Meteor showers were first postulated by Olmsted and Twining in 1834. H. Olbers suggested a period of 33 years for the Leonid shower. H. A. Newton came independently to the same conclusion, and traced the Leonids back to AD 902; he predicted another major shower for 1866, and it duly occurred.

The Perseid shower was identified by Quetelet in 1836. The shower has been nicknamed 'the tears of St. Lawrence'.

The richest showers on record have probably been the Leonids, which on 17 November 1966 reached a rate of 60 000 per hour over a period of 40 minutes. There were major showers also in 1799, 1833 and 1866, but not in 1899 or 1933, because of perturbations by Jupiter and Saturn.

The first radiant point to be measured was that of the Leonids, on 12–13 November 1833.

The first suggestion of an association between meteors and comets was made by D. Kirkwood in 1861; he believed meteors to be the debris of disintegrated comets. In 1862 G. V. Schiaparelli showed the association of the Perseids with comet P/Swift-Tuttle. The November Andromedids were subsequently shown to be associated with the defunct comet P/Biela. On 27 November 1872 the hourly rate of the Andromedids (sometimes termed Bieliids) reached 100 per hour, but in recent years the shower has become very feeble.

The greatest 'cometary' shower of meteors in the present century was that of 9 October 1933, associated with Comet P/Giacobini-Zinner. The rate reached 350 meteors per minute for a brief period.

The first measures of meteor heights by the triangulation method (i.e. observing the same meteors from widely separated points – at least two observers, at two different stations) were made by two German students, Brandes and Benzenberg, in 1798. They concluded that most meteors penetrate down to an altitude of 80 km above the ground.

The earliest meteor photograph (of an Andromedid) was taken on 27 November 1885 by L. Weinek, at Prague.

The first radar measurements of

METEORS

Shower	Begins	Maximum	Ends	Max. ZHR	Position of radiant: R.A. h	m	Dec. °	Associated Comet
Quadrantids	1 January	3 January	6 January	110	15	28	+50	
Corona Australids	14 March	16 March	18 March	5	16	20	−48	
Lyrids	19 April	22 April	24 April	12	18	08	+32	1861 I (Thatcher)
Eta Aquarids	2 May	4 May	7 May	20	22	24	00	P/Halley?
June Lyrids	10 June	15 June	21 June	8	18	32	+35	
Ophiuchids	17 June	20 June	26 June		17	20	−20	
Capricornids	10 July	26 July	15 August	6	21	00	−15	
Delta Aquarids	15 July	28 July	15 August	35	22	36	−10	
Piscis Australids	15 July	30 July	20 August	8	22	40	−30	
Alpha Capricornids	15 July	2 August	25 August	8	20	36	−10	
Iota Aquarids	15 July	6 August	25 August	6	22	15	− 9	
Perseids	25 July	12 August	18 August	68	03	04	+58	P/Swift-Tuttle
Kappa Cygnids	19 August	20 August	22 August	4	19	20	+55	
Draconids	10 October	10 October	10 October	var.	18	00	+54	P/Giacobini-Zinner
Orionids	16 October	21 October	26 October	30	06	24	+15	P/Halley?
Taurids	20 October	4 November	25 November	12	03	44	+22	P/Encke
Cepheids	7 November	9 November	11 November	8?	23	30	+63	
Leonids	15 November	17 November	19 November	var.	10	08	+22	P/Tempel-Tuttle
Andromedids	15 November	20 November	6 December	low	00	50	+55	P/Biela
Phœnicids	4 December	4/5 December	5 December	5	01	00	−55	
Geminids	7 December	14 December	15 December	58	07	28	+32	
Ursids	17 December	22 December	24 December	12	14	28	+76	P/Tuttle

Permanent daytime streams include the Arietids (29 March to June 17), the Xi Perseids (1 to 15 June) and the Beta Taurids (23 June to 7 July). The Beta Taurids seem to be associated with Encke's Comet.

The greatest meteor shower ever recorded, the Leonids, over Arizona, USA on 17 November 1966

meteors were made during the last war; pioneer work was done by the British radio astronomers Hey and Stewart, who in 1945 measured the Delta Aquarids by radar. Nowadays, radar has largely superseded the old method of visual observation.

The most reliable annual shower is that of the August Perseids, which never fail to produce a spectacular display. It is noteworthy that so far as meteor observation is concerned, the northern hemisphere has the best of matters!

Noctilucent clouds, which occur at heights of from 80 to 100 km, are thought to be due to ice crystals forming along meteor trails, though upon this point there is still no universal agreement among astronomers and meteorologists.

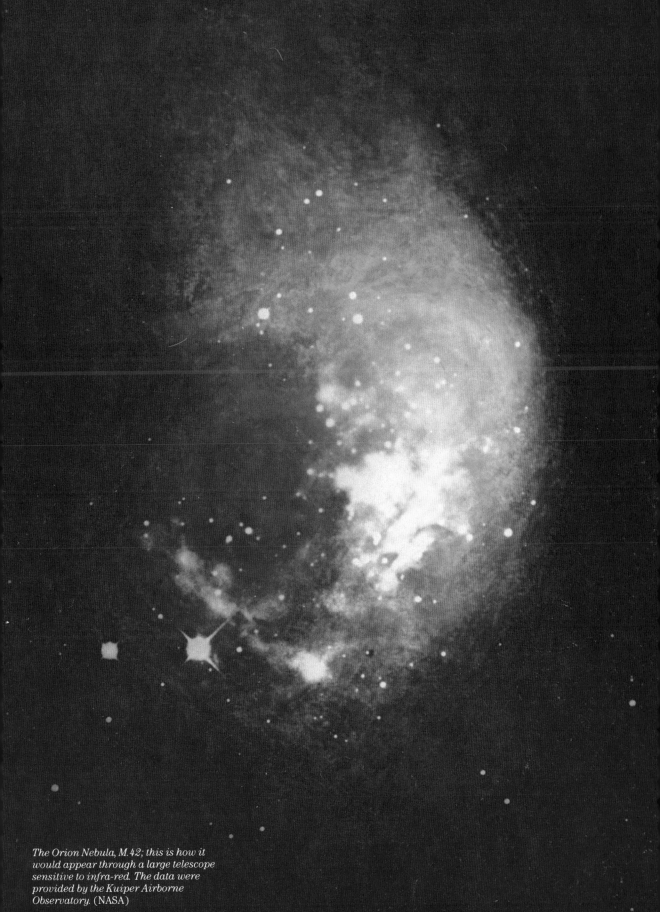

The Orion Nebula, M.42; this is how it would appear through a large telescope sensitive to infra-red. The data were provided by the Kuiper Airborne Observatory. (NASA)

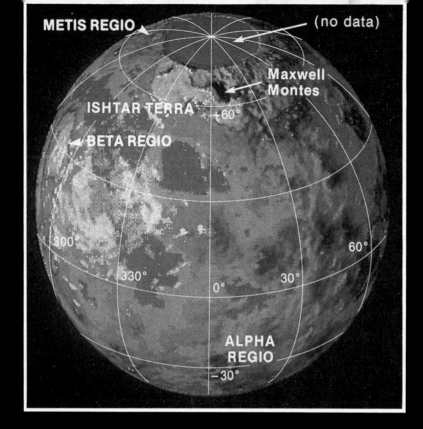

METIS REGIO

(no data)

Maxwell
Montes

ISHTAR TERRA +60°

BETA REGIO

300°

330°

0°

30°

60°

60°

ALPHA
REGIO

−30°

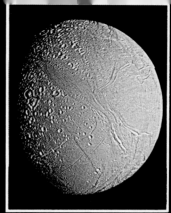

Above: *Enceladus, from Voyager 2; the
surface is icy, with small craters and
areas relatively lacking in detail.*
(Science Photo Library)

Left: *False-colour hemisphere of Venus;
from Pioneer, 1978–9.* (NASA)

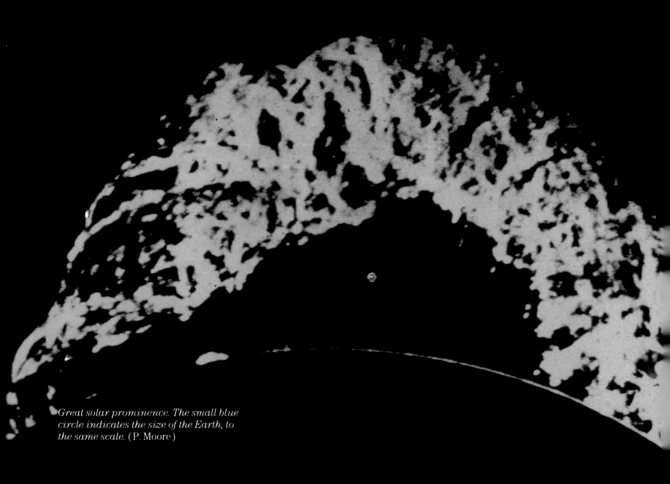

*Great solar prominence. The small blue
circle indicates the size of the Earth, to
the same scale.* (P. Moore)

Aurora Borealis; 1978 April, from Dundee. (K. Kennedy)

The supernova 1987A in the Large Cloud of Magellan, photographed on 5 May 1987 by Akira Fujii from Australia.

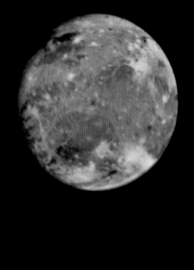
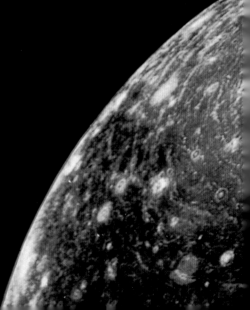

Composite of Jupiter and its large satellites (Voyager 1); bottom right: *Callisto;* lower left: *Ganymede;* centre: *Europa;* upper left: *Io.* (NASA)

West's Comet, photographed from Cayenne on 1 July 1975 by the discoverer, Richard West.

The 'Diamond Ring' effect during the total solar eclipse of 1982. (Photograph by Donald Trombino)

Gosse's Bluff. This is not far from Alice Springs in Australia, and appears to be a very denuded meteorite crater; the Bluff we see today is only the central portion of a ringed complex over 10 miles across. (P. Moore)

Below left: *Surface of Mars; Viking 1 lander (Chryse), July 1976.* (NASA)

Below right: *The crater Daedalus, 80 km in diameter, on the Moon's far side; taken from Apollo 2, 31 July 1969.* (NASA)

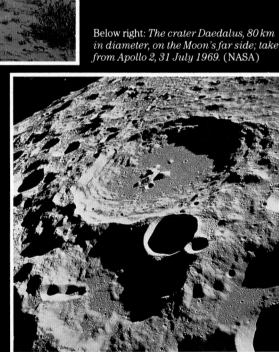

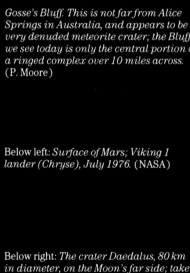

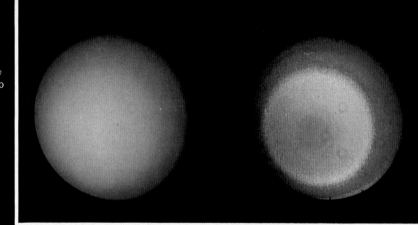

Uranus; 17 January 1986, Voyager 2. Left: true colour. Right: false colour view to bring out polar details. (Science Photo Library)

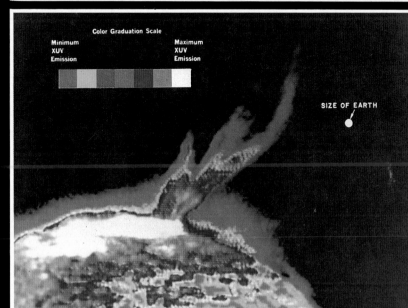

Solar eruption, 21 August 1973 (Skylab). Colour electronically processed from an original black-and-white picture in the light of ionized helium. (NASA)

Color Graduation Scale

Minimum XUV Emission

Maximum XUV Emission

SIZE OF EARTH

The first infra-red image of Pluto ever constructed using IRAS data. (University of Arizona)

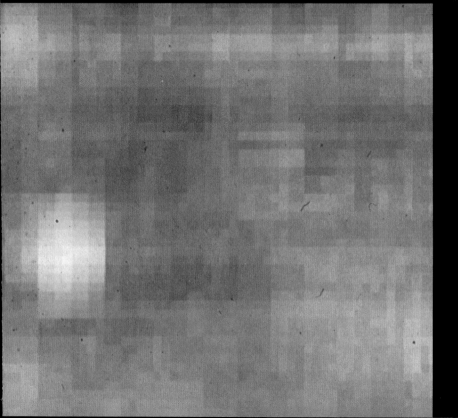

The Whirlpool Galaxy, M.51; US Naval Observatory, 1984. (NASA)

METEORITES

Meteorites reach ground level without being destroyed, and are not simply large meteors. They do not belong to showers, and any association with comets is dubious. Meteorites may in fact be more closely associated with asteroids, and it has been claimed that there is no distinction between a small asteroid and a large meteorite!

One important sub-group consists of the carbonaceous chondrites, which contain carbon compounds and organic matter (such as hydrocarbons). One famous example is the Orgueil Meteorite, which fell in France on 14 May 1864. In 1964 B. Nagy, at Columbia University together with his colleague Claus, suggested that this meteorite contained 'organized elements'– leading to the suggestion that there might be evidence of past living material. However, others maintain that these substances are due to terrestrial contamination. (In 1908 Svante Arrhenius suggested that life was brought to Earth by a meteorite; this 'panspermia' theory never met with much support, but has recently been revived in modified form, by F. Hoyle and C. Wickramasinghe.)

The earliest reports of meteoritic phenomena are recorded on Egyptian papyrus, around 2000 BC. Early meteorite falls are, naturally, poorly documented, but it has been suggested that a meteorite fell in Crete in 1478 BC; stones near Orchomenos in Boetia in 1200 BC, and an iron on Mount Ida in Crete in 1168 BC. According to Livy, 'stones' fell on Alban Hill in 634 BC, and there is evidence that in 416 BC a meteorite fell at Ægospotamos in Greece. The Sacred Stone in Mecca is certainly a meteorite. **The oldest meteorite which can be positively dated** is the Ensisheim Meteorite, which fell on 16 November 1492 and is now on show in Ensisheim Church (Switzerland).

In India, it is said that the Emperor Jahangir ordered two sword-blades, a dagger and knife to be made from the Jalandhar Meteorite of 10 April 1621. A sword was made from a meteorite which fell in Mongolia in 1670, and in the 19th century part of a South African meteorite was used to make a sword for the Emperor Alexander of Russia. **Meteorite ages** are usually estimated at between 4 and 4·6 æons.

The most famous terrestrial meteorite crater is in Arizona (lat. 35°·3 N., long. 111°·2 W.), 1265 metres in diameter and 175 metres deep. It is well-preserved, and there is no reasonable doubt about its origin – even though G. K. Gilbert, who was strongly of the opinion that the main lunar craters were due to impact, regarded the Arizona crater as volcanic! The crater is certainly more than 20 000 years old, and is a well-known tourist attraction. Other craters which are probably meteoritic include those at Wolf Creek (Australia) and Waqar (Arabia).

List of Meteorite Craters

Name	Diameter, metres (largest crater)	Date of discovery	Notes
Meteor Crater, Arizona	1265	1891	Many metallic meteorites found.
Wolf Creek, Australia	850	1947	
Henbury, Australia	200×110	1931	Thirteen craters.
Boxhole, Australia	175	1937	
Odessa, Texas	170	1921	
Waqar, Arabia	100	1932	Two craters.
Oesel, Estonia	100	1927	Six craters.
Campo del Cielo, Argentina	75	1933	Many craters.
Dalgaranga, Australia	70	1928	
Sikhote-Alin, Siberia	28	1947	Total of 106 craters. (Fall observed.)

Various other craters have been attributed to meteoric impact. In some cases this may be true, but there are serious doubts about, for instance, the New Quebec crater in Canada (discovered 1950; diameter 3400 metres) and others in the Canadian shield, which appear to be associated with geological features. The Vredefort Ring near Pretoria, in South Africa, is often listed as an impact crater, but geologists who have studied it are unanimous in stating that it is of volcanic origin. The huge Gosse's Bluff formation in Northern Territory, Australia, is almost certainly an ancient impact structure.

Tektites are small glassy objects, found only in a few regions. They seem to have been heated twice, and are usually aerodynamically shaped. Four main groups are known:

The largest tektite known was found in 1932 at Muong Nong, Laos; it weighs 3·2 kg. Whether or not tektites are extraterrestrial is still unknown. In 1897 R. O. M. Verbeek even suggested that they might come from the Moon, but this theory has now been rejected, and we have to confess that tektites remain a complete enigma.

The first suggestion that objects could 'fall from the sky' was made by E. F. Chladni in 1794, but his idea was met with considerable scepticism. (As recently as 1807 Thomas Jefferson, President of the United States, is quoted as saying 'I could more easily believe that two Yankee professors would lie than that stones would fall from heaven.')

The first scientific proof of meteorite falls was obtained by the French scientist J. B. Biot, who investigated the meteoritic shower at L'Aigle on 26 April 1803.

The classification of meteorites is, broadly speaking, into three types: stones (aerolites), stony-irons (siderolites) and irons (siderites).

IRONS (SIDERITES)

Hexahedrites contain between 4 and 6 per cent of nickel, and consist of the cubic mineral kamacite. With increased nickel another cubic mineral, taenite, is formed. These two minerals orient themselves parallel to octahedral planes, and when the surface is etched with acid after polishing the characteristic Widmanstätten patterns are re-

Region	Name	Geological age
Australasia	Australites	Middle/Late Pleistocene
Ivory Coast	Ivory Coast tektites	Lower Pleistocene
Czechoslovakia	Moldavites	Miocene
United States	Bediasites (Texas)	
	Georgiaites (Georgia)	Oligocene

Meteorite (Sample 79001), discovered in 1981 in Antarctica. (NASA)

vealed (**octahedrites**). The **nickel-rich ataxites**, with more than 12 per cent nickel, do not show the Widmanstätten patterns, and consist of plessite.

STONY IRONS (SIDEROLITES)
Pallasites consist of a network of nickel-iron enclosing crystals of olivine. **Mesosiderites** are heterogeneous aggregates of silicate minerals and a nickel-iron alloy. There are two further groups, **siderophytes** and **lodranites**, but only one example of each is known.

STONES (AEROLITES)
Chondrites contain spherical particles called chondrules. These chondrules are fragments of minerals, and show a radiating structure. Chondrites account for 85 per cent of known specimens. In **achondrites**, chondrules are absent; the achondrites are of coarser structure, with little free iron. They may be calcium-rich or calcium-poor.

The early systems of classification were due to G. Rose (1863), G. Tschermak (1883) and A. Brezina (1904); they were extended by G. Prior (1920) and latterly by G. J. H. McCall (1973). Stones are more commonly found than

irons in the ratio of 96 per cent to 4 per cent, but this is misleading, as irons are much more durable and are more likely to survive. Over 10 000 meteorites have now been located, though few have actually been seen to fall. Among famous falls which have resulted in meteorite recovery are those of the Přibram fireball (Czechoslovakia, which was recorded as magnitude −19!) on 7 April 1959, the Lost City meteorite (Oklahoma) in 1970, the Barwell Meteorite (Leicestershire) of 24 December 1965, and the Sikhote-Alin fall in Siberia on 12 February 1947 – the greatest of the century, apart from that of 1908 – which was an iron, and produced many craters, since it broke up during its descent.

Of special interest are eight meteorites known as the **SNC Meteorites** after the regions in which they were found (Shergotty in India, Nakhla in Egypt and Chassigny in France). They seem to have crystallized only 1·3 thousand million years ago, and their crynposition and texture indicates that they formed on or in a planet which had a strong gravitational

field. They have concentration of volatile elements, and glassy incursions which were presumably formed in the extreme heat of whichever process ejected them from a parent body. These glassy incursions have trapped gases such as Ar, Kr, Xe and N. It has been suggested that they are of Martian origin, though this is of course highly speculative.

There is no record of any human death due to a meteorite. Reports that a monk was killed at Cremona in 1511, and another monk at Milan in 1650, are unsubstantiated. However, in 1954 a woman in Alabama, USA, had a narrow escape when a meteorite fell through the roof of her house, and she suffered a minor injury to her arm. The only casualty has been a dog, which was in the wrong place at the wrong time when a meteorite landed in Egypt.

Meteorites are not always easy to identify on sight, but when etched with acid irons will show the characteristic **Widmanstätten patterns**, not found elsewhere.

The largest known meteorite is still lying where it fell, in prehistoric times, at Hoba West, near Grootfontein in South-West Africa; it weighs at least 60 tons. All the known meteorites weighing more than 10 tons are irons. They are:

Meteorite	Weight, tonnes
Hoba West, Grootfontein, S.W. Africa	60
Ahnighito (The Tent), Cape York, W. Greenland	30·4
Bacuberito, Mexico	27
Mbosi, Tanganyika	26
Agpalik, Cape York, W. Greenland	20·1
Armanty, Outer Mongolia	20 (estimated)
Willamette, Oregon, USA	14
Chupaderos, Mexico	14
Campo del Cielo, Argentina	13
Mundrabilla, Western Australia	12
Morito, Mexico	11

There is a report of an iron found by W. A. Cassidy in Argentina, in September 1969, which is said to have a weight of about 18 tonnes.

The largest stone meteorite fell in Kirin Province, Manchuria, on 8 March 1976. It weighs 1766 kg.

The largest meteorite in a museum is the Ahnighito, found by Robert Peary in Greenland in 1897; it is now in the Hayden Planetarium, New York, along with two other meteorites found at the same time and on the same site (known as The Woman and The Dog). Apparently the local Eskimos were rather reluctant to let them go! The Willamette meteorite is also in the Hayden Planetarium. (This meteorite was the subject of a lawsuit. It was found in 1902 on property belonging to the Oregon Iron and Steel Company. The discoverer moved it to his own property, and exhibited it; the Company sued him for possession, but the Court ruled in favour of the discoverer.)

The most famous fall of recent times was that of 30 June 1908, in the Tunguska region of Siberia. As seen from Kansk, 600 km away, the descending object was said to outshine the Sun, and detonations were heard 1000 km away; reindeer were killed and pine-trees blown flat over a large area. The first expedition to the site was not dispatched until 1927, led by L. Kulik. No fragments have been found, and it has been suggested that the object was the nucleus of a small comet, or else a fragment from Encke's Comet – which, if icy in nature, would presumably evapo-

rate during the descent and landing. (Inevitably, flying saucer enthusiasts have suggested that it was a space-ship in trouble!)

A second major fall in Siberia occurred on 12 February 1947, in the Sikhote-Alin area. The fall was observed, and more than 100 craters were located, of which thirty were of appreciable size. It is fortunate that both these Siberian falls occurred in uninhabited territory. If a meteorite of this size had hit a city, the death-toll would inevitably have been high.

Twenty-two meteorite falls in the British Isles are known:

1623	January 10	Stretchleigh, Devon. 12 kg.
1628	April 9	Hatford, Berkshire. Three stones; about 33 kg.
1779	–	Pettiswood, West Meath.
1795	December 13	Wold Cottage, Yorkshire. 25·4 kg.
1804	April 5	High Possil, Strathclyde (Lanarkshire). 4·5 kg.
1810	August ?	Mooresfort, Tipperary. 3·2 kg.
1813	September 10	Limerick. 48 kg (shower).
1830	February 15	Launton, Oxfordshire. 0·9 kg.
1830	May 17	Perth. About 11 kg.
1835	August 4	Aldsworth, Gloucestershire. Small shower, over 0·5 kg.
1844	April 29	Killeter, Tyrone. Small shower; small amount preserved.
1865	August 12	Dundrum, Tipperary. 1·8 kg.
1876	April 20	Rowton, Shropshire. 3·2 kg. (Iron meteorite.)
1881	March 14	Middlesbrough, Cleveland (Yorkshire). 1·4 kg.
1902	September 13	Crumlin, Antrim. 4·1 kg.
1914	October 13	Appley Bridge, Lancashire. 33 kg.
1917	December 3	Strathmore, Tayside (Perthshire). 13 kg (four stones).
1923	March 9	Ashdon, Essex. 0·9 kg.
1931	April 14	Pontlyfni, Gwynedd (Caernarvon). 120 gr.
1949	September 21	Beddgelert, Gwynedd (Caernarvon). 723 gr.
1965	December 24	Barwell, Leicestershire. 46 kg (total).
1969	April 25	Bovedy, N. Ireland. Main mass presumably fell in the sea.

Both these latter falls were well observed. Many fragments of the Barwell Meteorite were found; one was detected some time later nestling in a vase of artificial flowers on the window sill of a house in Barwell village. Though it broke up during the descent, it is the

Fragment found by Patrick Moore of the 46 kg Barwell meteorite (Patrick Moore)

largest stone known to have fallen over the British Isles.

The largest meteorites to have been found in different regions are:

Region	Meteorite	Weight
Africa	Hoba West, Grootfontein	60 tonnes
USA	Willamette, Oregon	14 tonnes
Asia	Armanty, Outer Mongolia	20 tonnes
South America	Campo del Cielo, Argentina	13 tonnes
Australia	Mundrabilla	12 tonnes
Europe	Magura, Czechoslovakia	1·5 tonnes
Ireland	Limerick	48 kg
England	Barwell, Leicestershire	±46 kg (total)
Scotland	Strathmore, Tayside (Perthshire)	10·1 kg
Wales	Beddgelert, Gwynedd (Caernarvon)	723 g

The only known case of a meteorite entering the Earth's atmosphere and leaving it again is that of the object of 10 August 1972. It seems to have approached the Earth 'from behind' at a relative velocity of 10 km/s, which increased to 15 km/s as the Earth's gravity accelerated it. The object entered the atmosphere at a slight angle, becoming detectable at a height of 76 km above Utah and reaching its closest point to the ground at 58 km above Montana. It then began to move outward, and became undetectable at just over 100 km above Alberta after a period of visibility of 1 minute 41 seconds; the magnitude was estimated by eye-witnesses to be at least −15, and the diameter of the object may have been as much as 80 m. After emerging from the Earth's atmosphere it re-entered a solar orbit, admittedly somewhat modified by its encounter, and presumably it is still orbiting the Sun.

GLOWS AND ATMOS- PHERIC EFFECTS

AURORÆ

Aur, or polar lights (Aurora Borealis in the northern hemisphere, Aurora Australis in the southern) must have been observed in ancient times, since they may often become extremely brilliant. There is excellent evidence that the Roman emperor Tiberius, who reigned from AD 14 to 37, once dispatched his fire-fighters to the port of Ostia on account of a red glow in the sky, seen from Rome, which proved to be auroral.

The first use of the term 'aurora' was due to the French astronomer P. Gassendi, following the brilliant display of 12 November 1621. However, **the first really good description of a display of Aurora Borealis** was given by K. Gesner of Zürich for the aurora of 27 December 1560. **The first Antarctic aurora** was described by Captain Cook on 20 February 1773.

The first major book dealing with auroræ was the *Traité physique et historique de l'aurora boréale*, written by J. J. de Mairan in 1731.

The first suggestion of an association between auroræ and electrical discharges may have been due to Halley in 1716, when he linked auroral displays with discharges associated with the Earth's magnetic field. (It must be remembered that Halley saw his first aurora in this year, following the end of the solar Maunder Minimum.)

The first suggestion of a connection between auroræ and magnetic effects was made by Hjorter in 1741.

The first suggestion of a connection between auroræ and sunspot phenomena was made by E. Loomis, of Yale, in 1870. The solar association was defined in more detail by the Italian astronomer G. Donati in 1872, following the great display of 4–5 February. In 1896 Birkeland reproduced 'miniature auroræ' by using a spherical electro-

magnet and a beam of cathode rays. In 1929 S. Chapman and V. Ferraro suggested that the cause of auroræ was solar plasma. It is now known that auroral displays are indeed due to particles emitted by the Sun, and it is assumed that these electrified particles enter the Van Allen zones round the Earth and 'overload' them, so that particles cascade downwards and produce the auroral glows in the upper atmosphere. **The maximum activity of auroræ** is seen along geomagnetic latitude 68 degrees N. or S.

On average, auroræ of some kind or other are seen in 240 nights per year in North Alaska, North Canada, Iceland, North Norway and Novaya Zemlya; 25 nights per year along the Canada/ United States border and in Central Scotland; and 1 night per year in Central France. Obviously, however, this varies according to the state of the solar cycle, and in general auroræ are most common 2 years after a spot-maximum.

Exceptionally brilliant auroræ occur now and then. Among many cases are the displays of 24–5 October 1870, 4–5 February 1872 and January 25–26 1938. The latter display was spectacular in Southern England and elsewhere, but was not seen from Tromsø in Norway! In 1909 an aurora was seen from Singapore (latitude N. 1 degree 25 minutes), but auroræ are extremely rare from latitudes as low as this.

The heights of auroræ vary. In general, the sharp lower boundary is at about 98 km above ground level, the maximum region of activity at about 110 km, and the normal upper boundary 300 km. In the cases of sunlit activity the upper limit may reach 700 km, or even 1000 km in extreme cases. Extremely low auroræ have been reported now and then, but with no certainty. There have also been many reports of noise accompanying auroral displays – either hissing or crackling. Many cases were listed by S. Tromholt in Norway in 1885 (reported in *Nature*, vol. 32, p. 499); it is extremely hard to explain the noise, but there does seem to be considerable weight of evidence that it occurs, even though final proof is still lacking.

Auroræ may be seen in various forms.

There are glows; arcs, with or without rays; bands, more diffuse and irregular; draperies, or curtains made up of very long rays; individual rays or streamers; coronæ, or radiating systems of rays converging at the zenith; 'surfaces', which are diffuse patches, sometimes pulsating; flaming auroræ, composed of quickly-moving sheets of light; and ghost arcs, which may persist long after the main display has ended.

THE ZODIACAL LIGHT

The Zodiacal Light may be seen as a faint cone of light rising from the horizon after sunset or before sunrise. It extends away from the Sun, and is generally observable for a fairly short period only after the Sun has disappeared or before it rises. On a clear, moonless night, under ideal conditions it contributes about one-third of the total sky light, and may be brighter than the average Milky Way region. It is due to particles scattered in the Solar System along and near the main plane of the system. The diameters of the particles are of the order of 0·1 to 0·2 micron. (One micron is equal to one-millionth of a metre.) Since the Zodiacal Light extends along the ecliptic, it is best seen when the ecliptic is most nearly vertical to the horizon–i.e. February to March and again in September to October.

The discovery of the Zodiacal Light was due to G. D. Cassini in 1683. He correctly suggested that it was caused by sunlight reflected by interplanetary 'dust'.

THE GEGENSCHEIN

The Gegenschein or Counterglow is seen as a faint patch of radiance in the position in the sky exactly opposite to the Sun. It is extremely elusive, and is generally visible only under near-perfect conditions. The best opportunities come when the anti-sun position is well away from the Milky Way (i.e. in February/April, and September/ November) and when the anti-sun position is high, at local midnight.* It has

*From England I have seen it only–in March 1942, when the whole country was blacked out as a precaution against German air-raids.

generally an oval shape, measuring 10 by 20 degrees, so that its maximum diameter is roughly 40 times that of the full moon.

The discovery of the Gegenschein seems to have been due to Esprit Pézénas, who reported it to the Paris Academy in 1731. **It was named** by Humboldt, who saw it on 16 March 1803. It was described in more detail by the Danish astronomer Theodor Brorsen, who saw it in 1854 and wrote about it in 1863, but (contrary to statements in many books) Brorsen did not claim the discovery for himself, as he was familiar with Humboldt's description.

The Zodiacal Band is a very faint, parallel-sided band of radiance, which may extend to either side of the Gegenschein, or be prolonged from the apex of the Zodiacal Light cone, or join the Zodiacal Light with the Gegenschein. It may be from 5 to 10 degrees wide, and is extremely faint. Like the Zodiacal Light and the Gegenschein, it is due to sunlight being reflected from interplanetary particles in the main plane of the Solar System.

Aurora, 1978 13 April, photographed from Dundee by K. Kennedy.

OTHER GLOWS AND ATMOSPHERIC EFFECTS

The Kordylewski Clouds were reported by the Polish astronomer K. Kordylewski in 1961. He believed them to be due to collections of interplanetary debris lying at the 'Lagrangian points' of the Moon's orbit – that is to say, moving in the same path as the Moon, but at distances of 60 degrees ahead and 60 degrees behind respectively. (The Trojan asteroids behave in this way with respect to Jupiter.) It has been claimed that the clouds have been seen with the naked eye as excessively faint patches of light, but their existence has yet to be properly confirmed.

The Airglow (dayglow and nightglow) was named by O. Struve in 1950, who suggested the name in correspondence with C. T. Elvey. Nightglow prevents the sky from being completely dark at any time, quite apart from the diffusion of starlight. There are various causes for it; chemical reactions of the neutral constituents of the upper atmosphere, with light emission; reactions from ionized constituents, again producing light emission; the influx of incoming particles from space, and so on. There also seems to be a sort of 'geo-corona' in the exosphere (1000 to 10 000 km altitude), due to emissions from hydrogen and helium atoms in this highly rarefied region of the atmosphere.

Diffused starlight is quite appreciable – coming as it does from the thousand million stars above magnitude 20, of which approximately one million are above magnitude 11·2, 4850 above magnitude 6, and 1620 above magnitude 5. Over a full visible hemisphere of the sky (20 626 square degrees) the total starlight is approximately equal to 51 stars the brilliance of Sirius or 4 planets the brilliance of Venus at maximum. It is, however, a tiny fraction of the light sent by the Moon.

The Green Flash (or Green Ray) is an atmospheric effect. As the Sun sinks below the horizon, the last segment of it may flash brilliant green for an instant. There are vague references to this in Egyptian and Celtic folklore, but **the first scientific reference to it** was made by W. Swan, who saw it on 13 September 1865. However, Swan's account was not published until 1883. **The first published reference** to the phenomenon was due to J. P. Joule, in 1869. Much more recently, some superb photographs of it have been taken by D. J. K. O'Connell at the Vatican Observatory; these were published in book form in 1958.

The Green Flash has also been seen with the planet Venus. The first published reference to it seems to be due to Admiral Murray, who saw it from HMS *Cornwall*, off Colombo, at 13 50 (1½ hours after sunset) on 28 November 1939. Venus was setting over a sea horizon; Admiral Murray was using binoculars, and described the flash as 'emerald'.

THE STARS

THE STARS: DISTANCES AND MOVEMENTS

The number of stars visible with the naked eye is approximately 5780, though this value is bound to be arbitrary. This means that it is seldom or never possible to see more than 2500 stars with the naked eye at any one time.

It has long been known that the stars are suns, and are very remote, but it was not until 1838 that the first star-distance was measured. This was achieved by F. W. Bessel, using the method of trigonometrical parallax; the star concerned was 61 Cygni. At about the same time, also using the parallax method, T. Henderson measured the distance of Alpha Centauri, and F. Struve measured the distance of Vega.

The nearest star (excluding the Sun) is Proxima Centauri, at 4·28 light-years; it is a member of the α Centauri system. The two bright components of α Centauri are only slightly further away. Stars within 13 light-years of the Sun are as follows:

In 1976 O. J. Eggen reported the dis-covery of a red star, magnitude 10·8, in Sculptor (near τ Sculptoris) which was regarded as very close; later work showed that it is well beyond the 13 light-year limit of the above table of these.

If placed at the standard distance, 10 parsecs (32·6 light-years) only α Centauri, Sirius, Procyon and τ Ceti would be visible with the naked eye; it is interesting to note that if our Sun could be observed from Sirius it would be a second-magnitude star at R.A., 18 h 44 m, declination −16°4′. Most of the stars in the table are faint red dwarfs; thus the very feeble Wolf 359 has only 0·000 02 of the luminosity of the Sun. The companions of Sirius and Procyon are white dwarfs. There are four flare stars; UV Ceti B (Luyten L 726−8 B), Proxima, Krüger 60 B and Ross 154. Lalande 21185 and Groombridge 34 A are excessively close binaries (too close to be separated visually) and there is a third body, of low mass, in the 61 Cygni system.

Radial velocities (the towards or away motion of the star relative to the Sun) are given as − for a velocity of approach, + for a velocity of recession.

The star with the greatest known proper motion is Barnard's Star, dis-covered in June 1916 by E. E. Barnard at the Yerkes Observatory. The annual proper motion is 10″·31, so that in 180 years it crosses the sky by a distance equal to the apparent diameter of the Full Moon. It is approaching us at 108 km/s, so that its distance is de-creasing at the rate of 0·036 light-years per century. The proper motion is in-creasing by about 0″·0013 per year, and will reach 25″ by AD 11800, when the star will be at its closest to us − 3·85 light-years, closer than Proxima; the parallax will be 0″·87. The apparent magnitude will then have increased to 8·5. Subsequently the star will begin to recede. Barnard's Star is also known as Munich 15040.

The star with the largest proper motion is Barnard's Star (Munich 15040); this is the only star with an annual proper motion exceeding 10 seconds of arc. Some large values are:

Star	Proper motion per year, seconds of arc
Barnard's Star	10·27
Lalande 21185	4·75
61 Cygni	4·12
ε Indi	3·93
Wolf 359	3·84
Proxima	3·75
α Centauri	3·61
Lalande 8760	3·53
Van Maanen's Star	3·01
Groombridge 34	2·87
τ Ceti	1·72
Sirius	1·21
Procyon	1·03
ε Eridani	0·97
Krüger 60	0·80

Star	R.A. h m (epoch 1950)		Dec. °		Spec.	Apparent Mag	Absolute Mag	Parallax ″	Radial Velocity, km/s	Distance l/y
Proxima	14	26·3	−62	28	Me	10·7	15·1	0·760		4·3
α Centauri A	14	36·2	−60	38	G4	0·0	4·4	0·760	−25	4·3
α Centauri B	14	36·2	−60	38	K5	1·7	5·8	0·760	−21	4·3
Barnard's Star	17	55·4	+04	33	M5	9·5	13·2	0·552	−108	5·8
Wolf 359	10	54·1	+07	20	M8	13·5	16·7	0·431	+13	7·6
Lalande 21185	11	00·7	+36	18	M2	7·5	10·5	0·402	−87	8·1
Sirius A	06	42·9	−16	39	AO	−1·4	1·4	0·377	−8	8·7
Sirius B	06	42·9	−16	39	dA	8·5	11·4	0·377	−8	8·7
UV Ceti A	01	36·4	−18	13	M6e	12·5	15·3	0·365	+29	8·9
UV Ceti B	01	36·4	−18	13	M6e	13·0v	15·8v	0·365	+29	8·9
Ross 154	18	46·1	−23	53	M6	10·6	13·3	0·345	−4	9·5
Ross 248	23	39·5	+43	56	M6	12·2	14·7	0·317	−81	10·3
ε Eridani	03	30·6	−09	38	KO	3·8	6·2	0·305	+15	10·7
Ross 128	11	43·5	+01	06	M5	11·1	13·5	0·301	−13	10·8
Luyten 789−6	22	35·7	−15	36	M7	11·3	13·3	0·302	−60	10·8
61 Cygni A	21	04·7	+38	30	K5	5·2	7·5	0·292	−64	11·2
61 Cygni B	21	04·7	+38	30	K7	6·0	8·4	0·292	−64	11·2
ε Indi	21	59·6	−57	00	K5	4·7	7·0	0·291	−40	11·2
Procyon A	07	36·7	+05	21	F5	0·4	2·7	0·287	−3	11·4
Procyon B	07	36·7	+05	21	−	10·8	13·1	0·287	−3	11·4
Σ 2393 A	18	42·2	+59	33	M4	8·9	11·1	0·284	+1	11·5

Star	R.A. h m (epoch 1950)		Dec. °		Spec.	Apparent Mag +	Absolute Mag	Parallax ″	Radial Velocity, km/s	Distance l/y
Σ 2398 B	18	42·2	+59	33	M5	9·7	11·9	0·284	+1	11·5
Groombridge 34 A	00	15·5	+43	44	M1	8·1	10·3	0·282	+14	11·6
Groombridge 34 B	00	15·6	+43	44	M6	11·0	13·3	0·282	+21	11·6
Lacaille 9352	23	02·6	−36	09	M2	7·4	9·6	0·279	+10	11·7
τ Ceti	01	41·7	−16	12	K0	3·5	5·7	0·273	−16	11·9
Luyten's Star	07	24·7	+05	29	M5	10·9	13·0	0·266	+26	12·3
Lacaille 8760	21	14·3	−39	04	M1	6·7	8·8	0·260	+23	12·5
Kapteyn's Star	05	09·7	−45	00	M0	8·8	12·7	0·256	+242	12·7
Krüger 60 A	22	26·3	+57	27	M4	9·8	11·9	0·254	−24	12·8
Krüger 60 B	22	26·3	+57	27	M6	11·3	13·3	0·254	−28	12·8

M.31, the Andromeda Galaxy – the most remote object clearly visible with the naked eye.

The brightest star within 13 light-years, both apparently and absolutely, is of course Sirius A, with 26 times the luminosity of the Sun. The **least massive star** is UV Ceti B, with a mass of only 0·035 that of the Sun. Apart from the two white dwarfs, all the stars within 13 light-years belong to the Main Sequence; there are no giants.

The first predictions of the positions of invisible stars were made by F. W. Bessel in 1844, when he forecast the positions of the companions to Sirius and Procyon; he based his calculations upon irregularities in the proper motions of the primaries. In 1862 Clark discovered Sirius B in almost the position predicted by Bessel; the orbital period of the system is 50 years. The companion to Procyon was found in 1898 by Schaeberle, again almost in Bessel's position; the orbital period of the system is 39 years.

Efforts have been made to detect planets of other stars by astrometric methods. At the Sproule Observatory in Pennsylvania P. van de Kamp has studied the proper motion of Barnard's Star ever since 1937, and has announced the probable existence of two planets, each rather less massive than Jupiter; there are various other 'suspect cases', though it is fair to say that the accuracy of van de Kamp's measurements has been questioned by some authorities. Obviously any extra-solar planets must be very faint (not above magnitude +30) but it is just possible that some may be detectable with the Hubble Space Telescope.

The nearest stars which are at all like the Sun are τ Ceti and ε Eridani. From 4 April 1960 to March 1961 astronomers at the Green Bank Radio Astronomy Observatory in West Virginia, led by Dr Frank Drake, carried out a 'listening operation' at a wavelength of 21 cm, using the 80 ft radio telescope, in the hope of picking up rhythmical signals which could be interpreted as artificial. The project was known officially as Project Ozma, though unofficially as Operation Little Green Men. Not surprisingly, the results were negative, but since 1969 operations of rather similar type have been carried out in the USSR.

Stellar populations were first noted by W. Baade in the early 1950s.

Population I stars are metal-rich, with about 2 per cent of their mass being made up of elements heavier than helium. The most brilliant Population I stars are of types W, O and B. The disk and arms of the Galaxy (and other spirals) are mainly Population I.

Population II stars are metal-poor, and are clearly older, so that their most brilliant members are red giants which have already left the Main Sequence. They are found in the halo and nucleus of the Galaxy (and other galaxies) and in globular clusters. It must, however, be stressed that there is no hard and fast boundary between Population I and Population II regions.

CLASSIFICATION AND EVOLUTION

The stars are of many different types, and show tremendous range in size and luminosity, though rather less in mass. The Sun is a normal star, and the only one sufficiently close to be studied in great detail.

The first attempt to classify the stars according to their spectra was made by the Italian Jesuit astronomer, Angelo Secchi, in 1863–7. He divided the stars into four types:

I. White or bluish stars, with broad, dark spectral lines of hydrogen but obscure metallic lines. Example: Sirius.

II. Yellow stars; hydrogen less prominent, metals more so. Examples: Capella, the Sun.

III. Orange stars: complicated banded spectra. Examples: Betelgeux, Mira. The class included many long-period variables.

IV. Red stars, with prominent carbon lines: all below magnitude 5. Example: R Cygni. This class also included many variables.

Secchi's catalogue included over 500 stars. His system was, however, superseded by that worked out at Harvard in the United States.

The Harvard system was intro-

duced by E. C. Pickering in 1890, and was extended by two famous women astronomers, Miss A. Cannon and Mrs W. Fleming. It has been further modified into what is now called the MKK (after Morgan, Keenan and Kellerman of the USA) or Yerkes system.

Most of the stars lie in the sequence from B to M. The very hot W stars are (as noted) very rare; O stars are also rare. Types R, N and S consist entirely of giants, and are regarded as no 'later' than Type M. R and N are now often combined into a simple type, C.

Originally it had been planned to make the sequence alphabetical, beginning at A, representing the spectra with the strongest hydrogen lines. When it was realized that the spectra represented star temperatures, major modifications had to be introduced, and the final result was alphabetically chaotic. A good and famous mnemonic is 'Wow! Oh Be A Fine Girl Kiss Me Right Now Sweetie' (or, if you prefer it, 'Smack'!)

In addition, P is used for gaseous nebulæ, and Q for novæ.

The following additional letters are used: e (emission lines present), n (nebulous lines, i.e. stars rotating rapidly), s (sharp lines), k (interstellar lines), m (metallic lines), v (variable), and p or pec (peculiar).

Stellar Evolution. It was originally believed that a star began its career as a red giant, condensing out of nebular material (type M; giant), shrank and heated up to become a Main Sequence star (type B or A), and then cooled while continuing to shrink, ending its career as a red dwarf of type M. This was – it was thought – shown by the famous Hertzsprung-Russell or H/R diagram.

The H/R diagram was originally due to Ejnar Hertzsprung, a Danish astronomer, in 1911; it was proposed independently in 1913 by H. W. Russell of the United States. In an H/R diagram, a star is plotted according to its luminosity (or equivalent, such as absolute magnitude) against spectral type (or surface temperature). The Main Sequence, running diagonally from hot O and B stars at the top left, down to faint M-type dwarfs at the lower right, is very marked; so are the giant and supergiant areas, to the upper right. White dwarfs

Type	Spectrum	Surface Temperature	Examples	Notes
W	Many bright lines: divided into WN (nitrogen sequence) and WC (carbon sequence)	Up to 80 000 °C	Rare: about 150 have been found in our galaxy, and 50 in the Large Magellanic Cloud. 16 cases are known as planetary nebulæ in which the central star is a Wolf-Rayet; one of these is NGC 7009, the 'Saturn Nebula' in Aquarius.	Known as Wolf-Rayet stars. Have expanding shells, moving outwards at up to 3000 km/sec. All very remote, and appear faint despite their considerable luminosity
O	Both bright and dark lines	40 000 °C–35 000 °C	γ Velorum	Represent a transition between W and B stars, though this does not imply any evolutionary sequence
B	Bluish-white (B0) to white (B9). No emission lines, but dominant absorption lines of hydrogen and (particularly) helium.	Over 25 000 °C for B0 12 000 °C for B9	Rigel B8 (i.e. 8/10 of way from B0 to A0)	Rigel is particularly luminous, with peculiarites in its spectrum
A	White stars: spectra dominated by hydrogen lines	10 000 °C–8000 °C	Sirius, Vega, Altair	
F	Yellowish hue. Calcium very conspicuous, with less prominent hydrogen lines	7500 °C–6000 °C	Procyon, Polaris	Yellow hue so elusive that to the naked eye most F-stars will be regarded as white
G	Yellow; weaker hydrogen lines, numerous conspicuous metallic lines	5500 °C–4200 °C (giants) 6000 °C–5000 °C (dwarfs)	Capella (giant) Sun (dwarf)	Beginning of division into dwarf or Main Sequence stars and the giants
K	Orange; weak hydrogen lines, strong metallic lines	4000 °C 3000 °C (giants) 5000 °C 4000 °C (dwarfs)	Arcturus, Aldebaran, Pollux	K-stars are more numerous than any other type
M	Orange-red; very complicated spectra with many bands due to molecules	3400 °C (giants) 3000 °C (dwarfs)	Mira Ceti (variable), Betelgeux, Antares (giants), Proxima Centauri (dwarf)	Strong differences between giants and Main Sequence dwarfs. Proxima Centauri closest star beyond the Solar System
R*	Reddish	2600 °C	V Arietis, T Lyræ	Remote, and appear faint
N*	Reddish; strong carbon lines	2500 °C	R Leporis, V Aquilæ	Remote, with many variable examples
S	Reddish; prominent bands of titanium oxide and zirconium oxide	2600 °C	χ Cygni, R Cygni	Examples given are long period variables

* Types R and N are now often combined as Type C.

(unrecognized in 1913) are to the lower left.

The key to stellar energy was discovered in 1939 by H. Bethe (and, at about the same time, by G. Gamow). Bethe actually worked it out during a train journey from Washington to Cornell University! The stars – or, at least, the normal ones – shine by means of nuclear transformations. Thus inside the Sun, hydrogen is being converted into helium. It takes four hydrogen nuclei to make one helium nucleus; each time this happens, a little mass is lost and a little energy is released. Each second, the Sun converts 600 million tonnes of hydrogen into helium – and loses 4 million tonnes in mass. This may sound a great deal, but the Sun is certainly older than the Earth, and will continue in virtually its present state for 5000 million years in the future – perhaps rather longer.

The career of a star depends upon its initial mass, and the H/R diagram itself is not indicative of a definite evolutionary sequence. In every case the star begins by condensing out of nebular material; this star formation is certainly going on in M.42, the Orion Nebula. We can also observe fairly small dark 'globules' against some bright nebulæ, known as Bok globules in honour of their discoverer, Bart J. Bok, who came originally from Holland but who settled in the United States. Bok drew attention to the globules in the 1930s. Some of them are over 6 light-years in diameter, though others are much smaller. They may well indicate regions of star formation by gravitational contraction, though it is unlikely that all stars begin

as Bok globules. There are many known cases in which many stars of similar type, and presumably similar age, are concentrated in a limited area; these are **stellar associations**.

Moreover, it is logical to assume that the stars in any cluster (such as the Pleiades) had a common origin in the same nebular cloud. Not all open clusters are similar in age: for instance the Pleiades cluster is relatively young, while M.67 in Cancer is comparatively old.

The first brown dwarf to be definitely identified was detected by B. Zuckerman (University of California) and E. Becklin (University of Hawaii) in 1987. They found that the spectrum of a white dwarf star, Giclas 29–38, shows an infra-red excess in the region between 2 and 5 micrometres. This is unusual, and indicates the presence of an object with a diameter 15 per cent that of the Sun and a temperature of about 900 °C. An earlier reported detection of a brown dwarf (1985, associated with the faint star Van Biesbroeck 8) was not confirmed.

(**a**) **Stars of less than 0·1 solar masses**.–Condensation out of nebular material causes a rise in temperature, but the cores never become hot enough for nuclear reactions to begin (about 10 000 000 degrees is the 'starting point'). The star therefore turns into a dim red dwarf of type M, and then simply fades until it has lost all its energy.

(**b**) **Stars of between 0·1 and 1·4 solar masses**.–The star begins by condensing gravitationally out of nebular material. The 'protostar' will be large and red, though by no means the same as the red giants such as Betelgeux. As contraction continues, the surface temperature remains the same, so that the star becomes less luminous. On the H/R diagram, it passes along what are known as the Hayashi and Henyey tracks (in honour of the astronomers who first described them) and then joins the Main Sequence. The T Tauri stars, named after the most famous member of the class, are still contracting toward the Main Sequence, and are irregularly variable. Condensation to the Main Sequence takes several million years, and a young star will also have to 'blow away' a surrounding cocoon of dust. It will then brighten, and cases have been found; thus the star V.1057 Cygni brightened up from magnitude 16 to above magnitude 11, in 1969, over a period of only about 250 days. Not surprisingly, stars of this type are strong emitters of infra-red radiation.

With the Sun and similar stars, the core temperature is about 14 000 000 °C. Hydrogen is being converted into helium, admittedly by a rather roundabout process known as the **proton-proton cycle**. (In hotter stars the so-called **carbon-nitrogen cycle** is dominant; the end result is much the same, with four hydrogen nuclei forming one helium nucleus, but carbon and nitrogen are used as catalysts.) Obviously, the core is being depleted of hydrogen and enriched with helium. When the

supply of hydrogen in the core runs low, energy production ceases, and the core contracts under the influence of gravitation, though hydrogen conversion continues in the shell surrounding the core. Eventually the star leaves the Main Sequence, and enters the giant branch of the H/R diagram; the core now has a temperature of at least 100 million degrees, and this is enough to start reactions involving the conversion of helium into carbon (the 'helium flash'). It has been calculated that when the Sun reaches this stage in its evolution it will have a diameter of at least 25 000 000 miles, and will be 100 times as luminous as it is now, though the surface temperature will be lower; the Sun will have become a red giant. Life on Earth cannot survive this ordeal, and it is very likely that the Earth itself will be destroyed.

For a star such as the Sun, the red giant stage may last for 100 million years, which is not long on the cosmical scale. A whole series of reactions takes place inside the star, and heavier elements are built up in succession, but at last all the nuclear energy is exhausted. There is now nothing to halt gravitational shrinkage, and the star contracts into what is termed a **white dwarf**. The dim Companion of Sirius is of this type.

In a white dwarf, the atoms are broken up and packed tightly together with little waste space. This leads to amazing density values, up to a million times the density of water in extreme cases; a tablespoonful of white dwarf material might well weigh a thousand tonnes. For instance, the Companion of Sirius is as massive as the Sun, but smaller than planets such as Uranus and Neptune; we even know of white dwarfs which are smaller than the Moon. They have been described as 'bankrupt stars', though their surface temperatures are still high (10 000 to 40 000 °C). We must assume that eventually a white dwarf will fade away into a cold, dead 'black dwarf' sending out no energy at all.

The Indian astronomer S. Chandrasekhar has shown that a star more than 1·4 times as massive as the Sun cannot become a white dwarf unless it sheds some of its mass. Some stars appear to do this during the preceding

giant stage, when the outer part of the envelope is blown away as a stellar wind. When the wind becomes visible later it produces a **planetary nebula**. This may dispose of 10 to 20 per cent of the star's mass. The shell will not fall back, but will expand and dissipate. Planetary nebulæ, as such, are therefore rather short-lived, and do not last for more than 30 000 to 40 000 years.

(**c**) **Stars with initial mass appreciably more than 1·4 times that of the Sun**. Here, everything happens at an accelerated pace, and if the proto-star is, say, 10 to 40 times as massive as the Sun it may reach the main sequence in a few thousand years. Things proceed much as before, though more quickly, until the end of helium conversion. The carbon-rich core reaches a temperature of between 600 and 700 million °C, and carbon conversion begins, producing elements such as magnesium, sodium, neon and oxygen. When the carbon is exhausted, the core contracts once more; oxygen nuclei are then fused to produce sulphur and phosphorus, though shells nearer the surface are still using up carbon, helium and hydrogen. The situation is very complicated, and when the central temperature has reached 3000 million °C or so even iron nuclei are being produced.

This, however, is the beginning of the end. When enough iron has accumulated in the core, nuclear reactions stop. As the core collapses in seconds there is an 'implosion' (the opposite of an explosion), and the outer shells, where reactions are still going on, are heated to incredible temperatures. Over the next hour, a violent shock wave disrupts the star, producing a supernova outburst. Much of the star's material is hurled into space – and since this material contains many heavy elements, the supernova enriches the interstellar medium. The end product is a cloud of expanding material, while the remnant of the original star's core becomes a **neutron star** or **pulsar**. The collapse is so dramatic that the protons (positively charged) and the electrons (negatively charged) are fused into neutrons (no electrical charge).

The concept of neutron stars was proposed in 1932 by the Russian physicist Landau, and again in 1934 by F. Zwicky and W. Baade at Caltech (USA) but the first neutron star was not detected until 1967 – not by visible light, but by its radio emissions. A neutron star is indeed an amazing object. The diameter is of the order of 10 km, and the density perhaps a thousand million million tonnes per cubic metre, or 100 million million times that of water. A pin's head of neutron star material would 'weigh' more than the liner *QE2*.

According to theory (which may or may not be correct!) the outer crust of a neutron star is crystalline and iron-rich, composed of what we may term 'normal' matter. Cracks occur in it, producing 'starquakes' which affect the object's behaviour. Below the crust comes the neutron-rich liquid mantle; below again, a superfluid core composed mainly of neutrons; and at the centre, material made up of 'hyperons', about which we can only speculate. If one could stand on the surface of a neutron star, one's weight would be 10 000 million times greater than on Earth. There are strong magnetic fields – perhaps a hundred million tesla, as against 30 millionths of a tesla for the Earth.

The first neutron star was discovered in 1967 by Miss Jocelyn Bell (now Dr Jocelyn Bell-Burnell) working with the Cambridge team led by Professor A. Hewish. During radio surveys of the sky, she found a weak source which was fluctuating quickly as though 'ticking'. The radio pulses were very regular and rapid, so that the object was called a **pulsar** (CP 1919). The official announcement was made on 29 February 1968. More pulsars were soon found, and today over 300 are known. Their periods have a limited range; that of CP 1919 is 1·3373 seconds.

The slowest pulsar is NP 0527; period 3·745491 seconds. In general all pulsars seem to be slowing down by very small amounts; thus the period of CP 1919 is lengthening by a thousand millionth of a second each month, so that in 3000 years' time the period will be 1·3374 second instead of the present-day 1·3373 second. However, some pulsars show irregular changes in period, or **glitches**; thus on 1 March 1969 the pulsar PSR 0833–45 speeded

up a full quarter of a millionth of a second!

It was some time before astronomers decided upon the nature of pulsars. There was even the short-lived LGM or 'Little Green Men' theory – that the radio signals were artificial! Pulsating or rotating white dwarfs were also suggested, but it became apparent that a pulsar must be smaller than even a white dwarf, and the idea of pulsars as neutron stars is now universally accepted. The pulsar is spinning rapidly, and the more normal surface protons and electrons are accelerated outward at the north and south magnetic poles of the star in rather the manner of searchlight beams; every time the Earth passes through the beam, we receive a pulse. Glitches are due to the disturbances in rotational period caused by starquakes.

The first pulsar to be seen visually was NP 0532 in the Crab Nebula. This was achieved in January 1969 by a team at the Steward Observatory in Arizona, using a 36-in reflector. They identified a faint, flashing object whose mean magnitude was about 17, and whose period was the same as that of the pulsar. Its period is 0·0339 second, so that it pulses 30 times a second. Later in 1969 the pulsar was photographed from Kitt Peak (Arizona). It is undoubtedly the remnant of the supernova of 1054. (R. Minkowski and W. Baade had observed it in 1942, and had suspected that it was the centre of activity in the Crab.) The second visual identification was that of the pulsar PSR 0833–45 in the southern constellation of Vela; the pulsar lies in the Gum Nebula (named after its discoverer, the late Dr C. S. Gum), and it may be assumed that the nebula is a supernova remnant. The Vela pulsar itself was found in 1968 at the Molonglo Radio Observatory in Australia. The period is 0·089 second – the third shortest known. In 1977 a team working with the 3·9 m Anglo-Australian telescope at Siding Spring (Australia) identified the pulsar as a faint flashing object with a mean magnitude of 24·2.

The first-known millisecond pulsar was PSR 1937 + 215 in Vulpecula. It pulses at a record 642 flashes per second, regularly spaced at intervals of 1·557 milliseconds. It was identified in

October 1982 by a team at Berkeley, California, led by D. Backer. Subsequent investigations with the 305-metre radio telescope at Arecibo indicated a distance of about 7000 light-years, and it has also been detected visually at Lick Observatory. In view of its very fast pulses, the object should be young, but it is not an X-ray emitter – at any rate, it was not detected by the Einstein X-ray Observatory – and it shows no evidence of slowing down. Both these facts indicate greater age – at least 100 000 years. C. Heiles has suggested that it is a neutron star, a supernova remnant, which is now collecting material from its companion and is being speeded up, so that it may be unstable.

A second millisecond pulsar in a globular cluster was detected by A. Brinklow, A. Lyne, J. Biggs and M. Ashworth in 1987, with the Lovell Telescope at Jodrell Bank. The pulsar is in the cluster M.4; the period is 11·0757 milliseconds, with a single radio pulse repeating 90 times per second.

It had always been thought that a supernova of Type II was due to the destruction of a red supergiant, leaving a neutron star remnant. The recent supernova in the Large Cloud of Magellan shows, however, that this is not always true, since in this case the progenitor was a blue supergiant and not a red supergiant; the mass of the progenitor was much the same, but the smaller size meant that the outburst was underluminous by supernova standards. Of course, a supernova of Type I, involving the destruction of the white dwarf member of a binary system, will not produce a neutron star. Also, some supernovæ appear to leave different kinds of remnants which have been named *scintars*. The first to be detected was near the border of Aquila and Serpens, which is 50 000 light-years away. It was originally known as W.50, but contained a radio-emitting star, SS 433, in its centre; X-rays from it were also detected, and the spectrum is not only peculiar but also variable. The best interpretation is that SS 433 is not a single star, but a binary system. One member of the pair is ejecting streamers of gas in diametrically opposite directions; these jets are not fixed in position, but are behaving in the man-

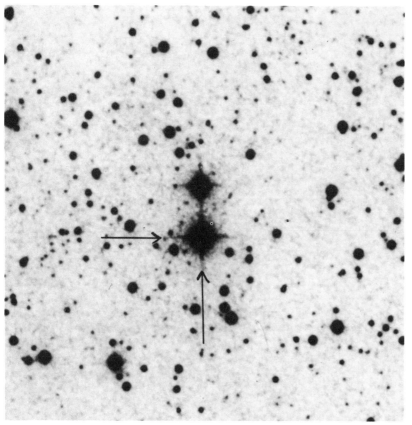

Optical counterpart of the X-ray source Cygnus X-1. The visible star is arrowed. (Royal Greenwich Observatory. Science Photo Lib.)

black hole concept is questioned by some astronomers – and we have as yet no decisive proof that they exist.

Note, also, that the basic principle was forecast as long ago as 1798 by the great French mathematician Laplace. Like Newton, he believed light to consist of a stream of particles, and commented that if a body were sufficiently small and dense it would be invisible, since the light-particles would not be able to travel fast enough to escape from it.

Meanwhile, all we can really say is that black holes are theoretical possibilities, and that the concept has been widely accepted. If valid, then black holes represent the end products of the most massive stars in the universe.

The first identification of a 'disk-star' which may be in the process of forming its own planets was made in 1977 by R. Thompson, P. Stritmatter, E. Erickson, P. Witteborn and D. Strecker at the Steward Observatory, University of Arizona. Observations were made with the 91 cm infra-red telescope of the Kuiper Airborne Observatory and with the Steward 2·3 m infra-red telescope. (Infra-red observations were essential to see through the dusty veil surrounding the star.) The star is MWC

ner of a rotating lawn sprinkler, with the star orbiting its companion in a period of 12 days. A second possible scintar is Circinus X-1, in the southern sky.

(d) Stars too massive initially to become supernovæ. Here we are by no means certain of the course of events, but it is suggested that once the final collapse starts nothing can stop it. Eventually the star has become so small and so dense that its escape velocity becomes greater than the velocity of light. Nothing can therefore escape, and to all intents and purposes the collapsed star or **collapsar** is cut off from the rest of the universe.

For obvious reasons, these **black holes** can be detected only by their effects upon objects which emit detectable radiation. One candidate is Cygnus X-1, so called because it is an X-ray source. The system consists of a B0 type supergiant, HDE 226868, with about 30 times the mass of the Sun and a diameter 23 times that of the Sun (18 000 000 km), together with an in-

visible secondary with 14 times the mass of the Sun. The orbital period is 5·6 days, the distance 6500 light-years, and the magnitude of the primary star 9 (it lies at RA 19 h 56 m 29s, dec. + 35° 03′55″, near η Cygni). The secondary would certainly be visible if it were a normal star, and it may be a black hole.

The size of a black hole will depend upon the mass of the collapsed star. The critical radius of a non-rotating black hole is called the Schwarzschild radius, after the German astronomer K. Schwarzschild, who investigated the problem mathematically in 1916; the boundary around the mass having this radius is the **event horizon**. For a body the mass of the Sun, the Schwarzschild radius would be about 3 km; for the Earth, less than 1 cm.

Conditions inside the event horizon are so different from our everyday experience that we cannot visualize them; it may even be that the old star finally crushes itself out of existence. However, it is important to add that the whole

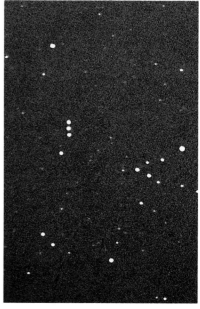

Barnard's Star; a composite of photographs taken by John Sabford in 1975, 1977 and 1979. The proper motion of Barnard's Star is very evident. (Science Photo Library)

349, in Cygnus. The surrounding disk of intensely glowing gas seems to have a diameter 20 times that of the central star, and to emit 10 times as much light. At the outer edge, the disk is about as thick as the star's diameter. The star itself is 10 times the diameter and 30 times the mass of the Sun. The luminous disk is believed to be the inner part of a surrounding larger disk of non-luminous gas in which outer planets may already have formed; this disk has a diameter greater than that of the orbit of Pluto, while the luminous disk has a diameter slightly greater than that of the orbit of the Earth. The luminous disk is wedge-shaped in cross-section, and where it joins the star's surface the thickness is 1/40 of the star's diameter.

The disk is estimated to be only 1000 years old, and is fading at about 1 per cent per month, as luminous material from the disk spirals into the central star. Therefore, the luminous disk may have vanished in a century from now.

The evidence rests upon several facts. The star is much brighter in visible light than it should be; it has declined since first identified forty years ago, and the spectrum is not that of a hot star – it is that of a hot, glowing disk. The distance from us is about 10 000 light-years.

The first star found to have a marked infra-red excess was Vega. The discovery was made in 1983 by H. Aumann and F. Gillett, from results obtained from IRAS (the Infra-Red Astronomical Satellite). The infra-red excess was found when the IRAS instruments were being calibrated. It was presumably due to solid particles coming from an extended region round Vega, stretching out some 80 astronomical units from the star. The temperature of the material was about − 185 °C, and it was suggested that the material might be planet-forming. Other similar cases followed, notably with Fomalhaut. An even greater excess was found with Beta Pictoris, an A-type star 78 light-years away and 60 times as luminous as the Sun. The material round Beta Pictoris was subsequently detected optically by B. Terrile and Bradford Smith, from the Las Campanas Observatory in Chile. Apparently Beta Pictoris is surrounded by a disk of material extending nearly 80 000 000 000 km from the star; we see it almost edge-on, and it may be no more than a few hundred million years old. So far as its composition is concerned, ices, silicates and carbonaceous substances are strong candidates, and it may be that the interior particles in the disk have been swept away by orbiting planets. Of course, this interpretation is highly speculative, but it cannot be ruled out.

The most luminous star known is probably η Carinæ, which is unique in many ways. Telescopically it has been described as looking more like a 'red blob' than a normal star, and it is associated with nebulosity; the distance is estimated at 6400 light-years, and the maximum luminosity may amount to several million times that of the Sun (one estimate gives 6 000 000 Suns). For some years in the 1830s and early 1840s it was the brightest star in the sky apart from Sirius, but for a century now it has been below naked-eye visibility, though it emits strongly in the infra-red and the decline in real luminosity is not so dramatic as might be thought. There have been speculative suggestions that it may suffer a supernova outburst in the foreseeable future (that is to say, within the next few tens of thousands of years); it has also been regarded as an exceptionally slow nova, or as an extraordinary type of variable star. The spectrum can only by classified as 'peculiar'.

Excluding η Carinæ, the most luminous star known is S. Doradûs, in the Large Magellanic Cloud; the absolute magnitude is −8·9, so that it is about a million times as powerful as the Sun (it is somewhat variable). From Earth, it cannot be seen with the naked eye.

The least luminous star known is RG 0050−2722 in Sculptor, with an absolute magnitude of + 19. Its distance is about 80 light-years, its surface temperature 2600 °C and its mass 0·023 that of the Sun – ten times the mass of Jupiter. Its nearest rival is VB 10 Aquilæ, absolute magnitude + 18·57.

The most massive star known is Plaskett's Star, HD 47129 Monocerotis, identified by K. Plaskett in 1922, when he discovered the binary nature of the O7-type supergiant. The orbital period is 14 days. The mass of each component is thought to be 55 times that of the Sun, and the primary has a radius of 25 times that of the Sun; the absolute magnitude is − 7.

The largest star known was formerly believed to be the invisible secondary of the eclipsing binary ε Aurigæ, with a diameter of about 5 700 000 000 km. Eclipses occur every 27 years; the latest eclipse began in 1982. However, it is now thought more likely that the secondary is a relatively small, hot star surrounded by an immense shell of gas, as was originally proposed by M. Hack. The system of α Herculis, made up of a red variable supergiant with a double companion, is enveloped in a huge cloud of gas which could have a diameter of 250 000 000 000 km. Some red supergiants, such as Betelgeux, have diameters exceeding 400 000 000 km. In 1978 it was announced that Betelgeux is surrounded by an extremely tenuous 'shell' of potassium, about 11 000 astronomical units in diameter.

The smallest known star, excluding neutron stars, is the white dwarf LP 327−16, discovered in May 1962 at Minneapolis. The diameter is about 1700 km (half that of the Moon), and the distance from us is 100 light-years. It is possible that another white dwarf, LP 768−500, is even smaller (less than 1600 km).

The first measures of stellar diameters using an intensity interferometer have been made by J. Hanbury Brown, from Australia. Some of the values found are as follows:

Star	Diameter, km
Canopus	115 000 000
β Carinæ	12 400 000
Vega	7 700 000
Regulus	6 600 000
α Gruis	6 100 000
Procyon	5 700 000
Sirius	4 800 000
Altair	4 700 000
Fomalhaut	4 600 000
ζ Puppis	4 400 000
Spica	2 200 000

Luminosity and absolute magnitude

Many people prefer to visualize a star's luminosity in terms of the Sun as unity rather than as absolute magnitude. The following conversion table is approximate, but good enough for most purposes.

CONVERSION OF ABSOLUTE MAGNITUDE
(M) TO SOLAR LUMINOSITIES (L)

M	L		
−20	8 300 000 000	+0·5	52·5
−19·5	5 250 000 000	+1	33·1
−19	3 300 000 000	+1·5	20·9
−18·5	2 000 000 000	+2	13·2
−18	1 300 000 000	+2·5	8·3
−17·5	832 000 000	+3	5·2
−17	525 000 000	+3·5	3·3
−16·5	330 000 000	+4	2·1
−16	209 000 000	+4·5	1·3
−15·5	132 000 000	+5	0·8
−15	83 200 000	+5·5	0·5
−14·5	52 500 000	+6	0·3
−14	33 100 000	+6·5	0·2
−13·5	20 900 000	+7	0·1
−13	13 200 000	+7·5	0·08
−12·5	8 320 000	+8	0·05
−12	5 250 000	+8·5	0·03
−11·5	3 310 000	+9	0·02
−11	2 090 000	+9·5	0·01
−10·5	1 320 000	+10	0·008
−10	832 000	+10·5	0·005
−9·5	525 000	+11	0·003
−9	331 000	+11·5	0·002
−8·5	209 000	+12	0·001
−8	132 000	+12·5	0·000 8
−7·5	83 200	+13	0·000 5
−7	52 500	+13·5	0·000 3
−6·5	33 100	+14	0·000 2
−6	20 900	+14·5	0·000 1
−5·5	13 200	+15	0·000 08
−5	8 320	+15·5	0·000 05
−4·5	5 250	+16	0·000 03
−4	3 310	+16·5	0·000 02
−3·5	2 090	+17	0·000 01
−3	1 320	+17·5	0·000 008
−2·5	832	+18	0·000 005
−2	525	+18·5	0·000 003
−1·5	331	+19	0·000 002
−1	209	+19·5	0·000 001
−0·5	132	+20	0·000 000 8
0	83·2		

DOUBLE STARS

The term 'double star' was first used by Ptolemy, who wrote that η Sagittarii was 'διπλους'. There are, of course, several doubles which may be separated with the naked eye, so that they have presumably been known from antiquity; the most celebrated is ζ Ursæ Majoris (Mizar), which makes a naked-eye pair with 80 Ursæ Majoris (Alcor). The Arabs gave full descriptions of it – though it is true that they apparently regarded Alcor as rather a difficult object. (This is not true today, but it is unlikely that there has been any real change.)

The first double star discovered telescopically was Mizar itself, which is made up of two components 14"·5 apart. The discovery was made by Riccioli in 1651. Alcor is 700" from the main pair, which is rather too wide for a recognized 'double' as entered in the official catalogues. The duplicity of γ Arietis was discovered by Robert Hooke in 1665, while he was searching telescopically for a comet.

The first southern double star to be discovered was α Crucis, by Father Guy Tachard in 1685. Tachard was on his way to Siam, by sea, and stopped off at the Cape of Good Hope, where he was warmly welcomed by the Dutch settlers and set up a temporary observatory, mainly for navigational purposes. He recorded that 'the foot of the Crozier marked in Bayer is a Double Star, that is to say, consisting of two bright stars distant from one another about their own Diameter, only much like to the most northern of the Twins; not to speak of a third much less, which is also to be seen, but further from these two.' The Crozier is the Southern Cross; 'Bayer' refers to J. Bayer's famous star catalogue of 1603, and the 'northern of the Twins' is Castor, already known to be double. Like most of his contemporaries, Tachard believed the stars to show definite apparent diameters rather than being virtual point sources.

Other doubles discovered at an early stage included α Centauri (1689), γ Virginis, and the 'Trapezium', θ Orionis, in the Orion Nebula, which is a multiple system.

The first true list of doubles was published in 1871 by C. Mayer of Mannheim. His list included γ Andromedæ, ζ Cancri, α Herculis and β Cygni. Mayer used an 8 ft mural quadrant, with magnifications of 60 and 80.

The first comment upon possible physically-associated or binary pairs was made by the Rev. John Michell in 1767, who wrote: 'It is highly probable in particular, and next to a certainty in general, that such double stars as appear to consist of two or more stars placed very near together, do really consist of stars placed near together, and under the influence of some general law.' Michell repeated this view in 1784. However, in 1782 William Herschel had commented that it was 'much too soon to form any theories of small stars revolving round large ones'.

The first proof of the existence of binary pairs was given by Herschel in 1802. From 1779 he had been attempting to measure the parallaxes of stars, and he had concentrated upon the doubles, since if one member of the pair were more remote than the other it followed that the closer member should show an annual parallax relative to the more distant member. He failed, because his equipment was not sufficiently sensitive, but he made the fortuitous discovery that some of the pairs under study (such as Castor) showed orbital motion, and by 1802 he was confident enough to publish his findings. His classic paper actually appeared in the *Philosophical Transactions* on 9 June 1803.

The first successful measurement of a star-distance was announced in 1838 by F. W. Bessel of Königsberg; the star concerned was 61 Cygni, which Bessel had selected because it had relatively large proper motion and was a wide binary – presumably indicating that by stellar standards it was comparatively close. At about the same time T. Henderson announced a parallax value for α Centauri, based upon measurements which he had made earlier (1831–33) at the Cape, while Struve in Russia gave a rather less accurate value for the parallax of Vega.

Rather surprisingly, binaries are more common than optical pairs, in which the components are not genuinely associated. It has even been estimated that more than 50 per cent of all stars are members of binary systems, with a mean separation from 10 to 20 astronomical units, though this may be too high a ratio. A visual binary will have an orbital period of at least 2 years. It is important to note that both components of a binary move round their common centre of gravity, and the orbits are not in general so dissimilar as might be thought, since in mass there is a far lower spread among the stars than there is in luminosity and in size.

The first reliable orbit for a binary pair (ξ Ursæ Majoris) was worked out by the French astronomer Felix Savary in 1830. (The period is 60 years.) Such calculations are of great importance, since they lead to a determination of the combined masses of the components – something which is much more difficult to calculate for a single star. In fact, our knowledge of stellar masses depends very largely upon the orbital movements of binaries.

Many important catalogues of double stars have been published since Mayer's. Among them are the catalogues of F. G. W. Struve (1822, with later additions), E. Dembowski (Naples, 1852; over 20 000 measures), S. W. Burnham (1870, and again in 1906, listing 13 665 pairs – he was personally responsible for the discovery of 1340 of them), R. Aitken (1932; 17 180 pairs) and the Lick Index Catalogue or IDS (1963; 65 000 pairs, of which 40 000 are binaries). Work of the greatest importance was carried out at the Republic Observatory, Johannesburg (formerly the Union Observatory) between 1917 and 1965, under the successive directorships of R. T. A. Innes (1917–27), H. E. Wood (1927–41), W. H. van den Bos (1941–1956) and W. S. Finsen (1957–65). The telescope used for most of the work was the 27 in refractor. It is a matter of great regret that the virtual closing of the Observatory, following the concentration of South African astronomy at Sutherland in Cape Province, has brought this work to an end – temporarily, one must hope.

The first spectroscopic binary

(Mizar A) was discovered in 1889 by E. C. Pickering at Harvard; another identification (β Aurigæ) soon followed. Spectroscopic binaries have too small a separation for the components to be seen individually, but the binary nature of the system betrays itself because of the Doppler shifts in the spectra. If both spectra are visible, the absorption lines will be periodically doubled; if one spectrum is too faint to be seen, the lines due to the primary will oscillate about a mean position. There are some 'borderline' cases; thus Capella was long known to be a spectroscopic binary, but the world's largest telescopes can just indicate that it is not a single star. The orbital period is 100 days.

The first astrometric binaries to be studied were Sirius and Procyon, by F. W. Bessel in 1844. In an astrometric binary, the presence of an invisible companion is inferred from slight 'wobblings' of the primary. In both these first cases, the companions were subsequently discovered, but with other astrometric binaries the secondary stars remain unseen.

The first White Dwarf binary companion to be discovered (in fact, the first white dwarf of any kind) was Sirius B, or the Companion of Sirius. It was first seen in 1862 by Clark, at Washington. The orbital period is 50 years, and the maximum separation is 11″·5 (as in 1975). For many years after its discovery the Companion was assumed to be large and red, but in 1915 W. S. Adams, at Mount Wilson, studied its spectrum and found that it was white; the surface temperature was at least 8000 °C. Since the luminosity was only 1/10 000 that of Sirius itself, the Companion had to be very small; smaller, in fact, than Uranus or Neptune, and with a diameter only 3 times that of the Earth. The resulting density turned out to be 70 000 times that of water.

It is worth noting that in ancient times Sirius was described as being red. There is no chance that the primary has changed in colour, and suggestions that the Companion was then going through its red giant stage, before collapsing into the white dwarf condition, seem to be untenable. We must therefore assume that there was some error in

observation or interpretation in the old records, though an element of mystery remains.

Binary stars are of many different kinds; sometimes the components are dissimilar (as with Sirius), sometimes they are identical twins – as with γ Virginis, which has a period of 180 years. Several decades ago it was wide and easy, but is now closing, and by 2016 will be single except in giant telescopes. This does not, of course, indicate any actual closing; everything depends upon the angle from which we see the pair. Another binary, formerly easy but now much less so, is Castor. In fact both components of the bright pair are spectroscopic binaries, and also associated with the system is Castor C or YY Geminorum, made up of two red dwarfs; it is an eclipsing variable. Castor therefore consists of six stars, four luminous and two dim.

The first eclipsing binary to be discovered was Algol, by G. Montanari in 1669 (it is not now thought that the 'winking' was known in ancient times, even though Algol was nicknamed the Demon Star). The cause of the variability was explained by J. Goodricke, the deaf-mute astronomer, in 1782.

The most extreme eclipsing binary is probably ε Aurigæ. The variability was discovered by Fritsch in 1821, but was then thought to be irregular. In fact the system is made up of a powerful supergiant together with an invisible secondary, detectable only in infra-red, which may be a hot star surrounded by an immense shell of dust and gas. Close to it in the sky is another extreme eclipsing binary ζ Aurigæ, with a K-type supergiant primary and a B-type secondary; when the B-star passes behind the supergiant, its light comes to us for a period via the outer layers of the primary, and the spectral effects are both complicated and highly informative. Another interesting system is β Lyræ. Here, and also with other stars of the same type, the two components are almost in contact; they are tidally distorted into elliptical shapes, and the whole system is enveloped in gas.

The binary with the shortest known period is X-1820–303, an X-ray star in the globular cluster NGC 6624, distant 30 000 light-years. Its

Alpha Centauri, a fine binary, with its distant companion Proxima.

period is 685 seconds, or 11 minutes. It was discovered in 1987 by the aptly-named Luigi Stella and collaborators with the satellite Exosat.

It is impossible to say which is the binary with the longest known period. With very large separations the period may amount to millions of years, and it is better merely to say that very widely separated components share a common binary motion through space.

The first binary White Dwarf was found by W. Luyten and P. Higgins in 1973. Its position is R.A. 9 h 42 m, dec. +23°41′. The separation is 13″, the position angle 052, and the period 12 000 years; at present the two components are 600 astronomical units apart.

The first binary neutron star was discovered in July 1974. The orbital period is 8 hours. The discovery was made by means of radio observations, and the object (pulsar PSR 1913 +16)

has not been seen optically, but it is assumed to be a binary made up of two neutron-star components.

The first X-ray binaries were reported following the launch of the first X-ray astronomical satellite, UHURU, in 1970. X-ray binaries are white dwarfs or neutron stars in orbit around another star – a B star, K star or another white dwarf. The X-ray star is pulling material off the companion, perhaps at 100 million million tonnes per second, and X-rays are being emitted.

The origin of binary systems is not known with certainty. The old theory – that a binary formed as a result of the fission of a single star – has been rejected by most astronomers. It has been suggested that binaries may be due to the mutual capture of the components, but in most cases it is believed that the components were formed from the same cloud of interstellar material in the same region of space. If the components are identical or near-identical twins, the spectra are generally the same. When there is a definite difference in brightness between the components, the spectra also differ. If both stars belong to the Main Sequence, the primary is usually of earlier type than the secondary, while if the primary is a giant the secondary is either a giant of earlier type or else a dwarf of similar spectral type. Novæ are binary systems (DQ Herculis 1934 is a good example; it is an eclipsing system with a period of 4 h 39 m made up of the old nova itself, probably so dense that it resembles a white dwarf, together with an invisible companion which is probably a red dwarf of type M). We also have the SS Cygni or U Geminorum stars, all of which seem to be binaries. In a star of this kind the primary is a white dwarf, the secondary a Main Sequence star, presumably a red dwarf. The components are very close together, and the white dwarf is pulling material away from the less massive secondary; gas moves across to the primary in a narrow stream, producing a disk or shell of gas round the white dwarf. Where the stream hits the disk there is a bright spot, and irregularities in the stream make the light we receive flicker rapidly with small amplitude; this flickering is detectable only with very sensitive

photoelectric equipment. The larger outbursts in the SS Cygni stars originate in the white dwarf component of the system. It is easy to see why these variables are sometimes termed dwarf novæ.

The most spectacular multiple stars in the sky are probably ε Lyræ and θ Orionis. ε Lyræ has two main components, making up a naked-eye pair; each component is again double, and all four may be distinguished with a fairly small telescope. θ Orionis, in the great nebula M.42, is nicknamed the 'Trapezium' for obvious reasons. It lies on the outskirts of the nebula, and is responsible for making the nebulosity luminous.

The position angle of a visual double star (either a binary or an optical pair) is measured according to the angular direction of the secondary (B) from the primary (A), reckoned from 000 at north round by east (090), south (180) and west (270) back to north. With rapid binaries the separations and position angles alter quickly; a good example is the fine binary ζ Herculis, where the magnitudes are 3 and 5·6, and the period is only 34 years. Some of the published catalogues are already out of date, and need revising. There is scope here for the skilful and well-equipped amateur as well as for the professional, and certainly double-star research is a fascinating branch of astronomy.

VARIABLE STARS

THE CLASSIFICATION OF VARIABLE STARS

Variable stars are of many kinds. Elaborate systems of classifying them have been proposed, and the notes given below are not intended to be more than a general guide.

ECLIPSING VARIABLES

Better termed 'eclipsing binaries', since there is no intrinsic variation. The light-changes are caused by one component of the binary system passing in front of the other.

Algol type (EA). A principal minimum and a very small (almost imperceptible) secondary minimum. Thus the star remains at its maximum brightness for most of the time.

Beta Lyræ type (EB). Components so close that they are distorted into elliptical shapes, and may be almost in contact. Light-variations continuous, with alternate deep and shallow minima.

W Ursæ Majoris type (EW). Dwarf pairs, with ellipsoidal components almost in contact; the periods are less than one day.

PULSATING VARIABLES

The variations are intrinsic; the star expands and contracts.

Mira variables (M), often called long-period variables. Named after Mira Ceti, the brightest and first-discovered member of the class. They are late-type giants of types M, R, N or S, with periods of from about 80 to almost 1000 days, though neither the periods nor the amplitudes are constant. The median absolute magnitudes are from $+2$ to -2, and the range may be very great – 11 magnitudes in the case of χ Cygni. The spectra show emission lines, and many of the stars may be binary systems.

Semi-regular variables. These are also of late spectral type. The periods are often so ill-defined as to be virtually unrecognizable, and the amplitudes are less than with the Mira stars. The periods range from around 30 to around 1000 days. There are various subdivisions. SRa: late-type supergiants with relatively stable cycles. SRb: also late-type, but less regular. SRc: young semi-regular supergiants. SRd: giants of types F, G or K.

Cepheids. Named after the prototype star, δ Cephei. They are yellow supergiants, and show the celebrated period-luminosity law which gives a key to their distances. The period of pulsation is the time required for a vibration to travel from the surface of the star to

NGC 6960, the Veil Nebula in Cygnus – part of a supernova remnant. (Palomar photograph)

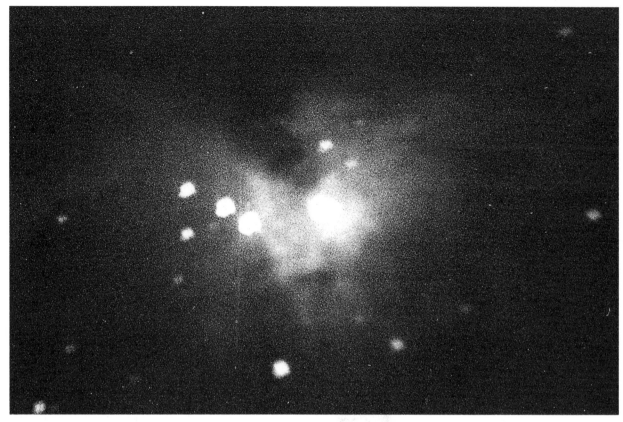

M.42, the Orion Nebula, showing the stars of the Trapezium (Theta Orionis). Photograph by Commander H. R. Hatfield, 30-cm reflector. Exposure 9 minutes.

the centre and back again, so that large stars have longer periods than smaller ones. *Classical Cepheids* (Cδ) are F, G or K supergiants, and are highly evolved, so that they have exhausted their available hydrogen and helium; the periods range from a few days to 70 days, with amplitudes of from 0·1 to 2 magnitudes. The rise to maximum is sharper than the subsequent decline, and the light-curves are symmetrical (or almost so). *W Virginis stars* or Population II Cepheids have lower masses, and are about 2 magnitudes fainter than classical Cepheids of equal periods. They are also less regular. The brightest example is the southern κ Pavonis.

RR Lyræ variables. Formerly known as cluster-Cepheids, though not all are found in clusters; RR Lyræ itself is not. The spectra are of type A or F. They are old, with masses lower than that of the Sun, but radii 4 to 5 times greater, and all are of about the same luminosity. The range is about 1 magnitude, and the periods are from 0·2 to 1·2

days. There are various subdivisions: RRa (RR Lyræ) have amplitudes around 1 magnitude, with periods around 0·5 day and asymmetrical light-curves. RRb: similar, but with smaller amplitudes (0·3 to 0·8 magnitude) and longer periods (around 0·7 day). The maxima are more rounded, and the asymmetry less pronounced. RRc: almost sinusoidal light-curves, with amplitude around 0·5 magnitude and period around 0·3 day. Some RR Lyræ stars, such as AR Herculis, have several pulsation periods, and this results in a continuous deformation of the light-curve (Blazhko effect).

RV Tauri variables. These are pulsating supergiants, usually of types F to K, but occasionally M. There are alternate deep and shallow minima; the interval between successive primary minima may be from 30 to 150 days. There are several subdivisions; RVa: the period is not constant, but is at least reasonably consistent; the brightest member of the class, R Scuti, is of this type. RVb: several superposed cycles,

with rough periods from 30 to 1000 days. RVc: very rough periods; these are exceptionally luminous supergiants of type M. RVd: yellow giants, with periods and amplitudes much less predictable.

Dwarf Cepheids, formerly classed with the RR Lyræ stars as RRs; the prototype is AI Velorum. They are of types A to F, with absolute magnitudes of from +1 to +5 and periods from 0·05 to 0·25 day.

δ Scuti variables (δ S), once confused with dwarf Cepheids. They have smaller amplitudes, often below 0 m·1, and periods from 0·02 to 0·25 day. They are young, with spectra of types A to F, and many are spectroscopic binaries. Subdwarfs showing the same characteristics are termed *SX Phœnicis stars*.

β Canis Majoris or β Cephei variables. B0 to B3 giants or subgiants, with periods of from 0·1 to 0·3 day, and amplitudes from 0·1 to 0·3 magnitude. They are relatively massive stars which have almost exhausted their core hydrogen. Ultra-short periods of a few hundredths of a day are suffixed s; such a star is χ Centauri.

α Cygni variables. Pulsating supergiants of types B or A, with short periods and amplitudes below 0 m·1.

ZZ Ceti variables. Pulsating white dwarfs with periods which may be as short as 30 seconds and amplitudes below 0 m·2. The longest periods are around 23 minutes. Flares of about a magnitude, due to interactions with a close companion, are sometimes seen.

Irregular variables. Slowly-varying stars denoted by L. Lb variables are of late spectral type, and Lc stars are late-type supergiants.

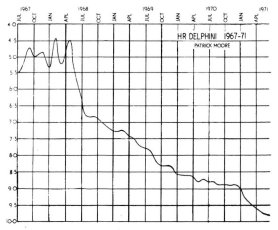

Variation of magnitude of HR Delphini 1967–71
(Patrick Moore)

ERUPTIVE VARIABLES

Novæ. Novæ can be spectacular in the extreme. A formerly faint star suddenly flares up, and becomes thousands of times brighter than its usual brilliance, remaining at maximum for a brief period before declining again. The rise may take only a few hours – as with the bright nova V1500 Cygni of 1975, which reached an absolute magnitude of −10 and an apparent magnitude 1·8, but within a week or so had fallen below naked-eye visibility. This was a fast nova; slow novæ, such as HR Delphini of 1967, are much more gradual in their decline.

A nova is a binary system, consisting of a low-density red star with a white dwarf companion. The white dwarf pulls material away from the red star, and this material produces a ring 9 r 'accretion disk' round the white dwarf. Eventually there is a nuclear outburst in the atmosphere of the white dwarf, and gas is ejected at high velocity; at the end of the outburst the system returns to its former state. Some stars, such as T Coronæ, have shown more than one outburst; these are the *recurrent novæ*.

Dwarf novæ, otherwise known as U Geminorum or SS Cygni stars, show minor outburst at roughly regular intervals. Of the two components, one is a K or M red dwarf and the other is a white dwarf. The amplitude is of the order of 3 to 5 magnitudes, and the interval between outbursts may be from 10 days to several years. The basic cause of variation is the same as with true novæ, but on a much reduced scale. *SU Ursæ Majoris* stars have both normal maxima and 'supermaxima' of greater amplitude. *Z Camelopardalis* stars show outbursts from 2 to 5 magnitudes every 10 to 40 days, but with unpredictable 'standstills' when variation is virtually suspended.

Nova-like variables. There are various types. γ *Cassiopeiæ* variables (γ C) are rapidly-rotating B-type stars with emission lines in their spectra; they are losing mass from their equatorial regions as they evolve off the Main Sequence. They have small amplitudes, usually less than 2 magnitudes. Stars of the *Z Andromedæ* or *Symbiotic* type are close binaries, often involved in nebulosity. One component is a cool red star, while the other is a hot star; the variations are caused by pulsations in the red star together with interactions between the two. *RR Telescopii* stars show slow increases which may be of very long duration, and it is suggested that they may be in the process of becoming planetary nebulæ. *S Doradûs variables* are very luminous, with absolute magnitudes of around −10 and spectra of B to F; they show variations from 1 to 3 magnitudes which may be either cyclic or irregular.

P Cygni variables, also very luminous, produce expanding gas-shells.

Nebular variables are very young. *In*: T Orionis stars, with spectra of from type B to K; they are of low mass, and fluctuate irregularly. *Is* stars, such as RW Aurigæ, are similar, but are not associated with nebulæ. They have amplitudes of from 0·5 to 1 magnitude, and vary over periods from a few hours to several days. *IsT*: T Tauri variables are extremely young, and show irregular fluctuations of small amplitude. *FU Orionis* stars have spectra of from type A to F; they may brighten up by several magnitudes over a period of several months, and stay at maximum for decades. At maximum, emission lines appear in their spectra.

Flare stars (UV Ceti type) are dwarfs of type K or M. They show sudden outbursts of from 1 to 6 magnitudes, lasting for several minutes. 'Flash' variables (UVa) are of earlier spectral type, and are more luminous; they are associated with nebulosity.

R Coronæ Borealis variables are deficient in hydrogen but rich in carbon. They are highly luminous, and of spectral type F, G, K or R. They remain at maximum for most of the time, but undergo sudden, unpredictable drops to minimum, taking from several weeks to several months to recover. The amplitudes are large – at least 10 magnitudes in the case of R Coronæ itself. The fadings are due to clouds of 'soot' accumulating in the star's atmosphere. R Coronæ variables are rare.

ROTATING VARIABLES

These have small amplitudes. *BY Draconis* stars are young, rapidly-rotating K or M dwarfs with emission lines in their spectra; they are quasi-periodic, with amplitudes of no more than 0·3 magnitude, and periods from a few hours to 120 days. The variations are due to their non-uniform surface brightness. *FK Comæ* stars are of type G; they have fast rotation and great chromospheric activity. The rapid span of FK Comæ itself may be due to the presence of a close, low-mass companion in a decaying orbit, leading possibly to eventual coalescence. *RS Canum Venaticorum* stars are binaries, of which one component is a rapidly-rotating F to K giant or subgiant; the periods range from a few hours up to several months. The variations are due to 'starspots' and, in some cases, eclipses. *Ellipsoidal variables* are also binaries, which are close together and gravitationally distorted; there are no eclipses, and the fluctuations are due to the differing amount of surface which is presented to us.

α^2 **Canum Venaticorum** stars are magnetic or spectrum variables, usually of type A, but with enhanced lines of some elements such as silicon, strontium, chromium or europium. The changes in visual light are small, usually below 0 m·1. *SX Arietis* stars are helium variables of type B, with intense helium and silicon lines in their spectra, and strong, variable magnetic fields.

Secular variables are stars which may have brightened or faded in historical times. Thus Ptolemy and others ranked β Leonis and θ Eridani as being of the first magnitude, whereas today they are respectively of magnitude 2 and below 3. δ Ursæ Majoris was ranked as equal to the other stars of the Plough, but is now at least a magnitude fainter; on the other hand α Ophiuchi was ranked as of magnitude 3, and is now 2. These variations must be regarded as suspect, since it would be unwise to trust the old estimates too far.

The first variable star to be positively identified as such was Mira (o Ceti), in 1638. It had been previously recorded by Fabricius (1596) and by Bayer (1603) and had even been allotted a Greek letter, so that it is surprising that its fluctuations were not tracked down earlier. In the latter part of the 17th century two more variables were identified – Algol, now known to be an eclipsing binary, and the long-period χ Cygni.

The following variables were identified between 1638 and 1850 (novæ are not included):

Star	Discoverer of variability	Date
Mira (o Ceti)	Holwarda	1638
Algol (β Persei)	Montanari	1669
χ Cygni	Kirch	1686
R Hydræ	Maraldi	1704
Rasalgethi (α Herculis)	W. Herschel	1759
μ Cephei	W. Herschel	1782
R Leonis	Koch	1782
δ Cephei	Goodricke	1784
β Lyræ	Goodricke	1784
η Aquilæ	Pigott	1784
R Scuti	Pigott	1795
R Coronæ Borealis	Pigott	1795
R Virginis	Harding	1809
R Aquarii	Harding	1811
ε Aurigæ	Fritsch	1821
R Serpentis	Harding	1826
η Carinæ	Burchell	1827
S Serpentis	Harding	1828
U Virginis	Harding	1831
δ Orionis	J. Herschel	1834
S Vulpeculæ	Rogerson	1837
Betelgeux (α Orionis)	J. Herschel	1840
β Pegasi	Schmidt	1847
λ Tauri	Baxendell	1848
R Orionis	Hind	1848
R Pegasi	Hind	1848
R Capricorni	Hind	1848
S Hydræ	Hind	1848
S Cancri	Hind	1848
S Geminorum	Hind	1848
R Geminorum	Hind	1848
T Geminorum	Hind	1848
R Tauri	Hind	1849
T Virginis	Boguslawsky	1849
T Cancri	Hind	1850
R Piscium	Hind	1850

Sir John Herschel regarded Alphard (α Hydræ) as a variable, but this has not been confirmed. It is now known that δ Orionis has a very small range, so that Herschel's discovery may not have been valid.

A CATALOGUE OF BRIGHT VARIABLE STARS

The following list includes variable stars with a maximum of magnitude 6 or brighter, and a range of at least 0 m·4.

MIRA VARIABLES

Star	Max.	Min.	Period, d.	Spectrum
R And	5·8	14·9	409	S
R Aql	5·5	12·0	284	M
R Car	3·9	10·5	309	M
S Car	4·5	9·9	149	K–M
R Cas	4·7	13·5	430	M
R Cen	5·3	11·8	546	M
o Cet	1·7	10·1	332	M Mira
S CrB	5·8	14·1	360	M
χ Cyg	3·3	14·2	407	S
U Cyg	5·9	12·1	462	N
R Gem	6·0	14·0	370	S
S Gru	6·0	15·0	401	M
R Hor	4·7	14·3	404	M
R Hya	4·0	10·0	390	M
R Leo	4·4	11·3	312	M
R Lep	5·5	11·7	432	N
V Mon	6·0	13·7	334	M
X Oph	5·9	9·2	334	M + K
U Ori	4·8	12·6	372	M
RU Sgr	6·0	13·8	240	M
RT Sgr	6·0	14·1	305	M
RR Sco	5·0	12·4	279	M
S Scl	5·5	13·6	365	M
R Ser	5·1	14·4	356	M
R Tri	5·4	12·6	266	M
SS Vir	6·0	9·6	355	N
R Vir	6·0	12·1	146	M

SEMI-REGULAR VARIABLES

Star	Max.	Min.	Period, d.	Spectrum
UU Aur	5·1	6·8	234	N
W Boö	4·7	5·4	450	M
VZ Cam	4·7	5·2	24	M
X Cnc	5·6	7·5	195	N
TU CVn	5·6	6·6	50	M
S Cen	7·0	65	0	N
T Cen	5·5	9·0	90	K–M
T Cet	5·0	6·9	159	M
FS Com	5·3	6·1	58	M
W Cyg	5·0	7·6	126	M
EU Del	5·8	6·9	59	M
R Dor	4·8	6·6	338	M
UX Dra	5·9	7·1	168	N
RY Dra	5·6	8·0	173	N
η Gem	3·2	3·9	233	M Propus
π¹ Gru	5·4	6·7	150	S
g Her	5·7	7·2	70	M
α Her	3	4	±100	M Rasalgethi
R Lyr	3·9	5·0	46	M
ε Oct	4·9	5·4	55	M
α Ori	0·1	0·9	2110	M Betelgeux
W Ori	5·9	7·7	212	N
CK Ori	5·9	7·1	120	K
Y Pav	5·7	8·5	233	N
SX Pav	5·4	6·0	50	M
β Peg	2·3	2·8	38	M Scheat
ρ Per	3	4	33–55	M
TV Psc	4·6	5·4	70	M
L² Pup	2·6	6·2	140	M
R Scl	5·8	7·7	370	N
RR UMi	6·0	6·5	40?	M

CEPHEIDS

Star	Max.	Min.	Period, d.	Spectrum
η Aql	3·5	4·4	7·2	F–G
RT Aur	5·0	5·8	3·7	F–G
ZZ Car	3·3	4·2	36·5	F–K
U Car	5·7	7·0	38·8	F–G
SU Cas	5·7	6·2	1·9	F
δ Cep	3·5	4·4	5·4	F–G
AX Cir	5·6	6·1	5·3	F–G
X Cyg	5·9	6·9	16·4	F–G
β Dor	3·7	4·1	9·8	F–G
ζ Gem	3·7	4·1	10·1	F–G
T Mon	6·0	6·6	27·0	F–K
S Mus	5·9	6·4	9·7	F
R Mus	5·9	6·7	7·5	F
Y Oph	5·9	6·4	17·1	F–G
κ Pav	3·9	4·7	9·1	F (W Virginis type)
S Sge	5·3	6·0	8·4	F–G
X Sgr	4·2	4·8	7·0	F
W Sgr	4·3	5·1	7·6	F–G
Y Sgr	5·4	6·1	5·8	F
AH Vel	5·5	5·9	4·2	F
T Vul	5·4	6·1	4·4	F–G

RV TAURI VARIABLE

Star	Max.	Min.	Period, d.	Spectrum
R Sct	4·4	8·2	140	G–K

SYMBIOTIC VARIABLES

	Max.	Min.	Period, d.	Spectrum
R Aqr	5·8	12·4	387	M + P
AG Peg	6·0	9·4	830	WN + M

RECURRENT NOVÆ

	Max.	Min.	Spectrum	Outbursts
T CrB	2·0	10·8	M + Q	1866, 1946 Blaze Star
RS Oph	5·3	12·3	O + M	1901, 1933, 1958, 1967

R CORONÆ VARIABLES

	Max.	Min.	Spectrum
R CrB	5·7	15	Fp
RY Sgr	6·0	15	Gp

ERUPTIVE VARIABLES

	Max.	Min.	Spectrum
U Ant	5·7	6·8	N
η Car	−0·8	7·9	Pec
ρ Cas	4·1	6·2	F–K (Occasional fades)
α Cas	2·1?	2·5?	K (Suspected variable)
γ Cas	1·6	3·3	B
μ Cen	2·9	3·5	B
μ Cep	3·4	5·1	M
θ Cir	5·0	5·4	B
P Cyg	3	4	Bp
T Cyg	5·0	5·5	K
BU Gem	5·7	7·5	M
RX Lep	5·0	7·0	M
S Mon	4	5	O7
BO Mus	6·0	7·7	M
χ Oph	4·2	5·0	B
λ Pav	3·4	4·3	B
X Per	6·0	7·0	O9·5 X-ray star
d Ser	4·9	5·9	G + A
VY UMa	5·9	6·5	N
BU Tau	4·8	5·5	Bp Pleione

ECLIPSING BINARIES

(Algol type)

	Max.	Min.	Period, d.	Spectrum
R Ara	6·0	6·9	4·4	B
WW Aur	5·8	6·5	2·5	A + A
R CMa	5·7	6·3	1·1	F
RS Cha	6·0	6·7	0·1	A + F
δ Lib	4·9	5·9	2·3	B
U Oph	5·9	6·6	1·7	B + B
β Per	2·2	3·4	2·9	B + G Algol
ζ Phe	3·9	4·4	1·7	B + B
RS Sgr	6·0	6·9	2·4	B + B
λ Tau	3·3	3·8	3·9	B + A
HU Tau	5·9	6·7	2·1	A

(Beta Lyræ type)

	Max.	Min.	Period, d.	Spectrum
UW CMa	4·0	5·3	4·3	O7
u Her	4·6	5·3	2·0	B + B
GG Lup	5·4	6·0	2·1	B + A
β Lyr	3·3	4·3	12·9	B + A Sheliak
V Pup	4·7	5·2	1·4	B + B

(Long period)

	Max.	Min.	Period, d.	Spectrum
ε Aur	2·9	3·8	9892	F
ζ Aur	3·7	4·1	972	K + B
VV Cep	4·8	5·4	7430	M + B

BRIGHT NOVÆ

The following list includes all novæ since 1600 to have attained magnitude 6·0 or brighter.

Nova	Year	Max. mag.	Discoverer
CK Vulpeculæ	1670	3	Anthelm
WY Sagittæ	1783	6	D'Agelet
V. 841 Ophiuchi	1848	4	Hind
Q Cygni	1876	3	Schmidt
T Aurigæ	1891	4·2	Anderson
V. 1059 Sagittarii	1898	4·9	Fleming
GK Persei	1901	0·0	Anderson
DM Geminorum	1903	5·0	Turner
OY Aræ	1910	6·0	Fleming
DI Lacertæ	1910	4·6	Espin
DN Geminorum	1912	3·3	Enebo
V. 603 Aquilæ	1918	−1·1	Bower
GI Monocerotis	1918	5·7	Wolf
V. 476 Cygni	1920	2·0	Denning
RR Pictoris	1925	1·1	Watson
XX Tauri	1927	6·0	Schwassmann and Wachmann
DQ Herculis	1934	1·2	Prentice
CP Lacertæ	1936	1·9	Gomi
V. 630 Sagittarii	1936	4·5	Okabayasi
BT Monocerotis	1939	4·3	Whippie and Wachmann
CP Puppis	1942	0·4	Dawson
DK Lacertæ	1950	6·0	Bertaud
V. 446 Herculis	1960	5·0	Hassell
V. 533 Herculis	1963	3·2	Dahlgren and Peltier
HR Delphini	1967	3·7	Alcock
LV Vulpeculæ	1968	4·9	Alcock
FH Serpentis	1970	4·4	Honda
V. 1500 Cygni	1975	1·8	Honda

SUPERNOVÆ

Supernovæ are the most colossal outbursts known in nature. Many have been seen in external galaxies, but the last supernova to be seen in our Galaxy was observed as long ago as 1604 – before the invention of the telescope. However, the 1987 supernova in the Large Cloud of Magellan attained naked-eye visibility.

Supernovæ are of two types, quite different in origin.

Type I. Here we are dealing with a binary system, of which one component (A) is initially more massive than its companion (B). A therefore evolves more quickly into the red giant stage. Material from it is pulled across to B, so that B grows in mass while A declines; eventually B becomes the more massive of the two. Meanwhile, A is reduced to a very small, dense core, made up chiefly of carbon. The situation is then reversed; B evolves in its turn, swells out, and starts to lose material back to

the shrunken A. The white dwarf builds up a gaseous layer, composed largely of hydrogen. But when the mass of the white dwarf exceeds the Chandrasekhar limit (around 1·4 times the mass of the Sun) the carbon detonates, and in a matter of a few seconds the white dwarf is completely destroyed. The resulting outburst takes a long time to die down. When the carbon white dwarf explodes, it creates other elements including neon, oxygen and silicon, ending up as nickel. Nickel decays to cobalt and then to iron, so that these elements can be detected spectroscopically. The energy produced is remarkable; the absolute magnitude of a Type I supernova at its peak can reach −19.

Type II. A supernova outburst of this type is due to the collapse of a very massive star – at least 8 times as massive as the Sun – which has used up its main nuclear fuel, and has produced a nickel-iron core with a mass about 1·5 times that of the Sun. The structure of the star prior to the outburst has been likened to that of an onion! Outside the iron core is a zone of silicon and sulphur; next comes a layer of neon and magnesium; then a layer of carbon, neon and oxygen; then a layer of helium, and finally an outer layer of hydrogen. Eventually the core can no longer support the pressure of the outer layers, and collapses to form a neutron star of immense density (one could pack over 2500 million tons of neutron star into a matchbox). The temperature now is of the order of 100 000 million degrees C, and vast numbers of neutrinos are produced, which pass straight through the star into space. The release of neutrons causes the core to 'bounce', and a shock-wave moves outward – to meet the material which is falling inward. The result is a catastrophic explosion. Most of the star's material is blown away, leaving only a small, incredibly dense remnant which may be a neutron star or – in extreme cases – a black hole. The peak luminosity is not so great as with a Type I supernova, but the absolute magnitude may reach a value of around −17, which is more than 500 million times as luminous as the Sun. A supernova remnant (SNR) may be detectable as a pulsar, the classic case being that of the Crab Nebula – known to be the remnant of the supernova observed in the year 1054.

The 1987 supernova in the Large Cloud of Magellan was of Type II. In this case the progenitor star (Sanduleak −69°202) was identifiable on pre-outburst photographs, and was found to have been a B3-type blue supergiant with a surface temperature of around 20 000 °C. It was a young star, only around 20 000 000 years old, with a mass 20 times that of the Sun. The fact that it was bluish came as a surprise, because it had been previously assumed that supernovæ were produced by old red supergiants; also, the peak luminosity was low by supernova standards – 'only' about 250 million times that of the Sun. At its brightest the supernova shone as a star of between magnitudes 2 and 3 even though it was 170 000 light-years away.

GALACTIC SUPERNOVÆ

Four definite galactic supernovæ have been observed in historical times: the stars of 1006, 1054, 1572 and 1604. Several other 'possibilities' have been listed by R. Stephenson (Newcastle University). The list is as follows:

1006	Supernova in Lupus.	
	Max. mag. about	−9·5
1054	Supernova in Taurus.	
	Max. mag. about	−4
1572	Supernova in Cassiopeia.	
	Max. mag. about	−4
1604	Supernova in Ophiuchus.	
	Max. mag. about	−3

185 December Discovered 7 December, near α and β Centauri; Chinese sources. Maximum brightness about −8. Possible identification with radio source G.315·4−2·3 and optical object RCW 86 (also an X-ray emitter). Probably observed by the Chinese for 20 months.

386 In Sagittarius. Chinese sources; visibility extended over 3 months. Several radio sources lie in the general area (near λ Sagittarii) but positive identification is impossible. This star may or may not have been a supernova.

393 February/March In Scorpius, near the tail (λ Scorpii). Visible for about 8 months according to Chinese

sources. Maximum magnitude about 0. Possible identification with radio source G.348·5+0·1 or G.348·7+0·3; distance ±20 000 light-years. Probably a supernova.

1006 April In Lupus, near β. Maximum brightness −9 to −10; according to Chinese and other sources, visible for 2 years. Definitely a supernova; probable identity with radio source G.327·6+14·5. Distance not over 3000 light-years.

1054 July In Taurus, near ζ. Chinese and Japanese sources. Maximum magnitude about the same as that of Venus (−4 to −4·5). Seen for 23 days in daylight and 21 months at night. Definite identification with the Crab Nebula; distance 6000 light-years.

1181 August In Cassiopeia, between ε and ι. Chinese and Japanese sources. Possible identification with radio source G.130·7+3·1. Distance about 25 000 light-years? Possible supernova; evidence by no means conclusive.

1572 November In Cassiopeia, near κ. Tycho's Star. Seen by W. Schüler of Wittenberg on 6 November, but may have been seen on 3 November (Tycho saw it on 11 November). Maximum magnitude slightly brighter than −4. Visible for 18 months. Identity with radio source G.120·1+1·4. The remnant is optically visible, and is an X-ray emitter. Distance about 20 000 light-years.

1604 October In Ophiuchus, near λ. Kepler's Star, discovered on 9 October; Kepler saw it on 11 October. (There may have been a first sighting on 7 October.) Maximum magnitude −2·5. Identified with radio source G.4·5+6·8.

The brightest of these galactic supernovæ was certainly the star of 1006. There has been another supernova about 1700 which has left the remnant which we know as the radio source Cassiopeia A, but the supernova was not observed, as it was too heavily obscured by interstellar material near the plane of the Galaxy.

Many supernovæ have been seen in external galaxies; the only one to have reached the fringe of naked-eye visibility, apart from the 1987 supernova in the Large Cloud of Magellan, was S Andromedæ (1885) in Messier 31.

STAR CLUSTERS AND NEBULÆ

Clusters and nebulæ are among the most striking of stellar objects. Several are easily visible with the naked eye. Few people can fail to recognize the lovely star-cluster of the Pleiades or Seven Sisters, which has been known since prehistoric times and about which there are many old legends. The nebula in the Sword of Orion, the Sword-Handle in Perseus, and Præsepe in Cancer are other objects visible without optical aid; keen-sighted people have little difficulty in locating the great Andromeda Spiral and the globular cluster in Hercules, while in the far south there are the two Clouds of Magellan, which cannot possibly be overlooked, as well as the bright globulars ω Centauri and 47 Tucanæ.

The most famous of all catalogues of nebulous objects was compiled by the French astronomer Charles Messier, and published in 1781. Ironically, Messier was not interested in the objects he listed; he was a comet-hunter, and merely wanted a quick means of identifying misty patches which were non-cometary in nature! In 1888 J. L. E. Dreyer, Danish by birth (though he spent much of his life in Ireland, and ended it in England), published his New General Catalogue (N.G.C.), augmented in 1895 and again in 1908 by his Index Catalogue (I. or I.C.).

The brightest nebular object is the Large Cloud of Magellan, which may be seen even in moonlight. Novæ have been seen in it, and one supernova (1987). A pulsar in it has been detected by J. Ables (Parkes, NSW). Next comes the Small Cloud of Magellan. The brightest cluster visible from England is the cluster of the Pleiades. (This claim may be disputed; the Hyades are also bright, but are somewhat drowned by the brilliant orange-red light of Aldebaran, which is not a bona-fide member of the cluster, but merely happens to lie roughly half-way between the Hyades and ourselves.)

Nebular objects may be divided into various types, all of which are represented in Messier's Catalogue and the N.G.C.:

Open or Loose Clusters. Aggregations of stars, with no particular shape. The Pleiades, the Hyades and Præsepe are good examples. These three (and many others!) are contained in our Galaxy, but other galaxies also include open clusters.

Globular Clusters. Symmetrical systems, which may contain up to a million stars, strongly concentrated towards the centre. In our Galaxy, the globulars lie around the edge of the main system, so that all are very remote. Leading examples are ω Centauri, 47 Tucanæ and M.13 Herculis. Again, other galaxies have globular clusters of their own.

Gaseous nebulæ, also termed **galactic nebulæ**. These may be described as 'stellar birth-places', since fresh stars are condensing out of the nebular gas and dust. The best known

example is M.42, the Sword of Orion. They are immensely rarefied; D. A. Allen has pointed out that if it were possible to take a 1-in core sample right through the Orion Nebula, which is of the order of 30 light-years in diameter, the total 'weight' of material collected would weigh no more than one new British penny. Some nebulæ (such as that in the Pleiades) shine by reflecting the light of stars in or very near them; others (such as the Orion Nebula) shine also because the very hot stars embedded in them ionize the hydrogen, and cause it to emit light on its own account. For this reason, bright emission nebulæ are termed H.II regions. If a nebula has no suitable stars in or near it to cause luminosity, it remains dark, and is detectable only because it blots out the light of stars beyond. The best-known dark nebula is the Coal Sack in Crux. The first exhaustive catalogue of dark nebulæ was published by E. E. Barnard in 1919. There is no basic difference between a bright nebula and a dark one; and indeed a nebula which appears bright to us could well seem

dark if viewed from elsewhere in the Galaxy. We cannot see the objects deep inside nebulæ, but they can be detected by their infra-red radiations; thus the Orion Nebula contains the so-called Becklin's Object, which may well be a remarkably powerful star permanently concealed from us.

Supernova remnants are also known. The classic example is M.1 Tauri (the Crab Nebula), identified with the supernova of 1054. This was first seen by Messier in August 1758; he had not known of its existence, though it had been recorded by J. Bevis in 1731 – and the observation led directly on to Messier's resolve to draw up his famous Catalogue.

Planetary nebulæ are neither planets nor nebulæ, but small, hot stars which are at an advanced stage of evolution, and are surrounded by shells of tenuous gas. The most famous example (though not actually the brightest) is M.57, the Ring Nebula in Lyra. The brightest planetary is NGC 7293, in Aquarius, which was not included in Messier's list. The faintest planetary in

The Pleiades. (Patrick Moore, 1987)

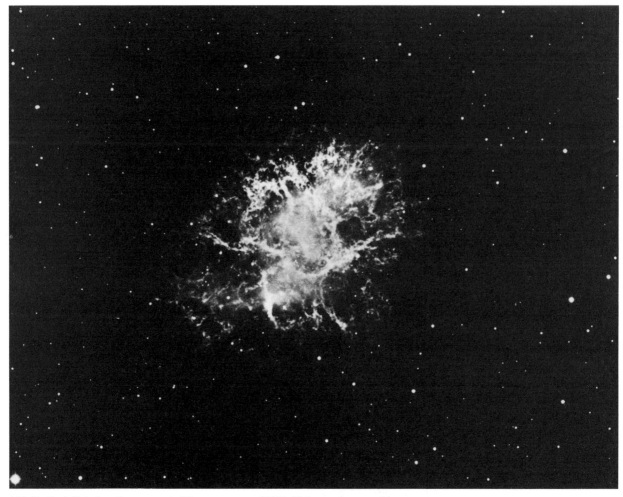

M.1, the Crab Nebula – the remnant of the supernova of 1054. (Palomar photograph)

Messier's catalogue is M.76 Persei, magnitude 12·2. (This is in fact the faintest of all Messier objects.)

Galaxies are external systems, beyond the limits of the Galaxy in which we live. Our Galaxy is often termed the Milky Way, though this name is better restricted to denoting the luminous band seen stretching across the night sky. Galaxies are of various types. In 1925 the American astronomer E. E. Hubble drew up a simple system of classification by shape, which is still widely used even though many more complex classifications have been proposed. Hubble's classes are as follows:

Spirals, resembling Catherine-wheels. Sa: conspicuous, often tightly-wound arms issuing from a well-defined nucleus. Sb: arms looser, nucleus less condensed. Sc: inconspicuous nucleus,

loose arms. Our Galaxy belongs to type Sb.

Barred spirals, in which the spiral arms issue from the ends of a kind of 'bar' through the nucleus. They are divided into types SBa, SBb and SBc in the same way as for the normal spirals.

Ellipticals, with no indications of spirality; they range from E7 (highly flattened) down to E0 (virtually spherical, looking superficially like globular clusters even though there is a great difference in mass).

Irregular galaxies, with no definite form.

It is thought that among the galaxies approximately 30 per cent are spiral, 60 per cent elliptical and 10 per cent irregular, though these values are very approximate.

The first spiral form was identified in 1845 by the third Earl of Rosse. Using

his newly-completed 72-in reflector at Birr Castle in Ireland, he saw the spiral shape of M.51 in Canes Venatici, now often nicknamed the Whirlpool. Others were subsequently found, though for some years the 72-in was the only telescope of aperture large enough to show the spiral forms clearly. Our Galaxy is an Sb spiral. It has an overall diameter of about 100 000 light-years (though some astronomers believe this to be an over-estimate), and contains approximately 100 000 million stars. The central bulge surrounding the galactic nucleus has a thickness of about 20 000 light-years. The Sun lies about 25 000 light-years from the centre of the Galaxy, on the inner edge of one of the spiral arms (known as the Carina-Cygnus arm). The Galaxy is in rotation; the Sun takes about 225 000 000 years to complete one circuit – a period sometimes termed the 'cosmic year', moving at about 250 km/s.

Other terms are also in use. **Seyfert galaxies**, first identified by Carl Seyfert in 1943, have bright, almost stellar nuclei which show variations in brightness; a good example is M.77 Ceti, the most massive galaxy in Messier's list. Many Seyferts are thought to be highly active. Some emit radio waves, infra-red and even X-rays which are detectable from Earth. The so-called N-galaxies also have bright nuclei, and may be closely associated with the Seyferts.

It is not now generally thought that spiral galaxies turn into ellipticals or vice versa, so that Hubble's classification does not indicate an evolutionary sequence. Ellipticals consist mainly of Population II objects, so that the most brilliant stars are old red giants, while Population I objects are dominant in the spirals, though the populations are always to some extent mixed, and no hard and fast boundaries can be drawn. Certainly the ellipticals contain relatively little interstellar matter – whereas there is plenty of it in the spirals, where star formation is still going on.

The first positive proof of interstellar matter in our Galaxy was obtained by Hartmann in 1904, when he was studying the spectrum of the star δ Orionis. δ Orionis is a spectroscopic binary, so that its lines show Doppler shifts corresponding to the orbital motion. However, some of the lines remained stationary, so that clearly they were associated not with the star itself, but with material between δ Orionis and ourselves. Further indications were obtained by J. Trümpler in 1930, because some of the Milky Way clusters appeared fainter than logically they should have done, and were presumably being dimmed. The gas between the stars is mainly made up of hydrogen and helium, with smaller amounts of carbon, nitrogen, oxygen and neon. In the 1930s astronomers at Mount Wilson found the first indications of interstellar molecules, but firm proof was postponed until 1963, when the hydroxyl radical, OH, was identified.

In 1968 C. Townes and his colleagues at Berkeley detected interstellar ammonia (NH_3) at a wavelength of 1·26 cm. The first organic molecules (that is to say, molecules containing carbon) were identified in 1969 by L. Snyder, D. Buhl, B. Zuckerman and P. Palmer, who detected formaldehyde (H_2CO). Others have since been found, including carbon monoxide, hydrogen cyanide and methanoic acid. Ethyl alcohol (CH_3CH_2OH) has also been detected. One 'cloud' contains enough of it to fill the entire globe of the Earth with alcohol!

Of supreme importance was the detection, in 1951, of cold hydrogen clouds emitting radiation at a wavelength of 21·1 cm. This had been predicted in 1945 by the Dutch astronomer H. C. van de Hulst; the 1951 identification was made by Ewen and Purcell in America. Studies of the 21-cm line have enabled astronomers to map the spiral arms of the Galaxy, since these are the regions in which the hydrogen clouds lie.

However, the average density of interstellar gas in our Galaxy is very low – about 500 000 hydrogen atoms per cubic metre, which is much lower than any laboratory vacuum. There are also dust grains, about 0·1 micron (10^{-7} metres) in radius, which are chiefly responsible for the reddening and dimming of distant objects. We cannot see through to the galactic centre, though

Lord Rosse's drawing of the Whirlpool Galaxy made in 1845, the first recognition of the spiral structure of a galaxy. The photograph reproduced in the frontispiece demonstrates the accuracy of Rosse's observation

as long ago as 1927 J. H. Oort established that the nucleus lies beyond the glorious star-clouds in Sagittarius. Dust is thought to make up about 1 per cent of the total mass of the Galaxy, though this figure is very arbitrary.

More molecules are being detected yearly. Also, in 1974 M. Cohen, who was engaged in studying the T Tauri variable HL Tauri, detected associated ice grains. HL Tauri is thought to be a very young star – perhaps no more than 300 000 years old.

As yet we have little information about intergalactic matter, but presumably it must exist, and certainly there seem to be streamers of tenuous gas linking the Magellanic Clouds with our Galaxy and with each other.

The first proof that the 'resolvable nebulæ' are independent galaxies was obtained by Hubble in 1923, when he was able to measure the periods of Cepheids in the Andromeda Spiral, M.31, and hence to find their distances. The idea was not new. It had been suggested as early as 1755 by Immanuel Kant, and had at one time been supported by Sir William Herschel. However, it had been very much of an open question, and was the subject of a great debate in 1920 between two American astronomers, Harlow Shapley and Heber D. Curtis. By studying short-period variables in the globular clusters, Shapley, in 1918, had been able to give the first reasonably accurate measurement of the size of our Galaxy, but he regarded the spirals as being contained in the Galaxy, whereas Curtis believed them to be external. Hubble's work, carried out with the Mount Wilson 100-in reflector (then the only telescope in the world of sufficient power) proved that Curtis had been right. It was thought that the Andromeda Spiral must be 900 000 light-years away, a figure subsequently modified to 750 000 light-years. In 1952 W. Baade found that there had been an error in the Cepheid scale; it was found that there were two kinds of Cepheids, one more luminous than the other, and those in the Spiral had been wrongly identified. They were more powerful than had been believed, and hence more remote. The distance of the Andromeda Spiral is now known to be

2 200 000 light-years. It is **the most remote object clearly visible with the naked eye**. It has been reported that the Triangulum Spiral, M.33, has been seen without optical aid, but this needs exceptional eyesight. The only external systems brighter than these two spirals are, of course, the Magellanic Clouds.

The first indication of an expanding universe was given by V. M. Slipher, who from 1912 to 1925 found that almost all the galaxies showed red shifts in their spectra, indicating recession. However, the significance of this was not immediately realized, since when Slipher began his work – and for more than a decade after – it was by no means certain that the 'resolvable nebulæ' were external galaxies, and men of the calibre of Shapley believed otherwise. Subsequently, Hubble showed a relationship between distance and velocity of recession; the further away the galaxy, the greater the recessional velocity. The value of what is termed Hubble's Constant is critical in any discussion of cosmology. One favoured figure is $55 \text{ km/s}^{-1}\text{M}^{-1}$ (M = megaparsecs; one megaparsec = 1 million parsecs).

Assuming that the spectral red shifts are true Doppler effects, the only galaxies not receding from us are the members of the **Local Group**. This consists of three or four large systems and more than two dozen dwarf galaxies. The whole Local Group is at least 5 000 000 light-years across.

The largest galaxy in the Local Group is M.31, the Andromeda Spiral, which is 1·5 times the size of our Galaxy. It contains over 300 globular clusters, and objects of all kinds (one supernova has been seen – S Andromedæ of 1885 – and numerous normal novæ). Second in size comes our Galaxy – possibly excluding Maffei 1, to be described below. Third is the Triangulum Spiral, M.33, half the size of our Galaxy and with 10 000 million stars. The Large Magellanic Cloud is a quarter the size of our Galaxy, and the Small Cloud one-sixth the size; both are classified as irregular, though it has been suggested that the Large Cloud shows traces of spirality.

In 1968 the Italian astronomer Maffei discovered two galaxies, only about half a degree from the plane of our Galaxy,

and therefore heavily obscured by interstellar matter. It now seems that the first of these, Maffei 1, is about 3 000 000 light-years away, and is a member of our Local Group; it appears to be a giant elliptical, but since we receive only about one-hundredth of its light our information about it is scanty. Maffei 2 was originally thought to be in the Local Group, but recent work indicates that it is well beyond, at a distance of about 15 000 000 light-years.

The other Local Group members are dwarfs, mainly less than 5000 light-years in diameter. Some of them, such as the Ursa Minor, Draco, Sculptor, Fornax, and Leo I and II systems are well within a million light-years of us, but are so sparse that they are inconspicuous. No doubt many more of these dwarfs exist, concealed by interstellar matter in the region of the galactic plane. **The smallest known galaxy** is GR-8, a dwarf irregular, a mere 1000 light-years in diameter. **The least massive Messier object** is M.32.

It is also worth noting that one globular cluster, NGC 5694, seems to be moving in a way that indicates its escape from our Galaxy. If so, it will become what is termed an 'intergalactic tramp'!

The following are the nearest members of the Local Group:

Name	Type	Abs. Mag.	Distance, thousands of light-years	Diameter, thousands of light-years
The Galaxy	Sb	−20·5	−	100
Large Magellanic Cloud	SBm	−18·5	170	30
Small Magellanic Cloud	SBm	−16·8	190	16
Ursa Minor dwarf	dE6	−8·8	250	2
Draco dwarf	dE3	−8·6	250	3
Sculptor dwarf	dE3	−11·7	280	5
Fornax dwarf	dE3	−13·6	420	7
Carina dwarf	dE	−	550	
Leo I	dE3	−11·0	750	2
Leo II	dE0	−9·4	750	3
NGC 6822 (Barnard's Galaxy)	Ir	−15·7	1700	5
M.31 (Andromeda Galaxy)	Sb	−21·1	2200	130
M.32	E2	−16·4	2200	6
NGC 205	E6	−16·4	2200	12
NGC 185	dE0	−11	2200	8
NGC 147	dE4	−14·9	2200	2
Andromeda I	dE0	−11	2200	2
Andromeda II	dE0	−11	2200	−
Andromeda III	dE2	−11	2200	−
NGC 1613	Ir	−14·8	2400	8
M.33 (Triangulum Galaxy)	Sc	−18·9	2900	52

Other clusters of galaxies are known, some of them much more populous than the Local Group. The most famous is the Virgo Cluster, which lies at a mean distance of 65 000 000 light-years and contains over 1000 members – notably the giant elliptical M.87, which is a strong radio source. According to E. Holmberg, its mass is as great as 790 000 million Suns, and the diameter is almost 40 kiloparsecs. Even more populous is the Coma cluster, 450 000 000 light-years away; the diameter is at least 10 million light-years, and there are two huge elliptical systems considerably larger than our Galaxy. Their combined absolute magnitude is about −24, as against −21 for our Galaxy.

The first quasar to be identified was 3C−273, in Virgo. (The prefix 3C indicates the third Cambridge catalogue of radio sources, published in 1962.) 3C−273 was known to be a strong radio emitter, but identifying it with a visual object proved to be difficult. On 5 August 1962, however, radio astronomers in Australia, working at the Parkes observatory, followed an occultation of the radio source by the Moon, and were able to fix its position very exactly. From this result, the source was identified with what seemed to be a faint bluish star of magnitude 12·8. In 1963 M. Schmidt, at Palomar in California, obtained an optical spectrum, and found that the object was not a star at all; its spectrum was quite different, and showed hydrogen lines which were tremendously red-shifted, indicating a high recessional velocity and therefore immense distance and luminosity. Other quasars were quickly found, together with quasar-like objects which were not radio emitters.

The brightest quasar is 3C−273. No others exceed magnitude 16.

The most remote object known is the quasar Q0051−279, discovered in 1987 by S. Warren and D. Hewitt. Its distance is of the order of 13 000 million light-years, and it is receding at over 93 per cent of the velocity of light.

Most astronomers believe that the red shifts in the spectra of quasars are 'cosmological', i.e. Doppler effects, but there are some dissentients. H. Arp claims that he has identified alignments of quasars and galaxies which are presumably associated, but which have different red shifts; F. Hoyle regards the quasars as reasonably local to our Galaxy. However, if the majority view is accepted, we may be observing quasars not so very far from the boundary of the observable universe, since an object receding at the full velocity of light would naturally be beyond our range. Evidence is growing that quasars are the nuclei of very active galaxies, probably containing massive central black holes.

The first quasar to be detected in infra-red was 3C−345, by P. Harvey at Texas in May 1982.

Associated with quasars are the BL Lacertæ objects, named after the brightest member of the class, identified as a radio source by Schmidt at Palomar in 1968. Most of them have featureless optical spectra; among early examples are AP Libræ and W Comæ, though some authorities have re-classified the latter as a quasar (it was originally thought to be a variable star within our own Galaxy!). In 1976 J. Wampler, at the Lick Observatory, was able to find some lines in the spectrum of the BL Lacertæ object PKS 0548−322, and measured the red shift. Subsequently R. Fosbury and M. Disney, using the 3·9 m Anglo-Australian telescope, obtained a spectrum of the 'fuzz' round the object, and found that it resembled the spectrum of an elliptical galaxy. They were able to show that the total luminosity was about the same as that of a faint quasar.

We cannot claim to known anything definite about the origin of the universe. The first modern-type theory – the 'big bang' – was proposed by the Belgian abbé Georges Lemaître in 1927; according to this theory all matter was created at one instant in a so-called primæval fireball, and expansion began, presumably to continue indefinitely. The Big Bang theory, elaborated by men such as George Gamow, was challenged in 1947 by H. Bondi and T. Gold, who proposed a 'steady-state' theory according to which the universe had no beginning and will have no end; as old galaxies die, new material in the form of hydrogen atoms is created spontaneously out of nothingness. There is also the oscillating universe theory, in which a phase of expansion is followed by a phase of contraction, and there is a new 'big bang' every 80 000 million years or so.

The steady-state theory has been abandoned by most authorities, since it does not fit the observed facts. It would mean that the distribution of galaxies would have been the same in the remote past as it is now (even though the galaxies themselves would be different from those which we know). Studying objects at distances of thousands of millions of light-years means, in effect, that we are looking back thousands of millions of years in time; and it seems that the distribution of remote radio sources is not the same as for objects closer to us, in which case the universe is not in a steady state.

We can even observe the remnants of the 'big bang'. In 1965 A. Penzias and R. Wilson, in America, were studying the sky with special radio equipment when

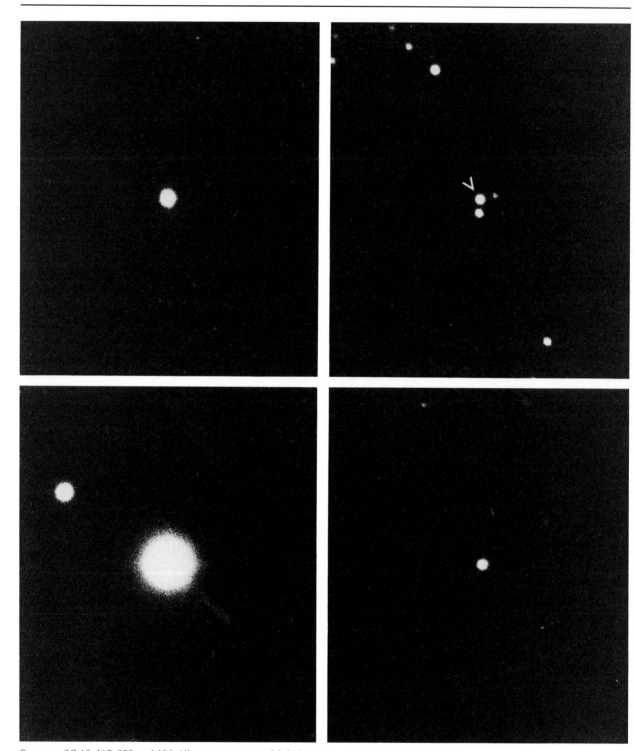

Quasars: 3C-48, 147, 273 and 196. All are very remote. 3C-273 is the brightest and nearest of the quasars, at rather over 2000 million light-years; it was also the first to be identified. (Science Photo Lib.)

they detected microwave radiation at 3·2 cm coming from all directions, and indicating a temperature of 3 °K (3 degrees Kelvin, that is to say, three degrees above absolute zero). At Princeton University, R. Dicke and his team had actually been busy making a radio telescope to search for this very radiation. Dicke had calculated that the temperature of the universe would have been immensely high at the moment of creation – the 'big bang' – and would by now have dropped to an overall value of 3 °K, which fitted in perfectly with the results by Penzias and Wilson.

Whether the overall expansion of the universe will continue indefinitely remains to be seen. Everything depends upon the amount of matter present. It has been calculated that the critical value is a mean of one hydrogen atom per cubic metre. If there is less matter than this, the expansion cannot be stopped. Present results indicate that the density is lower than the critical value, so that expansion will continue, though the problem cannot be regarded as definitely solved.

Note, however, that no theory can explain how any matter came to be created in the first place, so that we are not genuinely discussing the **origin** of the universe; we are dealing with its **evolution**, which is a very different thing!

THE FIRST KNOWN CLUSTERS AND NEBULÆ

It seems that a few nebulous objects have been known since prehistoric times. This certainly applies to the Pleiades, which can hardly be overlooked. In his great work the *Almagest*, Ptolemy (circa AD 120–180) records the Pleiades, and also the Sword-Handle in Perseus (NGC 869 and 884), M.44 (Præsepe) and, with almost certain identification, the open cluster M.7 in Scorpius; probably Ptolemy himself discovered the last of these. We may also assume that ancient men of the southern hemisphere knew the Magellanic Clouds, though they did not come to the notice of European astronomers until about 1520.

From the year 1745 more and more objects were found, mainly by astronomers such as De Chéseaux, Legentil, Lacaille and, of course, the two great Frenchmen, Messier and Méchain. By 1781, when Messier published his Catalogue, 138 nebulous objects were known. The following list includes all the objects found before 1745:

The globular cluster M13 in Hercules, photographed with the 60-inch reflector at Mount Wilson Observatory, California. M13 is the brightest globular cluster ever visible from Europe; the only brighter globulars (Omega Centauri and 47 Tucanæ) lie in the far south of the sky. M13 was discovered in 1714 by Edmond Halley, and is just visible with the naked eye on a clear night. It may contain as many as half a million stars. Its distance is 22 500 light-years. Some years ago radio astronomers in America sent a coded message to it in the hope that some astronomer in the cluster would receive it, but as the message will take 22 500 years to arrive we need not expect a rapid reply!

		Discoverer
M.45	Pleiades	(Prehistoric)
NGC 869/884	Sword-Handle in Perseus	(Listed by Ptolemy)
M.44	Præsepe	(Listed by Ptolemy)
M.7	Open cluster in Scorpio	Ptolemy, c. AD 140
M.31	Andromeda Spiral	Al-Sûfi, c. 964
o Velorum	Open cluster	Al-Sûfi, c. 964
Large Cloud of Magellan		1519
Small Cloud of Magellan		1519
M.42	Great Nebula in Orion	N. Pieresc, 1610
M.22	Globular in Sagittarius	A. Ihle, 1665
	ω Centauri	Halley, 1677
M.8	Lagoon Nebula	Flamsteed, 1680
M.11	Wild Duck Cluster	G. Kirch, 1681
NGC 2244	12 Monocerotis	Flamsteed, 1690
M.41	Open cluster in Canis Major	Flamsteed, 1702
M.5	Globular in Serpens	G. Kirch, 1702
M.50	Open cluster in Monoceros	G. D. Cassini, c. 1711
M.13	Hercules Globular	Halley, 1714
M.43	Part of Orion Nebula	de Mairan, c. 1731
M.1	Crab Nebula	J. Bevis, 1731

This makes in all 21 objects. Many others had been previously listed, but subsequently found to be mere groups of stars rather than true clusters or nebulæ. One remarkable fact relates to the Great Spiral in Andromeda, which was recorded by the Persian astronomer Al-Sûfi. It was not noted again until 1612, when Simon Marius described it. Amazingly, it was completely overlooked by the greatest observer of pre-telescopic times, Tycho Brahe.

MESSIER'S CATALOGUE OF NEBULOUS OBJECTS

Messier's original catalogue ends with M.103. The later numbers are objects which, apart from the last, were discovered by Méchain (M.110 may have been discovered by Messier himself, but it is still more generally known as NGC 205, one of the companions to the Great Spiral in Andromeda). M.104 was added to the Messier catalogue by Camille Flammarion in 1921, on the basis of finding a handwritten note about it in Messier's own copy of his 1781 catalogue. M.105 to 107 were listed by H. S. Hogg in the 1947 edition of the catalogue; M.108 and 109 by Owen Gingerich in 1960. The naming of NGC 205 as M.110 was suggested by K. G. Jones in 1968, but does not seem to have received general acceptance.

M.	NGC	Constellation	Type	
1	1952	Taurus	Supernova remnant	Crab Nebula
2	7089	Aquarius	Globular	
3	5272	Canes Venatici	Globular	
4	6121	Scorpius	Globular	
5	5904	Serpens	Globular	
6	6405	Scorpius	Open cluster	
7	6475	Scorpius	Open cluster	
8	6523	Sagittarius	Nebula	Lagoon Nebula
9	6333	Ophiuchus	Globular	
10	6254	Ophiuchus	Globular	
11	6705	Scutum	Open cluster	Wild Duck cluster
12	6218	Ophiuchus	Globular	
13	6205	Hercules	Globular	Great Hercules cluster
14	6402	Ophiuchus	Globular	
15	7078	Pegasus	Globular	
16	6611	Serpens	Nebula and embedded cluster	
17	6618	Sagittarius	Nebula	Omega or Horseshoe Nebula
18	6613	Sagittarius	Open cluster	
19	6273	Ophiuchus	Globular	
20	6514	Sagittarius	Nebula	Trifid Nebula
21	6531	Sagittarius	Open cluster	
22	6656	Sagittarius	Globular	
23	6494	Sagittarius	Open cluster	
24	6603	Sagittarius	Star-cloud in Milky Way	
25	I. 4725	Sagittarius	Open cluster	
26	6694	Scutum	Open cluster	
27	6853	Vulpecula	Planetary	Dumbbell Nebula
28	6626	Sagittarius	Globular	
29	6913	Cygnus	Open cluster	
30	7099	Capricornus	Globular	
31	224	Andromeda	Spiral galaxy	Great Spiral
32	221	Andromeda	Elliptical galaxy	Companion to M.31
33	598	Triangulum	Spiral galaxy	Triangulum Spiral
34	1039	Perseus	Open cluster	
35	2168	Gemini	Open cluster	
36	1960	Auriga	Open cluster	
37	2099	Auriga	Open cluster	
38	1912	Auriga	Open cluster	
39	7092	Cygnus	Open cluster	
40	–	–	Missing. Possibly a comet?	
41	2287	Canis Major	Open cluster	
42	1976	Orion	Nebula	Great Nebula
43	1982	Orion	Nebula	Part of Orion Nebula
44	2632	Cancer	Open cluster	Præsepe
45	–	Taurus	Open cluster	Pleiades
46	2437	Puppis	Open cluster	
47*	2422	Puppis	Open cluster	
48*	2548	Hydra	Open cluster	
49	4472	Virgo	Elliptical galaxy	
50	2323	Monoceros	Open cluster	
51	5194	Canes Venatici	Spiral galaxy	Whirlpool Galaxy
52	7654	Cassiopeia	Open cluster	
53	5024	Coma Berenices	Globular	
54	6715	Sagittarius	Globular	
55	6809	Sagittarius	Globular	
56	6779	Lyra	Globular	
57	6720	Lyra	Planetary	Ring Nebula
58	4579	Virgo	Spiral galaxy	
59	4621	Virgo	Elliptical galaxy	
60	4649	Virgo	Elliptical galaxy	
61	4303	Virgo	Spiral galaxy	
62	6266	Ophiuchus	Globular	
63	5055	Canes Venatici	Spiral galaxy	
64	4826	Coma Berenices	Spiral galaxy	
65	3623	Leo	Spiral galaxy	
66	3627	Leo	Spiral galaxy	
67	2682	Cancer	Open cluster	Famous old cluster
68	4590	Hydra	Globular	
69	6637	Sagittarius	Globular	
70	6681	Sagittarius	Globular	
71	6838	Sagitta	Globular	
72	6981	Aquarius	Globular	
73	6994	Aquarius	Four faint stars (Not a cluster)	
74	628	Pisces	Spiral galaxy	
75	6864	Sagittarius	Globular	
76	650	Perseus	Planetary	
77	1068	Cetus	Spiral galaxy	
78	2068	Orion	Nebula	

M.	NGC	Constellation	Type	
79	1904	Lepus	Globular	
80	6093	Scorpius	Globular	
81	3031	Ursa Major	Spiral galaxy	
82	3034	Ursa Major	Irregular galaxy	Companion to M.31
83	5236	Hydra	Spiral galaxy	
84	4374	Virgo	Spiral galaxy	
85	4382	Coma Berenices	Spiral galaxy	
86	4406	Virgo	Elliptical galaxy	
87	4486	Virgo	Elliptical galaxy	Seyfert galaxy
88	4501	Coma Berenices	Spiral galaxy	
89	4552	Virgo	Elliptical galaxy	
90	4569	Virgo	Spiral galaxy	
91	–	–	Not identified. Possibly a comet?	
92	6341	Hercules	Globular	
93	2447	Puppis	Open cluster	
94	4736	Canes Venatici	Spiral galaxy	
95	3351	Leo	Barred spiral galaxy	
96	3368	Leo	Spiral galaxy	
97	3587	Ursa Major	Planetary	Owl Nebula
98	4192	Coma Berenices	Spiral galaxy	
99	4254	Coma Berenices	Spiral galaxy	
100	4321	Coma Berenices	Spiral galaxy	
101	5457	Ursa Major	Spiral galaxy	
102	–	–	Missing. Possibly a faint spiral in Draco, or else identical with M.101.	
103	581	Cassiopeia	Open cluster	
104	4594	Virgo	Spiral galaxy	Sombrero Hat Galaxy
105	3379	Leo	Elliptical galaxy	
106	4258	Ursa Major	Spiral galaxy	
107	6171	Ophiuchus	Globular	
108	3556	Ursa Major	Spiral galaxy	
109	3992	Ursa Major	Spiral galaxy	
110	205	Andromeda	Elliptical galaxy	Companion to M.31

THE CONSTELLATIONS

THE CONSTELLATIONS

			Stars to mag:				
		1st mag	2.00	4.00	5.00	Area sq. deg.	Number of stars above mag. 5 per 100 square degrees (star density)
Andromeda	Andromeda	–	0	7	25	722	9·19
Antlia	The Airpump	–	0	0	4	239	1·67
Apus	The Bird of Paradise	–	0	2	6	206	2·91
Aquarius	The Water-bearer	–	0	7	31	980	3·16
Aquila	The Eagle	Altair	1	8	16	652	2·45
Ara	The Altar	–	0	7	10	237	4·21
Aries	The Ram	–	1	4	11	441	2·49
Auriga	The Charioteer	Capella	2	7	21	657	3·20
Boötes	The Herdsman	Arcturus	1	8	24	907	2·65
Cælum	The Graving Tool	–	0	0	2	125	1·60
Camelopardus	The Giraffe	–	0	0	11	757	1·45
Cancer	The Crab	–	0	1	6	506	1·19
Canes Venatici	The Hunting Dogs	–	0	1	7	465	1·51
Canis Major	The Great Dog	Sirius	4	10	26	380	6·84
Canis Minor	The Little Dog	Procyon	1	2	4	183	2·19
Capricornus	The Sea-Goat	–	0	5	16	414	3·86
Carina	The Keel	Canopus	3	14	40	494	8·10
Cassiopeia	Cassiopeia	–	0	7	23	598	3·85
Centaurus	The Centaur	Alpha Cen., Agena	2	14	49	1060	4·62
Cepheus	Cepheus	–	0	8	20	588	3·40
Cetus	The Whale	–	0	8	24	1232	1·95
Chamæleon	The Chameleon	–	0	0	6	132	4·55
Circinus	The Compasses	–	0	1	4	93	4·30
Columba	The Dove	–	0	4	9	270	3·33
Coma Berenices	Berenice's Hair	–	0	0	8	386	2·07
Corona Australis	The Southern Crown	–	0	0	7	128	5·47
Corona Borealis	The Northern Crown	–	0	3	10	179	5·59
Corvus	The Crow	–	0	5	6	184	3·26
Crater	The Cup	–	0	1	6	282	2·13
Crux Australis	The Southern Cross	Acrux, Beta Cru	3	5	13	68	19·12
Cygnus	The Swan	Deneb	1	11	43	804	5·35
Delphinus	The Dolphin	–	0	4	6	189	3·17
Dorado	The Swordfish	–	0	1	8	179	4·47

		1st mag	Stars to mag: 2.00	4.00	5.00	Area sq. deg.	Number of stars above mag. 5 per 100 square degrees (star density)
Draco	The Dragon	–	0	11	26	1083	2·40
Equuleus	The Foal	–	0	0	3	72	4·17
Eridanus	The River	Achernar	1	12	43	1138	3·78
Fornax	The Furnace	–	0	1	5	398	1·26
Gemini	The Twins	Pollux	3	13	23	514	4·47
Grus	The Crane	–	1	4	13	366	3·55
Hercules	Hercules	–	0	15	37	1225	3·02
Horologium	The Clock	–	0	1	2	249	0·80
Hydra	The Watersnake	–	1	9	32	1303	2·46
Hydrus	The Little Snake	–	0	3	9	243	3·70
Indus	The Indian	–	0	2	7	294	2·38
Lacerta	The Lizard	–	0	1	11	201	5·47
Leo	The Lion	Regulus	2	10	26	947	2·75
Leo Minor	The Little Lion	–	0	1	6	232	2·59
Lepus	The Hare	–	0	7	14	290	4·83
Libra	The Balance	–	0	5	13	538	2·42
Lupus	The Wolf	–	0	8	32	334	9·58
Lynx	The Lynx	–	0	2	12	545	2·20
Lyra	The Lyre	Vega	1	4	11	286	3·85
Mensa	The Table	–	0	0	0	153	0·00
Microscopium	The Microscope	–	0	0	4	210	1·90
Monoceros	The Unicorn	–	0	2	13	482	2·70
Musca Australis	The Southern Fly	–	0	4	11	138	7·97
Norma	The Rule	–	0	0	6	165	3·64
Octans	The Octant	–	0	1	4	291	1·37
Ophiuchus	The Serpent-bearer	–	0	12	36	948	3·80
Orion	Orion	Rigel, Betelgeux	5	15	42	594	7·07
Pavo	The Peacock	–	1	4	14	378	3·70
Pegasus	The Flying Horse	–	0	9	29	1121	2·59
Perseus	Perseus	–	1	10	34	615	5·52
Phœnix	The Phoenix	–	0	7	17	469	3·62
Pictor	The Painter	–	0	2	5	247	2·02
Pisces	The Fishes	–	0	3	24	889	2·70
Piscis Australis	The Southern Fish	Fomalhaut	1	1	7	245	2·86
Puppis	The Poop	–	0	11	42	673	6·24
Pyxis	The Compass	–	0	1	7	221	3·17
Reticulum	The Net	–	0	2	7	114	6·14
Sagitta	The Arrow	–	0	2	5	80	6·25
Sagittarius	The Archer	–	1	14	33	867	3·80
Scorpio	The Scorpion	Antares	3	17	38	497	7·64
Sculptor	The Sculptor	–	0	0	6	475	1·26
Scutum	The Shield	–	0	0	6	109	5·50
Serpens	The Serpent	–	0	9	17	637	2·67
Sextans	The Sextant	–	0	0	2	314	0·63
Taurus	The Bull	Aldebaran	2	14	44	797	5·52
Telescopium	The Telescope	–	0	1	4	252	1·59
Triangulum	The Triangle	–	0	2	3	132	2·27
Triangulum Australe	The Southern Triangle	–	1	3	6	110	5·45
Tucana	The Toucan	–	0	2	7	295	2·37
Ursa Major	The Great Bear	–	3	19	35	1280	2·73
Ursa Minor	The Little Bear	–	1	3	9	256	3·51
Vela	The Sails	–	2	10	30	500	6·00
Virgo	The Virgin	Spica	1	8	26	1294	2·01
Volans	The Flying Fish	–	0	3	7	141	4·96
Vulpecula	The Fox	–	0	0	10	268	3·73
Totals		21	50	455	1417		(Average: 3·8)

These counts do not include variable stars which can rise above the fifth magnitude, but whose average magnitude is below this limit. For instance, Mira Ceti is excluded even though its brightest maxima exceed magnitude 2.

FORMING THE CONSTELLATIONS

Ptolemy gave a list of 48 constellations: 21 northern, 12 Zodiacal and 15 southern, as follows:

Northern
Ursa Minor
Ursa Major
Draco
Cepheus
Boötes
Corona Borealis
Hercules
Lyra
Cygnus
Cassiopeia
Perseus
Auriga
Ophiuchus
Serpens
Sagitta
Aquila
Delphinus
Equuleus
Pegasus
Andromeda
Triangulum

Zodiacal
Aries
Taurus
Gemini
Cancer
Leo
Virgo
Libra
Scorpio
Sagittarius
Capricornus
Aquarius
Pisces

Southern
Cetus
Orion
Eridanus
Lepus
Canis Major
Canis Minor
Argo Navis
Hydra
Crater
Corvus
Centaurus
Lupus
Ara
Corona Australis
Piscis Australis

All these are still to be found in modern maps, though the huge, unwieldy Argo Navis has been divided up into Carina, Vela and Puppis.

Various constellations have been added since. With those that survive, the names have often been shortened; thus Piscis Volans, the Flying Fish, has become simply Volans, while Mons Mensæ, the Table Mountain, has become Mensa. There was also some confusion over two of Bayer's constellations, Apis (the Bee) and Avis Indica (the Bird of Paradise); modern maps give it as Apus. Surviving constellations were added as follows:

By Tycho Brahe, circa 1590:
Coma Berenices

By Bayer, 1603:
Pavo
Tucana
Grus
Phœnix
Dorado
Volans (originally Piscis Volans)
Hydrus
Chamæleon
Apus (originally Avis Indica)
Triangulum Australe
Indus

By Royer, 1679:
Columba (originally Columba Noachi, Noah's Dove)
Crux Australis

By Hevelius, 1690:
Camelopardus
Canes Venatici
Vulpecula (originally Vulpecula et Anser, the Fox and Goose)
Lacerta
Leo Minor
Lynx
Scutum Sobieskii
Monoceros
Sextans (originally Sextans Uraniæ, Urania's Sextant)

By La Caille, 1752:
Sculptor (originally Apparatus Sculptoris, the Sculptor's Apparatus)
Fornax (originally Fornax Chemica, the Chemical Furnace)
Horologium
Reticulum (originally Reticulus Rhomboidalis, the Rhomboidal Net)

Cælum (originally Cæla Sculptoris, the Sculptor's Tools)
Pictor (originally Equuleus Pictoris, the Painter's Easel)
Pyxis (originally Pyxis Nautica, the Mariner's Compass)
Antlia (originally Antlia Pneumatica, the Airpump)
Octans
Circinus
Norma (or Quadra Euclidis, Euclid's Square)
Telescopium
Microscopium
Mensa (originally Mons Mensæ, the Table Mountain)
The list of rejected constellations is very long. The most notable are as follows:

Proposed by Tycho Brahe, circa 1559:
Antinoüs

Proposed by Royer, 1679:
Nubes Major (the Great Cloud)
Nubes Minor (the Little Cloud)
Lilium (the Lily)

Proposed by Halley, circa 1680:
Robur Caroli (Charles' Oak)

Proposed by Flamsteed, circa 1700:
Mons Mænalus (the Mountain Mænalus)
Cor Caroli (Charles' Heart)

Proposed by Hevelius, 1690:
Triangulum Minor (the Little Triangle)
Cerberus

Proposed by Le Monnier, 1776:
Tarandus (the Reindeer)
Solitarus (the Solitaire)

Proposed by Lalande, 1776:
Messier

Proposed by Poczobut, 1777:
Taurus Poniatowski (Poniatowski's Bull)

Proposed by Hell, circa 1780:
Psalterium Georgianum (George's Lute)

Proposed by Bode, circa 1775:
Honores Frederici (the Honours of Frederick)
Sceptrum Brandenburgicum (the Sceptre of Brandenburg)
Telescopium Herschelii (Herschel's Telescope)

Globus Ærostaticus (the Balloon)
Quadrans Muralis (the Mural Quadrant)
Lochium Funis (the Log Line)
Machina Electrica (the Electrical Machine)
Officiana Typographica (the Printing Press)
Felis (the Cat).

There seem to have been two Muscas, one formed by La Caille to replace Bayer's Apis and the other (rejected) formed by Bode out of stars near Aries. One discarded group, Quadrans, has at least given its name to the Quadrantid meteor shower of early January.

LIST OF STARS OF MAGNITUDE 2·00 OR BRIGHTER

Stars of above magnitude 1·4 are conventionally termed 'first magnitude' stars – that is to say, down the list as far as Regulus.

Obviously the distances and luminosities of the stars, apart from the relatively close ones, are not known precisely, and in this list the values have been 'rounded off'. The distances are given in light-years, though in the main catalogue which follows it has seemed better to use parsecs. One parsec is equal to 3·2633 light-years.

Excluding the unique η Carinæ, there are two other stars which may exceed magnitude 2 though they are generally fainter. Mira Ceti attained 1·2 at the maximum of 1772, according to the Swedish astronomer Per Wargentin, though most maxima are well below 2; the official maximum magnitude is usually given as 1·7. γ Cassiopeiæ is an irregular variable; in 1936 it reached magnitude 1·6, but at present (1988) and for some years past its magnitude has been about 2·2.

According to a recent estimate, the most luminous star in the list is Canopus; next come Wezea, Deneb and Rigel. The difficulties of estimating very remote stars can be judged that in some other lists the luminosity of Canopus is reduced to only about 4000 times that of the Sun! The most remote star in the list is Wezea; the closest is, of course, α Centauri.

LIST OF STARS OF MAGNITUDE 2·00 OR BRIGHTER

Star		Name	Apparent magnitude	Luminosity Sun = 1	Spectrum	Distance light-years
1	α Canis Maj.	Sirius	−1·46	26	A1	8·8
2	α Carinæ	Canopus	−0·72	200 000	F0	1200
3	α Centauri		−0·27	1·5	K1 + G2	4·3
4	α Boötis	Arcturus	−0·04	115	K2	36
5	α Lyræ	Vega	0·03	52	A0	26
6	α Aurigæ	Capella	0·08	70	G8	42
7	β Orionis	Rigel	0·12v	60 000	B8	900
8	α Canis Min.	Procyon	0·38	11	F5	11·4
9	α Eridani	Achernar	0·46	780	B5	85
10	α Orionis	Betelgeux	var.	15 000	M2	310
11	β Centauri	Agena	0·61	10 500	B1	460
12	α Aquilæ	Altair	0·77	10	A7	16·6
13	α Crucis	Acrux	0·83	3200 + 2000	B1 + B3	360
14	α Tauri	Aldebaran	0·85	100	K5	68
15	α Scorpii	Antares	0·96v	7500	M1	330
16	α Virginis	Spica	0·98v	2100	B1	260
17	β Geminorum	Pollux	1·14	60	K0	36
18	α Pisc. Aust.	Fomalhaut	1·16	13	A3	22
19 =	α Cygni	Deneb	1·25	70 000	A2	1800
19 =	β Crucis		1·25v	8200	B0	425
21	α Leonis	Regulus	1·35	130	B7	85
22	ε Canis Maj.	Adhara	1·50	5000	B2	490
23	α Geminorum	Castor	1·58	45	A0	46
24 =	γ Crucis		1·63	160	M3	88
24 =	λ Scorpii	Shaula	1·63	1300	B2	275
26	γ Orionis	Bellatrix	1·64	2200	B2	360
27	β Tauri	Al Nath	1·65	470	B7	130
28	β Carinæ	Miaplacidus	1·68	130	A0	85
29	ε Orionis	Alnilam	1·70	26 000	B2	1200
30	α Gruis	Alnair	1·74	230	B5	69
31 =	ζ Orionis	Alnitak	1·77	19 000	O9·5	1100
31 =	ε Ursæ Maj.	Alioth	1·77	60	A0	62
33	γ Velorum		1·78	3800	WC7	520
34	α Ursæ Maj.	Dubhe	1·79	60	K0	75
35	α Persei	Mirphak	1·80	6000	F5	620
36	ε Sagittarii	Kaus Australis	1·85	100	B9	85
37 =	δ Canis Maj.	Wezea	1·86	130 000	F8	3000
37 =	ε Carinæ	Avior	1·86	600	K0	200
37 =	η Ursæ Maj.	Alkaid	1·86	450	B3	108
40	θ Scorpii	Sargas	1·87	14 000	G0	900
41	β Aurigæ	Menkarlina	1·90	50	A2	72
42	α Tri. Aust.	Atria	1·92	96	K2	55
43	γ Geminorum	Alhena	1·93	82	A0	85
44	α Pavonis		1·94	700	B3	230
45	δ Velorum	Koo She	1·96	50	A0	69
46 =	α Hydræ	Alphard	1·98	115	K3	85
46 =	β Canis Maj.	Mirzam	1·98v	7200	B1	710
48 =	γ Leonis	Algieba	1·99	60	K0 + G7	90
48 =	α Ursæ Min.	Polaris	1·99v	6000	F8	680
50	α Arietis	Hamal	2·00	96	K2	85

Star		Name	Apparent magnitude	Luminosity Sun = 1	Spectrum	Distance light-years
−	o Ceti	Mira	var.	140v	Md	95
−	γ Cassiopeiæ	Cih	2·2v	6000v	B0p	780

CONSTELLATIONS AND STARS

The largest constellation is Hydra, with an area of 1303 square degrees. (Formerly it was exceeded by Argo Navis, but Argo has now been unceremoniously chopped up into a keel, sails and a poop.)

The smallest constellation is Crux Australis, with an area of 68 square degrees.

The constellation with the greatest number of stars above the second magnitude is Orion (5 stars).

The constellation with the greatest number of stars above the fourth magnitude is Ursa Major (19 stars).

The constellation with the greatest number of stars above the fifth magnitude is Centaurus (49 stars).

The only constellation with no star as bright as the fifth magnitude is Mensa. Next in order of 'dimness' come Cælum, Horologium and Sextans, each with only 2 stars above the fifth magnitude.

For stars above the fifth magnitude, the greatest 'star density' is Crux Australis, which averages out at 19·12 stars per 100 square degrees. Its nearest rival, Lupus, has a figure of only 9·58.

The least star density, excluding Mensa, is for Sextans (0·63 star per 100 square degrees).

The brightest star is Sirius (α Canis Majoris), magnitude −1·46.

The brightest star in the northern hemisphere of the sky is Arcturus (α Boötis), magnitude −0·04.

The only naked-eye star which is said to have a greenish tint is β Libræ – though most observers will certainly class it as white!

The reddest naked-eye star is probably μ Cephei, called by Herschel 'the Garnet Star', though optical aid is needed to bring out its colour well. The reddest first-magnitude stars are Antares, Betelgeux and Aldebaran. Among the reddest telescopic stars are R Leporis and U Cephei.

The following data apply to the stars above magnitude 5·00, listed in the Catalogue which follows:

Most powerful stars, all of absolute magnitude −7·5 or brighter; Canopus (α Carinæ), s Carinæ, ε Aurigæ, ιScorpii, b Velorum, ρ Cassiopeiæ, β Doradûs, Deneb (α Cygni), ν Cephei and 3 Puppis.

Least powerful stars are ε Indi (absolute magnitude 7·0), ε Eridani (6·1), o^2 Eridani (6·0) and σ Draconis (5·9).

THE STAR CATALOGUE

This catalogue has been compiled from various sources; positions of the brightest stars (down to magnitude 4·75) are given for epoch 2000·0. The published values for absolute magnitudes and distances differ somewhat, and are subject to uncertainty; therefore the values given here have been in general 'rounded off' except for those stars for which the data are known with true precision. I have followed the data given in the authoritative *Cambridge Sky Catalogues* (1987).

The list of double stars, variable stars, and nebular objects does not pretend to be complete, and could be extended almost *ad infinitum*, but I have included most variable stars whose maximum magnitudes are 8·0 or brighter and where the range is at least half a magnitude. For double stars, I have given pairs which are within the range of modest telescopes together with some which may be regarded as test objects. For binary stars of reasonably short period, the values of position angle and distance are for approximate date 1990.

Data given are:

Variable Stars Range, type, period in days, and spectral type.

Double Stars Position angle in degrees, separation in seconds of arc, and magnitudes of the components.

Open Clusters Diameter in minutes of arc, magnitude, and approximate number of stars (though in many cases this is subject to great uncertainty).

Globular Clusters Diameter in minutes of arc, and approximate total magnitude.

Planetary Nebulæ Dimensions in seconds of arc, total magnitude, and magnitude of the central star.

Nebulæ Dimensions in minutes of arc, and, where appropriate, the magnitude of the illuminating star.

Galaxies Magnitude; dimensions, in minutes of arc; and type.

☀ ◉ ○		**VARIABLE STARS**
⊖ ⊖		**PLANETARY NEBULAE**
⁘ ⁙		**GLOBULAR CLUSTERS**
⬭ ○		**GALAXIES**
◌ ◌		**OPEN CLUSTERS**
✿ ✾		**GASEOUS NEBULA**

ANDROMEDA

(Abbreviation: And).

A large and important northern constellation; one of Ptolemy's 'originals'. It contains the Great Spiral, M.31. One of the leading stars in the constellation – Alpheratz or α Andromedæ – is actually a member of the Square of Pegasus, and was formerly known, more logically, as δ Pegasi.

In mythology, Andromeda was the beautiful daughter of King Cepheus and Queen Cassiopeia. Cassiopeia offended the sea-god Neptune by her boasting about Andromeda's beauty, which, she claimed, was greater than that of any sea-nymph. Neptune thereupon sent a sea-monster to ravage the kingdom, and the Oracle stated that the only solution was to chain Andromeda to a rock by the shore where she would be devoured by the monster. This was duly done, but the situation was redeemed by the hero Perseus, who was on his way home after killing the Gorgon, Medusa. Mounted upon his winged sandals, Perseus arrived in the nick of

BRIGHTEST STARS

Star		R.A.			Dec.			Mag.	Abs. mag.	Spectrum	Dist.	
		h	m	s	°	′	″				pc	
1	o	23	01	55·1	+42	19	34	3·6v	1·4	B6+A2	35	
7		23	12	32·9	+49	24	33	4·52	2·6	F0	23	
16	λ	23	37	33·7	+46	27	30	3·82v	1·8	G8	24	
17	ι	23	38	08·0	+43	16	05	4·29	−0·2	B8	79	
19	κ	23	40	24·4	+44	20	02	4·14	−0·2	B8	60	
21	α	00	08	23·2	+29	05	26	2·06	−0·1	A0p	22	Alpheratz
24	θ	00	17	05·4	+38	40	54	4·61	1·4	A2	44	
25	σ	00	18	19·6	+36	47	07	4·52	1·4	A2	42	
29	π	00	36	52·8	+33	43	09	4·36	−1·1	B5	120	
30	ε	00	38	33·3	+29	18	42	4·37	0·3	G8	41	
31	δ	00	39	19·6	+30	51	40	3·27	−0·2	K3	49	
34	ξ	00	47	20·3	+24	16	02	4·06v	−2·2	K1	48	
35	ν	00	49	48·8	+41	04	44	4·53	−1·1	B5	130	
37	μ	00	56	45·1	+38	29	58	3·87	2·1	A5	25	
38	η	00	57	12·4	+23	25	04	4·42	1·8	G8	35	
42	φ	01	09	30·1	+47	14	31	4·25	−1·2	B8	120	
43	β	01	09	43·8	+35	37	14	2·06	−0·4	M0	27	Mirach
50	υ	01	36	47·8	+41	24	20	4·09	4·0	F8	13	
51		01	37	59·5	+48	37	42	3·57	−0·2	K3	57	
57	γ	{02	03	53·9	+42	19	47	2·18	−0·1	K2	37	Almaak
		{02	03	54·7	+42	19	51	5·03		A0		
65		02	25	37·3	+50	16	43	4·71	−0·3	K4	82	

o is a β-Lyræ variable with a small range (3·5–3·7), period 1·6 days.
ζ is an ellipsoidal variable with a small range, 4·06–4·20.
λ is slightly variable (3·7–4·1) and is probably a spectroscopic binary.

Also above mag. 5:

		Mag.	Abs. mag.	Spectrum	Dist.	
3		4·65	0·2	K0	71	
5		5·70	3·4	F5	29	
8		4·85	−0·5	M2	100	
46	ξ	4·88	1·7	K0	35	Adhil
48	ω	4·83	2·0	F4	36	
52	χ	4·98	0·3	G8	86	

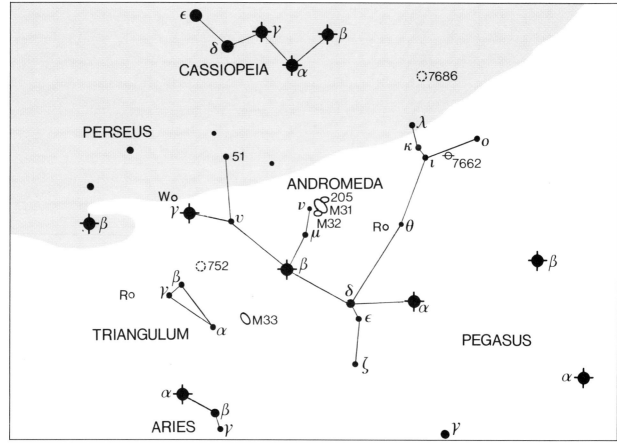

ANDROMEDA *(continued)*

time, turned the monster to stone by the simple expedient of showing it Medusa's head, and then, in the best story-book tradition, married Andromeda. Cepheus, Cassiopeia and of course Perseus are to be found in the sky; the Gorgon's head is marked by the 'Demon Star' Algol, and the sea-monster has sometimes been identified with the constellation of Cetus.

Andromeda has no first-magnitude star, but there are seven above magnitude 4:

β	2·06
α	2·06
γ	2·14
δ	3·27
51	3·57
o	3·6v
μ	3·87

		Mag.	Abs. mag.	Spectrum	Dist.
53	τ	4·94	−0·6	B8	130
58		4·80	1·9	A4	38
60		4·83	−0·3	K4	97

VARIABLE STARS

	R.A.		Dec.		Range	Type	Period, d.	Spectrum
	h	m	°	'				
Z	23	33·7	+48	49	8·0–12·4	Z And	–	M
ST	23	38·8	+35	46	7·7–11·8	Semi-reg.	328	R
SV	00	04·3	+40	07	7·7–14·3	Mira	325·2	M
KU	00	06·9	+43	05	6·5–10·5	Mira	750	M
VX	00	19·9	+44	43	7·8–9·3	Semi-reg.	369	N
T	00	22·4	+27	00	7·7–14·5	Mira	280·8	M
R	00	24·0	+38	35	5·8–14·9	Mira	409·3	S
TU	00	32·4	+26	02	7·8–13·1	Mira	316·8	M
RW	00	47·3	+32	41	7·9–15·7	Mira	430·3	M
W	02	17·6	+44	18	6·7–14·6	Mira	395·9	S

DOUBLE STARS

	R.A.		Dec.		P.A.	Sep.	Mags.	
	h	m	°	'	°	"		
π	00	36·9	+33	43	173	35·9	4·4, 8·6	
γ	02	03·9	+42	20	063	9·8	2·3, 4·8	
γ²					106	0·5	5·5, 6·3	Binary, period 61 years

OPEN CLUSTERS

M	NGC	R.A.		Dec.		Diameter	Mag.	No. of stars
		h	m	°	'	'		
	7686	23	30·2	+49	08	15	5·6	20
	752	01	57·8	+37	41	50	5·7	60

PLANETARY NEBULA

M	NGC	R.A.		Dec.		Dimensions	Mag.	Mag. of central star
		h	m	°	'	"		
	7662	23	25·9	+42	33	20×130	9·2	13·2

GALAXIES

M	NGC	R.A. h m	Dec. ° '	Mag.	Dimensions '	Type	
	205	00 40·4	+41 41	8·0	17·4×9·8	E6	Companion to M.31
31	224	00 42·7	+41 16	3·5	178×63	Sb	
32	221	00 42·7	+40 52	8·2	7·6×5·8	E2	Companion to M.31
	891	02 22·6	+42 21	9·9	13·5×2·8	Sb	

ANTLIA

(Abbreviation: Ant).

A small southern constellation, covering 239 square degrees, added to the sky by Lacaille in 1752; its original name was Antlia Pneumatica. It contains no star brighter than magnitude 4·4, and no mythological legends are associated with it.

See chart for Carina

BRIGHTEST STARS

Star	R.A. h m s	Dec. ° ' "	Mag.	Abs. mag.	Spectrum	Dist. pc
ε	09 29 14·7	−35 57 05	4·51	−0·4	M0	96
α	10 27 09·1	−31 04 04	4·25	−0·4	M0	85
ι	10 56 43·0	−37 08 16	4·60	0·3	F2	79

Also above mag. 5:

	Mag.	Abs. mag.	Spectrum	Dist.
θ	4·79	3·8	F7	14

VARIABLE STARS

	R.A. h m	Dec. ° '	Range	Type	Period, d.	Spectrum
S	09 32·3	−28 38	6·4–6·9	W Uma	0·65	A
U	10 35·2	−39 34	5·7–6·8	Irr.	−	N

DOUBLE STARS

	R.A. h m	Dec. ° '	P.A. °	Sep. "	Mags.
θ	09 44·2	−27 46	005	0·1	5·4, 5·6
δ	10 29·6	−30 36	226	11·0	5·6, 9·6

GALAXIES

M	NGC	R.A. h m	Dec. ° '	Mag.	Dimensions '	Type
	2997	09 45·6	−31 11	10·6	8·1×6·5	Sc
	3223	10 21·6	−34 16	11·8	4·1×2·6	Sb
	3347	10 42·8	−36 22	12·5	4·4×2·6	SBb

APUS

(Abbreviation: Aps).

Originally Avis Indica; it was introduced by Bayer in 1603. It lies in the far south, and has only two stars brighter than magnitude 4:

α 3·83   γ 3·89

See chart for Musca

BRIGHTEST STARS

Star	R.A. h m s	Dec. ° ' "	Mag.	Abs. mag.	Spectrum	Dist. pc
α	14 47 51·6	−79 02 41	3·83	−0·3	K5	67
δ	16 20 20·7	−78 41 44	4·68	−0·5	M4+K5	110
γ	16 33 27·1	−78 53 49	3·89	3·2	K0	14
β	16 43 04·5	−77 31 02	4·24	0·2	K0	42
ζ	17 21 59·3	−67 46 13	4·78	−0·3	gK5	100

The magnitude of δ is given as the combination of δ¹ and δ², which are separated by 103″ and make up a very wide physically connected pair.

Also above mag. 5:

	Mag.	Abs. mag.	Spectrum	Dist.
η	4·91		A2p	

VARIABLE STARS

	R.A. h m	Dec. ° '	Range	Type	Period, d.	Spectrum
θ	14 05·3	−76 48	6·4–8·6	Semi-reg.	119	M
S	15 04·3	−71 53	9·5–15	R CrB	−	R
DW	17 23·5	−67 56	7·9–9·1	Algol	2·31	B

(R Apodis, magnitude 5·3, has been suspected of variability.)

DOUBLE STAR

	R.A. h m	Dec. ° '	P.A. °	Sep. "	Mags.
δ	16 20·3	−78 42	012	102·9	4·7, 5·1

GLOBULAR CLUSTERS

M	NGC	R.A. h m	Dec. ° '	Diameter '	Mag.
	IC 4499	15 00·3	−82 13	7·6	10·6
	6101	16 25·8	−72 12	10·7	9·3

APUS (continued)

GALAXY

M	NGC	R.A. h m	Dec. ° '	Mag.	Dimensions '	Type
	5967	15 48·1	−75 40	12·5	2·9 × 1·8	SBc

AQUARIUS

(Abbreviation: Aqr).

A Zodiacal constellation, and of course one of Ptolemy's 'originals'. Oddly enough there are no well-defined legends attached to it, though it has been associated with Ganymede, cup-bearer of the Olympian gods. There are 9 stars brighter than magnitude 4·00:

β 2·91
α 2·96
δ 3·27
ζ 3·6
c² 3·66
λ 3·74
ε 3·77
γ 3·84
b¹ 3·97

BRIGHTEST STARS

Star		R.A. h m s	Dec. ° ' "	Mag.	Abs. mag.	Spectrum	Dist. pc	
2	ε	20 47 40·3	−09 29 45	3·77	1·2	A1	33	Albali
3	k	20 47 44·0	−05 01 40	4·42	−0·5	M3	91	
6	μ	20 52 39·0	−08 59 00	4·73	1·0	A8	30	
13	ν	21 09 35·4	−11 22 18	4·51	0·3	G8	70	
22	β	21 31 33·3	−05 34 16	2·91	−4·5	G0	300	Sadalsuud
23	ξ	21 37 44·9	−07 51 15	4·69	2·4	A7	36	
31	o	22 03 18·7	−02 09 19	4·69	−0·2	B8	95	
34	α	22 05 46·8	−00 19 11	2·96	−4·5	G2	290	Sadalmelik
33	ι	22 06 26·1	−13 52 11	4·27	−0·2	B8	78	
43	θ	22 16 49·9	−07 47 00	4·16	1·8	G8	26	Ancha
48	γ	22 21 39·2	−01 23 14	3·84	0·6	A0	28	Sadachiba
52	π	22 25 16·4	+01 22 39	4·66	−4·1	B0	450	
55	ζ¹	22 28 49·5	−00 01 13	4·53	0·6	F2	30	
	ζ²	22 28 49·9	−00 01 12	4·31		F2		
62	η	22 35 21·2	−00 07 03	4·02	−0·2	B8	46	
71	τ	22 49 35·3	−13 35 33	4·01	−0·4	M0	74	
73	λ	22 52 36·6	−07 34 47	3·74	−0·5	M2	71	
76	δ	22 54 38·8	−15 49 15	3·27	−0·2	A2	30	Scheat
88	c²	23 09 26·6	−21 10 21	3·66	0·2	K0	33	
90	φ	23 14 19·2	−06 02 56	4·22	−0·5	M2	75	
91	ψ¹	23 15 53·4	−09 05 16	4·21	1·8	K0	30	
93	ψ²	23 27 54·1	−09 10 57	4·39	−1·1	B5	130	
98	b¹	23 22 58·0	−20 06 02	3·97	0·2	K0	49	
99	b²	23 26 02·5	−20 38 31	4·39	−0·3	K5	87	
101	b³	23 33 16·4	−20 54 32	4·71	—	A0	25	
105	ω²	23 42 43·2	−14 32 42	4·49	0·4	B9·5	43	

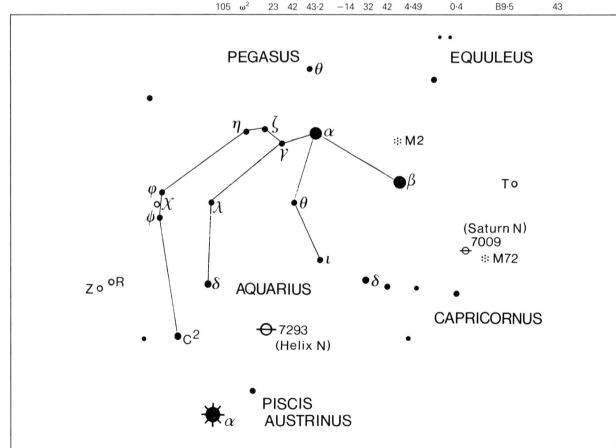

AQUARIUS (continued)

Also above mag. 5:

		Mag.	Abs. mag.	Spectrum	Dist.
57	σ	4·82	0·0	A0	92
95	ψ³	4·98	0·6	A0	75
102	ω¹	5·00	—	A5	18
104	A²	4·82	−2·0	G0	220

VARIABLE STARS

	R.A. h m	Dec. ° '	Range	Type	Period, d.	Spectrum
W	20 46·4	−04 05	8·4−14·9	Mira	381·1	M
V	20 46·8	+02 26	7·6−9·4	Semi-reg.	244	M
T	20 49·9	−05 09	7·2−14·2	Mira	202·1	M
X	22 18·7	−20 54	7·5−14·8	Mira	311·6	S
S	22 57·1	−20 21	7·6−15·0	Mira	297·3	M
R	23 43·8	−15 17	5·8−12·4	Symb.	387	M + Pec

DOUBLE STARS

	R.A. h m	Dec. ° '	P.A. °	Sep. "	Mags.	
β	21 31·6	−05 34	AB 321	35·4	2·9, 10·8	
			AC 186	57·2	11·4	
41	22 14·3	−21 04	AB 114	5·0	5·6, 7·1	
			AC 043	212·1	9·0	
51	22 24·1	−04 50	AB 324	0·5	6·5, 6·5	
			AB + D 191	116·0	10·1	
			AC 342	54·4	10·2	
			AE 133	132·4	8·6	
ζ	22 28·8	−00 01	200	2·0	4·3, 4·5	Binary, p. 856 y
89	23 09·9	−22 27	007	0·4	5·1, 5·9	
107	23 46·0	−18 41	136	6·6	5·7, 6·7	

ASTERISM

M	NGC	R.A. h m	Dec. ° '	
73	6994	20 58·9	−12 38	Four stars; not a true cluster.

GLOBULAR CLUSTERS

M	NGC	R.A. h m	Dec. ° '	Diameter '	Mag.
72	6981	20 53·5	−12 32	5·9	9·3
2	7089	21 33·5	−00 49	12·9	6·5

PLANETARY NEBULÆ

M	NGC	R.A. h m	Dec. ° '	Dimensions "	Mag.	Mag. of central star	
	7009	21 04·2	−11 22	2·5×100	8·3	11·5	Saturn Nebula
	7293	22 29·6	−20 48	770	6·5	13·5	Helix Nebula

GALAXIES

M	NGC	R.A. h m	Dec. ° '	Mag.	Dimensions '	Type
	7184	22 02·7	−20 49	12·0	5·8×1·8	Sb
	7606	23 19·1	−08 29	10·8	5·8×2·6	Sb
	7723	23 38·9	−12 58	11·1	3·6×2·6	Sb
	7727	23 39·9	−12 18	10·7	4·2×3·4	SBap

AQUILA

(Abbreviation: Aql).

One of the most distinctive of all the northern constellations, and, of course, an 'original'. Mythologically it represents an eagle which was sent by Jupiter to collect a Phrygian shepherd-boy, Ganymede, who was destined to become cup-bearer of the Gods – following an unfortunate episode in which the former holder of the office, Hebe, tripped and fell during a particularly solemn ceremony.

The leading star is Altair. Altogether

BRIGHTEST STARS

Star		R.A. h m s	Dec. ° ' "	Mag.	Abs. mag.	Spectrum	Dist. pc	
13	ε	18 59 37·2	+15 04 06	4·02	−0·1	K2	65	
12	i	19 01 40·7	−05 44 20	4·02	0·0	K1	64	
17	ζ	19 05 24·4	+13 51 48	2·99	0·2	B9	e2	Dheneb
16	λ	19 06 14·7	−04 52 57	3·44	0·0	B8·5	30	Althalimain
30	δ	19 25 29·7	+03 06 53	3·36	2·1	F0	16	
32	ν	19 26 30·9	+00 22 44	4·66	−4·6	F2	530	
38	μ	19 34 05·2	+07 22 44	4·45	−0·2	K3	39	
41	ι	19 36 43·1	−01 17 11	4·36	−2·2	B5	180	
50	γ	19 46 15·4	+10 36 48	2·72	−2·3	K3	87	Tarazed
53	α	19 50 46·8	+08 52 06	0·77	2·2	A7	5·1	Altair
55	η	19 52 28·1	+01 00 20	var.	−4·5v	G0v	440	
59	ξ	19 54 14·7	+08 27 41	4·71	0·2	K0	73	
60	β	19 55 18·5	+06 24 24	3·71	3·2	G8	11	Alshain
65	θ	20 11 18·1	−00 49 17	3·23	−0·8	B9	61	
71		20 38 20·1	−01 06 19	4·32	0·3	G8	64	

AQUILA *(continued)*

there are eight stars above magnitude 4:

α 0·77
γ 2·72
ζ 2·99
θ 3·23
δ 3·36
λ 3·44
η 3·7 (max)
β 3·71

Also above mag. 5:

		Mag.	Abs. mag.	Spectrum	Dist.
39	κ	4·95		B0·5	
67	ρ	4·95	1·4	A2	51
69		4·91	−0·1	K2	100
70		4·89	−2·3	K5	240

VARIABLE STARS

	R.A. h m	Dec. ° '	Range	Type	Period, d.	Spectrum
V	19 04·4	−05 41	6·6−8·4	Semi-reg.	353	N
R	19 06·4	+08 14	5·5−12·0	Mira	284·2	M
TT	19 08·2	+01 18	6·4−7·7	Cepheid	13·75	F−G
W	19 15·4	−07 03	7·3−14·3	Mira	490·4	S
U	19 29·4	−07 03	6·1−6·9	Cepheid	7·02	F−G
X	19 51·5	+04 28	8·3−15·5	Mira	347·0	M
η	19 52·5	+01 00	3·5−4·4	Cepheid	7·18	F−G
RR	19 57·6	−01 53	7·8−14·5	Mira	394·8	M

DOUBLE STARS

	R.A. h m	Dec. ° '	P.A. °	Sep. "	Mags.
ε	18 59·6	+15 04	187	131·1	4·0, 9·9
23	19 18·5	+01 05	005	3·1	5·3, 9·3
31	19 25·0	+11 57	343	105·6	5·2, 8·7
δ	19 25·5	+03 07	271	108·9	3·4, 10·9
ν	19 26·5	+00 20	288	201·0	4·7, 8·9
U	19 29·4	−07 03	228	1·5	var., 11·7
χ	19 42·6	+11 50	077	0·5	5·6, 6·8
γ	19 46·3	+10 37	258	132·6	2·7, 10·7
π	19 48·7	+11 49	110	1·4	6·1, 6·9
α	19 50·8	+08 52	301	165·2	0·8, 9·5
57	19 54·6	−08 14	170	35·7	5·8, 6·5

OPEN CLUSTERS

M	NGC	R.A. h m	Dec. ° '	Diameter '	Mag.	No. of stars
	6709	18 51·5	+10 21	13	6·7	40
	6755	19 07·8	+04 14	15	7·5	100

PLANETARY NEBULÆ

M	NGC	R.A. h m	Dec. ° '	Diameter "	Mag.	Mag. of central star
	6741	19 02·6	−00 27	6	10·8	14·7
	6751	19 05·9	−06 00	20	12·5	13·9
	6790	19 23·2	+01 31	7	10·2	13·5
	6803	19 31·3	+10 03	6	11·3	15·2

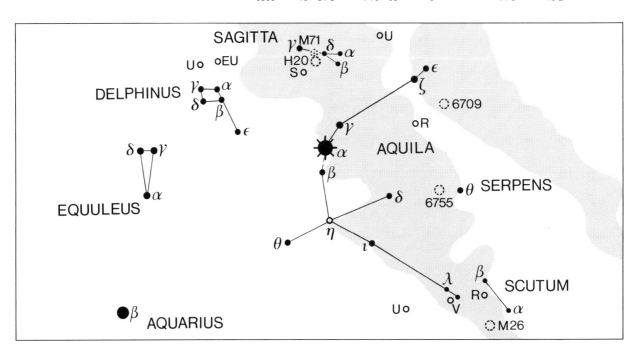

ARA

(Abbreviation: Ara).

An original constellation, though apparently without any definite legends attached to it. There are 7 stars above magnitude 4·00:

β 2·85
α 2·95
ζ 3·13
γ 3·34
δ 3·62
θ 3·66
η 3·76

BRIGHTEST STARS

Star	R.A. h	m	s	Dec. °	′	″	Mag.	Abs. mag.	Spectrum	Dist. pc	
η	16	49	47·0	−59	02	29	3·76	−0·3	K5	58	
ζ	16	58	37·1	−55	59	24	3·13	−0·3	K5	42	
ε¹	16	59	34·9	−53	09	38	4·06	−0·5	M1	82	
β	17	25	17·9	−55	31	47	2·85	−4·4	K3	240	
γ	17	25	23·5	−56	22	39	3·34	−4·4	B1	330	
δ	17	31	05·8	−60	41	01	3·62	−0·2	B8	29	
α	17	31	50·3	−49	52	34	2·95	−1·7	B3	58	Choo
σ	17	35	39·4	−46	30	20	4·59	—	A0	47	
θ	18	06	37·6	−50	05	30	3·66	−5·1	B1	480	

Also above mag. 5:

	Mag.	Abs. mag.	Spectrum	Dist.
λ	4·77	3·4	dF5	19

VARIABLE STARS

	R.A. h	m	Dec. °	′	Range	Type	Period, d.	Spectrum
X	16	36·4	−55	24	8·0–13·5	Mira	175·8	M
R	16	39·7	−57	00	6·0–6·9	Algol	4·42	B
U	17	53·6	−51	41	7·7–14·1	Mira	225·2	M

DOUBLE STARS

	R.A. h	m	Dec. °	′	P.A. °	Sep. ″	Mags.
γ	17	25·4	−56	23	AB 328	17·9	3·3, 10·3
					AC 066	41·6	11·8

OPEN CLUSTERS

M	NGC	R.A. h	m	Dec. °	′	Diameter ′	Mag.	No. of stars
	6193	16	41·3	−48	46	15	5·2	—
	6204	16	46·5	−47	01	5	8·2	45
	6208	16	49·5	−53	49	16	7·2	60
	6250	16	58·0	−45	48	8	5·9	60
	6253	16	59·1	−52	43	5	10·2	30
	H.13	17	05·4	−48	11	15	—	15
	IC 4651	17	24·7	−49	57	12	6·9	80

GLOBULAR CLUSTERS

M	NGC	R.A. h	m	Dec. °	′	Diameter ′	Mag.
	6352	17	25·5	−48	25	7·1	8·1
	6362	17	31·9	−67	03	10·7	8·3
	6397	17	40·7	−53	40	25·7	5·6

GALAXIES

M	NGC	R.A. h	m	Dec. °	′	Mag.	Dimensions ′	Type
	6215	16	51·1	−58	59	11·8	2·0 × 1·6	Sc
	6221	16	52·8	−59	13	11·5	3·2 × 2·3	SBc

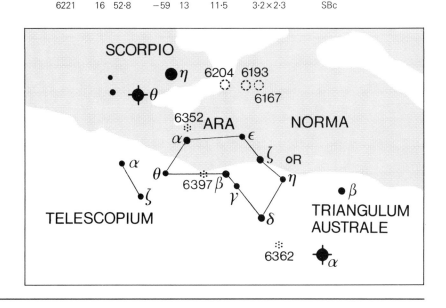

ARIES

(Abbreviation: Ari).

The first constellation of the Zodiac – though since the vernal equinox has shifted into Pisces, Aries should logically be classed as second! In mythology it represents a ram with a golden fleece, sent by the god Mercury (Hermes) to rescue the two children of the king of Thebes from an assassination plan by their stepmother. The ram, which had the remarkable ability to fly, carried out its mission; unfortunately, the girl (Helle) lost her balance and fell to her death in that part of the sea now called the Hellespont, but the boy (Phryxus) arrived safely. After the ram's death, its golden fleece was hung in a sacred grove, from which it was later removed by the Argonauts commanded by Jason.

Aries is moderately conspicuous. There are four stars above the fourth magnitude:

α 2·00 β 2·64 c 3·63 γ 3·9

Of these, γ is an excellent double with two equal components.

See chart for Andromeda

BRIGHTEST STARS

Star		R.A.			Dec.			Mag.	Abs. mag.	Spectrum	Dist.	
		h	m	s	°	′	″				pc	
5	γ	{ 01	53	31·7	+19	17	45	4·68	0·4	A9	36	Mesartim
		{ 01	53	31·8	+19	17	37	4·59	0·2	B9	36	
6	β	01	54	38·3	+20	48	29	2·64	2·1	A5	14	Sheratan
13	α	02	07	10·3	+23	27	45	2·00	−0·1	K2	26	Hamal
35		02	43	27·0	+27	42	26	4·66	−1·7	B3	180	
39		02	47	54·5	+29	14	50	4·51	0·0	K1	58	
41	c	02	49	58·9	+27	15	38	3·63	−0·2	B8	36	
48	ε	02	59	12·6	+21	20	25	4·63	1·4	A2	48	
57	δ	03	11	37·7	+19	43	36	4·35	−0·1	K2	78	Boteïn

Also above mag. 5:

		Mag.	Abs. mag.	Spectrum	Dist.
9	λ	4·79	1·7	F0	42
14		4·98	0·6	F2	75
58	ζ	4·89	0·0	A0	91

VARIABLE STARS

	R.A.		Dec.		Range	Type	Period, d.	Spectrum
	h	m	°	′				
R	02	16·1	+25	03	7·4–13·7	Mira	186·8	M
T	02	48·3	+17	31	7·5–11·3	Semi-reg.	317 (var.)	M
U	03	11·0	+14	48	7·2–15·2	Mira	371·1	M

DOUBLE STARS

	R.A.		Dec.		P.A.	Sep.	Mags.
	h	m	°	′	°	″	
γ	01	53·5	+19	18	000	7·8	4·8, 4·8
π	02	49·3	+17	28	AB 120	3·2	5·2, 8·7
					AC 110	25·2	10·8
ε	02	59·2	+21	20	191	1·5	5·2, 5·5

GALAXIES

M	NGC	R.A.		Dec.		Mag.	Dimensions	Type	
		h	m	°	′		′		
	772	01	59·3	+19	01	10·3	7·1×4·5	Sb	Arp 78
	976	02	34·0	+20	59	12·4	1·7×1·5	Sb	

AURIGA

(Abbreviation: Aur).

One of the most brilliant northern constellations, with Capella outstanding. Mythologically it honours Erechthonius, son of Vulcan, the blacksmith of the gods, who became King of Athens and also invented the four-horse chariot.

There are eight stars brighter than the fourth magnitude:

α 0·08
β 1·90
θ 2·62
ι 2·69
ε 2·99v
η 3·17
δ 3·72
ζ 3·75v

In addition, Al Nath used to be called γ Aurigæ, but has now been given a free transfer, and is included in Taurus as β Tauri.

BRIGHTEST STARS

Star		R.A.			Dec.			Mag.	Abs. mag.	Spectrum	Dist.	
		h	m	s	°	′	″				pc	
3	ι	04	56	59·5	+13	09	58	2·69	−2·3	K3	82	Hassaleh
7	ε	05	01	58·1	+43	49	24	2·99v	−8·5	F0	1400	
8	ζ	05	02	28·6	+41	04	33	3·75v	−2·3	K4	160	Sadatoni
10	η	05	06	30·8	+41	14	04	3·17	−1·7	B3	61	
13	α	05	16	41·3	+45	59	53	0·08	0·3	G8	13	V. close double Cape
15	λ	05	19	08·4	+40	05	57	4·71	4·4	G0	13	
29	τ	05	49	10·4	+39	10	52	4·52	0·3	G8	70	
31	υ	05	51	02·4	+37	18	20	4·74	−0·5	gM1	110	
32	ν	05	51	29·3	+39	08	55	3·97	0·2	K0	45	
33	δ	05	59	31·6	+54	17	05	3·72	0·2	K0	50	
34	β	05	59	31·7	+44	56	51	1·90	0·6	A2	22	Menkarlina
37	θ	05	59	43·2	+37	12	45	2·62	0·1	A0p	25	
35	π	05	59	56·1	+45	56	12	4·26	−2·4	M3	200	
44	κ	06	15	22·6	+29	29	53	4·35	0·3	G8	46	

Also above mag. 5:

		Mag.	Abs. mag.	Spectrum	Dist.
4	ω	4·94	0·6	A0	69
2		4·78	−0·2	K3	190
9		5·00	2·6	F0	30
11	μ	4·86	1·8	A3	22
21	σ	4·89	−0·3	K4	110
25	χ	4·76	−6·3	B5	930
30	ξ	4·99	0·8	A2	
46	ψ¹	4·91	−5·7	M0	600

VARIABLE STARS

	R.A.		Dec.		Range	Type	Period, d.	Spectrum
	h	m	°	′				
RX	05	01·4	+39	58	7·3–8·0	Cepheid	11·62	F–G
ε	05	02·0	+43	49	2·9–3·8	Eclipsing	9892	F
ζ	05	02·5	+41	05	3·7–4·1	Eclipsing	972·1	K+B

AURIGA (continued)

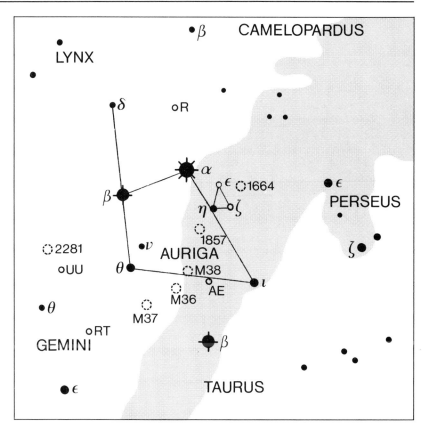

VARIABLE STARS (continued)

	R.A.		Dec.		Range	Type	Period, d.	Spectrum
	h	m	°	'				
R	05	17·3	+53	35	6·7–13·9	Mira	457·5	M
UV	05	21·8	+32	31	7·4–10·6	Mira	394·4	M
U	05	42·1	+32	02	7·5–15·5	Mira	408·1	M
X	06	12·2	+50	14	8·0–13·6	Mira	163·8	M
RT	06	28·6	+30	30	5·0–5·8	Cepheid	3·73	F–G
WW	06	32·5	+38	27	5·8–6·5	Algol	2·53	A+A
UU	06	36·5	+38	27	5·1–6·8	Semi-reg.	234	N

DOUBLE STARS

	R.A.		Dec.		P.A.	Sep.	Mags.
	h	m	°	'	°	"	
ω	04	59·3	+37	53	359	5·4	5·0, 8·0
R	05	17·3	+53	35	339	47·5	var., 8·6
ν	05	51·5	+39	09	206	54·6	4·9, 9·3
δ	05	59·5	+54	17	AB 271	115·4	3·7, 9·5
					AC 067	197·1	9·5
θ	05	59·7	+37	13	AB 313	3·6	2·6, 7·1
					AC 297	50·0	10·6

OPEN CLUSTERS

M	NGC	R.A.		Dec.		Diameter	Mag.	No. of stars
		h	m	°	'	'		
	1664	04	51·1	+43	42	18	7·6	–
	1778	05	08·1	+37	03	7	7·7	25
	1857	05	20·2	+39	21	6	7·0	40
	1893	05	22·7	+33	24	11	7·5	60
38	1912	05	28·7	+35	50	21	6·4	100
36	1960	05	36·1	+34	08	12	6·0	60
37	2099	05	52·4	+32	33	24	5·6	150
	2126	06	03·0	+49	54	6	10·2	40
	2281	06	49·3	+41	04	15	5·4	30

NEBULA

M	NGC	R.A.		Dec.		Dimensions	Mag. of illuminating star
		h	m	°	'	'	
	IC 405	05	16·2	+34	16	30×19	6v AE Aurigæ;
							Flaming Star Nebula

BOÖTES

(Abbreviation: Boö).

An original constellation, dominated by Arcturus. There are various myths attached to it, but none is very definite. According to one version, Boötes was a herdsman who invented the plough drawn by two oxen, for which service he was transferred to the sky.

There are eight stars above magnitude 4·00:

α −0·04
ε 2·37
η 2·68
γ 3·03
δ 3·47
β 3·50
ρ 3·58
ζ 3·78

BRIGHTEST STARS

Star		R.A. h	m	s	Dec. °	′	″	Mag.	Abs. mag.	Spectrum	Dist. pc	
4	τ	13	47	15·6	+17	27	24	4·50	3·8	F7	16	
5	υ	13	49	28·6	+15	47	52	4·06	−0·3	K5	72	
8	η	13	54	41·0	+18	23	51	2·68	2·7	G0	9·8	
17	κ	14	13	27·6	+51	47	15	4·40	1·5	A7	38	
16	α	14	15	39·6	+19	10	57	−0·04	−0·2	K2	11	Arcturus
21	ι	14	16	09·8	+51	22	02	4·75	2·4	A7	28	
19	λ	14	16	22·9	+46	05	18	4·18	1·8	A0p	29	
23	θ	14	25	11·7	+51	51	02	4·05	3·8	F7	13	
25	ρ	14	31	49·7	+30	22	17	3·58	−0·2	K3	56	
27	γ	14	32	04·6	+38	13	30	3·03	0·5	A7	32	Seginus
28	σ	14	34	40·7	+29	44	42	4·46	3·0	F2	17	
30	ζ	14	41	08·8	+13	43	42	3·78	−0·2	A2	63	
36	ε	14	44	59·1	+27	04	27	2·37	−0·9	K0	46	Izar
35	o	14	45	14·4	+16	57	51	4·60	0·2	K0	29	
37	ξ	14	51	23·2	+19	06	04	4·55	5·5	G8	6·8	
42	β	15	01	56·6	+40	23	26	3·50	0·3	G8	42	Nekkar
43	ψ	15	04	26·6	+26	56	51	4·54	−0·1	K2	75	
49	δ	15	15	30·1	+33	18	53	3·47	0·3	G8	43	
51	μ¹	15	24	29·3	+37	22	38	4·31	2·6	F0	18	Alkalurops

Also above mag. 5:

		Mag.	Abs. mag.	Spectrum	Dist.
6		4·91	−0·3	K4	110
12		4·83	2·4	F8	29
20		4·86	−0·2	K3	100
29	π¹	4·93	−0·3	B9	40
31		4·86	0·3	G8	75
34		4·81		gM0	30
41	ω	4·81	−0·3	K4	91
44		4·76	4·4	G0	12
45		4·93	3·4	F5	19
A		4·83	0·7	K1	

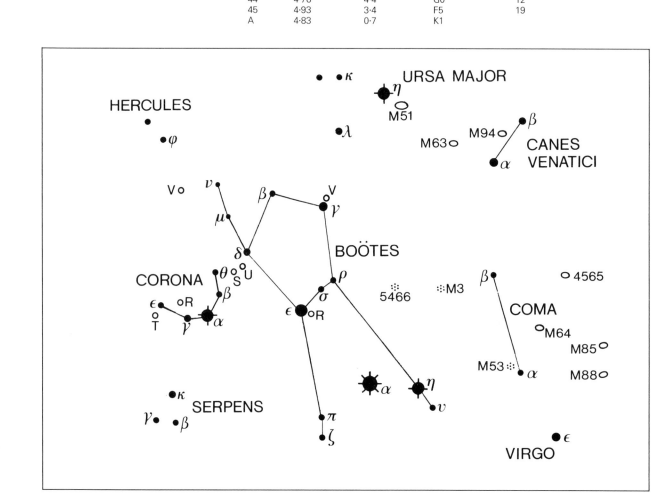

BOÖTES (continued)

VARIABLE STARS

	R.A. h m	Dec. ° '	Range	Type	Period, d.	Spectrum
S	14 22·9	+53 49	7·8–13·8	Mira	270·7	M
V	14 29·8	+38 52	7·0–12·0	Semi-reg.	258	M
R	14 37·2	+26 44	6·2–13·1	Mira	223·4	M
W	14 43·4	+26 32	4·7–5·4	Semi-reg.	450	M
i (44)	15 03·8	+47 39	6·5–7·1	W UMa	0·27	G+G

DOUBLE STARS

	R.A. h m	Dec. ° '	P.A. °	Sep. "	Mags.	
κ	14 13·5	+51 47	236	13·4	4·6, 6·6	
ι	14 16·2	+51 22	033	38·5	4·9, 7·5	
π	14 40·7	+16 25	108	5·6	4·9, 5·8	
ζ	14 41·1	+13 44	AB 303	1·0	4·5, 4·6	Binary, 123·3 y
			AC 259	99·3	10·9	
i (44)	15 03·8	+47 39	040	1·0	5·3v, 6·2	Binary, 225 y
μ	15 24·5	+37 23	171	108·3	4·3, 7·0	
ε	14 45·0	+27 04	339	2·8	2·5, 4·9	

GLOBULAR CLUSTER

M	NGC	R.A. h m	Dec. ° '	Diameter "	Mag.
	5466	14 05·5	+28 32	11	9·1

GALAXIES

M	NGC	R.A. h m	Dec. ° '	Mag.	Dimensions '	Type
	5248	13 37·5	+08 53	10·2	6·5×4·9	Sc
	5676	14 32·8	+49 28	10·9	3·9×2·0	Sc

CÆLUM

(Abbreviation: Cae).

This entirely unremarkable constellation was introduced in 1752 by Lacaille, under the name of Cæla Sculptoris. It has no star brighter than the fourth magnitude, and only two which are brighter than the fifth.

See chart for Columba

BRIGHTEST STARS

Star	R.A. h m s	Dec. ° ' "	Mag.	Abs. mag.	Spectrum	Dist. pc
α	04 40 33·6	−41 51 50	4·45	3·0	F2	20
γ	05 04 24·3	−35 29 00	4·55	0·2	gK0	52

VARIABLE STARS

	R.A. h m	Dec. ° '	Range	Type	Period, d.	Spectrum
R	04 40·5	−38 14	6·7–13·7	Mira	390·9	M
T	04 47·3	−36 13	7·0–9·8	Semi-reg.	156	N

DOUBLE STARS

	R.A. h m	Dec. ° '	P.A. °	Sep. "	Mags.
α	04 40·6	−41 52	121	6·6	4·5, 12·5
γ	05 04·4	−35 29	308	2·9	4·6, 8·1

CAMELOPARDUS

(Abbreviation: Cam).

A very barren northern constellation. It was introduced to the sky by Hevelius in 1690, and some historians have maintained that it represents the camel which carried Rebecca to Isaac! It is interesting to note that several of the apparently faint stars are in fact highly luminous and remote; for instance α Cam, which is below the fourth magnitude, is well over 20 000 times more luminous than the Sun.

There are no stars in Camelopardus above the fourth magnitude.

See chart for Cassiopeia

BRIGHTEST STARS

Star		R.A. h m s	Dec. ° ' "	Mag.	Abs. mag.	Spectrum	Dist. pc
2	H	03 29 04·1	+59 56 25	4·21	−7·1	B9	1100
	γ	03 50 21·5	+71 19 57	4·63	0·9	A3	56
9	α	04 54 03·0	+66 20 34	4·29	−6·2	09·5	860
7		04 57 17·1	+53 45 08	4·47	1·2	A1	45
10	β	05 03 25·1	+60 26 32	4·03	−4·5	G0	460

Also above mag. 5:

	Mag.	Abs. mag.	Spectrum	Dist.
22	4·80	0·6	A0	66

VARIABLE STARS

	R.A. h m	Dec. ° '	Range	Type	Period, d.	Spectrum
U	03 37·5	+62 29	7·7–8·9	Semi-reg.	400	N
RV	04 30·7	+57 25	7·1–8·2	Semi-reg.	101	M
T	04 40·1	+66 09	7·3–14·4	Mira	373·2	M
X	04 45·7	+75 06	7·4–14·2	Mira	143·6	K–M
S	05 41·0	+68 48	7·7–11·6	Semi-reg.	327	R
V	06 02·5	+74 30	7·7–16·0	Mira	522·4	M

CAMELOPARDUS (continued)

VARIABLE STARS (continued)

	R.A. h m	Dec. ° ′	Range	Type	Period, d.	Spectrum
VZ	07 31·1	+82 25	4·8–5·2	Semi-reg.	24	M
R	14 17·8	+83 50	7·0–14·4	Mira	270·2	S

OPEN CLUSTER

M	NGC	R.A. h m	Dec. ° ′	Diameter ′	Mag.	No. of stars
	1502	04 07·7	+62 20	8	5·7	45

PLANETARY NEBULA

M	NGC	R.A. h m	Dec. ° ′	Diameter ″	Mag.	Mag. of central star
	IC 3568	12 32·9	+82 33	6	11·6	12·3

GALAXIES

M	NGC	R.A. h m	Dec. ° ′	Mag.	Dimensions ′	Type
	IC 342	03 46·8	+68 06	9·2	17·8×17·4	SBc
	1961	05 42·1	+69 23	11·1	4·3×3·0	Sb
	2146	06 18·7	+78 21	10·5	6·0×3·8	SBb
	2366	07 28·9	+69 13	10·9	7·6×3·5	Irr.
	2403	07 36·9	+65 36	8·4	17·8×11·0	Sc
	2460	07 56·9	+60 21	11·7	2·9×2·2	Sb
	2655	08 55·6	+78 13	10·1	5·1×4·4	SBa
	2715	09 08·1	+78 05	11·4	5·0×1·9	Sc

CANCER

(Abbreviation: Cnc).

Cancer is an obscure constellation, redeemed only by the presence of two famous star-clusters, Præsepe and M.67. However, it lies in the Zodiac, and it has a legend attached to it. It represents a sea-crab which Juno, queen of

BRIGHTEST STARS

Star		R.A. h m s	Dec. ° ′ ″	Mag.	Abs. mag.	Spectrum	Dist. pc	
16	ζ	08 12 12·6	+17 38 52	4·67	3·2, 4·1	F7, G2	16	Tegmine
17	β	08 16 30·9	+09 11 08	3·52	−0·3	K4	52	
43	γ	08 43 17·1	+21 28 06	4·66	1·2	A1	48	Asellus Borealis
47	δ	08 44 41·0	+18 09 15	3·94	0·2	K0	47	Asellus Australis
48	ι¹	08 46 41·8	+28 45 36	4·02	−2·1	G8	130	
65	α	08 58 29·2	+11 51 28	4·25	1·9	A3	23	Acubens

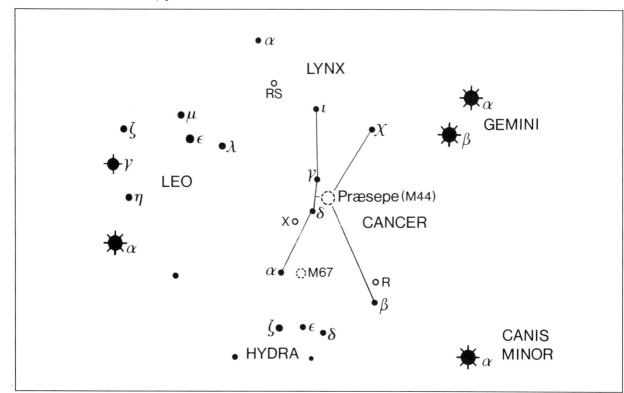

CANCER (continued)

Olympus, sent to the rescue of the multi-headed hydra which was doing battle with Hercules. Not surprisingly, Hercules trod on the crab, but as a reward for its efforts Juno placed it in the sky!

There are only two stars above the fourth magnitude; β (3·52) and δ (3·94).

VARIABLE STARS

	R.A. h m	Dec. ° '	Range	Type	Period, d.	Spectrum
R	08 16·6	+11 44	6·1–11·8	Mira	361·6	M
V	08 21·7	+17 17	7·5–13·9	Mira	272·1	S
X	08 55·4	+17 14	5·6–7·5	Semi-reg.	195	N
T	08 56·7	+19 51	7·6–10·5	Semi-reg.	482	R–N
W	09 09·9	+25 15	7·4–14·4	Mira	393·2	M
RS	09 10·6	+30 58	6·2–7·7	Semi-reg.	120	M

DOUBLE STAR

	R.A. h m	Dec. ° '	P.A. °	Sep. "	Mags.	
ζ	08 12·2	+17 39	AB + C 088	5·7	5·0, 6·2	Binary, 1150 y
			AB 182	0·6	5·3, 6·0	Binary, 59·7 y
			AB + D 108	287·9	9·7	

OPEN CLUSTERS

M	NGC	R.A. h m	Dec. ° '	Diameter '	Mag.	No. of stars	
44	2632	08 40·1	+19 59	95	3·1	50	Præsepe
67	2682	08 50·4	+11 49	30	6·9	200	

GALAXY

M	NGC	R.A. h m	Dec. ° '	Mag.	Dimensions '	Type
	2775	09 10·3	+07 02	10·3	4·5 × 3·5	Sa

CANES VENATICI

(Abbreviation: CVn).

One of Hevelius' constellations, dating only from his maps of 1690, and evidently representing two hunting dogs (Asterion and Chara) which are being held by the herdsman Boötes – possibly to stop them from chasing the two Bears round and round the celestial pole. The name of Cor Caroli was given to α² CVn by Edmond Halley in honour of King Charles I. Cor Caroli (mag. 2·90) is the only star above the fourth magnitude. It is also the prototype 'magnetic variable'.

See chart for Ursa Major

BRIGHTEST STARS

Star		R.A. h m s	Dec. ° ' "	Mag.	Abs. mag.	Spectrum	Dist. pc	
8	β	12 33 44·4	+41 21 26	4·26	4·5	G0	9·2	Chara
12	α²	12 56 01·6	+38 19 06	2·90	0·1	A0p	20	Cor Caroli
20		13 17 32·5	+40 34 21	4·73	−0·7	F0	52	
24		13 34 27·2	+49 00 57	4·70	1·9	A4	35	

Also above mag. 5:

	Mag.	Abs. mag.	Spectrum	Dist.
5	4·80	0·3	G7	38
25	4·82	0·5	A7	73

VARIABLE STARS

	R.A. h m	Dec. ° '	Range	Type	Period, d.	Spectrum	
T	12 30·2	+31 30	7·6–12·6	Mira	290·1	M	
Y	12 45·1	+45 26	7·4–10·0	Semi-reg.	157	N	La Superba
U	12 47·3	+38 23	7·2–11·0	Mira	345·6	M	
TU	12 54·9	+47 12	5·6–6·6	Semi-reg.	50	M	
V	13 19·5	+45 32	6·5–8·6	Semi-reg.	192	M	
R	13 49·0	+39 33	6·5–12·9	Mira	328·5	M	

DOUBLE STAR

	R.A. h m	Dec. ° '	P.A. °	Sep. "	Mags.
α²	12 56·0	+38 19	22·9	19·4	2·9, 5·5

GLOBULAR CLUSTER

M	NGC	R.A. h m	Dec. ° '	Diameter '	Mag.
3	5272	13 42·2	+28 23	16·2	6·4

GALAXIES

M	NGC	R.A. h m	Dec. ° '	Mag.	Dimensions '	Type	
	4111	12 07·1	+43 04	10·8	4·8 × 1·1	S0	
	4138	12 09·5	+43 41	12·3	2·9 × 1·9	E4	
	4145	12 10·0	+39 53	11·0	5·8 × 4·4	Sc	
	4151	12 10·5	+39 24	10·4	5·9 × 4·4	Pec.	(Strong UV source)
	4618	12 41·5	+41 09	10·8	4·4 × 3·8	Sc	
	4214	12 15·6	+36 20	9·8	7·9 × 6·3	Irr.	
	4217	12 15·8	+47 06	11·9	5·5 × 1·8	Sb	
	4242	12 17·5	+45 37	11·0	4·8 × 3·8	S	
106	4258	12 19·0	+47 18	8·3	18·2 × 7·9	Sb	
	4395	12 25·8	+33 33	10·1	12·9 × 11·0	S	
	4449	12 28·2	+44 06	9·4	5·1 × 3·7	Irr.	
	4490	12 30·6	+41 38	9·8	5·9 × 3·1	Sc	

CANES VENATICI *(continued)*

GALAXIES *(continued)*

M	NGC	R.A. h	m	Dec. °	'	Mag.	Dimensions '	Type
	4631	12	42·1	+32	32	9·3	15·1×3·3	Sc
	4656–7	12	44·0	+32	10	10·4	13·8×3·3	Sc
94	4736	12	50·9	+41	07	8·2	110×9·1	Sb
	5005	13	10·9	+37	03	9·8	5·4×2·7	Sb
	5033	13	13·4	+36	36	10·1	10·5×5·6	Sb
63	5055	13	15·8	+42	02	8·6	12·3×7·6	
	5112	13	21·9	+38	44	11·9	3·9×2·9	Sc
51	5194	13	29·9	+47	12	8·4	11·0×7·8	Sc Whirlpool
	5195	13	30·0	+47	16	9·6	5·4×4·3	Pec. Companion to M.51
	5371	13	55·7	+40	28	10·7	4·4×3·6	Sb

CANIS MAJOR

(Abbreviation: CMa).

An original constellation, representing one of Orion's hunting dogs. Though dominated by Sirius, it contains several other bright stars, of which one (Adhara) is only just below the first magnitude. Altogether there are 11 stars brighter than the fourth magnitude:

α	−1·46
ε	1·50
δ	1·86
β	1·98v
η	2·44
ζ	3·02
o^2	3·03
o^1	3·86
ω	3·86
ν^2	3·95
κ	3·96

It is interesting to note that three of these stars–Wezea, Aludra and o^2–are extremely luminous. Compared with them, Sirius is extremely feeble, as it has only 26 times the Sun's luminosity.

BRIGHTEST STARS

Star		R.A. h	m	s	Dec. °	'	"	Mag.	Abs. mag.	Spectrum	Dist. pc	
1	ζ	06	20	18·7	−30	03	48	3·02	−1·7	B3	88	Phurad
2	β	06	22	41·9	−17	57	22	1·98v	−4·8	B1	220	Mirzam
	λ	06	28	10·1	−32	34	48	4·48	−0·5	B5	17	
4	ξ^1	06	31	51·2	−23	25	06	4·34	−3·9	B1	440	
5	ξ^2	06	35	03·3	−22	57	53	4·54	0·6	A0	61	
7	ν^2	06	36	41·0	−19	15	22	3·95	0·0	K1	72	
8	ν^3	06	37	53·3	−18	14	15	4·43	0·0	K1	72	
9	α	06	45	08·9	−16	42	58	−1·46	1·4	A1	2·7	Sirius
13	κ	06	49	50·4	−32	30	31	3·96	−2·5	B2	200	
16	o^1	06	54	07·8	−24	11	02	3·86	−6·0	K3	520	
14	θ	06	54	11·3	−12	02	19	4·07	−0·3	K4	73	
19	π	06	55	37·3	−20	08	11	4·68	0·6	gF2	65	
20	ι	06	56	08·1	−17	03	14	4·38	−3·9	B3	410	
21	ε	06	58	37·5	−28	58	20	1·50	−4·4	B2	150	Adhara
22	σ	07	01	43·1	−27	56	06	3·46	−5·7	M0	460	
24	o^2	07	03	01·4	−23	50	00	3·03	−6·8	B3	860	
23	γ	07	03	45·4	−15	38	00	4·11	−3·4	B8	320	Muliphen
25	δ	07	08	23·4	−26	23	36	1·86	−8·0	F8	940	Wezea
27		07	14	15·1	−26	21	09	4·66	−2·9	B3	320	
28	ω	07	14	48·6	−26	46	22	3·86	−2·3	B3	170	
29	UW	07	18	40·3	−24	33	32	4·98v	—	O7·8		
30	τ	07	18	41·4	−24	57	15	4·39	−6·0	O9	1100	
31	η	07	24	05·6	−29	18	11	2·44	−7·0	B5	760	Aludra

Sirius has a white dwarf companion, mag. 8; binary, period 50 years. Though the companion does not appear faint, it is difficult to see because of the overpowering glare of the primary. The separation ranges between 3" to 11"·5. It was widest in 1975, so that it is now closing up. β (Mirzam) is variable over a very small range, and is also a spectrum variable. ξ^1 is also a spectrum variable, and optically variable over a very small range of less than 0·1 magnitude.

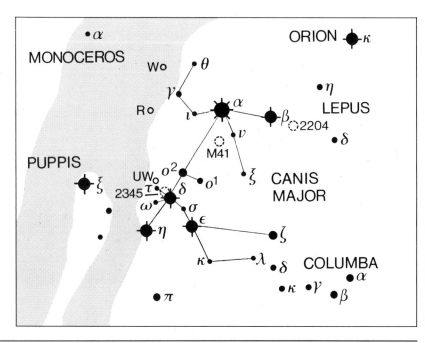

CANIS MAJOR *(continued)*

Also above mag. 5:

		Mag.	Abs. mag.	Spectrum	Dist.
15	EY	4·82	−3·9	B1	560
29	UW	4·98v	(Lyræ type, 4·4d)		

VARIABLE STARS

	R.A. h m	Dec. ° '	Range	Type	Period, d.	Spectrum
W	07 08·1	−11 55	6·4–7·9	Irr.	--	N
RY	07 16·6	−11 29	7·7–8·5	Cepheid	4·68	F–G
UW	07 18·4	−24 34	4·0–5·3	Beta Lyræ	4·39	07
R	07 19·5	−16 24	5·7–6·3	Algol	1·14	F

DOUBLE STARS

	R.A. h m	Dec. ° '	P.A. °	Sep. "	Mags.	
ν¹	06 36·4	−18 40	262	17·5	5·8, 8·5	
α	06 45·1	−16 43	005	4·5	−1·5, 8·5	Binary, 50 y
17	06 55·0	−20 24	AB 147	44·4	5·8, 9·3	
			AC 184	50·5	9·0	
			AD 186	129·9	9·5	
π	06 55·6	−20 08	018	11·6	4·7, 9·7	
μ	06 56·1	−14 03	AB 340	3·0	5·3, 8·6	
			AC 288	88·4	10·5	
			AD 061	101·3	10·7	
ε	06 58·6	−28 58	161	7·5	1·5, 7·4	

OPEN CLUSTERS

M	NGC	R.A. h m	Dec. ° '	Diameter '	Mag.	No. of stars	
	2204	06 15·7	−18 39	13	8·6	80	
41	2287	06 47·0	−20 44	38	4·5	80	
	2345	07 08·3	−13 10	12	7·7	20	
	2360	07 17·8	−15 37	13	7·2	80	
	2362	07 17·8	−24 57	8	4·1	60	τ CMa

GALAXIES

M	NGC	R.A. h m	Dec. ° '	Mag.	Dimensions '	Type
	2207	06 16·4	−21 22	10·7	4·3×2·9	Sc
	2217	06 21·7	−27 14	10·4	4·8×4·4	SBa
	2223	06 24·6	−22 50	11·4	3·3×3·0	SBb
	2280	06 44·8	−27 38	11·8	5·6×3·2	Sb

CANIS MINOR

(Abbreviation: CMi).

The second of Orion's two dogs. There are two stars above the fourth magnitude:

α 0·38
β 2·90

BRIGHTEST STARS

Star	R.A. h m s	Dec. ° ' "	Mag.	Abs. mag.	Spectrum	Dist. pc	
3 β	07 27 09·0	+08 17 21	2·90	−0·2	B8	42	Gomeisa
4 γ	07 28 09·7	+08 55 33	4·32	−0·2	K3	65	
6	07 29 47·7	+12 00 24	4·54	−0·1	K2	70	
10 α	07 39 18·1	+05 13 30	0·38	2·6	F5	3·5	Procyon
HD 66141	08 02 15·8	+02 20 04	4·39	−0·1	K2	70	

Procyon has a white dwarf companion; magnitude 13, period 40 years. It is very difficult to observe because of the glare from the primary.

Also above mag. 5:

		Mag.	Abs. mag.	Spectrum	Dist.
2	ε	4·99	0·3	G8	78

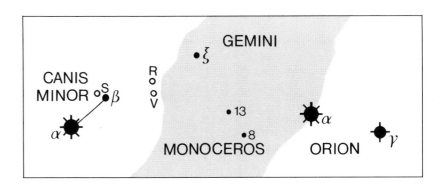

CANIS MINOR *(continued)*

VARIABLE STARS

	R.A.		Dec.		Range	Type	Period, d.	Spectrum
	h	m	°	′				
V	07	07·0	+08	53	7·4–15·1	Mira	366·1	M
R	07	08·7	+10	01	7·3–11·6	Mira	337·8	S
S	07	32·7	+08	19	6·6–13·2	Mira	332·9	M

DOUBLE STAR

	R.A.		Dec.		P.A.	Sep.	Mags.	
	h	m	°	′	°	″		
α	07	39·3	+05	14	021	5·2	0·4, 12·9	Binary, 40·7 y

CAPRICORNUS

(Abbreviation: Cap).

A Zodiacal constellation, though by no means brilliant. It has been identified with the demigod Pan, but there seems to be no well-defined mythological associations. There are five stars above magnitude 4:

δ 2·87v
β 3·08
α² 3·57
γ 3·68
ζ 3·74

α¹ and α² make up a naked-eye pair, separated by 376″ (position angle 291°), but the two components are not genuinely associated, and each is itself double; the fainter component of α² is also double!

BRIGHTEST STARS

Star		R.A.			Dec.			Mag.	Abs. mag.	Spectrum	Dist.	
		h	m	s	°	′	″				pc	
5	α¹	20	17	38·6	−12	30	30	4·24	−4·5	G3	490	} Al Giedi
6	α²	20	18	03·1	−12	32	42	3·57	0·2	G9	36	
9	β	20	21	00·5	−14	46	53	3·08	4·0	F8	32	Dabih
16	ψ	20	46	05·5	−25	16	16	4·14	3·4	F5	12	
18	ω	20	51	49·1	−26	55	09	4·11	−0·3	K5	64	
23	θ	21	05	56·6	−17	13	58	4·07	0·6	A0	49	
24	A	21	07	07·5	−25	00	21	4·50	−0·5	M1	95	
32	ι	21	22	14·6	−16	50	05	4·28	0·3	G8	63	
34	ζ	21	26	39·9	−22	24	41	3·74	−4·5	G4	450	
36	b	21	28	43·2	−21	48	26	4·51	0·3	gG5	67	
39	ε	21	37	04·7	−19	27	58	4·68	−2·3	B3	250	
40	γ	21	40	05·2	−16	39	45	3·68	1·2	F0p	18	Nashira
43	κ	21	42	39·3	−18	51	59	4·73	0·3	G8	77	
49	δ	21	47	02·3	−16	07	38	2·87v	2·0	A5	15	Deneb al Giedi

Also above mag. 5:

		Mag.	Abs. mag.	Spectrum	Dist.	
8	ν	4·76	−0·3	B9	99	Alshat
11	ρ	4·78	0·6	F2	32	
22	η	4·84	2·6	A4	12	

VARIABLE STARS

	R.A.		Dec.		Range	Type	Period, d.	Spectrum
	h	m	°	′				
RT	20	17·1	−21	19	6·5–8·1	Semi-reg.	393	N
RR	21	02·3	−27	05	7·8–15·5	Mira	277·5	M
RS	21	07·2	−16	25	7·0–9·0	Semi-reg.	340	M

DOUBLE STARS

	R.A.		Dec.		P.A.	Sep.	Mags.	
	h	m	°	′	°	″		
α	20	18·1	−12	33	291	377·7	5·6, 4·2	(α¹–α²)
α¹	20	17·6	−12	30	AB 182	44·3	4·2, 13·7	
					AC 221	45·4	9·2	
α²	20	18·1	−12	33	AB 172	6·6	3·6, 11·0	
					AD 156	154·6	9·3	
					BC 240	1·2	11·3	

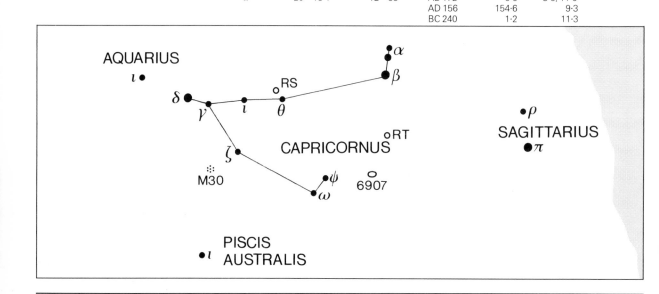

CAPRICORNUS (continued)

DOUBLE STARS (continued)

	R.A. h m	Dec. ° ′	P.A. °	Sep. ″	Mags.	
σ	20 19·6	−19 07	179	55·9	5·5, 9·0	
π	20 27·3	−18 13	148	3·2	5·3, 8·9	
ρ	20 28·9	−17 49	158	0·5	5·0, 10·0	
ε	21 37·1	−19 28	047	68·1	4·7, 9·5	
τ	20 39·3	−14 57	118	0·3	5·8, 6·3	Binary, 200 y

GLOBULAR CLUSTER

M	NGC	R.A. h m	Dec. ° ′	Diameter ′	Mag.
30	7099	21 40·4	−23 11	11·0	7·5

GALAXY

M	NGC	R.A. h m	Dec. ° ′	Mag.	Dimensions ′	Type
	6907	20 25·1	−24 49	11·3	3·4×3·0	SBb

CARINA

(Abbreviation: Car).

The keel of the ship Argo, in which Jason and his companions sailed upon their successful if somewhat unprincipled expedition to remove the Golden Fleece of the sacred ram (Aries) from its grove. Carina is the brightest part of Argo, and contains Canopus, which is second only to Sirius in brilliance. Canopus is a highly luminous star, but estimates of its power and distance vary considerably. A recent estimate makes it over 200 000 times as luminous as the Sun; other estimates are much lower. Also in this constellation is η Carinæ, which during part of the last century rivalled Sirius, but which is now well below naked-eye visibility. It is unique, and at its peak may have been the most luminous star in the Galaxy. Excluding η, there are 15 stars in Carina above the fourth magnitude:

α	−0·72
β	1·68
ε	1·86
ι	2·25
θ	2·76
υ	2·97
ρ	3·32
ω	3·32
q	3·40
a	3·44
χ	3·47
u	3·78
R	3·8v
c	3·84
x	3·91

BRIGHTEST STARS

Star	R.A. h m s	Dec. ° ′ ″	Mag.	Abs. mag.	Spectrum	Dist. pc	
α	06 23 57·1	−52 41 44	−0·72	−8·5	F0	360	Canopus
N	06 34 58·5	−52 58 32	4·39	−0·8	B9	100	
A	06 49 51·3	−53 37 20	4·40	0·4	gG3	52	
χ	07 56 46·7	−52 58 56	3·47	−3·0	B2	180	
d	08 40 37·0	−59 45 40	4·33	−3·6	B2	350	
ε	08 22 30·8	−59 30 34	1·86	−2·1	K0	62	Avior
f	08 46 42·6	−56 46 11	4·49	−2·5	B2	250	
c	08 55 02·8	−60 38 41	3·84	−1·0	B8	93	
G	09 05 38·4	−70 32 20	4·71	−2·5	B2	280	
a	09 10 57·9	−58 58 01	3·44	−3·0	B2	190	
i	09 11 16·7	−62 19 02	3·97	−2·3	B3	180	
β	09 13 12·2	−69 43 02	1·68	−0·6	A0	26	Miaplacidus
g	09 16 12·2	−57 32 28	4·34	−0·3	gK5	70	
ι	09 17 05·4	−59 16 31	2·25	−4·7	F0	250	Tureis (or
R	09 32 14·7	−62 47 19	3·8–10	var.	M5	250	Aspidiske)
h	09 34 26·6	−59 13 46	4·08	−3·7	B5	350	
m	09 39 20·9	−61 19 40	4·52	0·2	B9	73	
l (ZZ)	09 45 14·8	−62 30 28	4·1v	4·4	G0	16	
υ	09 47 06·1	−65 04 18	2·97	−2·0	A7	99	
ω	10 13 44·3	−70 02 16	3·32	−1·0	B7	70	
q	10 17 04·9	−61 19 56	3·40	−4·4	K5	280	
1	10 24 23·7	−74 01 54	4·00	0·6	F2	17	
s	10 27 24·4	−57 38 19	4·68	−8·5	F0	3600	
p	10 32 01·4	−61 41 07	3·32v	−1·7	B3	96	
r	10 35 35·2	−57 33 27	4·45	−0·4	gM0	92	
t²	10 38 45·0	−59 10 58	4·66	−5·9	cK	1300	
θ	10 42 57·4	−64 23 39	2·76	−4·1	B0	230	
W	10 43 32·1	−60 33 59	4·57	−0·3	K5	75	
u	10 53 29·6	−58 51 12	3·78	1·7	K0	26	
x	11 08 35·3	−58 58 30	3·91	−8·0	G0	1400	
y	11 12 36·0	−60 19 03	4·60	−8·5	F0	3300	

Also above mag. 5:

	Mag.	Abs. mag.	Spectrum
Q	4·92	−0·8	M0
D¹	4·96	−2·4	B3
B	4·80	3·5	F5
e²	4·80	0·2	G6
p	4·94		F1
k	4·87	1·5	G4
K	4·94	−0·5	A2
z¹	4·76	1·1	G5

VARIABLE STARS

	R.A. h m	Dec. ° ′	Range	Type	Period, d.	Spectrum	
V	08 28·7	−60 07	7·1–7·8	Cepheid	6·70	F–H	
X	08 31·3	−59 14	7·9–8·6	Beta Lyræ	1·08	A+A	
R	09 32·2	−62 47	3·9–10·5	Mira	308·7	M	
ZZ	09 45·2	−62 30	3·3–4·2	Cepheid	35·53	F–K	(l Carinæ)
S	10 09·4	−61 33	4·5–9·9	Mira	149·5	K–M	
η	10 45·1	−59 41	−0·8–7·9	Irreg.	−	Pec.	
BO	10 45·8	−59 29	7·2–8·5	Irreg.	−	M	
U	10 57·8	−59 44	5·7–7·0	Cepheid	38·77	F–G	

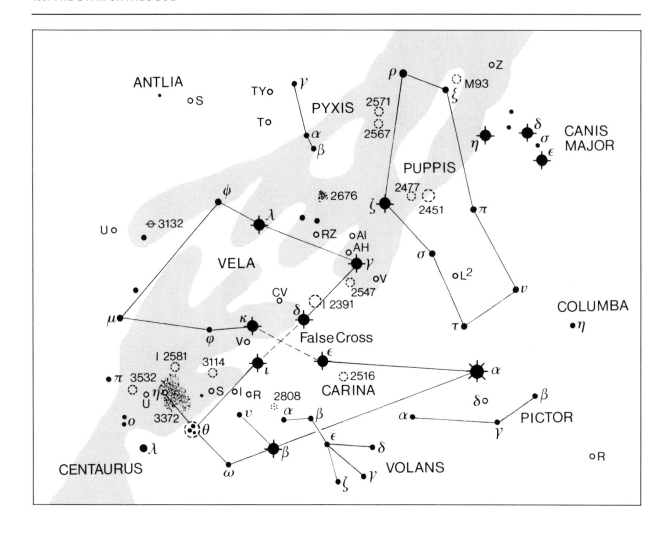

CARINA (continued)

DOUBLE STARS

	R.A.		Dec.		P.A.	Sep.	Mags.
	h	m	°	'	°	"	
η	10	45·1	−59	41	195	0·2	var., 8·6
υ	09	47·1	−65	04	127	5·0	3·1, 6·1

OPEN CLUSTERS

M	NGC	R.A.		Dec.		Diameter	Mag.	No. of stars	
		h	m	°	'	'			
	2516	07	58·3	−60	52	30	3·8	80	
	3114	10	02·7	−60	07	35	4·2	–	
	IC 2581	10	27·4	−57	38	8	4·3	25	
	IC 2602	10	43·2	−64	24	50	1·9	60	θ Carinæ cluster
	3532	11	06·4	−58	40	55	3·0	150	
	Mel 101	10	42·1	−65	06	14	8·0	50	
	3572	11	10·4	−60	14	7	6·6	35	
	3590	11	12·9	−60	47	4	8·2	25	
	Mel 105	11	19·5	−63	30	4	8·5	70	
	IC 2714	11	17·9	−62	42	12	8·2	100	
	3680	11	25·7	−43	15	12	7·6	30	

GLOBULAR CLUSTER

M	NGC	R.A.		Dec.		Diameter	Mag.
		h	m	°	'	'	
	2808	09	12·0	−64	52	13·8	6·3

NEBULA

M	NGC	R.A.		Dec.		Mag.	Mag. of illuminating star
		h	m	°	'		
	3372	10	43·8	−59	52	6·2	var. η Carinæ nebula

CARINA (continued)

PLANETARY NEBULÆ

M	NGC	R.A.		Dec.		Diameter	Mag.	Mag. of central star
		h	m	°	'	"		
	IC 2448	09	07·1	−69	57	8	11·5	12·9
	2867	09	21·4	−58	19	11	9·7	13·6
	IC 2501	09	38·8	−60	05	25	11·3	−
	3211	10	17·8	−62	40	12	11·8	−
	IC 2621	11	00·3	−65	15	5	−	13·6

CASSIOPEIA

(Abbreviation: Cas).

An original constellation – one of the most distinctive in the sky. Mythologically, Cassiopeia was Andromeda's mother and wife of Cepheus; it was her boasting which led to the unfortunate contretemps with Neptune's sea-monster.

There are seven stars above magnitude 4. Of these, γ (Cih) is variable, and it is also suspected that α (Shedir) is variable over a small range – possibly 2·1 to 2·4, though opinions differ. κ, which looks obscure, is extremely luminous.

There are 8 stars brighter than magnitude 4:

γ 2·2v
α 2·23v?
β 2·27
δ 2·68
ε 3·38
η 3·44
ζ 3·67
ι 3·98

BRIGHTEST STARS

Star		R.A.			Dec.			Mag.	Abs. mag.	Spectrum	Dist.	
		h	m	s	°	'	"				pc	
7	ρ	23	54	22·9	+57	29	58	var.	−8·0	F8	1500	
11	β	00	09	10·6	+59	08	59	2·27	1·9	F2	13	Chaph
14	λ	00	31	46·3	+54	31	20	4·73	−0·9	B8	13	
15	κ	00	32	59·9	+62	55	55	4·16	−6·6	B1	930	
17	ζ	00	36	58·2	+53	53	49	3·67	−2·5	B2	170	
18	α	00	40	30·4	+56	32	15	2·23v?	−0·9	K0	37	Shedir
22	ο	00	44	43·4	+48	17	04	4·54	−2·5	B2	210	
24	η	00	49	05·9	+57	48	58	3·44	4·6	G0	5·9	Achird
28	ν²	00	56	39·7	+59	10	52	4·63	1·8	G8	29	
27	γ	00	56	42·4	+60	43	00	2·2v	−4·6	B0p	240	Cih
33	θ	01	11	06·1	+55	09	00	4·33	2·4	A7	37	Marfak
37	δ	01	25	48·9	+60	14	07	2·68	2·1	A5	19	Ruchbah
36	ψ	01	25	55·9	+68	07	48	4·74	0·2	K0	74	
39	χ	01	33	50·8	+59	13	56	4·71	0·2	K0	80	
45	ε	01	54	23·6	+63	40	13	3·38	−2·9	B3	160	Segin
48		02	01	57·3	+70	54	26	4·48	1·9	A4	33	
50	ι	02	03	26·0	+72	25	17	3·98	1·2	A1	36	

Also above mag. 5:

		Mag.	Abs. mag.	Spectrum	Dist.
1		4·85	−4·4	B1	570
5	τ	4·87	0·0	K1	92
4		4·97	−0·3	K5	86
8	σ	4·88	−3·5	B1	400
19	ξ	4·80	−2·5	B2	260
25	ν	4·89		B9	
26	υ¹	4·83	−0·1	K2	91
46	ω	4·99		B8	

VARIABLE STARS

	R.A.		Dec.		Range	Type	Period, d.	Spectrum
	h	m	°	'				
V	23	11·7	+59	42	6·9–13·4	Mira	228·8	M
ρ	23	54·4	+58	30	4·1–6·2	?	−	F–K
R	23	58·4	+51	24	4·7–13·5	Mira	430·5	M
T	00	23·2	+55	48	6·9–13·0	Mira	444·8	M
TU	00	26·3	+51	17	6·9–8·1	Cepheid	2·14	F
α	00	40·5	+56	22	?2·1–2·5	Suspected	−	K
U	00	46·4	+48	15	8·0–15·7	Mira	277·2	S
RV	00	52·7	+47	25	7·3–16·1	Mira	331·7	M
W	00	54·9	+58	34	7·8–12·5	Mira	405·6	N
γ	00	56·7	+60	43	1·6–3·3	Irreg.	−	B
S	01	19·7	+72	37	7·9–16·1	Mira	612·4	S
SU	02	52·0	+68	53	5·7–6·2	Cepheid	1·95	F
RZ	02	48·9	+69	38	6·2–7·7	Algol	1·19	A

DOUBLE STARS

	R.A.		Dec.		P.A.	Sep.	Mags.	
	h	m	°	'	°	"		
λ	00	31·8	+54	31	176	0·5	5·3, 5·6	
η	00	49·1	+57	49	293	12·2	3·4, 7·5	Binary, 480 y
ψ	01	25·9	+68	08	113	25·0	4·7, 9·6	
ι	02	29·1	+67	24	232	2·4	4·6, 6·9	Binary, 840 y
σ	23	59·0	+55	45	326	3·0	5·0, 7·1	

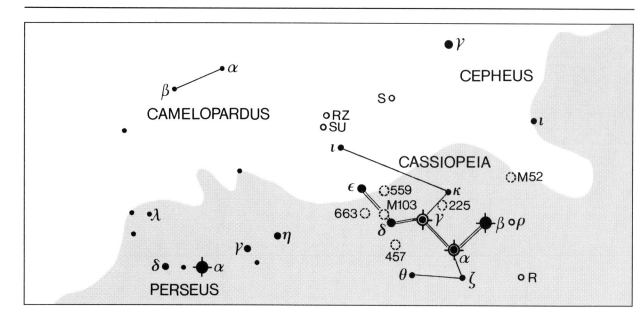

CASSIOPEIA (continued)

OPEN CLUSTERS

M	NGC	R.A. h m	Dec. ° '	Diameter '	Mag.	No. of stars	
52	7654	23 24·2	+61 35	13	6·9	100	
	7788	23 56·7	+61 24	9	9·4	20	
	7789	23 57·0	+56 44	16	6·7	300	
	H.21	23 54·1	+61 46	4	9·0	6	
	129	00 29·9	+60 14	21	6·5	35	Contains DL Cas.
	133	00 31·2	+63 22	7	9·4	5	
	146	00 33·1	+63 18	7	9·1	20	
	381	01 08·3	+61 35	6	9·3	50	
	436	01 15·6	+58 49	6	8·8	30	
	457	01 19·1	+58 20	13	6·4	80	φ Cas cluster.
	559	01 29·5	+63 18	4·4	9·5	60	
	IC 1805	02 32·7	+61 27	22	6·5	40	
103	581	01 33·2	+60 42	6	7·4	25	
	637	01 42·9	+64 00	3·5	8·2	20	
	1027	02 42·7	+61 33	20	6·7	40	
	654	01 44·1	+61 53	5	6·5	60	
	659	01 44·2	+60 42	5	7·9	40	
	663	01 46·0	+61 15	16	7·1	80	

NEBULÆ

M	NGC	R.A. h m	Dec. ° '	Dimensions '	Mag. of illuminating star	
	7635	23 20·7	+61 12	15×8	7	Bubble Nebula
	281	00 52·8	+56 36	35×30	8	
	IC 1805	02 33·4	+61 26	60×60	−	
	IC 1848	02 51·3	+60 25	60×30	−	

GALAXIES

M	NGC	R.A. h m	Dec. ° '	Mag.	Dimensions '	Type
	147	00 33·2	+48 30	9·3	12·9×8·1	dE4
	185	00 39·0	+48 20	9·2	11·5×9·8	dE0

CENTAURUS

(Abbreviation: Cen).

A brilliant southern constellation (one of Ptolemy's originals), containing many important objects, including the nearest of all the bright stars (α) and the superb globular cluster ω. There are 19 stars brighter than the fourth magnitude.

α	−0·27
β	0·61
θ	2·06
γ	2·17
ε	2·30
η	2·31v
ζ	2·55
δ	2·60v
ι	2·75
μ	3·04v
κ	3·13
λ	3·13
ν	3·41
φ	3·83
τ	3·86
υ^1	3·87
d	3·88
π	3·89
σ	3·91

BRIGHTEST STARS

Star	*R.A.* h	m	s	*Dec.* °	′	″	*Mag.*	*Abs. mag.*	*Spectrum*	*Dist.* pc	
π	11	21	00·4	−54	29	27	3·89	−1·1	B5	99	
λ	11	35	46·8	−63	01	11	3·13	−0·8	B9	57	
65G	11	46	30·7	−61	10	12	4·11	−2·0	G0	140	
j	11	49	41·0	−45	10	25	4·32	−1·1	B5	120	
B	11	51	08·5	−45	43	05	4·46	−0·3	K4	90	
δ	12	08	21·5	−50	43	20	2·60v	−2·5	B2	100	
ρ	12	11	39·1	−52	22	07	3·96	−1·4	B4	120	
τ	12	37	42·1	−48	32	28	3·86	1·4	A2	30	
σ	12	28	02·4	−50	13	51	3·91	−1·7	B3	130	
γ	12	41	30·9	−48	57	34	2·17	−0·6	A0	34	Menkent
w	12	42	35·3	−48	48	47	4·66	0·0	gK1	86	
e	12	53	06·8	−48	56	35	4·33	−0·1	gK2	54	
n	12	53	26·1	−40	10	44	4·27	0·5	A7	57	
ξ^2	13	06	54·5	−49	54	22	4·27	−3·0	B2	280	
ι	13	20	35·8	−36	42	44	2·75	1·4	A2	16	
J	13	22	37·8	−60	59	18	4·53	−1·1	B5	130	
m	13	24	0·05	−64	32	09	4·53	1·8	G5	35	
d	13	31	02·6	−39	24	27	3·88	0·3	G8	36	
ε	13	39	53·2	−53	27	58	2·30	−3·5	B1	150	
1	13	45	43·0	−33	02	30	4·23	0·8	F2	28	
M	13	46	39·3	−51	25	58	4·65	0·2	gK0	78	
2	13	49	26·6	−34	27	03	4·19	−0·5	M1	87	
ν	13	49	30·2	−41	41	16	3·41	−2·5	B2	150	
μ	13	49	36·9	−42	28	25	3·04	−1·7	B3	89	
3	13	51	50·0	−32	59	41	4·32	−1·6	B5	91	
4	13	53	12·4	−31	55	39	4·73	−1·6	B5	190	
ζ	13	55	32·3	−47	17	17	2·55	−3·0	B2	110	
294G	13	57	38·9	−63	41	11	4·71	−0·3	K4	100	
φ	13	58	16·2	−42	06	02	3·83	−2·5	B2	190	
υ^1	13	58	40·7	−44	48	13	3·87	−2·3	B3	170	
υ^2	14	01	43·3	−45	36	12	4·34	−2·0	F5	160	
β	14	03	49·4	−60	22	22	0·61	−5·1	B1	140	Agena
χ	14	16	02·7	−41	10	47	4·36	−2·5	B2	240	
5 θ	14	06	40·9	−36	22	12	2·06	1·7	K0	14	
ν	14	20	19·4	−56	23	12	4·33	−3·7	B5	310	
ψ	14	20	33·3	−37	53	07	4·05	0·0	A0	65	
a	14	23	02·1	−39	09	34	4·44	−1·9	B6	180	
η	14	35	30·3	−42	09	28	2·31v	−2·9	B3	110	
α^2	14	39	35·4	−60	50	13	1·39	5·7	K1	1·3	
α^1	14	39	36·7	−60	50	02	0·00	4·4	G2	1·3	
b	14	41	57·5	−37	47	37	4·00	−1·7	B3	140	
c^1	14	43	39·3	−35	10	25	4·05	−0·3	K5	71	
κ	14	59	09·6	−42	06	15	3·13	−2·5	B2	130	

Strangely, α has no official proper name. It has been called Al Rijil, and also Toliman, but astronomers in general prefer to call it simply α Centauri.

Also above mag. 5:

	Mag.	*Abs. mag.*	*Spectrum*	*Dist.*
	4·85	0·6	A0	67
1	4·79	0·5	B8p	230
o^1	4·96	1·1	G4	
4 n	4·76	−1·3	B7	
f	4·96	−2·1	B3	

VARIABLE STARS

	R.A. h	m	*Dec.* °	′	*Range*	*Type*	*Period, d.*	*Spectrum*
RS	11	20·5	−61	52	7·7–14·1	Mira	164·4	M
X	11	49·2	−41	45	7·0–13·8	Mira	315·1	M
W	11	55·0	−59	15	7·6–13·7	Mira	201·6	M
S	12	24·6	−49	26	6·0–7·0	Semi-reg.	65	N
U	12	33·5	−54	40	7·0–14·0	Mira	220·3	M
RV	13	37·5	−56	29	7·0–10·8	Mira	446·0	N
XX	13	40·3	−57	37	7·3–8·3	Cepheid	10·95	F–G
T	13	41·8	−33	36	5·5–9·0	Semi-reg.	60	K–M
μ	13	49·6	−42	28	2·9–3·5	Irr.	–	B
R	14	16·6	−59	55	5·3–11·8	Mira	546·2	M
V	14	32·5	−56	53	6·4–7·2	Cepheid	5·49	F–G

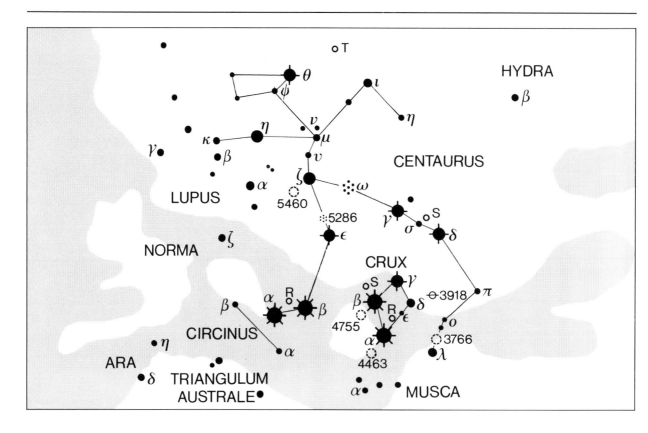

CENTAURUS (continued)

DOUBLE STARS

	R.A. h m	Dec. ° '	P.A. °	Sep. "	Mags.	
γ	12 41·5	−48 58	353	1·4	2·9, 2·9	Binary, 84·5 y
ε	13 39·9	−53 28	158	36·0	2·3, 12·7	
3	13 51·8	−33 00	108	7·9	4·5, 6·0	
4	13 53·2	−31 56	185	14·9	4·8, 8·4	
β	14 03·8	−60 22	251	1·3	0·7, 3·9	
η	14 35·5	−42 09	270	5·0	2·6, 13·5	
α	14 39·6	−60 50	215	19·7	0·0, 1·2	Binary, 79·9 y

OPEN CLUSTERS

M	NGC	R.A. h m	Dec. ° '	Diameter '	Mag.	No. of stars	
	3766	11 36·1	−61 37	12	5·3	100	
	IC 2944	11 36·6	−63 02	15	4·5	30	λ Cen cluster
	3960	11 50·9	−55 42	7	8·3	45	
	5138	13 27·3	−59 01	8	7·6	40	
	5281	13 46·6	−62 54	5	5·9	40	
	5316	13 53·9	−61 52	14	6·0	80	
	5460	14 07·6	−48 19	25	5·6	40	
	5617	14 29·8	−60 43	10	6·3	80	
	5662	14 35·2	−56 33	12	5·5	70	

GLOBULAR CLUSTERS

M	NGC	R.A. h m	Dec. ° '	Diameter '	Mag.	
	5139	13 26·8	−47 29	36·3	3·6	ω Centauri
	5286	13 46·4	−51 22	9·1	7·6	

NEBULA

M	NGC	R.A. h m	Dec. ° '	Dimensions '	
	5367	13 57·7	−39 59	4·3	Includes IC 4347. Double nucleus

CENTAURUS (continued)

GALAXIES

M	NGC	R.A. h m	Dec. ° '	Mag.	Dimensions '	Type	
	4603	12 40·9	−40 59	12·0	3·8×2·5	Sc	
	4696	12 48·8	−41 19	10·7	3·5×3·2	E1p	
	4945	13 05·4	−49 28	9·5	20·0×4·4	SBc	
	4976	13 08·6	−49 30	10·2	4·3×2·6	E4p	
	5128	13 25·5	−43 01	7·0	18·2×14·3	S0p	Centaurus A
	5253	13 39·9	−31 39	10·6	4·0×1·7	E5	
	5483	14 10·4	−43 19	12·0	3·1×2·8	Sc	
	3557	11 10·0	−37 32	10·4	4·0×2·7	E3	

CEPHEUS

(Abbreviation: Cep).

A rather undistinguished constellation, though it contains δ Cephei – the prototype Cepheid. Mythologically, Cepheus was Andromeda's father and Cassiopeia's husband. There are 8 stars above the 4th magnitude – though one of these, the 'Garnet Star' μ, is an irregular variable which is generally rather below this limit:

α 2·44
γ 3·21
β 3·23v
ζ 3·35
η 3·43
ι 3·52
μ 3·6v
δ 3·7v

BRIGHTEST STARS

Star		R.A. h m s	Dec. ° ' "	Mag.	Abs. mag.	Spectrum	Dist. pc	
1	κ	20 08 53·1	+77 42 41	4·39	−0·8	B9	100	
2	θ	20 29 34·7	+62 59 39	4·22	1·8	A0	34	
3	η	20 45 17·2	+61 50 20	3·43	3·2	K0	14	
5	α	21 18 34·6	+62 35 08	2·44	1·9	A7	14	Alderamin
8	β	21 28 39·4	+70 33 39	3·23v	−3·6	B2	230	Alphirk
9		21 37 55·0	+62 04 55	4·73	−5·7	B2	640	
11		21 41 55·1	+72 18 42	4·56	0·2	K0	60	
	μ	21 43 30·2	+58 46 48	var.	−7·0	M2	480	The 'Garnet Star'
10	ν	21 45 26·8	+61 07 15	4·29	−7·5	A2	1200	
17	ξ	22 03 45·7	+64 37 42	4·29	2·3	F7+G	37	Kurdah
21	ζ	22 10 51·1	+58 12 05	3·35	−4·4	K1	220	
23	ε	22 15 01·8	+57 02 37	4·19	1·7	F0	30	
27	δ	22 29 10·1	+58 24 55	var.	−4·6v	F8	410	Prototype Cepheid
32	ι	22 49 40·6	+66 12 02	3·52	0·0	K1	39	
33	π	23 07 53·7	+75 23 16	4·41	0·4	G2	59	
35	γ	23 39 20·7	+77 37 57	3·21	2·2	K1	16	Alrai
	43H	01 08 44·6	+86 15 26	4·25	−0·1	K2	68	

Also above mag. 5:

		Mag.	Abs. mag.	Spectrum	Dist.
24		4·79	0·3	G8	360
34	o	4·90	1·2	G7	

VARIABLE STARS

	R.A. h m	Dec. ° '	Range	Type	Period, d.	Spectrum
T	21 09·5	+68 29	5·2–11·3	Mira	388·1	M
VV	21 56·7	+63 38	4·8–5·4	Eclipsing	7430	M+B
S	21 35·2	+78 37	7·4–12·9	Mira	486·8	N
μ	21 43·5	+58 47	3·4–5·1	Irreg.?	–	M
δ	22 29·2	+58 25	3·5–4·4	Cepheid	5·37	F–G
W	22 36·5	+58 26	7·0–9·2	Semi-reg.	Long	K–M
U	01 02·3	+81 53	6·7–9·2	Algol	2·49	B+G

DOUBLE STARS

	R.A. h m	Dec. ° '	P.A. °	Sep. "	Mags.	
κ	20 08·9	+72 43	122	7·4	4·4, 8·4	
β	21 28·7	+70 34	249	13·3	3·2, 7·9	
ξ	22 03·8	+64 38	277	7·7	4·4, 6·5	Binary, 3800 y
δ	22 29·2	+58 25	191	41·0	var., 7·5	
π	23 07·9	+75 23	346	1·2	4·6, 6·6	Slow binary
o	23 18·6	+68 07	220	2·9	4·9, 7·1	Binary, 796 y

OPEN CLUSTERS

M	NGC	R.A. h m	Dec. ° '	Diameter '	Mag.	No. of stars
	IC 1396	21 39·1	+57 30	50	3·5	50
	7160	21 53·7	+62 36	7	6·1	12
	7235	22 12·6	+57 17	4	7·7	30
	7261	22 20·4	+58 05	6	8·4	30
	7510	23 11·5	+60 34	4	7·9	60
	188	00 44·4	+85 20	14	8·1	120

PLANETARY NEBULA

M	NGC	R.A. h m	Dec. ° '	Dimensions "	Mag.	Mag. of central star
	40	00 13·0	+72 32	37	10·7	11·6

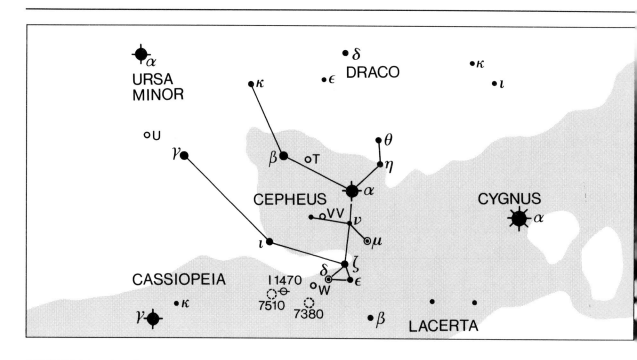

CEPHEUS (continued)

NEBULÆ

M	NGC	R.A. h m	Dec. ° '	Dimensions "	Mag.
	7023	21 01·8	+68 12	18×18	6·8
	Sh2–155	22 56·8	+62 37	50×30	7·7 Cave Nebula

CETUS

(Abbreviation: Cet).

One of the largest of all constellations; sometimes associated with the sea-monster of the Perseus legend, at others relegated to the status of a harmless whale. It contains the prototype long-period variable Mira, which can occasionally rise to magnitude 1·7, but which spends most of its period below naked-eye visibility. Cetus abounds in faint galaxies.

Excluding Mira, there are 8 stars above the fourth magnitude:

β 2·04
α 2·53
η 3·45
γ 3·47
τ 3·50
ι 3·56
θ 3·60
ζ 3·73

BRIGHTEST STARS

Star		R.A. h m s	Dec. ° ' "	Mag.	Abs. mag.	Spectrum	Dist. pc	
2		00 03 44·3	−17 20 10	4·55	−0·3	B9	91	
7		00 14 38·4	−18 55 58	4·44	−0·5	gM1	86	
8	ι	00 19 25·6	−08 49 26	3·56	−0·1	K2	50	Baten Kaitos Shemali
16	β	00 43 35·3	−17 59 12	2·04	0·2	K0	21	Diphda
17	φ¹	00 44 11·3	−10 36 34	4·75	0·2	K0	81	
31	η	01 08 35·3	−10 10 56	3·45	−0·1	K2	36	
45	θ	01 24 01·3	−08 11 01	3·60	0·2	K0	35	
53	χ	01 49 35·0	−10 41 11	4·67	1·9	F2	30	
52	τ	01 44 04·0	−15 56 15	3·50	5·7	G8	3·6	
55	ζ	01 51 27·6	−10 20 06	3·73	−0·1	K2	58	Baten Kaitos
59	υ	02 00 00·2	−21 04 40	4·00	−0·5	M1	79	
65	ξ¹	02 12 59·9	+08 50 48	4·37	−2·1	G8	200	
68	ο	02 19 20·6	−02 58 39	var.	−0·5v	Md	29	Mira
73	ξ²	02 28 09·5	+08 27 36	4·28	−0·8	B9	98	
76	σ	02 32 05·1	−15 14 41	4·75	2·1	F5	33	
82	δ	02 39 28·9	+00 19 43	4·07	−3·0	B2	260	
86	γ	02 43 18·0	+03 14 09	3·47	1·4	A2	23	Alkaffaljidhina
89	π	02 44 07·3	−13 51 32	4·25	−0·6	B7	93	
87	μ	02 44 56·4	+10 06 51	4·27	1·7	F0	30	
91	λ	02 59 42·8	+08 54 27	4·70	−2·2	B5	220	
92	α	03 02 16·7	+04 05 23	2·53	−0·5	M2	40	Menkar

Also above mag. 5:

		Mag.	Abs. mag.	Spectrum	Dist.
6		4·89	3·7	F6	18
20		4·77	−0·4	M0	110
46		4·90	−0·2	K3	110
72	ρ	4·89	0·2	B9	85
78	ν	4·86	0·3	G8	82
83	ε	4·84	2·8	F5	22
96	κ	4·83	5·0	G5	9·3

CETUS *(continued)*

VARIABLE STARS

	R.A. h m	Dec. ° '	Range	Type	Period, d.	Spectrum
W	00 02·1	−14 41	7·1−14·8	Mira	351·3	S
T	00 21·8	−20 03	5·0−6·9	Semi-reg.	159	M
S	00 24·1	−09 20	7·6−14·7	Mira	320·5	M
U	02 33·7	−13 09	6·8−13·4	Mira	234·8	M
UV	01 38·8	−17 58	6·8−13·0	Flare	−	dM
o	02 19·3	−02 59	1·7−10·1	Mira	332·0	M

DOUBLE STARS

	R.A. h m	Dec. ° '	P.A. °	Sep. "	Mags.	
37	01 14·4	−07 55	331	49·7	5·2, 8·7	
χ	01 49·6	−10 41	250	183·8	4·9, 6·9	
66	02 12·8	−02 24	AB 234	16·5	5·7, 7·5	
			AC 061	172·7	11·4	
o	02 19·3	−02 59	085	0·3	var., 12·0	(B is VZ Ceti)
ν	02 35·9	+05 36	081	8·1	4·9, 9·5	
ε	02 39·6	−11 52	039	0·1	5·8, 5·8	Binary, 2·7 y
γ	02 43·3	+03 14	294	2·8	3·5, 7·3	

PLANETARY NEBULA

M	NGC	R.A. h m	Dec. ° '	Diameter "	Mag.	Mag. of central star
	246	00 47·0	−11 53	225	8·0	11·9

GALAXIES

M	NGC	R.A. h m	Dec. ° '	Mag.	Dimensions '	Type
	45	00 14·1	−23 11	10·4	8·1×5·8	S
	247	00 47·1	−20 46	8·9	20·0×7·4	S
	428	01 12·9	+00 59	11·3	4·1×3·2	Scp
	578	01 30·5	−22 40	10·9	4·8×3·2	Sc
	584	01 31·3	−06 52	10·3	3·8×2·4	E4
	720	01 53·0	−13 44	10·2	4·4×2·8	E3
	864	02 15·5	+06 00	11·0	4·6×3·5	Sc
	895	02 21·6	−05 31	11·8	3·6×2·8	Sb
	908	02 23·1	−21 14	10·2	5·5×2·8	Sc
	936	02 27·6	−01 09	10·1	5·2×4·4	SBa
	1042	02 40·4	−08 26	10·9	4·7×3·9	Sc
	1055	02 41·8	+00 26	10·6	7·6×3·0	Sb
77	1068	02 42·7	−00 01	8·8	6·9×5·9	SBp
	1073	02 43·7	+01 23	11·0	4·9×4·6	SBc
	1087	02 46·4	−00 30	11·0	3·5×2·3	Sc

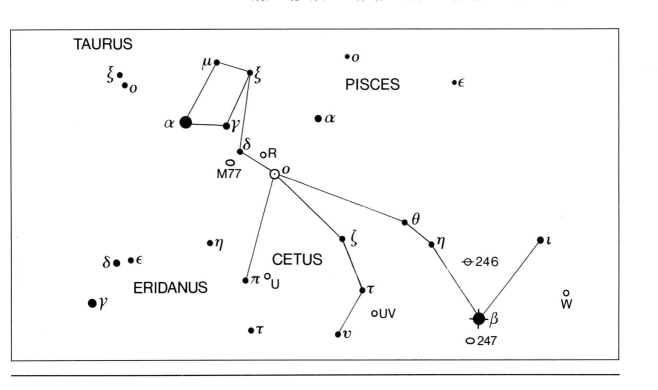

CHAMÆLEON

(Abbreviation: Cha).

A small southern constellation of no particular note; there are no legends attached to it, and there are no stars above the fourth magnitude.

See chart for Musca

BRIGHTEST STARS

Star	R.A. h	m	s	Dec. °	′	″	Mag.	Abs. mag.	Spectrum	Dist. pc
α	08	18	31·7	−76	55	10	4·07	2·2	F6	24
θ	08	20	38·7	−77	29	04	4·35	1·7	K0	24
γ	10	35	28·1	−78	36	27	4·11	−0·4	M0	77
δ²	10	45	46·7	−80	32	24	4·45	−1·7	B3	170
β	12	18	20·7	−79	18	43	4·26	−0·9	B6	110

Also above mag. 5:

	Mag.	Abs. mag.	Spectrum	Dist.
ε	4·91	0·2	B9	88

VARIABLE STARS

	R.A. h	m	Dec. °	′	Range	Type	Period, d.	Spectrum
R	08	21·8	−76	21	7·5–14·2	Mira	334·6	M
RS	08	43·2	−79	04	6·0–6·7	Algol + δ Scuti	1·67	A–F

DOUBLE STARS

	R.A. h	m	Dec. °	′	P.A. °	Sep. ″	Mags.
δ	10	45·3	−80	28	076	0·6	6·1, 6·4
ε	11	59·6	−78	13	188	0·9	5·4, 6·0

CIRCINUS

(Abbreviation: Cir).

A very small southern constellation, in the area of α and β Centauri. It is associated with no legends. The only star above magnitude 4 is α (3·19).

See chart for Centaurus

BRIGHTEST STARS

Star	R.A. h	m	s	Dec. °	′	″	Mag.	Abs. mag.	Spectrum	Dist. pc
α	14	42	28·0	−64	58	43	3·19	2·6	F0	14
β	15	17	30·8	−58	48	04	4·07	1·7	A3	22
γ	15	23	22·6	−59	19	14	4·51	−1·0	B5	84

Also above mag. 5:

	Mag.	Abs. mag.	Spectrum	Dist.
ε	4·86	0·3	K4	110

VARIABLE STARS

	R.A. h	m	Dec. °	′	Range	Type	Period, d.	Spectrum
AX	14	52·6	−63	49	5·6–6·1	Cepheid	5·27	F–G
θ	14	56·7	−62	47	5·0–5·4	Irreg.	–	B

DOUBLE STARS

	R.A. h	m	Dec. °	′	P.A. °	Sep. ″	Mags.
α	14	42·5	−64	59	232	15·7	3·2, 8·6
δ	15	16·9	−60	57	270	50·0	5·1, 13·4
γ	15	23·4	−59	19	033	0·6	5·1, 5·5 Binary, 180 y

OPEN CLUSTER

M	NGC	R.A. h	m	Dec. °	′	Diameter ′	Mag.	No. of stars
	5823	15	05·7	−55	36	10	7·9	100

COLUMBA

(Abbreviation: Col).

A 'modern' constellation dating from 1679, when it was introduced to the sky by Royer. Apparently it represents the dove which Noah released from the Ark; it was originally called Columba Noachi. There are four stars above magnitude 4:

α 2·64
β 3·12
δ 3·85
ε 3·87

δ Columbæ was formerly called 3 Canis Majoris.

BRIGHTEST STARS

Star	R.A.			Dec.			Mag.	Abs. mag.	Spectrum	Dist.	
	h	m	s	°	′	″				pc	
ε	05	31	12·7	−35	28	15	3·87	0·2	gK0	42	
α	05	39	38·9	−34	04	27	2·64	−0·2	B8	37	Phakt
β	05	50	57·5	−35	46	06	3·12	−0·1	K2	44	Wazn
γ	05	57	32·2	−35	17	00	4·36	−2·3	B3	210	
η	05	59	08·8	−42	48	55	3·96	0·2	K0	44	
κ	06	16	33·0	−35	08	26	4·37	0·3	G8	58	
δ	06	22	06·7	−33	26	11	3·85	0·5	gG1	36	

μ (mag. 5·16), position RA 05·44·1 dec. −32°19′, is one of the three 'runaway stars' which seem to be moving away from the nebulous region of Orion (the others are 53 Arietis and AE Aurigæ). μ Colurnbæ has an annual proper motion of 0″·025. Its spectral type is 09·5.

Also above mag. 5:

	Mag.	Abs. mag.	Spectrum	Dist.
ξ	4·97	0·0	K1	97
λ	4·87	−1·1	B5	160
o	4·83	3·2	K0	21

VARIABLE STARS

	R.A.		Dec.		Range	Type	Period, d.	Spectrum
	h	m	°	′				
T	05	19·3	−33	42	6·6−12·7	Mira	225·9	M
R	05	50·5	−29	12	7·8−15·0	Mira	327·6	M

DOUBLE STARS

	R.A.		Dec.		P.A.	Sep.	Mags.
	h	m	°	′	°	″	
α	05	39·6	−34	04	359	13·5	2·6, 12·3
γ	05	57·5	−35	17	110	33·8	4·4, 12·7
π²	06	07·9	−42	09	150	0·1	6·2, 6·3

GLOBULAR CLUSTER

M	NGC	R.A.		Dec.		Diameter	Mag.	
		h	m	°	′	′		
	1851	05	14·1	−40	03	11·0	7·3	X-ray source

GALAXIES

M	NGC	R.A.		Dec.		Mag.	Dimensions	Type
		h	m	°	′		′	
	1792	05	05·2	−37	59	10·2	4·0×2·1	Sb
	1808	05	07·7	−37	32	9·9	7·2×4·1	SBa
	2090	05	47·0	−34	14	11·7	4·5×2·3	Sc

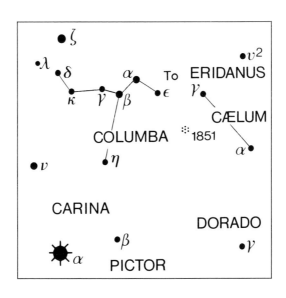

COMA BERENICES

(Abbreviation: Com).

At first glance this constellation gives the impression of being a vast, dim cluster. Coma has no star above magnitude 4·3, but it abounds in faint ones, and there are many telescopic galaxies.

Though the constellation is not 'original', there is a legend attached to it. When Ptolemy Euergetes, King of Egypt, set out in an expedition against the Assyrians, his wife Berenice vowed that if he returned safely she would cut off her lovely hair and place it in the temple of Venus. The King returned; the Queen kept her vow, and Jupiter placed the shining tresses in the sky.

See chart for Boötes

BRIGHTEST STARS

Star		R.A. h	m	s	Dec. °	′	″	Mag.	Abs. mag.	Spectrum	Dist. pc	
15	γ	12	26	56·2	+28	16	06	4·35	−0·2	K1	31	
42	α	13	09	59·2	+17	31	45	4·32	3·4	F5	18	Diadem
43	β	13	11	52·3	+27	52	41	4·26	4·7	G0	8·3	

α is a very close binary, with a period of 26 years; the separation is never more than 0″·3. The components are approximately equal.

Also above mag. 5:

	Mag.	Abs. mag.	Spectrum	Dist.
12	4·79	0·7	F8	27
7	4·94	0·2	K0	89
11	4·74	0·3	G8	69
14	4·93		F0	24
16	4·99		A4	21
23	4·81	−0·6	A0	110
31	4·94	0·6	G0	74
36	4·78	−0·5	M1	110
37	4·90		K1	27
41	4·80	−0·3	K5	110

VARIABLE STARS

	R.A. h	m	Dec. °	′	Range	Type	Period, d.	Spectrum
R	12	04·0	+18	49	7·1–14·6	Mira	362·8	M
FS	13	06·4	+22	37	5·3–6·1	Semi-reg.	58	M

GALAXIES

M	NGC	R.A. h	m	Dec. °	′	Mag.	Dimensions ′	Type	
	4136	12	09·3	+29	56	11·7	4·1×3·9	Sc	
98	4192	12	13·8	+14	54	10·1	9·5×3·2	Sb	
	4251	12	18·1	+28	10	11·6	4·2×1·9	E7	
99	4254	12	18·8	+14	25	9·8	5·4×4·8	Sc	
	4278	12	20·1	+29	17	10·2	3·6×3·5	E1	
	4314	12	22·6	+29	53	10·5	4·8×4·3	SBa	
100	4321	12	22·9	+15	49	9·4	6·9×6·2	Sc	
	4448	12	28·2	+28	37	11·1	4·0×1·6	Sb	
	4450	12	28·5	+17	05	10·1	4·8×3·5	Sb	
	4459	12	29·0	+13	59	10·4	3·8×2·8	E2	
	4473	12	29·8	+13	26	10·2	4·5×2·6	E4	
	4477	12	30·0	+13	38	10·4	4·0×3·5	SBa	
88	4501	12	32·0	+14	25	9·5	6·9×3·9	SBb	
	4548	12	35·4	+14	30	10·2	5·4×4·4	SBb	
	4559	12	36·0	+27	58	9·8	10·5×4·9	Sc	
	4565	12	36·3	+25	59	9·6	16·2×2·8	Sb	
	4651	12	43·7	+16	24	10·7	3·8×2·7	Sop	
	4689	12	47·8	+13	46	10·9	4·0×3·5	Sb	
	4725	12	50·4	+25	30	9·2	11·0×7·9	SBb	
64	4826	12	56·7	+21	41	8·5	9·3×5·4	Sb	Black-Eye Galaxy

OPEN CLUSTER

M	NGC	R.A. h	m	Dec. °	′	Diameter ′	Mag.	No. of stars	
	Mel 111	12	25	+26		275	4	80	Coma Berenices

GLOBULAR CLUSTER

M	NGC	R.A. h	m	Dec. °	′	Diameter ′	Mag.
53	5024	13	12·9	+18	10	12·6	7·7

CORONA AUSTRALIS

(Abbreviation: CrA).

An original constellation. It has no bright stars, but is easy to recognize because of its distinctive shape.

See chart for Sagittarius

BRIGHTEST STARS

Star	R.A. h	m	s	Dec. °	′	″	Mag.	Abs. mag.	Spectrum	Dist. pc
θ	18	33	29·9	−42	18	45	4·64	0·3	G5	60
ζ	19	03	06·7	−42	05	43	4·75	0·6	A0	68
γ	19	06	24·9	−37	03	48	4·21	4·0	F8	12
δ	19	08	20·6	−40	29	48	4·59	0·2	gK0	63
α	19	09	28·2	−37	54	16	4·11	1·6	A2	14
β	19	10	01·5	−39	20	27	4·11	0·3	gG5	34

Also above mag. 5:

	Mag.	Abs. mag.	Spectrum	Dist.
ε	4·8v	2·6	F0	28

CORONA AUSTRALIS (continued)

DOUBLE STARS

	R.A. h m	Dec. ° ′	P.A. °	Sep. ″	Mags.	
κ	18 33·4	−38 44	359	21·6	5·9, 5·9	
λ	18 43·8	−38 19	214	29·2	5·1, 9·7	
γ	19 06·4	−37 04	109	1·3	4·8, 5·1	Binary, 120·4 y

GLOBULAR CLUSTER

M	NGC	R.A. h m	Dec. ° ′	Diameter ′	Mag.
	6541	18 08·0	−43 42	13·1	6·6

NEBULA

M	NGC	R.A. h m	Dec. ° ′	Dimensions ′	Mag.	Mag. of illuminating star
	6729	19 01·9	−36 57	1 (var.)	var.	9·7v (R Coronæ Australis)

PLANETARY NEBULA

M	NGC	R.A. h m	Dec. ° ′	Diameter ″	Mag.	Mag. of central star
	IC 1297	19 17·4	−39 37	7	—	12·9v RU Coronæ Australis

CORONA BOREALIS

(Abbreviation: CrB).

A small but very distinctive constellation representing a crown given by Bacchus to Ariadne, the daughter of King Minos of Crete. There are three stars above magnitude 4; α is an eclipsing binary with a very small range:

α 2·23v
β 3·68
γ 3·84v

See chart for Boötes

BRIGHTEST STARS

Star	R.A. h m s	Dec. ° ′ ″	Mag.	Abs. mag.	Spectrum	Dist. pc	
3 β	15 27 49·7	+29 06 21	3·68	1·2	F0p	18	Nusakan
4 θ	15 32 55·7	+31 21 32	4·14	−1·1	B5	110	
5 α	15 34 41·2	+26 42 53	2·23v	0·6	A0	24	Alphekka
7 ζ²	15 39 22·6	+36 38 09	4·7	−0·6	B7	130	Wide binary with ζ¹, mag. 6·0
8 γ	15 42 44·5	+26 17 44	3·84v	−0·3	A0	64	
10 δ	15 49 35·6	+26 04 06	4·63	1·8	G5	38	
13 ε	15 57 35·2	+26 52 40	4·15	−0·2	K3	74	

α is an eclipsing binary with a very small range (0 m·1).
γ is a δ Scuti variable; range only 0·06 mag.
β is a magnetic spectrum variable.

Also above mag. 5:

	Mag.	Abs. mag.	Spectrum	Dist.
2 η	4·98	4·4	G0	14
ζ	5·0+6·0	−0·6	B6+B7	130
17 σ	5·22	4·4	G0	21
14 ι	4·99	−0·6	A0	130
16 τ	4·76	0·2	K0	52
19 ξ	4·85	0·2	K0	85
11 κ	4·82	1·7	K0	33

VARIABLE STARS

	R.A. h m	Dec. ° ′	Range	Type	Period, d.	Spectrum	
U	15 18·2	+31 39	7·7–8·8	Algol	3·45	B+F	
S	15 21·4	+31 22	5·8–14·1	Mira	360·3	M	
R	15 48·6	+28 09	5·7–15	R CrB	—	F8p	
V	15 49·5	+39 34	6·9–12·6	Mira	357·6	N	
T	15 59·5	+25 55	2·0–10·8	Recurrent nova		M+Q	(1866, 1946)
W	16 15·4	+37 48	7·8–14·3	Mira	238·4	M	

DOUBLE STARS

	R.A. h m	Dec. ° ′	P.A. °	Sep. ″	Mags.	
ο	15 20·1	+29 37	337	147·3	5·5, 9·4	
η	15 23·2	+30 17	AB 030	1·0	5·6, 5·9	Binary, 41·6 y
			AC 012	57·7	12·5	
			AB+D 047	215·0	10·0	
ζ	15 39·4	+36 38	305	6·3	5·1, 6·0	
γ	15 42·7	+26 18	118	0·6	4·1, 5·5	Binary, 91 y
ε	15 57·6	+26 53	003	1·8	4·2, 12·6	
ρ	16 01·0	+33 18	071	89·8	5·5, 8·7	
σ	16 14·7	+33 52	234	7·0	5·6, 6·6	Binary, 1000 y

CORVUS

(Abbreviation: Crv).

An original group. When the god Apollo became enamoured of Coronis, mother of the great doctor Æsculapius, he sent a crow to watch her and report on her behaviour. To be candid, the crow's report was decidedly adverse; but Apollo rewarded the bird with a place in the sky!

Corvus is distinctive, since its leading stars form a quadrilateral. There are four stars above magnitude 4:

γ 2·59   δ 2·95
β 2·65   ε 3·00

Curiously, the star lettered α is more than a magnitude fainter than any of these.

See chart for Hydra

BRIGHTEST STARS

Star		R.A. h	m	s	Dec. °	′	″	Mag.	Abs. mag.	Spectrum	Dist. pc	
1	α	12	08	24·7	−24	43	44	4·02	1·9	F2	21	Alkhiba
2	ε	12	10	07·4	−22	37	11	3·00	−0·1	K2	32	
4	γ	12	15	48·3	−17	32	31	2·59	−1·2	B8	57	Minkar
7	δ	12	29	51·8	−16	30	55	2·95	0·2	B9	36	Algorel
8	η	12	32	04·1	−16	11	46	4·31	1·7	F0	29	
9	β	12	34	23·2	−23	23	48	2·65	−2·1	G5	89	Kraz

VARIABLE STARS

	R.A. h	m	Dec. °	′	Range	Type	Period, d.	Spectrum
R	12	19·6	−19	15	6·7–14·4	Mira	317·0	M
SV	12	49·8	−15	05	6·8–7·6	Semi-reg.	70	M

DOUBLE STAR

	R.A. h	m	Dec. °	′	P.A. °	Sep. ″	Mags.
δ	12	29·9	−16	31	214	24·2	3·0, 9·2

PLANETARY NEBULA

M	NGC	R.A. h	m	Dec. °	′	Dimensions ″	Mag.	Mag. of central star
	4631	12	24·5	−18	48	45×110	10·3	13·2

CRATER

(Abbreviation: Crt).

Like Corvus, a small constellation adjoining Hydra; it has been identified with the wine-goblet of Bacchus. The only star above the fourth magnitude is δ (3·56).

See chart for Hydra

BRIGHTEST STARS

Star		R.A. h	m	s	Dec. °	′	″	Mag.	Abs. mag.	Spectrum	Dist. pc	
7	α	10	59	46·4	−18	17	56	4·08	0·2	K0	37	Alkes
11	β	11	11	39·4	−22	49	33	4·48	0·2	A2	72	
12	δ	11	19	20·4	−14	46	43	3·56	1·8	G8	22	
15	γ	11	24	52·8	−17	41	02	4·08	2·1	A5	24	
21	θ	11	36	40·8	−09	48	08	4·70	0·2	B9	79	
27	ζ	11	44	45·7	−18	21	03	4·73	0·3	G8	74	

Also above mag. 5:

	Mag.	Abs. mag.	Spectrum	Dist.
ε	4·83	−0·3	K5	96

DOUBLE STAR

	R.A. h	m	Dec. °	′	P.A. °	Sep. ″	Mags.
γ	11	24·9	−17	41	096	5·2	4·1, 9·6

GALAXIES

M	NGC	R.A. h	m	Dec. °	′	Mag.	Dimensions ′	Type
	3511	11	03·4	−23	05	11·6	5·4×2·2	Sc
	3513	11	03·8	−23	15	12·0	2·8×2·3	SBc
	3571	11	11·5	−18	17	12·8	3·3×1·3	Sa
	3672	11	25·0	−09	48	11·5	4·1×2·1	Sb
	3887	11	47·1	−16	51	11·0	3·3×2·7	Sc
	3981	11	56·1	−19	54	12·4	3·9×1·5	Sb

CRUX AUSTRALIS

(Abbreviation: Cru).

Though Crux is the smallest constellation in the entire sky, it is also one of the most famous. Before Royer introduced it, in 1679, it had been included in Centaurus. Strictly speaking it is more like a kite than a cross. As well as its brilliant stars it contains the glorious 'Jewel Box' cluster, and also the dark nebula known as the Coal Sack.

BRIGHTEST STARS

Star	R.A. h	m	s	Dec. °	′	″	Mag.	Abs. mag.	Spectrum	Dist. pc	
θ¹	12	03	01·6	−63	18	46	4·33	−0·5	A0	17	
θ²	12	04	19·2	−63	09	56	4·72	−3·0	B2	290	
η	12	06	52·8	−64	36	49	4·15	0·6	F0	33	
δ	12	15	08·6	−58	44	55	2·80	−3·0	B2	79	
ζ	12	18	26·1	−64	00	11	4·04	−2·3	B3	150	
ε	12	21	21·5	−60	24	04	3·59	0·0	K2	18	
α ⎰	12	26	35·9	−63	05	56	1·41	−3·9	B1	110	Acrux
⎱	12	26	36·5	−63	05	58	1·88	−3·4	B3		
γ	12	31	09·9	−57	06	47	1·63	−0·5	M3	27	
ι	12	45	37·8	−60	58	52	4·69	0·0	K1	87	
β	12	47	43·2	−59	41	19	1·25v	−5·0	B0	130	(Mimosa)
μ¹	12	54	35·6	−57	10	40	4·03	−2·3	B3	190	

β is variable over a small range (below 0 m·1); β Canis Majoris type.

CRUX AUSTRALIS (continued)

There are five stars above the fourth magnitude:

α 0·83
β 1·25v
γ 1·63
δ 2·80v
ε 3·59

Of the four main stars, any casual glance will show the difference between γ, a red giant, and the other three, which are hot and bluish-white.

See chart for Centaurus

VARIABLE STARS

	R.A. h m	Dec. ° ′	Range	Type	Period, d.	Spectrum
BH	12 16·3	−56 17	7·2–10·0	Mira	421	S
T	12 21·4	−62 17	6·3–6·8	Cepheid	6·73	F
R	12 23·6	−61 38	6·4–7·2	Cepheid	5·83	F–G
S	12 54·4	−58 26	6·2–6·9	Cepheid	4·69	F–G

DOUBLE STARS

	R.A. h m	Dec. ° ′	P.A. °	Sep. ″	Mags.
θ¹	12 03·0	−63 19	325	4·5	4·3, 13·6
η	12 06·9	−64 37	299	44·0	4·2, 11·7
α	12 26·6	−63 06	AB 115	4·4	1·4, 1·9
			AC 202	90·1	1·0, 4·9
γ	12 31·2	−57 07	AB 031	110·6	1·6, 6·7
			AC 082	155·2	9·5
ι	12 45·6	−60 59	022	26·9	4·7, 9·5
μ¹	12 54·6	−57 11	017	34·9	4·0, 5·2

OPEN CLUSTERS

M	NGC	R.A. h m	Dec. ° ′	Diameter ′	Mag.	No. of stars	
	4052	12 01·9	−63 12	8	8·8	80	
	4103	12 06·7	−61 15	7	7·4	45	
	4337	12 23·9	−58 08	3·5	8·9	–	
	4349	12 24·5	−61 54	16	7·4	30	
	H.5	12 29·0	−60 46	5	9·0	40	
	4439	12 28·4	−60 06	4	8·4	–	
	4609	12 42·3	−62 58	5	6·9	40	
	4755	12 53·6	−60 20	10	4·2	50+	Jewel Box (κ Crucis)

DARK NEBULA

M	NGC	R.A. h m	Dec. ° ′	Dimensions	Area, sq. °	
		12 53	−63	400×300	26·2	Coal Sack

CYGNUS

(Abbreviation: Cyg).

Cygnus is one of the richest constellations in the sky; it is often nicknamed the Northern Cross – certainly it is much more nearly cruciform than is Crux. Various legends are associated with it. According to one, the group was placed in the sky to honour a swan into which Jupiter once transformed himself when on a visit to the wife of the King of Sparta!

There are 14 stars above the fourth magnitude. To these must be added the red variable χ, which can rise above magnitude 4 at times.

α	1·25	τ	3·72
γ	2·20	ι	3·79
ε	2·46	κ	3·77
δ	2·87	o¹	3·79
β	3·08	η	3·89
ζ	3·20	ν	3·94
ξ	3·72	o²	3·98

BRIGHTEST STARS

Star		R.A. h m s	Dec. ° ′ ″	Mag.	Abs. mag.	Spectrum	Dist. pc	
1	κ	19 17 06·0	+53 22 07	3·77	0·2	K0	52	
10	ι	19 29 42·1	+51 43 47	3·79	2·1	A5	41	
6	β	19 30 43·1	+27 57 35	3·08	−2·3	K5	120	Albireo
8		19 31 46·1	+34 27 11	4·74	−2·3	B3	240	
13	θ	19 36 26·2	+50 13 16	4·48	2·1	F5	19	
12	φ	19 39 22·4	+30 09 12	4·69	1·8	G8	34	
18	δ	19 44 58·4	+45 07 51	2·87	−0·6	A0	49	
	χ	19 50 33·7	+32 54 51	var.	0·2v	S	25	
21	η	19 56 18·2	+35 05 00	3·89	0·2	K0	52	
33		20 13 23·7	+56 34 04	4·30	1·3	A3	39	
31	o¹	20 13 33·7	+46 44 29	3·79	−2·2	K2	160	
32	o²	20 15 28·1	+47 42 51	3·98	−3·4	K3	280	
34	P	20 17 47·0	+38 01 59	var.	var.	B1p	1400	
37	γ	20 22 13·5	+40 15 24	2·20	−4·6	F8	230	Sadr
39		20 23 51·4	+32 11 35	4·43	−0·2	K3	78	
41		20 29 23·6	+30 22 07	4·01	−2·0	F5	160	
47		20 33 54·0	+35 15 03	4·61	−4·4	K2	570	
50	α	20 41 25·8	+45 16 49	1·25	−7·5	A2	560	Deneb
52		20 45 39·6	+30 43 11	4·22	0·2	K0	59	
53	ε	20 46 12·5	+33 58 13	2·46	0·2	K0	25	Gienah
54	λ	20 47 24·3	+36 29 27	4·53	−1·1	B5	130	
58	ν	20 57 10·2	+41 10 02	3·94	0·6	A0	45	
59	f¹	20 59 49·3	+47 31 16	4·74v	−3·9	B1	480	(V.832 Cyg)
62	ξ	21 04 55·7	+43 55 40	3·72	−4·4	K5	290	
63		21 06 35·9	+47 38 54	4·55	−2·3	K4	220	
64	ζ	21 12 56·0	+30 13 37	3·20	−2·1	G8	120	
65	τ	21 14 47·3	+38 02 44	3·72	1·7	F0	21	
67	σ	21 17 24·7	+39 23 41	4·23	−7·1	B9	1600	
66	υ	21 17 54·9	+34 53 48	4·43	−2·5	B2	240	
73	ρ	21 33 58·7	+45 35 30	4·02	0·3	G8	56	
80	π¹	21 42 05·5	+51 11 23	4·67	−1·7	B3	180	Azelfafage
78	μ²	21 44 08·2	+28 44 35	6·14	3·7	dF3	17	Combined
	μ¹	21 44 08·4	+28 44 34	4·78		F6		mag. 4·4
81	π	21 46 47·4	+49 18 35	4·23	−2·9	B3	250	

CYGNUS *(continued)*

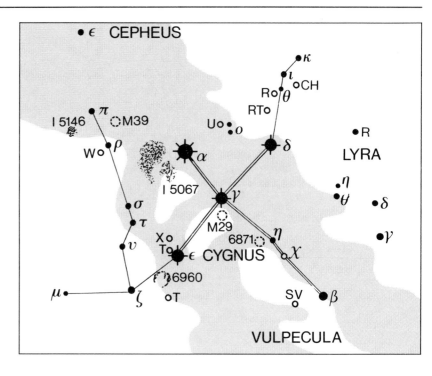

Also above mag. 5:

		Mag.	Abs. mag.	Spectrum	Dist.	
29	b³	4·97	1·4	A2	33	
30		4·83	0·0	A3	92	
28	b²	4·93	−1·7	B3	210	
22		4·94	−1·9	B6	220	
24	ψ	4·92	1·3	A3	52	
17		4·99		F5		
15		4·89	0·3	G8	83	
2		4·97	−2·3	B3	260	
45	ω¹	4·95	−2·5	B2	270	
55		4·84	−6·8	B3	1000	
57		4·78	−1·1	B5	150	
68		5·00		O8		
72		4·90	0·0	K1	96	
80	π¹	4·78	−1·3	B3	180	Azelfafage

VARIABLE STARS

	R.A.		Dec.		Range	Type	Period, d.	Spectrum
	h	m	°	′				
CH	19	24·5	+50	14	6·4–8·7	Z Andromedæ	±97	M+B
R	19	36·8	+50	12	6·1–14·2	Mira	426·4	M
RT	19	43·6	+48	47	6·4–12·7	Mira	190·2	M
SU	19	44·8	+29	16	6·5–7·2	Cepheid	3·84	F
χ	19	50·6	+32	55	3·3–14·2	Mira	406·9	S
Z	20	01·4	+50	03	7·4–14·7	Mira	263·7	M
RS	20	13·4	+38	44	6·5–9·3	Semi-reg.	417	N
P	20	17·8	+38	02	3–6	Irreg.	–	Bp
CN	20	17·9	+59	48	7·3–14·0	Mira	198·5	M
U	20	19·6	+47	54	5·9–12·1	Mira	462·4	N
V	20	41·3	+48	09	7·7–13·9	Mira	421·4	N
X	20	43·4	+35	35	5·9–6·9	Cepheid	16·39	F–G
T	20	47·2	+34	22	5·0–5·5	Irreg.	–	K
W	21	36·0	+45	22	5·0–7·6	Semi-reg.	126	M
SS	21	42·7	+43	35	8·4–12·4	SS Cygni	±50	A–G
WY	21	48·7	+44	15	7·5–14·0	Mira	304·5	M

DOUBLE STARS

	R.A.		Dec.		P.A.	Sep.	Mags.	
	h	m	°	′	°	″		
β	19	30·7	+27	-58	054	34·4	3·1, 5·1	Yellow, blue
δ	19	45·0	+45	07	225	2·4	2·9, 6·3	Binary, 828 y
ψ	19	55·6	+52	26	178	3·2	4·9, 7·4	
γ	20	22·2	+40	15	196	41·2	2·2, 9·9	B is a close dble
61	21	06·9	+38	45	148	29·9	5·2, 6·0	

CYGNUS (continued)

DOUBLE STARS (continued)

	R.A. h m	Dec. ° ′	P.A. °	Sep. ″	Mags.	
τ	21 14·8	+38 03	015	0·5	3·8, 6·4	Binary, 50 y
μ	21 44·1	+28 45	300	1·6	4·8, 6·1	Binary, 507 y

OPEN CLUSTERS

M	NGC	R.A. h m	Dec. ° ′	Diameter ′	Mag.	No. of stars
	6811	19 38·2	+46 34	13	6·8	70
	6819	19 41·3	+40 11	5	7·3	–
	6834	19 52·2	+29 25	5	7·8	50
	6866	20 03·7	+44 00	7	7·6	80
	6871	20 05·9	+35 47	20	5·2	15
	6910	20 23·1	+40 47	8	7·4	50
29	6913	20 23·9	+38 32	7	6·6	50
	6939	20 31·4	+60 38	8	7·8	80
	7067	21 24·2	+48 01	3	9·7	20
39	7092	21 32·2	+48 26	32	4·6	30

PLANETARY NEBULÆ

M	NGC	R.A. h m	Dec. ° ′	Dimensions ″	Mag.	Mag. of central star	
	6826	19 44·8	+50 31	30×140	9·8	10·4	Blinking Nebula
	7048	21 14·2	+46 16	61	11·3	18	

NEBULÆ

M	NGC	R.A. h m	Dec. ° ′	Dimensions ′	Mag. of illuminating star	
	6888	20 12·0	+38 21	20×10	7·4	Crescent Nebula
	6960	20 45·7	+30 43	70×6	–	Filamentary Nebula, 52 Cygni
	IC 5067/70	20 50·8	+44 21	80×70	–	Pelican Nebula
	6992/5	20 56·4	+31 43	60×8	–	Veil Nebula: SNR
	7000	20 58·8	+44 20	120×100	6	North America Nebula
	IC 5146	21 53·5	+47 16	12×12	10	Cocoon Nebula, with sparse cluster

DELPHINUS

(Abbreviation: Del).

A small but compact constellation; one of Ptolemy's originals. It honours the dolphin which carried the great singer Arion to safety, after he had been thrown overboard by the crew of the ship carrying him home after winning all the prizes in a competition. The curious names of α and β were allotted by one Nicolaus Venator, for reasons which are obvious!

There are 3 stars above the fourth magnitude:

β 3·54
α 3·77
γ 3·9 (4·5+5·5)

See chart for Aquila

BRIGHTEST STARS

Star		R.A. h m s	Dec. ° ′ ″	Mag.	Abs. mag.	Spectrum	Dist. pc	
2	ε	20 33 12·6	+11 18 12	4·03	−1·9	B6	150	
4	ζ	20 35 18·4	+14 40 27	4·68	1·7	A3	38	
6	β	20 37 32·8	+14 35 43	3·54	0·7	F5	33	Rotanev
9	α	20 39 38·1	+15 54 43	3·77	0·2	B9	52	Svalocin
11	δ	20 43 27·3	+15 04 28	4·43	0·5	A7	56	
12	γ	20 46 39·3	+16 07 27	3·9	3·2	G5	23	

VARIABLE STARS

	R.A. h m	Dec. ° ′	Range	Type	Period, d.	Spectrum
R	20 14·9	+09 05	7·6–13·8	Mira	284·9	M
EU	20 37·9	+18 16	5·8–6·9	Semi-reg.	59	M
HR	20 42·3	+19 10	3·7–12·7	Nova	–	Q
U	20 45·5	+18 05	7·6–8·9	Semi-reg.	110	M
S	20 43·1	+17 05	8·3–12·4	Mira	277·2	M
V	20 47·8	+19 20	8·1–16·0	Mira	533·5	M

DOUBLE STARS

	R.A. h m	Dec. ° ′	P.A. °	Sep. ″	Mags.	
1	20 30·3	+10 54	AB 346	0·9	6·1, 8·1	
			AC 349	16·8	14·1	
β	20 37·5	+14 36	167	0·3	4·0, 4·9	Binary, 26·7 y
α	20 39·6	+15 55	AB 224	29·5	3·8, 13·3	
			AC 272	43·4	11·8	
κ	20 39·1	+10 05	286	28·8	5·1, 11·7	
γ	20 46·7	+16 07	268	9·6	4·5, 5·5	
13	20 47·8	+06 00	194	1·6	5·6, 9·2	

GLOBULAR CLUSTER

M	NGC	R.A. h m	Dec. ° ′	Diameter ′	Mag.
	6394	20 34·2	+07 24	5·9	8·9

DELPHINIUS (continued)

PLANETARY NEBULA

M	NGC	R.A. h m	Dec. ° '	Dimensions "	Mag.	Mag. of central star
	6891	20 15·2	+12 42	12×74	11·7	12·4

DORADO

(Abbreviation: Dor).

A 'modern' southern constellation. The only star above the fourth magnitude is α (3·27), but Dorado contains part of the Large Magellanic Cloud (LMC), in which is the magnificent Tarantula looped nebula round 30 Doradûs.

See chart for Reticulum

BRIGHTEST STARS

Star	R.A. h m s	Dec. ° ' "	Mag.	Abs. mag.	Spectrum	Dist. pc
γ	04 16 01·6	−51 29 12	4·25	2·6	F0	45
α	04 33 59·8	−55 02 42	3·27	−0·6	A0	59
ζ	05 05 30·6	−57 28 22	4·72	4·0	F8	13
β	05 33 37·5	−62 29 24	var.	−8·0	F9v	2300
δ	05 44 46·5	−65 44 08	4·35	2·4	A7	450
HD 40409	05 54 06·1	−63 05 23	4·65	1·4	K0	13

Also above mag. 5:

	Mag.	Abs. mag.	Spectrum	Dist.
θ	4·83	−0·1	K2	80

VARIABLE STARS

	R.A. h m	Dec. ° '	Range	Type	Period, d.	Spectrum
R	04 36·8	−62 05	4·8−6·6	Semi-reg.	338	M
β	05 33·6	−62 29	3·7−4·1	Cepheid	9·84	F−G

DOUBLE STAR

	R.A. h m	Dec. ° '	P.A. °	Sep. "	Mags.
α	04 34·0	−55 03	AB 182	0·2	3·8, 4·3
			AB+C 101	77·7	9·8

GALAXIES

M	NGC	R.A. h · m	Dec. ° '	Mag.	Dimensions '	Type
	1549	04 15·7	−55 36	9·9	3·7×3·2	E0
	1553	04 16·2	−55 47	9·5	4·1×2·8	S0
	1596	04 27·6	−55 02	11·0	3·9×1·2	S0
	1617	04 31·7	−54 36	10·4	4·7×2·4	SBa
	1672	04 45·7	−59 15	11·0	4·8×3·9	SBb
	LMC	05 23·6	−69 45	0·1	650×550	Large Cloud of Magellan. Contains 30 Doradûs and 3 planetary nebulæ, NGC 1714, 1722 and 1743
	1947	05 26·8	−63 46	10·8	3·0×1·6	S0p

NEBULA

M	NGC	R.A. h m	Dec. ° '	Dimensions '	
	2070	05 38·7	−69 06	40×25	30 Doradûs. In the LMC

DRACO

(Abbreviation: Dra).

A long, sprawling northern group. In mythology it has been identified either with the dragon killed by Cadmus before the founding of the city of Bœotia, or with the dragon which guarded the golden apples in the Garden of the Hesperides. Thuban (α) was the pole star in ancient times.

There are 12 stars above the fourth magnitude:

γ	2·23	δ	3·07	χ	3·57	ε	3·83
η	2·74	ζ	3·17	α	3·65	λ	3·84
β	2·79	ι	3·29	ξ	3·75	κ	3·87

BRIGHTEST STARS

Star	R.A. h m s	Dec. ° ' "	Mag.	Abs. mag.	Spectrum	Dist. pc	
HD 81817	09 37 05·2	+81 19 35	4·29	−0·2	K3	57	
1 λ	11 31 24·2	+69 19 52	3·84	−0·4	M0	65	Giansar
5 κ	12 33 28·9	+69 47 17	3·87	0·2	B7	22	
10 i	13 51 25·8	+64 43 33	4·66	−0·5	gM3	110	
11 α	14 04 23·2	+64 22 33	3·65	0·2	A0	71	Thuban
12 ι	15 24 55·6	+58 57 58	3·29	−0·1	K2	48	Edasich
13 θ	16 01 53·2	+58 33 55	4·01	3·2	F8	16	
14 η	16 23 59·3	+61 30 50	2·74	0·3	G8	25	Aldhibain
22 ζ	17 08 47·0	+65 42 53	3·17	−1·9	B6	97	Aldhibah
23 β	17 30 25·8	+52 18 05	2·79	−2·1	G2	82	Alwaid
31 ψ	{ 17 41 56·1	+72 08 56	4·58	2·8	F5	23 }	Dziban
	{ 17 41 57·7	+72 09 24	5·79	4·0	F0	23 }	
32 ξ	17 53 31·5	+56 52 21	3·75	−0·1	K2	58	Juza
33 γ	17 56 36·2	+51 29 20	2·23	−0·3	K5	31	Eltamin
43 φ	18 20 45·2	+71 20 16	4·22	−0·6	A0p	33	
44 χ	18 21 03·0	+72 43 58	3·57	4·1	F7	7·8	
47 o	18 51 12·0	+59 23 18	4·66	−0·9	K0	67	
57 δ	19 12 33·1	+67 39 41	3·07	0·2	G9	36	Taïs
60 τ	19 15 32·8	+73 21 20	4·45	−0·2	K3	70	

DRACO (continued)

BRIGHTEST STARS (continued)

Star		R.A.			Dec.			Mag.	Abs. mag.	Spectrum	Dist.	
		h	m	s	°	′	″				pc	
58	π	19	20	39·9	+65	42	52	4·59	0·6	A2	63	
61	σ	19	32	21·5	+69	39	40	4·68	5·9	K0	5·7	Alrakis
63	ε	19	48	10·2	+70	16	04	3·83	0·3	G8	51	Tyl
67	ρ	20	02	48·9	+67	52	25	4·51	−0·2	K3	82	

Also above mag. 5:

		Mag.	Abs. mag.	Spectrum	Dist.	
42		4·82	−0·1	K2	96	
39		4·98	1·2	A1	53	
28	ω	4·80	3·4	F5	22	
24	ν¹	4·88		A8		} Kuma
25	ν²	4·87		A4		
21	μ	4·92		F5		
4		4·95	−0·5	M4	120	
6		4·94	−0·1	K2	84	
18	g	4·83	−4·5	K1		
19		4·89	3·7	F6	17	
45		4·77	−4·6	F7	450	
52	υ	4·82	0·2	K0	64	
54		4·99	−0·1	K2	100	

VARIABLE STARS

	R.A.		Dec.		Range	Type	Period, d.	Spectrum
	h	m	°	′				
RY	12	56·4	+66	00	5·6–8·0	Semi-reg.	173	N
R	16	32·7	+66	45	6·7–13·0	Mira	245·5	M
T	17	56·4	+58	13	7·2 13·5	Mira	421·2	N
UW	17	57·5	+54	40	7·0–8·0	Irreg.	–	K
UX	19	21·6	+76	34	5·9–7·1	Semi-reg.	168	N

DOUBLE STARS

	R.A.		Dec.		P.A.	Sep.	Mags.	
	h	m	°	′	°	″		
η	16	24·0	+61	31	142	5·2	2·7, 8·7	
μ	17	05·3	+54	28	020	1·9	5·7, 5·7	Binary, 482 y
ν	17	32·2	+55	11	312	61·9	4·9, 4·9	
ψ	17	41·9	+72	09	015	30·3	4·9, 6·1	
ε	19	48·2	+70	16	015	3·1	3·8, 7·4	

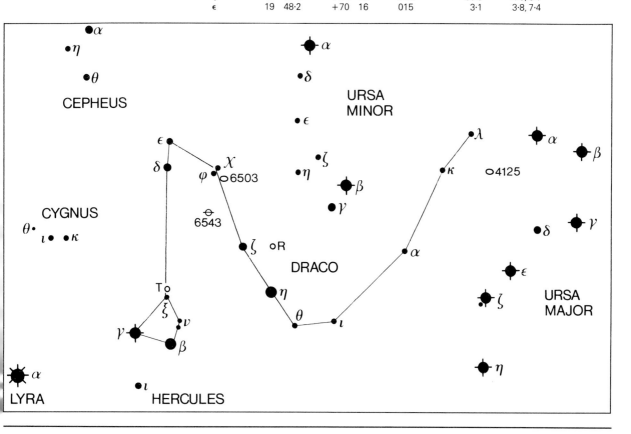

DRACO *(continued)*

PLANETARY NEBULA

M	NGC	R.A.		Dec.		Dimensions	Mag.	Mag. of central star
		h	m	°	'	"		
	6543	17	58·6	+66	38	18×350	8·8	11·4

GALAXIES

M	NGC	R.A.		Dec.		Mag.	Dimensions	Type
		h	m	°	'		'	
	3147	10	16·9	+73	24	10·6	4·0×3·5	Sb
	4125	12	08·1	+65	11	9·8	5·1×3·2	E5p
	4236	12	16·7	+69	28	9·7	18·6×6·9	Sb
	5866	15	06·5	+55	46	10·0	5·2×2·3	E6p
	5879	15	09·8	+57	00	11·5	4·4×1·7	Sb
	5907	15	15·9	+56	19	10·4	12·3×1·8	Sb
	5985	15	39·6	+59	20	11·0	5·5×3·2	Sb
	6015	15	51·4	+62	19	11·2	5·4×2·3	Sc
	5907	15	15·9	+56	19	10·4	12·3×1·8	Sb
	5985	15	39·6	+59	20	11·0	5·5×3·2	Sb
	6503	17	49·4	+70	09	10·2	6·2×2·3	Sb

EQUULEUS

(Abbreviation: Eql).

A very obscure and small constellation, but one of the 'originals'. It represents a foal given by Mercury to Castor, one of the Heavenly Twins. There is only one star above the fourth magnitude: α (3·92). The name often applied to α has a decidedly modern flavour!

See chart for Aquila

BRIGHTEST STARS

Star		R.A.			Dec.			Mag.	Abs. mag.	Spectrum	Dist.	
		h	m	s	°	'	"				pc	
5	γ	21	10	20·3	+10	07	53	4·69	1·0	F0p	64	
7	δ	21	14	28·7	+10	00	25	4·49	4·0	F8	15	
8	α	21	15	49·3	+05	14	52	3·92	0·6	G0	46	Kitalpha

VARIABLE STAR

	R.A.		Dec.		Range	Type	Period, d.	Spectrum
	h	m	°	'				
S	20	57·2	+05	05	8·0–10·1	Algol	3·44	B+F

DOUBLE STARS

	R.A.		Dec.		P.A.	Sep.	Mags.	
	h	m	°	'	°	"		
ε	20	59·1	+04	18	AB 285	1·0	6·0, 6·3	Binary, 101·4 y
					AB+C 070	10·7	7·1	
					AD 280	74·8	12·4	
γ	21	10·4	+10	08	AB 268	1·9	4·7, 11·5	
					AC 005	47·7	12·5	
δ	21	14·5	+10	00	029	0·3	5·2, 5·3	Binary, 5·7 y
β	21	22·9	+06	49	257	34·4	5·2, 13·7	

ERIDANUS

(Abbreviation: Eri).

An immensely long constellation, extending from Achernar in the far south as far as Kursa, near Orion. Achernar is the only brilliant star, though θ (Acamar) was ranked as the first magnitude in ancient times, and there are suggestions that it has faded. Mythologically, Eridanus is the river Po – and this was the river into which the youth Phaëthon was plunged when he had obtained permission to drive the Sun-chariot for a day, and had lost control of it, so that Jupiter was forced to strike him down with a thunderbolt.

BRIGHTEST STARS

Star		R.A.			Dec.			Mag.	Abs. mag.	Spectrum	Dist.	
		h	m	s	°	'	"				pc	
	α	01	37	42·9	−57	14	12	0·46	−1·6	B5	26	Achernar
	χ	01	55	57·5	−51	36	32	3·70	3·2	G5	15	
	κ	02	26	59·1	−47	42	14	4·25	−2·2	B5	190	
	φ	02	16	30·6	−51	30	44	3·56	−0·2	B8	37	
	ι	02	40	40·0	−39	51	19	4·11	0·2	K0	61	
1	τ¹	02	45	06·1	−18	34	21	4·47	3·7	F6	15	
2	τ²	02	51	02·2	−21	00	15	4·75	0·2	K0	81	
3	η	02	56	25·6	−08	53	54	3·89	0·9	K1	23	Azha
	θ	02	58	15·6	−40	18	17	2·92	0·6+1·7	A3+A2	17	Acamar
11	τ³	03	02	23·4	−23	37	28	4·09	2·1	A5	23	
13	ζ	03	15	49·9	−08	49	11	4·80	1·8	A3	16	Zibal
16	τ⁴	03	19	30·9	−21	45	28	3·69	−0·5	gM3	69	Angetenar
	e	03	19	55·7	−43	04	10	4·27	5·3	G5	6·2	
17	υ	03	30	37·0	−05	04	30	4·73	−0·2	B8	97	
18	ε	03	32	55·8	−09	27	30	3·73	6·1	K2	3·3	
19	τ⁵	03	33	47·2	−21	37	58	4·27	−0·2	B8	78	
	y	03	37	05·6	−40	16	29	4·58	0·2	K0	89	
	h	03	42	50·0	−37	18	49	4·59	−0·3	gK5	95	
	δ	03	43	14·8	−09	45	48	3·54	3·8	K0	9	Rana
26	π	03	46	08·4	−12	06	06	4·42	−1·1	gMa	50	
27	τ⁶	03	46	50·8	−23	14	59	4·23	3·1	F3	17	
	g	03	48	35·3	−37	37	20	4·27	0·6	A0	50	
33	τ⁸	03	53	42·6	−24	36	45	4·65	−1·1	B5	140	
32	w	03	54	17·4	−02	57	17	4·46	0·3	G8	68	
34	γ	03	58	01·7	−13	30	31	2·95	−0·4	M0	44	Zaurak
36	τ⁹	03	59	55·4	−24	00	59	4·66	−0·6	A0	110	
38	o¹	04	11	51·8	−06	50	15	4·04	−0·7	F2	85	Beid

ERIDANUS (continued)

There are 15 stars above magnitude 4:

α	0·46	χ	3·70
β	2·79	ε	3·73
θ	2·92	υ²	3·82
γ	2·95	53	3·87
δ	3·54	η	3·89
υ⁴	3·56	ν	3·93
φ	3·56	υ³	3·96
τ⁴	3·69		

BRIGHTEST STARS (continued)

Star		R.A. h	m	s	Dec. °	′	″	Mag.	Abs. mag.	Spectrum	Dist. pc	
40	o²	04	15	16·2	−07	39	10	4·43	6·0	K1	4·9	Keid
41	υ⁴	04	17	53·6	−33	47	54	3·56	0·0	B8·5	40	
43	υ³	04	24	02·1	−34	01	01	3·96	−0·5	M1	78	
50	υ¹	04	33	30·6	−29	46	00	4·51	0·3	gG6	63	
52	υ²	04	35	33·0	−30	33	45	3·82	0·2	K0	53	Theemini
48	ν	04	36	19·1	−03	21	09	3·93	−3·6	B2	350	
53	l	04	38	10·7	−14	18	15	3·87	−0·1	K2	44	Sceptrum
54		04	40	26·4	−19	40	18	4·32	−0·5	gM4	92	
57	μ	04	45	30·1	−03	15	17	4·02	−1·6	B5	130	
61	ω	04	52	53·6	−05	27	10	4·39	1·6	A9	36	
67	β	05	07	50·9	−05	05	11	2·79	0·0	A3	28	Kursa
69	λ	05	09	08·7	−08	45	15	4·27	−3·0	B2	280	

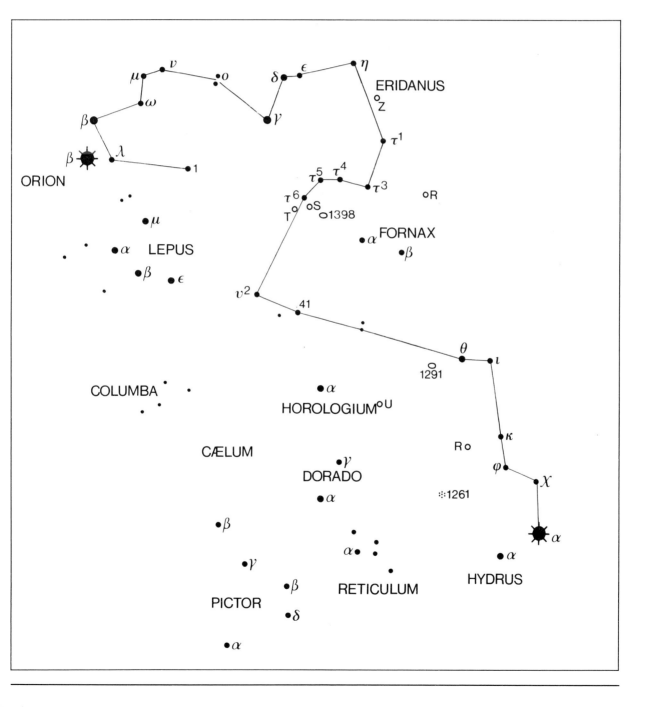

ERIDANUS (continued)

Also above mag. 5:

		Mag.	Abs. mag.	Spectrum	Dist.
45		4·91	−1·3	K3	170
39		4·87	−0·2	K3	100
15		4·88	0·3	G6	82
64	S	4·79	1·7	F0	42
65	ψ	4·81	−2·3	B2	290

VARIABLE STARS

	R.A. h m	Dec. ° ′	Range	Type	Period, d.	Spectrum
Z	02 47·9	−12 28	7·0–8·6	Semi-reg.	80	M
RR	02 52·2	−08 16	7·4–8·6	Semi-reg.	97	M
T	03 55·2	−24 02	7·4–13·2	Mira	252·2	M
W	04 11·5	−25 08	7·5–14·5	Mira	376·7	M
RZ	04 43·8	−10 41	7·8–8·7	Algol	39·28	A+G

DOUBLE STARS

	R.A. h m	Dec. ° ′	P.A. °	Sep. ″	Mags.	
χ	01 56·0	−51 37	202	5·0	3·7, 10·7	
p	01 39·8	−56 12	194	11·2	5·5, 5·8	Binary, 484 y
θ	02 58·3	−40 18	088	8·2	3·4, 4·5	
τ⁴	03 19·5	−21 45	AB 288	5·7	3·7, 9·2	
			AC 112	39·2	10·7	
υ⁴	04 17·9	−33 48	013	49·2	3·6, 11·8	A is a close dble
ρ²	03 02·7	−07 41	075	1·8	5·3, 9·5	

PLANETARY NEBULA

M	NGC	R.A. h m	Dec. ° ′	Dimensions ″	Mag.	Mag. of central star
	1535	04 14·2	−12 44	18×44	9·6	12·2

GALAXIES

M	NGC	R.A. h m	Dec. ° ′	Mag.	Dimensions ′	Type
	1084	02 46·0	−07 35	10·6	2·9×1·5	Sc
	1179	03 02·6	−18 54	11·8	4·6×3·9	Sp
	1187	03 02·6	−22 52	10·9	5·0×4·1	SBc
	1291	03· 17·3	−41 08	8·5	10·5×9·1	SBa
	1300	03 19·7	−19 25	10·4	6·5×4·3	SBb
	1332	03 26·3	−21 20	10·3	4·6×1·7	E7
	1337	03 28·1	−08 23	11·7	6·8×2·0	S
	1395	03 38·5	−23 02	11·3	3·2×2·5	E3
	1407	03 40·2	−18 35	9·8	2·5×2·5	E0
	1532	04 12·1	−32 52	11·1	5·6×1·8	Sb
	1637	04 41·5	−02 51	10·9	3·3×2·9	Sc

FORNAX

(Abbreviation: For).

A southern group, originally Fornax Chemica (the Chemical Furnace). It has no bright stars, but is notable for containing a large number of faint galaxies.

The only star above the fourth magnitude is α (3·87).

See chart for Eridanus

BRIGHTEST STARS

Star	R.A. h m s	Dec. ° ′ ″	Mag.	Abs. mag.	Spectrum	Dist. pc
ν	02 04 29·4	−29 17 49	4·69	−0·6	A0	110
β	02 49 05·4	−32 24 22	4·46	0·3	G6	61
α	03 12 04·2	−28 59 13	3·87	3·3	F8	14

Also above mag. 5:

	Mag.	Abs. mag.	Spectrum	Dist.
δ	5·00		B5	
ω	4·90	0·2	B9	

VARIABLE STARS

	R.A. h m	Dec. ° ′	Range	Type	Period, d.	Spectrum
R	02 29·3	−26 06	7·5–13·0	Mira	387·9	N
ST	02 44·4	−29 12	7·7–9·0	Semi-reg.	277	M
S	03 46·2	−24 24	?5·6–8·5?	Suspected	?	M

DOUBLE STARS

	R.A. h m	Dec. ° ′	P.A. °	Sep. ″	Mags.	
ω	02 33·8	−28 14	244	10·8	5·0, 7·7	
γ¹	02 49·8	−24 34	AB 145	12·0	6·1, 12·5	
			AC 143	40·9	10·5	
η²	02 50·2	−35 51	014	5·0	5·9, 10·1	
α	03 12·1	−28 59	298	4·0	4·0, 7·0	Binary, 314 y
χ³	03 28·2	−35 51	248	6·3	6·5, 10·5	

FORNAX (continued)

PLANETARY NEBULA

M	NGC	R.A.		Dec.		Diameter	Mag.	Mag. of central star
		h	m	°	′	″		
	1360	03	33·3	−25	51	390	−	11·3

GALAXIES

M	NGC	R.A.		Dec.		Mag.	Dimensions	Type
		h	m	°	′		′	
	986	02	33·6	−39	02	11·0	3·7×2·8	SBb
	1097	02	46·3	−30	17	9·2	9·3×6·6	SBb
	1201	03	04·1	−26	04	10·6	4·4×2·8	Sa
	1255	03	13·5	−25	44	11·1	4·1×2·8	Sa
	1302	03	19·9	−26	04	11·5	4·4×4·2	SBa
	1316	03	22·7	−37	12	8·8	7·1×5·5	SB0p
	1326	03	23·9	−36	28	10·5	4·0×3·0	SB0
	1344	03	28·3	−31	04	10·3	3·9×2·3	E3
	1350	03	31·1	−33	38	10·5	4·3×2·4	SBb
	1365	03	33·6	−36	08	9·5	9·8×5·5	SBb
	1371	03	35·0	−24	56	11·5	5·4×4·0	SBa
	1380	03	36·5	−34	59	11·1	4·9×1·9	S0
	1385	03	37·5	−24	30	11·2	3·0×2·0	Sc
	1399	03	38·5	−35	27	9·9	3·2×3·1	E1p
	1398	03	38·9	−26	20	9·7	6·6×5·2	SBb
	1404	03	38·9	−35	35	10·2	2·5×2·3	E1
	1425	03	42·2	−29	54	11·7	5·4×2·7	Sb

GEMINI

(Abbreviation: Gem).

A brilliant Zodiacal constellation. In mythology, Castor and Pollux were the twin sons of the King and Queen of Sparta. Pollux was immortal, but Castor was not. When Castor was killed, Pollux pleaded with the gods to be allowed to share his immortality; so Castor was brought back to life, and both youths placed in the sky.

Today Pollux is the brighter of the two stars, but in ancient times Castor was recorded as being the more brilliant. If any change has occurred (and this is by no means certain) it is more likely to have been in the late-type Pollux than in Castor, which is a multiple system. There are 13 stars in Gemini above the fourth magnitude:

β 1·14
α 1·58
γ 1·93
μ 2·88v
ε 2·98
η 3·1 (max)
ξ 3·36
δ 3·53
κ 3·57
λ 3·58
θ 3·60
ζ 3·7 (max)
ι 3·79

BRIGHTEST STARS

Star		R.A.			Dec.			Mag.	Abs. mag.	Spectrum	Dist.	
		h	m	s	°	′	″				pc	
1		06	04	07·2	+23	15	48	4·16	0·3	gG5	59	
7	η	06	14	52·6	+22	30	24	var.	−0·5v	M3	57	Propus
13	μ	06	22	57·6	+22	30	49	2·88v	−0·5	M3	71	Tejat
18	ν	06	28	57·7	+20	12	43	4·15	−1·0	B7	110	
24	γ	06	37	42·7	+16	23	57	1·93	0·0	A0	26	Alhena
27	ε	06	43	55·9	+25	07	52	2·98	−4·5	G8	210	Mebsuta
30		06	43	59·2	+13	13	40	4·49	0·0	K1	70	
31	ξ	06	45	17·3	+12	53	44	3·36	0·7	F5	23	Alzirr
34	θ	06	52	47·3	+33	57	40	3·60	0·0	A3	51	
38	e	06	54	38·6	+13	10	40	4·65	2·6	F0	26	
43	ζ	07	04	06·5	+20	34	13	var.	−4·5	G0	430	Mekbuda
46	τ	07	11	08·3	+30	14	43	4·41	−0·1	K2	67	

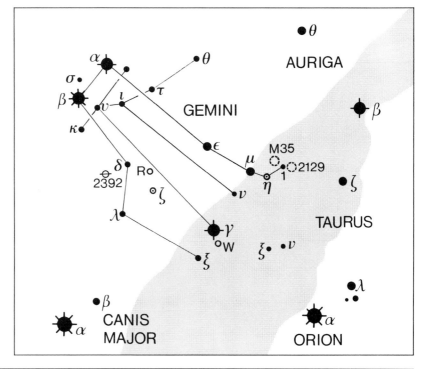

GEMINI *(continued)*

BRIGHTEST STARS *(continued)*

Star		R.A. h	m	s	Dec. °	′	″	Mag.	Abs. mag.	Spectrum	Dist. pc	
54	λ	07	18	05·5	+16	32	25	3·58	1·7	A3	25	
55	δ	07	20	07·3	+21	58	56	3·53	1·9	F2	18	Wasat
60	ι	07	25	43·5	+27	47	53	3·79	0·2	K0	50	
62	ρ	07	29	06·6	+31	47	03	4·18	2·6	F0	19	
66	α	07	34	35·9	+31	53	18	1·58	1·2	A0	14	Castor
69	ν	07	35	55·3	+26	53	45	4·06	−0·4	M0	78	
75	σ	07	43	18·7	+28	53	01	4·28	0·0	K1	40	
77	κ	07	44	26·8	+24	23	52	3·57	0·3	G8	45	
78	β	07	45	18·9	+28	01	34	1·14	0·2	K0	11	Pollux

Also above mag. 5:

		Mag.	Abs. mag.	Spectrum	Dist.
51		5·00	−0·5	M4	120
71	o	4·90	0·6	F3	70
81		4·88	−0·3	K5	110
83	φ	4·97	1·7	A3	45

VARIABLE STARS

	R.A. h	m	Dec. °	′	Range	Type	Period, d.	Spectrum
BU	06	12·3	+22	54	5·7–7·5	Irreg.	–	M
η	06	14·9	+22	30	3·2–3·9	Semi-reg.	233	M
W	06	35·0	+15	20	6·5–7·4	Cepheid	7·91	F–G
X	06	47·1	+30	17	7·5–13·6	Mira	263·7	M
ζ	07	04·1	+20	34	3·7–4·1	Cepheid	10·15	F–G
R	07	07·4	+22	42	6·0–14·0	Mira	369·8	S
V	07	23·2	+13	06	7·8–14·9	Mira	275·1	M
T	07	49·3	+23	44	8·0–15·0	Mira	287·8	S
U	07	55·1	+22	00	8·2–14·9	SS Cygni	±103	M+WD

DOUBLE STARS

	R.A. h	m	Dec. °	′	P.A. °	Sep. ″	Mags.	
η	06	14·9	+22	30	266	1·4	var., 8·8	Binary, 474 y
μ	06	22·9	+22	31	077	72·7	3·0, 9·8	
ν	06	29·0	+20	13	329	112·5	4·2, 8·7	
ε	06	43·9	+25	08	094	110·3	3·0, 9·0	
38	06	54·6	+13	11	147	7·0	4·7, 7·7	Binary, 3190 y
ζ	07	04·1	+20	34	AB 084	87·0	var., 10·5	
					AC 350	96·5	8·0	
λ	07	18·1	+16	32	033	9·6	3·6, 10·7	
δ	07	20·1	+21	59	223	6·0	3·5, 8·2	Binary, 1200 y
α	07	34·6	+31	53	AB 088	2·5	1·9, 2·9	Binary, 420 y
					AC 164	72·5	1·6, 8·8	
κ	07	44·4	+24	24	240	7·1	3·6, 8·1	

OPEN CLUSTERS

M	NGC	R.A. h	m	Dec. °	′	Diameter ′	Mag.	No. of stars	
	2129	06	01·0	+23	18	7	6·7	40	
	IC 2157	06	05·0	+24	00	7	8·4	20	
	2169	06	08·4	+13	57	7	5·9	30	
35	2168	06	08·9	+24	20	28	5·0	200	
	2266	06	43·2	+26	58	7	9·5	30	
	2355	07	16·9	+13	47	9	9·7	40	
	2395	07	27·1	+13	35	12	8·0	30	Asterism?

PLANETARY NEBULA

M	NGC	R.A. h	m	Dec. °	′	Dimensions ″	Mag.	Mag. of central star	
	2392	07	29·2	+20	55	13×44	10	10·5	Eskimo Nebula

GRUS

(Abbreviation: Gru).

The most distinctive of the 'Southern Birds'. The contrast in colour between α and β is very marked.

BRIGHTEST STARS

Star		R.A. h	m	s	Dec. °	′	″	Mag.	Abs. mag.	Spectrum	Dist. pc	
γ		21	53	55·6	−37	21	54	3·01	−1·2	B8	70	
λ		22	06	06·7	−39	32	36	4·46	−0·4	M0	94	
α		22	08	13·8	−46	57	40	1·74	−1·1	B5	21	Alnair
δ¹		22	29	15·9	−43	29	45	3·97	0·3	gG5	43	
δ²		22	29	45·3	−43	44	58	4·11		M4	27	
β		22	42	39·9	−46	53	05	2·11v	−2·4	M3	53	Al Dhanab
ε		22	48	33·1	−51	19	01	3·49	1·4	A2	25	
ζ		23	00	52·6	−52	45	15	4·12	0·3	G5	50	

GRUS *(continued)*

There are six stars above the fourth magnitude:

α 1·74
β 2·11v
γ 3·01
ε 3·49
ι 3·90
δ¹ 3·97

β has a very small range (2·11 to 2·23), so that its fluctuations are not detectable with the naked eye.

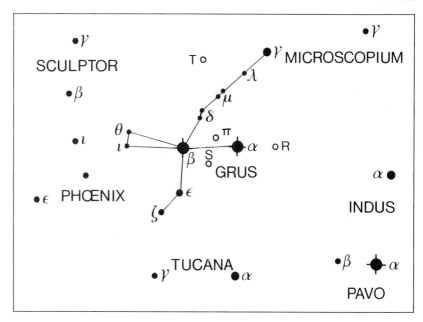

BRIGHTEST STARS *(continued)*

Star	R.A.			Dec.			Mag.	Abs. mag.	Spectrum	Dist.
	h	m	s	°	′	″				pc
θ	23	06	52·6	−43	31	14	4·28	2·2	F6	26
ι	23	10	21·4	−45	14	48	3·90	0·2	K0	53

Also above mag. 5:

	Mag.	Abs. mag.	Spectrum	Dist.
μ	4·79	0·3	G4	79
ρ	4·85	0·2	K0	83
η	4·85	0·2	K0	62

VARIABLE STARS

	R.A.		Dec.		Range	Type	Period, d.	Spectrum
	h	m	°	′				
RS	21	43·1	−48	11	7·9–8·5	Delta Scuti	0·15	A–F
R	21	48·5	−46	55	7·4–14·9	Mira	331·9	M
π¹	22	22·7	−45	57	5·4–6·7	Semi-reg.	150	S
T	22	25·7	−37	34	7·8–12·3	Mira	136·5	M
S	22	26·1	−48	26	6·0–15·0	Mira	401·4	M

DOUBLE STARS

	R.A.		Dec.		P.A.	Sep.	Mags.
	h	m	°	′	°	″	
θ	23	06·9	−43	31	075	1·1	4·5, 7·0
υ	23	06·9	−38	54	211	1·1	5·7, 8·0

GALAXIES

M	NGC	R.A.		Dec.		Mag.	Dimensions	Type
		h	m	°	′		′	
	7144	22	52·7	−48	15	10·7	3·5×3·5	E0
	7213	22	09·3	−47	10	10·4	1·9×1·8	Sa
	7410	22	55·0	−39	40	10·4	5·5×2·0	SBa
	7412	22	55·8	−42	39	11·4	4·0×3·1	SBb
	7418	22	56·6	−37	02	11·4	3·3×2·8	SBc
	IC 1459	22	57·2	−36	28	10·0	−	E3
	IC 5267	22	57·2	−43	24	10·5	5·0×4·1	S0
	7424	22	57·3	−41	04	11·0	7·6×6·8	SBc
	IC 5273	22	59·5	−37	42	11·4	2·9×2·1	SBc
	7456	23	02·1	−39	35	11·9	5·9×1·8	Sc
	7496	23	09·8	−43	26	11·1	3·5×2·8	SBb
	7531	23	14·8	−43	36	11·3	3·5×1·5	Sb
	7552	23	16·2	−42	35	10·7	3·5×2·5	SBb
	7582	23	18·4	−42	22	10·6	4·6×2·2	SBG
	7599	23	19·3	−42	15	11·4	4·4×1·5	Sc

HERCULES

(Abbreviation: Her).

A large but by no means brilliant constellation, commemorating the great hero of mythology. The most interesting objects are the red supergiant α and the globular clusters M.13 and M.92; M.13 is the brightest globular in the northern hemisphere of the sky, and is surpassed only by the southern ω Centauri and 47 Tucanæ.

Hercules contains fifteen stars above the fourth magnitude:

β	2·77
ζ	2·81
α	3·00 (max)
δ	3·14
π	3·16
μ	3·42
η	3·53
ξ	3·70
γ	3·75
ι	3·80
o	3·83
109	3·84
θ	3·86
τ	3·89
ε	3·92

BRIGHTEST STARS

Star		R.A. h	m	s	Dec. °	′	″	Mag.	Abs. mag.	Spectrum	Dist. pc	
1	χ	15	52	40·4	+42	27	05	4·62	4·2	F9	15	
11	φ	16	08	46·0	+44	56	05	4·26	0·0	B9p	23	
22	τ	16	19	44·3	+46	18	48	3·89	−1·6	B5	130	
20	γ	16	21	55·1	+19	09	11	3·75	0·6	A9	42	
24	ω	16	25	24·8	+14	02	00	4·57	1·8	A0p	33	Cujam
27	β	16	30	13·1	+21	29	22	2·77	0·3	G8	31	Kornephoros
35	σ	16	34	06·0	+42	26	13	4·20	0·2	B9	61	
40	ζ	16	41	17·1	+31	36	10	2·81	3·0	G0	9·6	Rutilicus
44	η	16	42	53·7	+38	55	20	3·53	1·8	G8	21	
58	ε	17	00	17·2	+30	55	35	3·92	0·6	A0	26	
64	α	17	14	38·8	+14	23	25	var.	−2·3v	M5	67	Rasalgethi
65	δ	17	15	01·8	+24	50	21	3·14	0·9	A3	28	Sarin
67	π	17	15	02·6	+36	48	33	3·16	−2·3	K3	120	
68	u	17	17	19·4	+33	06	00	var.	−2·9v	B3	350	
69	e	17	17	40·1	+37	17	29	4·65	1·4	A2	45	
75	ρ	17	23	40·8	+37	08	45	4·17	0·6	A0	52	
76	λ	17	30	44·1	+26	06	39	4·41	−0·3	K4	85	Masym
85	ι	17	39	27·7	+46	00	23	3·80	−1·7	B3	130	
86	μ	17	46	27·3	+27	43	15	3·42	3·9	G5	8·1	
91	θ	17	56	15·1	+37	15	02	3·86	−2·2	K1	130	
92	ξ	17	57	45·7	+29	14	52	3·70	0·2	K0	50	
94	ν	17	58	30·0	+30	11	22	4·41	−2·0	F2	190	
93		18	00	03·2	+16	45	03	4·67	−0·9	K0	100	
95		18	01	30·3	+21	35	44	4·27	0·5	A7	48	
103	o	18	07	32·4	+28	45	45	3·83	0·2	B9	52	
102		18	08	45·4	+20	48	52	4·36	−2·5	B2	240	
109		18	23	41·7	+21	46	11	3·84	−0·1	K2	33	
110		18	45	39·6	+20	32	47	4·19	3·7	F6	15	
111		18	47	01·1	+18	10	53	4·36	1·7	A3	34	
113		18	54	44·7	+22	38	43	4·56	1·7	A3	36	

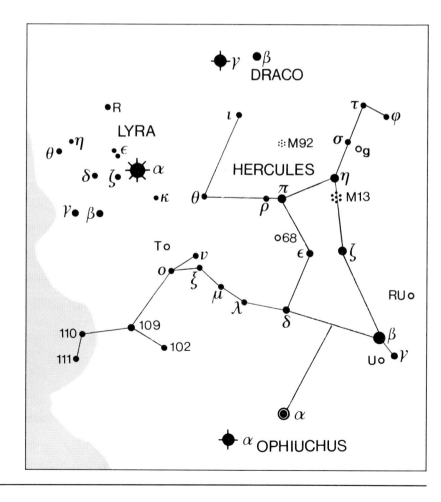

HERCULES (continued)

Also above mag. 5:

		Mag.	Abs. mag.	Spectrum	Dist.
6	υ	4·76		B9	
7	κ	5·00	−0·1	K2	87
29		4·84	−0·3	K5	70
g		5·0v	var.	M6	23
52		4·82	0·9	A2p	
60		4·91	0·9	A3	62
104		4·97	−0·5	M3	120
106		4·95	−0·4	M0	110

VARIABLE STARS

	R.A. h m	Dec. ° ′	Range	Type	Period, d.	Spectrum
X	16 02·7	+47 14	7·5–8·6	Semi-reg.	95	M
R	16 06·2	+18 22	7·8–15·0	Mira	318·4	M
RU	16 10·2	+25 04	6·8–14·3	Mira	485·5	M
U	16 25·8	+18 54	6·5–13·4	Mira	406·0	M
g (30)	16 28·6	+41 53	5·7–7·2	Semi-reg.	70	M
W	16 35·2	+37 21	7·6–14·4	Mira	280·4	M
S	16 51·9	+14 56	6·4–13·8	Mira	307·4	M
α	17 14·6	+14 23	3–4	Semi-reg.	±100?	M
u (68)	17 17·3	+35 06	4·6–5·3	Beta Lyræ	2·05	B+B
RS	17 21·7	+22 55	7·0–13·0	Mira	219·6	M
Z	17 58·1	+15 08	7·3–8·1	Algol	3·99	F+K
T	18 09·1	+31 01	6·8–13·9	Mira	165·0	M
AC	18 30·3	+21 52	7·4–9·7	RV Tauri	75·5	F+K
RX	18 30·7	+12 37	7·2–7·8	Algol	1·78	A+A

DOUBLE STARS

	R.A. h m	Dec. ° ′	P.A. °	Sep. ″	Mags.	
κ	16 08·1	+17 03	012	28·4	5·3, 6·5	
γ	16 21·9	+19 09	233	41·6	3·8, 9·8	
ω	16 25·4	+14 02	AB 223	1·0	4·6, 11·6	
			AC 096	28·4	11·1	
37	16 40·6	+04 13	230	69·8	5·8, 7·0	
ζ	16 41·3	+31 36	089	1·6	2·9, 5·5	Binary, 34·5 y
54	16 55·4	+18 26	183	2·5	5·4, 12·7	
α	17 14·6	+14 23	107	4·7	var., 5·4	Binary, 3600 y
δ	17 15·0	+24 50	236	8·9	3·7, 8·2	Optical pair
u (68)	17 17·3	+35 06	060	4·4	var., 10·2	
ρ	17 23·7	+37 09	316	4·1	4·6, 5·6	
μ	17 46·5	+27 43	247	33·8	3·4, 10·1	

GLOBULAR CLUSTERS

M	NGC	R.A. h m	Dec. ° ′	Diameter ′	Mag.
92	6341	17 17·1	+43 08	11·2	6·5
13	6205	16 41·7	+36 28	16·6	5·9

PLANETARY NEBULÆ

M	NGC	R.A. h m	Dec. ° ′	Diameter ″	Mag.	Mag. of central star
	6058	16 04·4	+40 41	23	13·3	13·8
	IC 4593	16 12·2	+12 04	12 × 120	10·9	11·3
	6210	16 44·5	+23 49	14	9·3	12·9

HOROLOGIUM

(Abbreviation: Hor).

A very obscure southern constellation. The only star above the fourth magnitude is α. The only other star above the fifth magnitude is δ; magnitude 4·93.

See chart for Eridanus

BRIGHTEST STAR

Star	R.A. h m s	Dec. ° ′ ″	Mag.	Abs. mag.	Spectrum	Dist. pc
α	04 14 00·0	−42 17 40	3·86	0·0	K1	59

Also above mag. 5:

	Mag.	Abs. mag.	Spectrum	Dist.
δ	4·93	2·6	F0	28

VARIABLE STARS

	R.A. h m	Dec. ° ′	Range	Type	Period, d.	Spectrum
R	02 53·9	−49 53	4·7–14·3	Mira	404·0	M
T	03 00·9	−50 39	7·2–13·7	Mira	217·7	M
V	03 03·5	−58 56	7·8–8·9	Semi-reg.	?	M
U	03 52·8	−45 50	7·8–15·1	Mira	348·4	M

HOROLOGIUM (continued)

GLOBULAR CLUSTER

M	NGC	R.A. h m	Dec. ° ′	Diameter ′	Mag.
	1261	03 12·3	−55 13	6·9	8·4

GALAXIES

M	NGC	R.A. h m	Dec. ° ′	Mag.	Dimensions ′	Type
	1249	03 10·0	−53 21	11·7	5·2×2·7	SBc
	1411	03 38·8	−44 05	11·9	2·8×2·3	S0
	1433	03 42·0	−47 13	10·0	6·8×6·0	SBa
	1448	03 44·5	−44 39	11·3	8·1×1·8	Sc
	1493	03 57·5	−46 12	11·8	2·6×2·3	SBc
	1512	04 03·9	−43 21	10·6	4·0×3·2	SBa

HYDRA

(Abbreviation: Hya).

The largest constellation in the sky, representing the hundred-headed monster which lived in the Lernæan marshes until it was killed by Hercules. It extends for over 6 hours of R.A. α (Alphard) is often known as 'the Solitary One', because there are no other bright stars anywhere in the neighbourhood.

Hydra contains 10 stars above the fourth magnitude:

α 1·98
γ 3·00
ζ 3·11
ν 3·11
π 3·27
ε 3·38
ξ 3·54
λ 3·61
μ 3·81
θ 3·88

BRIGHTEST STARS

Star		R.A. h m s	Dec. ° ′ ″	Mag.	Abs. mag.	Spectrum	Dist. pc	
4	δ	08 37 39·3	+05 42 13	4·16	0·6	A0	43	
5	σ	08 38 45·4	+03 20 29	4·44	−0·1	K2	75	
7	η	08 43 13·4	+03 23 55	4·30	−1·7	B3	160	
12	D	08 46 22·4	−13 32 52	4·32	0·3	G8	64	
11	ε	08 46 46·5	+06 25 07	3·38	0·6	G0	34	
13	ρ	08 48 25·9	+05 50 16	4·36	0·6	A0	57	
16	ζ	08 55 23·6	+05 56 44	3·11	0·2	K0	38	
22	θ	09 14 21·8	+02 18 51	3·88	0·6	A0	45	
30	α	09 27 35·2	−08 39 31	1·98	−0·2	K3	26	Alphard
31	τ¹	09 29 08·8	−02 46 08	4·60	3·7	F6	15	
32	τ²	09 31 58·9	−01 11 06	4·57	0·0	A3	82	
35	ι	09 39 51·3	−01 08 34	3·91	−0·2	K3	63	
39	υ¹	09 51 28·6	−14 50 48	4·12	0·3	G8	58	
40	υ²	10 05 07·4	−13 03 53	4·60	−1·2	B8	140	
41	λ	10 10 35·2	−12 21 15	3·61	0·2	K0	46	
42	μ	10 26 05·3	−16 50 11	3·81	−0·3	K4	59	
	ν	10 49 37·4	−16 11 37	3·11	−0·1	K2	39	
	ξ	11 33 00·1	−31 51 27	3·54	0·3	G7	44	
	o	11 40 12·9	−34 44 40	4·70	0·3	B9	?	
	β	11 52 54·5	−33 54 28	4·28	−0·3	B9	82	
46	γ	13 18 55·2	−23 10 17	3·00	0·3	G5	32	
R		13 29 42·7	−23 16 52	var.	var.	Md	100	Very red
49	π	14 06 22·2	−26 40 56	3·27	−0·1	K2	47	
58	E	14 50 17·2	−27 57 37	4·41	−0·3	gK4	88	

Also above mag. 5:

		Mag.	Abs. mag.	Spectrum	Dist.	
6		4·98	−0·3	K4	110	
9		4·88	0·0	K1	95	
18	ω	4·97	−1·1	K2	170	
26		4·79	0·3	G8	79	
27	P	4·80	1·8	G8	39	
F		4·70	−4·1	G4	(=31 Mon)	
HD 83953		4·77	−1·1	B5	150	
φ		4·91	0·2	K0	88	
χ¹		4·94	3·3	F4	21	
45	ψ	4·95	0·2	K0	81	
51	κ	4·77	−0·3	K5	48	
52		4·97	−0·6	B8	130	
54		4·94	0·6	F0	51	

VARIABLE STARS

	R.A. h m	Dec. ° ′	Range	Type	Period, d.	Spectrum
RT	08 29·7	−06 19	7·0−11·0	Semi-reg.	253	M
S	08 53·6	+03 04	7·4−13·3	Mira	256·4	M
T	08 55·7	−09 08	6·7−13·2	Mira	289·2	M
X	09 35·5	−14 42	8·0−13·6	Mira	301·4	M
U	10 37·6	−13 23	4·8−5·8	Semi-reg.	450	N
R	13 29·7	−23 17	4·0−10·0	Mira	389·6	M
TT	11 13·2	−26 28	7·5−9·5	Algol	6·95	A+G
HZ	11 26·3	−25 45	7·6−8·2	Semi-reg.	95	M
W	13 49·0	−28 22	7·7−11·6	Semi-reg.	397	M
RU	14 11·6	−28 53	7·2−14·3	Mira	333·2	M

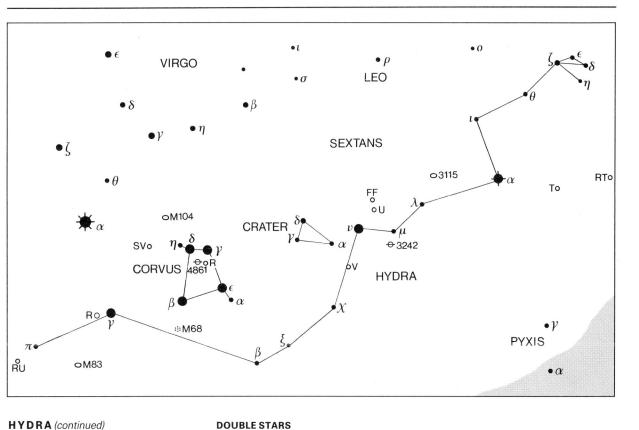

HYDRA (continued)

DOUBLE STARS

	R.A.		Dec.		P.A.	Sep.	Mags.	
	h	m	°	′	°	″		
ε	08	46·8	+06	25	AB 295	0·2	3·8, 4·7	Binary, 890 y
					AB+C 281	2·8	6·8	
θ	09	14·4	+02	19	197	29·4	3·9, 9·9	
α	09	17·6	−08	40	153	283·1	2·0, 9·5	
β	11	52·9	−33	54	008	0·9	4·7, 5·5	
R	13	29·7	−23	17	324	21·2	var., 12·0	
52	14	28·2	−29	30	AB 130	0·1	5·8, 5·8	
					AB+C 279	4·2	10·0	
					AB+D 282	140·8	12·0	
59	14	58·7	−27	39	335	0·8	6·3, 6·6	

OPEN CLUSTER

M	NGC	R.A.		Dec.		Diameter	Mag.	No. of stars
		h	m	°	′	′		
48	2548	08	13·8	−05	48	54	5·8	80

GLOBULAR CLUSTER

M	NGC	R.A.		Dec.		Diameter	Mag.
		h	m	°	′	′	
68	4590	12	39·5	−26	45	12·0	8·2

GALAXIES

M	NGC	R.A.		Dec.		Mag.	Dimensions	Type
		h	m	°	′		′	
	2784	09	12·3	−24	10	10·1	5·1×2·3	S0
	2835	09	17·9	−22	21	11·1	6·3×4·4	Sp
	3109	10	03·1	−26	09	10·4	14·5×3·5	Irr.
	3585	11	13·3	−26	45	10·0	2·9×1·6	E5
	3621	11	18·3	−32	49	9·9	10·0×6·5	Sc
	3923	11	51·0	−28	48	10·1	2·9×1·9	E3
	5078	13	19·8	−27	24	12·0	3·2×1·7	Sa
	5085	13	20·3	−24	26	11·9	3·4×3·0	Sb
	5061	13	18·1	−26	50	11·7	2·6×2·3	E2
	5101	13	21·8	−27	26	11·7	5·5×4·9	Sa
83	5236	13	37·0	−29	52	8·2	11·2×10·2	Sc

HYDRUS

(Abbreviation: Hyi).

A constellation in the far south – remarkably lacking in interesting objects. There are three stars above the fourth magnitude:

β 2·80
α 2·86
γ 3·24

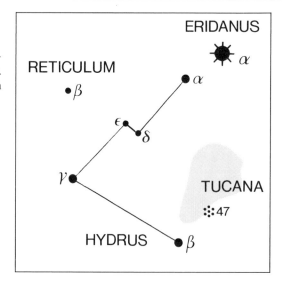

BRIGHTEST STARS

Star	R.A.			Dec.			Mag.	Abs. mag.	Spectrum	Dist.
	h	m	s	°	′	″				pc
β	00	25	46·0	−77	15	15	2·80	3·8	G1	6·3
η²	01	54	56·1	−67	38	50	4·69	0·3	G5	69
α	01	58	46·2	−61	34	12	2·86	2·6	F0	11
δ	02	21	45·0	−68	39	34	4·09	1·4	A2	35
ε	02	39	35·5	−68	16	00	4·11	−0·8	B9	91
ν	02	50	28·7	−75	04	00	4·75	−0·3	gK6	100
γ	03	47	14·5	−74	14	20	3·24	−0·4	M0	49

Also above mag. 5:

	Mag.	Abs. mag.	Spectrum	Dist.
ζ	4·84	−1·6	A2	

VARIABLE STAR

	R.A.		Dec.		Range	Type	Period, d.	Spectrum
	h	m	°	′				
VW	04	09·1	−71	18	8·4–14·4	SS Cygni	100	M

INDUS

(Abbreviation: Ind).

An undistinguished little constellation, but there are two stars above the fourth magnitude:

α 3·11
β 3·65

See chart for Grus

BRIGHTEST STARS

Star	R.A.			Dec.			Mag.	Abs. mag.	Spectrum	Dist.
	h	m	s	°	′	″				pc
α	20	37	33·9	−47	17	29	3·11	0·2	K0	38
η	20	44	02·2	−51	55	16	4·51	2·4	dA7	26
β	20	54	48·5	−58	27	15	3·65	0·2	K0	37
θ	21	19	51·1	−53	26	57	4·39	2·1	A5	28
δ	21	57	55·0	−54	59	34	4·40	1·7	F0	35
ε	22	03	21·5	−56	47	10	4·69	7·0	K5	3·4

Also above mag. 5:

	Mag.	Abs. mag.	Spectrum	Dist.
ζ	4·89	−0·5	M1	120

VARIABLE STARS

	R.A.		Dec.		Range	Type	Period, d.	Spectrum
	h	m	°	′				
S	20	56·4	−54	19	7·4–14·5	Mira	399·9	M
T	21	20·2	−45	01	7·7–9·4	Semi-reg.	320	N

DOUBLE STARS

	R.A.		Dec.		P.A.	Sep.	Mags.	
	h	m	°	′	°	″		
θ	21	19·9	−53	27	275	6·0	4·5, 7·0	
δ	21	57·9	−55	00	323	0·1	5·3, 5·3	Binary, 12 y

INDUS *(continued)*

GALAXIES

M	NGC	R.A. h	m	Dec. °	′	Mag.	Dimensions ′	Type
	7049	21	19·0	−48	34	10·7	2·8×2·2	S0
	7083	21	35·7	−63	54	11·8	4·5×2·9	Sb
	7090	21	36·5	−54	33	11·1	7·1×1·4	SBc
	7168	22	02·1	−51	45	12·6	2·0×1·6	E3
	7205	22	08·5	−57	25	11·4	4·3×2·2	Sb

LACERTA

(Abbreviation: Lac).

An obscure constellation with only one star above the fourth magnitude (α, 3·77).

BRIGHTEST STARS

Star		R.A. h	m	s	Dec. °	′	″	Mag.	Abs. mag.	Spectrum	Dist. pc
HD 211073	1H	22	13	52·5	+39	42	54	4·49	−0·2	K3	77
1		22	15	58·1	+37	44	56	4·13	−1·3	K3	100
2		22	21	01·4	+46	32	12	4·57	−1·3	B6	150
3	β	22	23	33·4	+52	13	44	4·43	0·2	G9	66
4		22	24	30·8	+49	28	35	4·57	−6·5	B9	1500
5		22	29	31·7	+47	42	25	4·36	−2·4	M0	230
6		22	30	29·1	+43	07	25	4·51	−3·0	B2	280
7	α	22	31	17·3	+50	16	57	3·77	1·4	A2	30
9		22	37	22·3	+51	32	43	4·63	1·5	A7	41
11		22	40	30·7	+44	16	35	4·46	−0·2	K3	84

Also above mag. 5:

	Mag.	Abs. mag.	Spectrum	Dist.
15	4·94	−0·4	M0	120
10	4·88	−4·8	O9	780

VARIABLE STARS

	R.A. h	m	Dec. °	′	Range	Type	Period, d.	Spectrum
S	22	29·0	+40	19	7·6–13·9	Mira	241·8	M
Z	22	40·9	+56	50	7·9–8·8	Cepheid	10·89	F–G

OPEN CLUSTERS

M	NGC	R.A. h	m	Dec. °	′	Diameter ′	Mag.	No. of stars
	7243	22	15·3	+49	53	21	6·4	40
	7296	22	28·2	+52	17	4	9·7	20

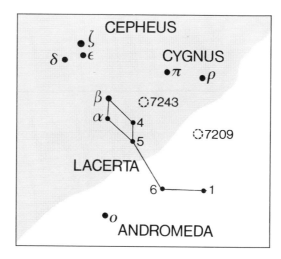

LEO

(Abbreviation: Leo).

An important Zodiacal constellation – mythologically, the Nemæan lion which was killed by Hercules. There are 12 stars above the fourth magnitude:

α 1·35
γ 1·99
β 2·14
δ 2·56
ε 2·98
θ 3·34
ζ 3·44
η 3·52
o 3·52
ρ 3·85
μ 3·88
ι 3·94

β (Denebola) was ranked as of the first magnitude by Ptolemy and others, and there is therefore a suspicion that it has faded, though the evidence is not conclusive.

BRIGHTEST STARS

Star		R.A. h	m	s	Dec. °	'	"	Mag.	Abs. mag.	Spectrum	Dist. pc	
1	κ	09	24	39·2	+26	10	56	4·46	−0·1	K2	73	
4	λ	09	31	43·1	+22	58	04	4·31	−0·3	K5	79	Alterf
14	o	09	41	09·0	+09	53	32	3·52	2·1	A5	17	Subra
17	ε	09	45	51·0	+23	46	27	2·98	−2·0	G0	95	Asad Australis
24	μ	09	52	45·8	+26	00	25	3·88	−0·1	K2	55	Rassalas
29	π	10	00	12·7	+08	02	39	4·70	−0·5	M2	110	
30	η	10	07	19·9	+16	45	45	3·52	−5·2	A0	560	
31	A	10	07	54·2	+09	59	51	4·37	−0·3	K4	81	
32	α	10	08	22·2	+11	58	02	1·35	−0·6	B7	26	Regulus
36	ζ	10	16	41·4	+23	25	02	3·44	0·6	F0	36	Adhafera
41	γ	10	19	58·3	+19	50	30	1·99	0·2	K0+G7	28	Algieba
47	ρ	10	32	48·6	+09	18	24	3·85	−5·7	B1	770	
54		10	55	37·2	+24	44	55	4·32	1·2	A1	42	
61		11	01	49·6	−02	29	04	4·74	−0·3	K5	86	
60	b	11	02	19·7	+20	10	47	4·42	0·4	A0	22	
63	χ	11	05	01·0	+07	20	10	4·63	1·3	F2	47	
68	δ	11	14	06·4	+20	31	25	2·56	1·9	A4	16	Zosma
70	θ	11	14	14·3	+15	25	46	3·34	1·4	A2	24	Chort
72		11	15	12·2	+23	05	43	4·63	−0·5	gM2	100	
74	φ	11	16	39·6	−03	39	06	4·47	1·5	A7	39	
77	σ	11	21	08·1	+06	01	45	4·05	0·2	B9	59	
78	ι	11	23	55·4	+10	31	45	3·94	1·9	F2	24	
91	υ	11	36	56·9	−00	49	26	4·30	0·2	G9	64	
93		11	47	59·0	+20	13	08	4·53	2·2	A0	22	
94	β	11	49	03·5	+14	34	19	2·14	1·7	A3	12	Denebola

Also above mag. 5:

		Mag.	Abs. mag.	Spectrum	Dist.
5	ξ	4·97	0·2	K0	84
10		5·00	0·0	K1	100
40		4·79	2·2	F6	29
58		4·84	0·0	K1	83
84	τ	4·95	−0·9	G8	150

VARIABLE STAR

	R.A. h m	Dec. ° '	Range	Type	Period, d.	Spectrum
R	09 47·6	+11 25	4·4–11·3	Mira	312·4	M

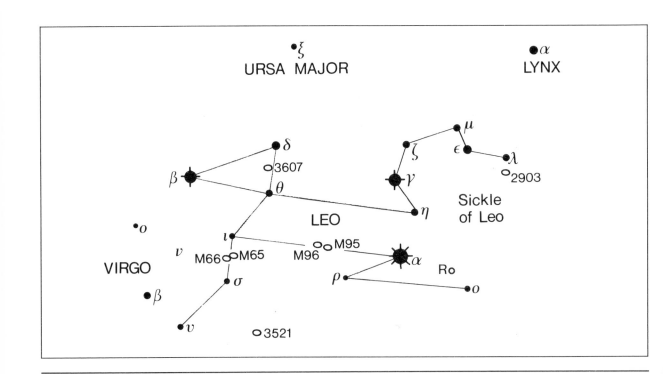

LEO *(continued)*

DOUBLE STARS

	R.A. h m	Dec. ° '	P.A. °	Sep. "	Mags.	
ω	09 28·5	+09 03	053	0·5	5·9, 6·5	Binary, 118 y
α	10 08·4	+11 58	307	176·9	1·4, 7·7	
γ	10 20·0	+19 51	AB 124	4·3	2·2, 3·5	Binary, 619 y
			AC 291	259·9	9·2	
			AD 302	333·0	9·6	
TX	10 35·0	+08 39	157	2·4	5·8, 8·5	
ι	11 23·9	+10 32	131	1·5	4·0, 6·7	Binary, 192 y
τ	11 27·9	+02 51	176	91·1	4·1, 8·0	

GALAXIES

M	NGC	R.A. h m	Dec. ° '	Mag.	Dimensions '	Type	
	3190	10 18·1	+21 50	11·0	4·6×1·8	SG	
95	3351	10 44·0	+11 42	9·7	7·4×5·1	SBb	
96	3368	10 46·8	+11 49	9·2	7·1×5·1	Sb	
	3377	10 47·7	+13 59	10·2	4·4×2·7	E5	
105	3379	10 47·8	+12 35	9·3	4·5×4·0	E1	
	3384	10 48·3	+12 38	10·0	5·9×2·6	E7	
	3412	10 50·9	+13 25	10·6	3·6×2·0	E5	
	3489	11 00·3	+13 54	10·3	3·7×2·1	E6	
	3521	11 05·8	−00 02	8·9	9·5×5·0	Sb	
	3593	11 14·6	+12 49	11·0	5·8×2·5	Sb	
	3596	11 15·1	+14 47	11·6	4·2×4·1	Sc	
	3607	11 16·9	+18 03	10·0	3·7×3·2	E1	
65	3623	11 18·9	+13 05	9·3	10·0×3·3	Sb	
	3626	11 20·1	+18 21	10·9	3·1×2·2	Sb	
66	3627	11 20·2	+12 59	9·0	8·7×4·4	Sb	
	3628	11 20·3	+13 36	9·5	14·8×3·6	Sb	Arp 317
	3630	11 20·3	+02 58	12·8	2·3×0·9	E7	
	3640	11 21·1	+03 14	10·3	4·1×3·4	E1	
	3646	11 21·7	+20 10	11·2	3·9×2·6	Sc	
	3686	11 27·7	+17 13	11·4	3·3×2·6	Sc	
	3810	11 41·0	+11 28	10·8	4·3×3·1	Sc	

LEO MINOR

(Abbreviation: LMi).

A small and obscure constellation. The only star above the fourth magnitude is 46 (mag. 3·83).

See chart for Ursa Major

BRIGHTEST STARS

Star		R.A. h m s	Dec. ° ' "	Mag.	Abs. mag.	Spectrum	Dist. pc	
10		09 34 13·3	+36 23 51	4·55	0·3	G8	71	
21		10 07 25·7	+35 14 41	4·48	2·4	A7	26	
30		10 25 54·8	+33 47 45	4·74	2·6	F0	27	
31	β	10 27 52·9	+36 42 26	4·21	1·8	G8	31	
37		10 38 43·1	+31 58 34	4·71	−2·1	G2	230	
46		10 53 18·6	+34 12 53	3·83	1·7	K0	23	Præcipua

VARIABLE STARS

	R.A. h m	Dec. ° '	Range	Type	Period, d.	Spectrum
R	09 45·6	+34 31	6·3–13·2	Mira	371·9	M
S	09 53·7	+34 55	7·9–14·3	Mira	233·8	M
RW	10 16·1	+30 34	6·9–10·1	Mira	?	N

DOUBLE STAR

	R.A. h m	Dec. ° '	P.A. °	Sep. "	Mags.	
β	10 27·9	+36 42	250	0·2	4·4, 6·1	Binary, 37·2 y

GALAXIES

M	NGC	R.A. h m	Dec. ° '	Mag.	Dimensions "	Type
	3003	09 48·6	+33 25	11·7	5·9×1·7	SBc
	3245	10 27·3	+28 30	10·8	3·2×1·9	E5
	3254	10 29·3	+29 30	11·5	5·1×1·9	Sb
	3294	10 36·3	+37 20	11·7	3·3×1·8	Sc
	3344	10 43·5	+24 55	9·9	6·9×6·5	Sc
	3414	10 51·3	+27 59	10·7	3·6×2·7	SBa
	3430	10 52·2	+32 57	11·5	3·9×2·3	Sc
	3432	10 52·5	+36 37	11·2	6·2×1·5	SB
	3486	11 00·4	+28 58	10·3	6·9×5·4	Sc

LEPUS

(Abbreviation: Lep).

An original constellation. In mythology, it was said that Orion was particularly fond of hunting hares – and so a hare was placed beside him in the sky.

Lepus is fairly distinctive. There are 8 stars above the fourth magnitude:

α 2·58
β 2·84
ε 3·19
μ 3·31
ζ 3·55
γ 3·60
η 3·71
δ 3·81

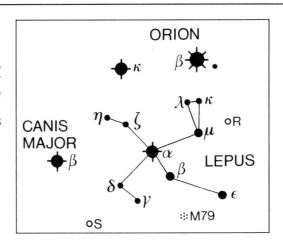

BRIGHTEST STARS

Star		R.A. h	m	s	Dec. °	′	″	Mag.	Abs. mag.	Spectrum	Dist. pc	
2	ε	05	05	27·6	−22	22	16	3·19	−0·3	K5	50	
3	ι	05	12	17·8	−11	52	09	4·45	−0·2	B8	85	
5	μ	05	12	55·8	−16	12	20	3·31	−0·8	B9	66	
4	κ	05	13	13·8	−12	56	30	4·36	−1·2	B8	130	
6	λ	05	19	34·4	−13	10	36	4·29	−4·5	B0·5	520	(Measure uncertain)
HD 34968		05	20	26·8	−21	14	23	4·71	0·6	A0	66	
9	β	05	28	14·7	−20	45	35	2·84	−2·1	G2	97	Nihal
11	α	05	32	43·7	−17	49	20	2·58	−4·7	F0	290	Arneb
13	γ	05	44	27·7	−22	26	55	3·60	4·1	F6	8·1	
14	ζ	05	46	57·2	−14	49	20	3·55	1·7	A3	24	
15	δ	05	51	19·2	−20	52	45	3·81	0·3	G8	48	
16	η	05	56	24·2	−14	10	04	3·71	1·7	F0	20	
18	θ	06	06	09·3	−14	56	07	4·67	0·8	A0	18	

Also above mag. 5:

	Mag.	Abs. mag.	Spectrum	Dist.
17	4·93		A2	

VARIABLE STARS

	R.A. h	m	Dec. °	′	Range	Type	Period, d.	Spectrum
R	04	59·6	−14	48	5·5−11·7	Mira	432·1	N
T	05	04·8	−21	54	7·4−13·5	Mira	368·1	M
RX	05	11·4	−11	51	5·0−7·0	Irreg.	–	M
S	06	05·8	−24	12	7·1−8·9	Semi-reg.	90	M

DOUBLE STARS

	R.A. h	m	Dec. °	′	P.A. °	Sep. ″	Mags.
ι	05	12·3	−11	52	337	12·7	4·5, 10·8
κ	05	13·2	−12	56	358	2·6	4·5, 7·4
β	05	28·2	−20	46	AB 330	2·5	2·8, 7·3
					AC 145	64·3	11·8
					AD 075	206·4	10·3
					AE 058	241·5	10·3
γ	05	44·5	−22	27	350	96·3	3·7, 6·3

GLOBULAR CLUSTER

M	NGC	R.A. h	m	Dec. °	′	Diameter ′	Mag.
79	1904	05	24·5	−24	33	8·7	8·0

PLANETARY NEBULA

M	NGC	R.A. h	m	Dec. °	′	Diameter ″	Mag.	Mag. of central star
	IC 418	05	27·5	−12	42	12	10·7	10·7

GALAXIES

M	NGC	R.A. h	m	Dec. °	′	Mag.	Dimensions ′	Type
	1744	05	00·0	−26	01	11·2	6·8×4·1	SBc
	1964	05	33·4	−21	57	10·8	6·2×2·5	Sb

LIBRA

(Abbreviation: Lib).

One of the Zodiacal constellations. It is, however, decidedly obscure. It was originally known as Chelæ Scorpionis (the Scorpion's Claws). Some Greek legends associate it, though rather vaguely, with Mochis, the inventor of weights and measures.

There are six stars above the fourth magnitude:

β 2·61
α² 2·75
σ 3·29
υ 3·58
τ 3·66
γ 3·91

BRIGHTEST STARS

Star		R.A. h	m	s	Dec. °	'	"	Mag.	Abs. mag.	Spectrum	Dist. pc	
9	α²	14	50	52·6	−16	02	30	2·75	1·2	A3	22	Zubenel-genubi
16		14	57	10·9	−04	20	47	4·49	1·7	F0	31	
20	σ	15	04	04·1	−25	16	55	3·29	−0·5	M4	51	Zubenal-gubi
24	ι	15	12	13·2	−19	47	30	4·54	−0·3	B9	93	
27	β	15	17	00·3	−09	22	58	2·61	−0·2	B8	37	Zubenel-chemale
38	γ	15	35	31·5	−14	47	23	3·91	1·8	G8	23	Zubenel-hakrabi
39	υ	15	37	01·4	−28	08	06	3·58	−0·3	K5	39	
40	τ	15	38	39·3	−29	46	40	3·66	−1·4	B4	100	
43	κ	15	41	56·7	−19	40	44	4·74	−0·3	K5	91	
46	θ	15	53	49·4	−16	43	46	4·15	1·8	G8	25	

σ Libræ was formerly known as γ Scorpii. β Libræ is said to be the only single star to show a greenish tint, though most people will certainly call it white!

Also above mag. 5:

		Mag.	Abs. mag.	Spectrum	Dist.
δ		4·8 (max)	0·6	A2	73
ν		4·83	−0·3	A0	120
31	ε	4·94	3·4	F5	23
42		4·96	−0·3	K4	110
48		4·88	−0·8	B9	120

VARIABLE STARS

	R.A. h	m	Dec. °	'	Range	Type	Period, d.	Spectrum
δ	15	01·1	−08	31	4·9–5·9	Algol	2·33	B
Y	15	11·7	−06	01	7·6–14·7	Mira	275·0	M
S	15	21·4	−20	23	7·5–13·0	Mira	192·4	M
RS	15	24·3	−22	55	7·0–13·0	Mira	217·7	M
RU	15	33·3	−15	20	7·2–14·2	Mira	316·6	M
RR	15	56·4	−18	18	7·8–15·0	Mira	277·0	M

DOUBLE STARS

	R.A. h	m	Dec. °	'	P.A. °	Sep. "	Mags.
μ	14	49·3	−14	09	355	1·8	5·8, 6·7
α	14	50·9	−16	02	314	231·0	2·8, 5·2
ι	15	12·2	−19	47	111	57·8	5·1, 9·4
κ	15	41·9	−19	41	279	172·0	4·7, 9·7

GLOBULAR CLUSTER

M	NGC	R.A. h	m	Dec. °	'	Diameter '	Mag.	
	5897	15	17·4	−21	01	12·6	8·6	H.IV. 19

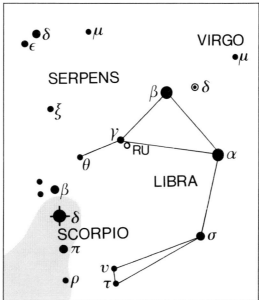

LUPUS

(Abbreviation: Lup).

An original constellation, though no definite legends seem to be attached to it. There are 11 stars above the fourth magnitude:

α	2·30
β	2·68
γ	2·78
δ	3·22
ε	3·37
ζ	3·41
η	3·41
φ¹	3·56
κ	3·72
π	3·89
χ	3·95

BRIGHTEST STARS

Star		R.A. h	m	s	Dec. °	′	″	Mag.	Abs. mag.	Spectrum	Dist. pc	
	ι	14	19	24·1	−46	03	28	3·55	−1·7	B3	110	
	τ¹	14	26	08·1	−45	13	17	4·56	−3·0	B2	310	
	τ²	14	26	10·7	−45	22	45	4·35	4·0	dF8	28	
	σ	14	32	36·8	−50	27	25	4·42	−2·5	B2	130	
	ρ	14	37	53·1	−49	25	32	4·05	−1·1	B5	110	
	α	14	41	55·7	−47	23	17	2·30	−4·4	B1	210	Men
	o	14	51	38·3	−43	34	31	4·32	−1·9	B6	180	
	β	14	58	31·8	−43	08	02	2·68	−2·5	B2	110	KeKouan
	π	15	05	07·1	−47	03	04	3·89	−1·6	B5	130	
	λ	15	08	50·5	−45	16	47	4·05	−2·3	B3	190	
	κ	15	11	56·0	−48	44	16	3·72	0·2	B9	39	
	ζ	15	12	17·0	−52	05	57	3·41	0·3	G8	42	
2	f	15	17	49·7	−30	08	55	4·34	0·1	gK0	55	
	μ	15	18	31·9	−47	51	30	4·27	−0·2	B8	77	
	δ	15	21	22·2	−40	38	51	3·22	−3·0	B2	180	
	φ¹	15	21	48·3	−36	15	41	3·56	−0·3	K5	56	
	ε	15	22	40·7	−44	41	21	3·37	−2·3	B3	140	
	φ²	15	23	09·2	−36	51	30	4·54	−2·3	B3	220	
	k	15	25	20·1	−38	44	01	4·60	0·0	A0	80	
	γ	15	35	08·4	−41	10	00	2·78	−1·7	B3	79	
	ω	15	38	03·1	−42	34	02	4·33	−0·4	M0	69	
3	ψ¹	15	39	45·9	−34	24	42	4·67	0·3	gG5	63	
	g	15	41	11·2	−44	39	40	4·64	2·8	F5	22	
4	ψ²	15	42	40·9	−34	42	37	4·75	−0·9	B6	140	
5	χ	15	50	57·4	−33	37	38	3·95	−0·3	B9	68	
	η	16	00	07·1	−38	23	48	3·41	−2·5	B2	150	
	θ	16	06	35·4	−36	48	08	4·23	−2·3	B3	200	

Also above mag. 5:

	Mag.	Abs. mag.	Spectrum	Dist.
1	4·91	−6·6	F0	1800
ν¹	5·00	3·4	F3	20
ξ	4·6 (5·1 + 5·6)		A + A	

VARIABLE STARS

	R.A. h	m	Dec. °	′	Range	Type	Period, d.	Spectrum
S	14	53·4	−46	37	7·8–13·5	Mira	342·7	S
GG	15	18·9	−40	47	5·4–6·0	Beta Lyræ	2·16	B + A

DOUBLE STARS

	R.A. h	m	Dec. °	′	P.A. °	Sep. ″	Mags.
τ¹	14	26·1	−45	13	204	148·2	4·6, 9·3
π	15	05·1	−47	03	073	1·4	4·6, 4·7
κ	15	11·9	−48	44	144	26·8	3·9, 5·8
μ	15	18·5	−47	53	AB 142	1·2	5·1, 5·2
					AC 130	23·7	7·2

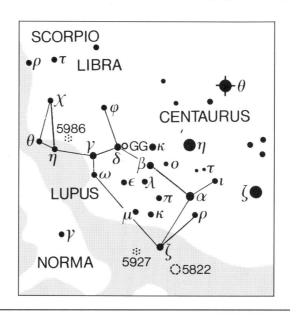

LUPUS (continued)

DOUBLE STARS (continued)

	R.A.		Dec.		P.A.	Sep.	Mags.
	h	m	°	'	°	"	
ε	15	22·7	−44	41	247	0·6	3·7, 7·2
υ	15	24·7	−39	43	038	1·4	5·4, 10·9
ξ	15	56·9	−33	58	049	10·4	5·3, 5·8
η	16	00·1	−38	24	020	15·0	3·6, 7·8

OPEN CLUSTERS

M	NGC	R.A.		Dec.		Diameter	Mag.	No. of stars
		h	m	°	'	'		
	5749	14	48·9	−54	31	8	8·8	30
	5822	15	05·2	−54	21	40	6·5	150

GLOBULAR CLUSTERS

M	NGC	R.A.		Dec.		Diameter	Mag.
		h	m	°	'	'	
	5824	15	04·0	−33	04	6·2	9·0
	5927	15	28·0	−50	40	12·0	8·3
	5986	15	46·1	−37	47	9·8	7·1

PLANETARY NEBULÆ

M	NGC	R.A.		Dec.		Diameter	Mag.	Mag. of central star
		h	m	°	'	"		
	IC 4406	14	22·4	−44	09	28	10·6	14·7
	5882	15	16·8	−45	39	7	10·5	12·0

GALAXY

M	NGC	R.A.		Dec.		Mag.	Dimensions	Type
		h	m	°	'		'	
	5643	14	32·7	−44	10	10·7	4·6×4·1	SB0

LYNX

(Abbreviation: Lyn).

A very ill-defined and obscure northern constellation. It was added to the sky by Hevelius, and has no mythological associations.

There are two stars above the fourth magnitude:

α 3·13
38 3·92

See chart for Ursa Major

BRIGHTEST STARS

Star		R.A.			Dec.			Mag.	Abs. mag.	Spectrum	Dist.
		h	m	s	°	'	"				pc
2		06	19	37·3	+59	00	39	4·48	1·4	A2	35
15		06	57	16·5	+58	25	21	4·35	1·8	G5	32
21		07	26	42·8	+49	12	42	4·64	0·3	A1	74
31		08	22	50·1	+43	11	17	4·25	−0·3	K5	76
HD 77912		09	06	31·7	+38	27	08	4·56	−3·3	G8	370
38		09	18	50·6	+36	48	09	3·92	1·7	A3	27
40	α	09	21	03·2	+34	23	33	3·13	−0·4	M0	51

Also above mag. 5:

	Mag.	Abs. mag.	Spectrum	Dist.
12	4·87	1·2	A2	
16	4·90	1·4	A2	50
24	4·99	0·0	A3	100
27	4·84	1·4	A2	60

VARIABLE STARS

	R.A.		Dec.		Range	Type	Period, d.	Spectrum
	h	m	°	'				
RR	06	26·4	+56	17	5·6–6·0	Algol	9·95	A
R	07	01·3	+55	20	7·2–14·5	Mira	378·7	S
Y	07	28·2	+45	59	7·8–10·3	Semi-reg.	110	M

DOUBLE STARS

	R.A.		Dec.		P.A.	Sep.	Mags.
	h	m	°	'	°	"	
4	06	22·1	+59	22	124	0·8	6·2, 7·7
12	06	45·2	+59	27	AB 070	1·7	5·4, 6·0
					AC 308	8·7	7·3
					AD 256	170·0	10·6
19	07	22·9	+55	17	AB 315	14·8	5·6, 6·5
					AD 003	214·9	8·9
					BC 287	74·2	10·9
38	09	18·8	+36	48	AB 229	2·7	3·9, 6·6
					BC 212	87·7	10·8
					BD 256	177·9	10·7

GALAXIES

M	NGC	R.A.		Dec.		Mag.	Dimensions	Type
		h	m	°	'		'	
	2541	08	14·7	+49	04	11·7	6·6×3·5	S
	2683	08	52·7	+33	25	9·7	9·3×2·5	Sb
	2776	09	12·2	+44	57	11·6	2·9×2·7	Sc

LYRA

(Abbreviation: Lyr).

A small constellation, but a very interesting one; it is graced by the presence of the brilliant blue Vega, as well as the prototype eclipsing binary β Lyræ, the quadruple ε Lyræ, and the 'Ring Nebula' M.57. Mythologically it represents the harp which Apollo gave to the great musician Orpheus.

There are several stars above the fourth magnitude. Of these, one (β) is the famous variable; the combined magnitude of the quadruple ε is about 3·9, though keen-sighted people can see the two main components as separated. The stars are:

α 0·03
γ 3·24
β 3·4 (max)
ε 3·9 (combined)

BRIGHTEST STARS

Star		R.A. h	m	s	Dec. °	′	″	Mag.	Abs. mag.	Spectrum	Dist. pc	
1	κ	18	19	51·5	+36	03	52	4·33	−0·1	K2	77	
3	α	18	36	56·2	+38	47	01	0·03	0·5	A0	8·1	Vega
4	ε¹	18	44	20·1	+39	40	15	4·67	1·7	A3	38	
5	ε²	18	44	22·7	+39	36	46	5·1	2·1	A5	38	
6	ζ¹	18	44	46·2	+37	36	18	4·36	1·2	A3	64	
7	ζ²	18	44	48·0	+37	35	40	5·73	2·8	F0	64	
10	β	18	50	04·6	+33	21	46	var.	−0·6	B7	92	Sheliak
12	δ²	18	54	30·0	+36	53	56	4·30v	−2·4	M4	220	
13	R	18	55	19·9	+43	56	46	var.	var.	M5	40	
14	γ	18	58	56·4	+32	41	22	3·24	−0·8	B9	59	Sulaphat
20	η	19	13	45·3	+39	08	46	4·39	−3·0	B2	270	Aladfar
21	θ	19	16	21·9	+38	08	01	4·36	−2·1	K0	170	

Also above mag. 5:

		Mag.	Abs. mag.	Spectrum	Dist.
15	λ	4·93	−2·3	K3	260

VARIABLE STARS

	R.A. h	m	Dec. °	′	Range	Type	Period, d.	Spectrum
W	18	14·9	+36	40	7·3–13·0	Mira	196·5	M
T	18	32·3	+37	00	7·8–9·6	Irreg.	–	R
β	18	50·1	+33	22	3·3–4·3	Beta Lyræ	12·94	B+A
R	18	55·3	+43	57	3·9–5·0	Semi-reg.	46	M
RR	19	25·5	+42	47	7·1–8·1	RR Lyræ	0·57	A–F

DOUBLE STARS

	R.A. h	m	Dec. °	′	P.A. °	Sep. ″	Mags.
ε	18	44·3	+39	40	AB+CD 173	207·7	4·7, 5·1
					ε¹=AB 357	2·6	5·0, 5·1
					ε²=CD 094	2·3	5·2, 5·5
ζ	18	44·8	+37	36	150	43·7	4·3, 5·9
β	18	50·1	+33	22	149	45·7	var., 8·6
δ¹	18	53·7	+36	58	020	174·6	5·6, 9·3
δ²	18	54·5	+36	54	349	86·2	4·5, 11·2
η	19	13·8	+39	09	082	28·1	4·4, 9·1

OPEN CLUSTER

M	NGC	R.A. h	m	Dec. °	′	Diameter ″	Mag.	No. of stars
	6791	19	20·7	+37	51	16	9·5	300

GLOBULAR CLUSTER

M	NGC	R.A. h	m	Dec. °	′	Diameter ″	Mag.
56	6779	19	16·6	+30	11	7·1	8·2

PLANETARY NEBULA

M	NGC	R.A. h	m	Dec. °	′	Dimensions ″	Mag.	Mag. of central star	
57	6720	18	53·6	+33	02	70×150	9·7	14·8	Ring Nebula

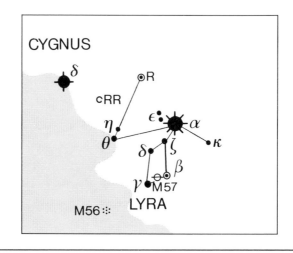

MENSA

(Abbreviation: Men).

A very dim constellation, introduced by Lacaille in 1752 under the name of Mons Mensæ (the Table Mountain). A small part of the Large Magellanic Cloud extends into it. There are no stars brighter than the fifth magnitude, and no objects to be listed. For the record, the brightest star is α (5·09). Next comes γ: R.A. 5 h 31 m 53 s·1, dec. −76° 20′ 28″, mag. 5·19, absolute mag. −0·3. Spectrum K4. Distance 130 pc. It has an optical companion of mag. 11, at P.A. 107°, distance 38″·2.

See chart for Musca

VARIABLE STARS

	R.A.		Dec.		Range	Type	Period, d.	Spectrum
	h	m	°	′				
U	04	09·6	−81	51	8·0–10·9	Mira	407	M
TY	05	26·9	−81	35	7·7–8·2	W UMa	0·46	A
TZ	05	30·2	−84	47	6·2–6·9	Algol	8·57	B

MICROSCOPIUM

(Abbreviation: Mic).

A small southern constellation. γ (4·67) is the brightest star.

See chart for Grus

BRIGHTEST STARS

Star	R.A.			Dec.			Mag.	Abs. mag.	Spectrum	Dist.
	h	m	s	°	′	″				pc
γ	21	01	17·3	−32	15	28	4·67	0·3	G4	70
ε	21	17	56·1	−32	10	21	4·71	2·1	A2p	

γ was formerly known as 1 PsA and ε as 4 PsA.

Also above mag. 5:

	Mag.	Abs. mag.	Spectrum	Dist.
α	4·90	0·3	G6	73
θ[1]	4·82	−0·6	A2p	

VARIABLE STARS

	R.A.		Dec.		Range	Type	Period, d.	Spectrum
	h	m	°	′				
T	20	27·9	−28	16	7·7–9·6	Semi-reg.	344	M
U	20	29·2	−40	25	7·0–14·4	Mira	334·2	M
S	21	26·7	−29	51	7·8–14·3	Mira	208·9	M

DOUBLE STARS

	R.A.		Dec.		P.A.	Sep.	Mags.
	h	m	°	′	°	″	
α	20	50·0	−33	47	166	20·5	5·0, 10·0
θ[2]	21	24·4	−41	00	AB 267	0·5	6·4, 7·0
					AC 066	78·4	10·5

GALAXIES

M	NGC	R.A.		Dec.		Mag.	Dimensions	Type
		h	m	°	′		′	
	6923	20	31·7	−30	50	12·1	2·5 × 1·4	Sb
	6925	20	34·3	−31	59	11·3	4·1 × 1·6	Sb

MONOCEROS

(Abbreviation: Mon).

Not an ancient constellation, and though it represents the fabled unicorn there are no definite legends attached to it. It is crossed by the Milky Way, and the general area is decidedly rich. There are 4 stars above the fourth magnitude; of these, β is a double, and the magnitude is combined. The stars are:

β 3·7   30 3·90   α 3·93   γ 3·98

BRIGHTEST STARS

Star		R.A.			Dec.			Mag.	Abs. mag.	Spectrum	Dist.	
		h	m	s	°	′	″				pc	
5	γ	06	14	51·3	−06	16	29	3·98	−0·2	K3	66	
8	ε	06	23	46·0	+04	35	34	4·33	0·3	A5	54	
11	β	06	28	48·9	−07	01	58	3·7	−2·6 / −1·7	B2 / B3	220	
13		06	32	54·2	+07	19	58	4·50	−5·2	A0	860	
15	S	06	40	58·6	+09	53	45	var.	−5·5	O7	920	
18		06	47	51·6	+02	24	44	4·47	0·2	K0	59	
22	δ	07	11	51·8	−00	29	34	4·15	0·0	A0	64	
26	α	07	41	14·8	−09	33	04	3·93	0·2	K0	54	
28		08	01	13·2	−01	23	33	4·68	−0·3	K4	88	
29	ζ	08	08	35·6	−02	59	02	4·34	−4·5	G2	560	
30		08	25	39·5	−03	54	23	3·90	0·6	A0	46	
31		08	43	40·4	−07	14	01	4·62	−4·5	G2	670	= F Hydræ

MONOCEROS (continued)

Also above mag. 5:

	Mag.	Abs. mag.	Spectrum	Dist.
17	4·77	−0·3	K4	100
19	4·99	−3·5	B1	500
27	4·93	−0·1	K2	97

VARIABLE STARS

	R.A. h m	Dec. ° ′	Range	Type	Period, d.	Spectrum
V	06 22·7	−02 12	6·0–13·7	Mira	333·8	M
T	06 25·2	+07 05	6·0–6·6	Cepheid	27·02	F–K
S	06 41·0	+09 54	4–5?	Irreg.	–	07
X	06 57·2	−09 04	6·9–10·0	Semi-reg.	156	M
RY	07 06·9	−07 33	7·7–9·2	Semi-reg.	466	N
U	07 30·8	−09 47	6·1–8·1	RV Tauri	92·3	F–K

DOUBLE STARS

	R.A. h m	Dec. ° ′	P.A. °	Sep. ″	Mags.
ε	06 23·8	+04 36	027	13·4	4·5, 6·5
β	06 28·8	−07 02	AB 132	7·3	4·7, 5·2
			AC 124	10·0	6·1
			AD 056	25·9	12·2
S (15)	06 41·0	09 54	AB 213	2·8	4·7v, 7·5
			AC 013	16·6	9·8
			AD 308	41·3	9·6
			AE 139	73·9	9·9
			AF 222	156·0	7·7
			AK 056	105·6	8·1

OPEN CLUSTERS

M	NGC	R.A. h m	Dec. ° ′	Diameter ′	Mag.	No. of stars	
	2215	06 21·0	−07 17	11	8·4	40	
	2244	06 32·4	+04 52	24	4·8	100	In Rosette Neb.
	2251	06 34·7	+08 22	10	7·3	30	
	2286	06 47·6	−03 10	15	7·5	50	
	2301	06 51·8	+00 28	12	6·0	80	
50	2323	07 03·2	−08 20	16	5·9	80	
	2335	07 06·6	−10 05	12	7·2	35	
	2343	07 08·3	−10 39	7	6·7	20	
	2353	07 14·6	−10 18	20	7·1	30	
	2506	08 00·2	−10 47	7	7·6	150	

NEBULÆ

M	NGC	R.A. h m	Dec. ° ′	Dimensions ′	Mag. of illuminating star
	2149	06 03·5	−09 44	3×2	9
	2237–9	06 32·3	+05 03	80×60	Rosette Nebula
	2261	06 39·2	+08 44	2×1	10v R Monocerotis
	2264	06 40·9	+09 54	60×30	4v S Monocerotis. Cone Nebula

MUSCA AUSTRALIS

(Abbreviation: Mus).

This small but fairly distinctive constellation, not far from Crux, is generally known simply as 'Musca'. There are five stars above the fourth magnitude:

α 2·69v
β 3·05
δ 3·62
λ 3·64
γ 3·87

α is actually variable over a very small range (2·66–2·73).

BRIGHTEST STARS

Star	R.A. h m s	Dec. ° ′ ″	Mag.	Abs. mag.	Spectrum	Dist. pc
λ	11 45 36·4	−66 43 43	3·64	2·1	A5	16
μ	11 48 14·4	−66 48 53	4·72	−0·5	gM2	110
ε	12 17 34·2	−67 57 38	4·11	2·2	gM6	12
γ	12 32 28·1	−72 07 58	3·87	−1·1	B5	59
α	12 37 11·0	−69 08 07	2·69v	−2·3	B3	100
β	12 46 16·9	−68 06 29	3·05	−1·7	B3	89
δ	13 02 16·3	−71 32 56	3·62	−0·1	K2	54

Also above mag. 5:

	Mag.	Abs. mag.	Spectrum	Dist.
η	4·80	−0·2	B8	100

VARIABLE STARS

	R.A. h m	Dec. ° ′	Range	Type	Period, d.	Spectrum
S	12 12·8	−70 09	5·9–6·4	Cepheid	9·66	F
B0	12 34·9	−67 45	6·0–6·7	Irreg.	–	M
R	12 42·1	−69 24	5·9–6·7	Cepheid	7·48	F
T	13 21·2	−74 27	7·0–9·0	Semi-reg.	93	N

MUSCA AUSTRALIS (continued)

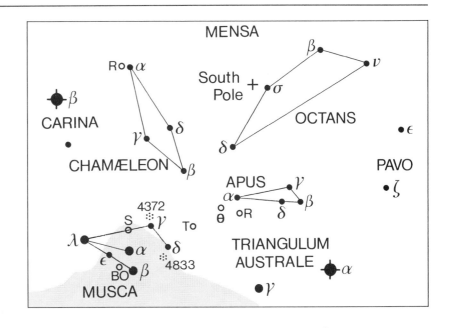

DOUBLE STARS

	R.A.		Dec.		P.A.	Sep.	Mags.
	h	m	°	′	°	″	
ζ²	12	22·1	−67	31	130	32·4	5·2, 10·6
α	12	37·2	−69	08	316	29·6	2·7, 12·8
β	12	46·3	−68	06	014	1·4	3·7, 4·0
θ	13	08·1	−65	18	187	5·3	5·7, 7·3

OPEN CLUSTER

M	NGC	R.A.		Dec.		Diameter	Mag.	No. of stars
		h	m	°	′	′		
	4463	12	30·0	−64	48	5	7·2	30

GLOBULAR CLUSTERS

M	NGC	R.A.		Dec.		Diameter	Mag.
		h	m	°	′	′	
	4372	12	25·8	−72	40	18·6	7·8
	4833	12	59·6	−70	53	13·5	7·3

PLANETARY NEBULÆ

M	NGC	R.A.		Dec.		Diameter	Mag.	Mag. of central star
		h	m	°	′	″		
	IC 4191	13	08·8	−67	39	5	12·0	—
	5189	13	33·5	−65	59	153	10	14 Gum 47

NORMA

(Abbreviation: Nor).

An obscure constellation, once known as Quadra Euclidis (Euclid's Quadrant); it was added to the sky by Lacaille. It contains no star above the fourth magnitude.

See charts for Ara and Lupus

BRIGHTEST STARS

Star	R.A.			Dec.			Mag.	Abs. mag.	Spectrum	Dist.
	h	m	s	°	′	″				pc
η	16	03	12·6	−49	13	47	4·65	0·3	gG4	64
ι¹	16	03	31·9	−57	46	31	4·63	2·1	A5	30
δ	16	06	29·3	−45	10	24	4·72	0·6	A0	18
γ²	16	19	50·3	−50	09	20	4·02	0·3	G8	40
ε	16	27	10·9	−47	33	18	4·47	−1·7	B3	150

Also above mag. 5:

	Mag.	Abs. mag.	Spectrum	Dist.
κ	4·94	0·3	G4	61
γ¹	4·99	−6·3	F8	1700

VARIABLE STARS

	R.A.		Dec.		Range	Type	Period, d.	Spectrum
	h	m	°	′				
R	15	36·0	−49	30	6·5–13·9	Mira	492·7	M
T	15	44·1	−54	59	6·2–13·6	Mira	242·6	M
S	16	18·9	−57	54	6·1–6·8	Cepheid	9·75	F–G

NORMA (continued)

DOUBLE STARS

	R.A.		Dec.		P.A.	Sep.	Mags.	
	h	m	°	′	°	″		
ι¹	16	03·5	−57	47	100	0·2	5·3, 5·5	Binary, 26·9 y
ε	16	27·2	−47	33	335	22·8	4·8, 7·5	

OPEN CLUSTERS

M	NGC	R.A.		Dec.		Diameter	Mag.	No. of stars	
		h	m	°	′	′			
	5925	15	27·7	−54	31	15	8·4	120	
	5999	15	52·2	−56	28	5	9·0	40	
	6031	16	07·6	−54	04	2	8·5	20	
	6067	16	13·2	−54	13	13	5·6	100	
	6087	16	18·9	−57	54	12	5·4	40	S Normæ cluster
	H.10	16	19·9	−54	59	30	−	30	
	6134	16	27·7	−49	09	7	7·2	−	
	6152	16	32·7	−52	37	30	8·1	70	
	6167	16	34·4	−49	36	8	6·7	−	

PLANETARY NEBULA

M	NGC	R.A.		Dec.		Diameter	Mag.	Mag. of central star
		h	m	°	′	″		
	Sp-1	15	51·7	−51	31	76	13·6	13·8

OCTANS

(Abbreviation: Oct).

The south polar constellation. It is very obscure, and contains only one star above the fourth magnitude: ν (3·76).

See chart for Musca

BRIGHTEST STARS

Star	R.A.			Dec.			Mag.	Abs. mag.	Spectrum	Dist.
	h	m	s	°	′	″				pc
δ	14	26	55·0	−83	40	04	4·32	−0·1	gK2	60
ν	21	41	28·6	−77	23	24	3·76	0·2	K0	32
β	22	46	03·1	−81	22	54	4·15	2·6	dF0	20

The south polar star is σ Octantis, R.A. 20 h 15·1 m, dec. −89° 08 (in 1983). The apparent magnitude is 5·46; spectrum F0; absolute magnitude 2·7; distance 37 parsecs. The polar distance was 45′ in 1900, but will have increased to 1° before the end of the century.

Also above mag. 5:

	Mag.	Abs. mag.	Spectrum	Dist.
θ	4·78	−0·1	K2	78

VARIABLE STARS

	R.A.		Dec.		Range	Type	Period, d.	Spectrum
	h	m	°	′				
R	05	26·1	−86	23	6·4−13·2	Mira	405·6	M
U	13	24·5	−84	13	7·1−14·1	Mira	302·6	M
S	18	08·7	−86	48	7·3−14·0	Mira	258·9	M
ε	22	20·0	−80	26	4·9−5·4	Semi-reg.	55	M

DOUBLE STARS

	R.A.		Dec.		P.A.	Sep.	Mags.
	h	m	°	′	°	″	
ι	12	55·0	−85	07	230	0·6	6·0, 6·5
μ²	20	41·7	−75	21	017	17·4	7·1, 7·6
λ	21	50·9	−82	43	070	3·1	5·4, 7·7

OPHIUCHUS

(Abbreviation: Oph).

This constellation is also sometimes known as Serpentarius. It commemorates Æsculapius, son of Apollo and Coronis, who became so skilled in medicine that he was even able to restore the dead to life. To avoid depopulation of the Underworld, Jupiter reluctantly disposed of Æsculapius with a thunderbolt, but relented sufficiently to place him in the sky!

Ophiuchus is a very large constellation, with 13 stars above the fourth

BRIGHTEST STARS

Star		R.A.			Dec.			Mag.	Abs. mag.	Spectrum	Dist.	
		h	m	s	°	′	″				pc	
1	δ	16	14	20·6	−03	41	39	2·74	−0·5	M1	43	Yed Prior
2	ε	16	18	19·1	−04	41	33	3·24	0·3	G8	32	Yed Post
4	ψ	16	24	06·0	−20	02	15	4·50	0·2	K0	72	
5	ρ	16	25	34·9	−23	26	46	4·59	−2·5	B2	230	
7	χ	16	27	01·3	−18	27	23	4v	−2·5v	B2p	150	
3	υ	16	27	48·1	−08	22	18	4·63	1·4	A2	21	
10	λ	16	30	54·7	+01	59	02	3·82	1·2	A1	33	Marfik
8	φ	16	31	08·2	−16	36	46	4·28	0·3	G8	63	
9	ω	16	32	08·0	−21	27	59	4·45	1·7	A7	19	
13	ζ	16	37	09·4	−10	34	02	2·56	−4·4	O9·5	170	Han
20		16	49	49·9	−10	46	59	4·65	0·7	F6	56	
25	ι	16	54	00·4	+10	09	55	4·38	−0·6	B8	99	
27	κ	16	57	40·0	+09	22	30	3·20	−0·1	K2	36	
35	η	17	10	22·5	−15	43	30	2·43	1·4	A2	18	Sabik
36		17	15	20·7	−26	36	04	4·31	6·4	K0	5·4	
41		17	16	36·5	−00	26	43	4·73	−0·1	K2	93	
40	ξ	17	21	00·0	−21	06	46	4·39	3·0	F2	19	

OPHIUCHUS (continued)

magnitude:

α 2·08
η 2·43
ζ 2·56
δ 2·74
β 2·77
κ 3·20
ε 3·24
θ 3·27
ν 3·34
72 3·73
γ 3·75
λ 3·82
67 3·97

BRIGHTEST STARS (continued)

Star		R.A. h	m	s	Dec. °	′	″	Mag.	Abs. mag.	Spectrum	Dist. pc	
42	θ	17	22	00·4	−24	59	58	3·27	−3·0	B2	180	
44	b	17	26	22·1	−24	10	31	4·17	2·5	A9	25	
49	σ	17	26	30·7	+04	08	25	4·34	−2·3	K3	190	
47		17	26	37·7	−05	05	12	4·54	3·1	F3	21	
45	d	17	27	21·1	−29	52	01	4·29	2·1	F5	27	
55	α	17	34	55·9	+12	33	36	2·08	0·3	A5	19	Rasalhague
57	μ	17	37	50·5	−08	07	08	4·62	−0·2	B8	73	
60	β	17	43	28·2	+04	34	02	2·77	−0·1	K2	37	Cheleb
62	γ	17	47	53·4	+02	42	26	3·75	0·6	A0	35	
64	ν	17	59	01·4	−09	46	25	3·34	0·2	K0	42	
66		18	00	15·5	+04	22	07	4·64	−2·5	B2	250	
67		18	00	38·5	+02	55	53	3·97	−5·7	B5	740	
68		18	01	45·0	+01	18	19	4·45	1·2	A1	45	
70	p	18	05	27·2	+02	29	58	4·03	5·7	K0	5·1	
71		18	07	18·2	+08	44	02	4·64	1·8	G8	34	
72		18	07	20·8	+09	33	50	3·73	1·9	A4	28	

Also above mag. 5:

		Mag.	Abs. mag.	Spectrum	Dist.
30		4·82	−0·3	K4	93
58		4·87	3·4	F5	20
69	τ	4·79	2·6	F0	22
74		4·86	0·3	G8	82

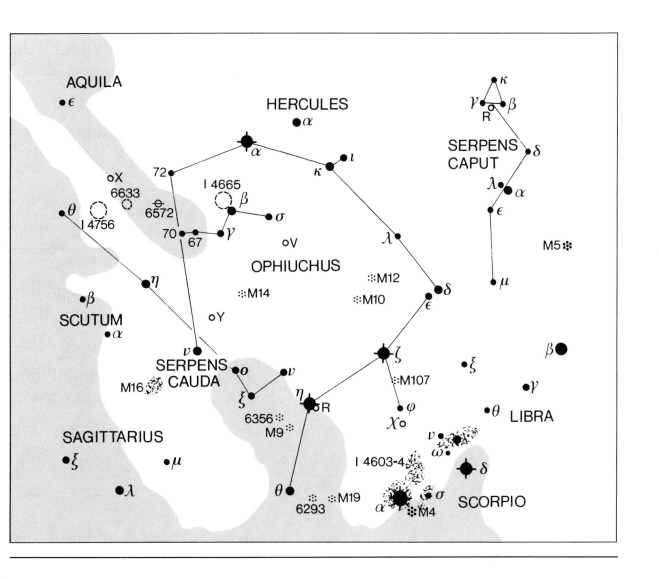

OPHIUCHUS *(continued)*

VARIABLE STARS

	R.A. h	m	Dec. °	′	Range	Type	Period, d.	Spectrum	
V	16	26·7	−12	26	7·3–11·6	Mira	298·0	N	
χ	16	27·0	−18	27	4·2–5·0	Irreg.	–	B	
SS	16	57·9	−02	46	7·8–14·5	Mira	180·0	M	
R	17	07·8	−16	06	7·0–13·8	Mira	302·6	M	
U	17	16·5	+01	13	5·9–6·6	Algol	1·68	B+B	
Z	17	19·5	+01	31	7·6–14·0	Mira	348·7	K–M	
RS	17	50·2	−06	43	5·3–12·3	Recurrent nova	–	O+M	(Outbursts 1933, 1958, 1987)
Y	17	52·6	−06	09	5·9–6·4	Cepheid	17·12	F–G	
RY	18	16·6	+03	42	7·5–13·8	Mira	150·5	M	
X	18	38·3	+08	50	5·9–9·2	Mira	334·4	M+K	

DOUBLE STARS

	R.A. h	m	Dec. °	′	P.A. °	Sep. ″	Mags.	
ρ	16	25·6	−23	27	344	3·1	5·3, 6·0	
υ	16	27·8	−08	22	095	1·0	4·6, 7·8	
λ	16	30·9	+01	59	AB 022	1·5	4·2, 5·2	Binary, 129·9 y
					AB+C 170	119·2	11·1	
					AD 246	313·8	9·9	
φ	16	31·1	−16	37	037	34·4	4·3, 12·8	
19	16	47·2	+02	04	089	23·4	6·1, 9·4	
η	17	10·4	−15	43	247	0·5	3·0, 3·5	Binary, 84·3 y
36	17	15·3	−26	36	150	4·7	5·1, 5·1	Binary, 549 y
41	17	16·6	−00	27	346	1·0	4·8, 7·8	
53	17	34·6	+09	35	191	41·2	5·8, 8·5	
τ	18	03·1	−08	11	AB 280	1·8	5·2, 5·9	Binary, 280 y
					AC 127	100·3	9·3	
70	18	05·5	+02	30	224	1·5	4·2, 6·0	Binary, 88·1 y
73	18	09·6	+04	00	300	0·4	6·1, 7·0	Binary, 270 y
X	18	38·3	+08	50	150	0·4	var., 8·6	Binary, 485 y

OPEN CLUSTERS

M	NGC	R.A. h	m	Dec. °	′	Diameter ′	Mag.	No. of stars
	IC 4665	17	46·3	+05	43	41	4·2	30
	6633	18	27·7	+06	34	27	4·6	30

GLOBULAR CLUSTERS

M	NGC	R.A. h	m	Dec. °	′	Diameter ′	Mag.
107	6171	16	32·5	−13	03	10·0	8·1
12	6218	16	47·2	−01	57	14·5	6·6
10	6254	16	57·1	−04	06	15·1	6·6
62	6266	17	01·2	−30	07	14·1	6·6
19	6273	17	02·6	−26	16	13·5	7·1
	6304	17	14·5	−29	28	6·8	8·4
	6316	17	16·6	−28	08	4·9	9·0
9	6333	17	19·2	−18	31	9·3	7·9
	6356	17	23·6	−17	49	7·2	8·4
	6355	17	24·0	−26	21	5·0	9·6
14	6402	17	37·6	−03	15	11·7	7·6
	6401	17	38·6	−23	55	5·6	9·5

PLANETARY NEBULA

M	NGC	R.A. h	m	Dec. °	′	Dimensions ″	Mag.	Mag. of central star
	6309	17	14·1	−12	55	14×66	10·8	14·4

GALAXY

M	NGC	R.A. h	m	Dec. °	′	Mag.	Dimensions ′	Type
	6384	17	32·4	+07	04	10·6	6·0×4·3	Sb

ORION

(Abbreviation: Ori).

One of the most magnificent constellations in the sky; it represents the mythological hunter who boasted that he could kill any creature on earth, but who was fatally stung by a scorpion. The two leading stars are Rigel, which is actually variable over a very small range (0·08 to 0·20) and the red variable Betelgeux – a name which may also be spelled Betelgeuse or Betelgeuze. The gaseous nebula M.42, in the Sword, is the most famous example of its type, and is easily visible with the naked eye. Altogether there are fifteen stars above the fourth magnitude:

β	0·12v
α	0·5v
γ	1·64
ε	1·70
ζ	1·77
κ	2·06
δ	2·23v
ι	2·76
π^3	3·19
η	3·36v
λ	3·39
τ	3·60
π^4	3·69
π^5	3·72v
σ	3·73

BRIGHTEST STARS

Star		R.A. h	m	s	Dec. °	′	″	Mag.	Abs. mag.	Spectrum	Dist. pc	
1	π^3	04	49	50·3	+06	57	41	3·19	3·8	F6	7·7	
2	π^2	04	50	36·6	+08	54	01	4·36	0·6	A0	55	
3	π^4	04	51	12·3	+05	36	18	3·69	−3·6	B2	280	
4	o^1	04	52	31·9	+14	15	02	4·74		M3	32	
8	π^5	04	54	15·0	+02	26	26	3·72v	−3·6	B2	290	
7	π^1	04	54	53·7	+10	09	03	4·65	0·9	A0p	83	
9	o^2	04	56	22·2	+13	30	52	4·07	−0·1	K2	68	
10	π^6	04	58	32·8	+01	42	51	4·47	−2·2	K2	190	
11		05	04	34·1	+15	24	14	4·68	0·0	A0p	26	
17	ρ	05	13	17·4	+02	51	40	4·46	−0·2	K3	86	
19	β	05	14	32·2	−08	12	06	0·12v	−7·1	B8	280	Rigel
20	τ	05	17	36·3	−06	50	40	3·60	−2·2	B5	130	
22	o	05	21	45·7	−00	22	57	4·73	−3·0	B2	340	
29	e	05	23	56·7	−07	48	29	4·14	0·3	G8	58	
28	η	05	24	28·6	−02	23	50	3·36	−3·5	B1	230	Algjebbah
24	γ	05	25	07·8	+06	20	59	1·64	−3·6	B2	110	Bellatrix
30	ψ	05	26	50·2	+03	05	44	4·59	−3·0	B2	330	
31	Cl	05	29	43·9	−01	05	32	4·71	−0·3	K5	93	
32	A	05	30	47·0	+05	56	53	4·20	−1·6	B5	150	
36	υ	05	31	55·8	−07	18	05	4·62	−4·1	B0	560	Thabit
34	δ	05	32	00·3	−00	17	57	2·23v	−6·1	09·5	720	Mintaka
37	ϕ^1	05	34	49·2	+09	29	22	4·41	−4·6	B0	570	
39	λ	05	35	08·2	+09	56	02	3·39	−5·1	08	550	Heka
42	c	05	35	23·1	−04	50	18	4·59	−3·6	B2	135	(near θ)
44	ι	05	35	25·9	−05	54	36	2·76	−6·0	09	570	Hatysa
46	ε	05	36	12·7	−01	12	07	1·70	−6·2	B0	370	Alnilam
40	ϕ^2	05	36	54·3	+09	17	27	4·09	0·2	K0	60	
48	σ	05	38	44·7	−02	36	00	3·73	−4·4	09·5	550	
47	ω	05	39	11·0	+04	07	17	4·57	−2·9	B3	310	
50	ζ	05	40	45·5	−01	56	34	1·77	−5·9	09·5	340	Alnitak
53	κ	05	47	45·3	−09	40	11	2·06	−6·9	B0·5	650	Saiph
54	χ^1	05	54	22·9	+20	16	34	4·41	4·4	G0	9·9	
58	α	05	55	10·2	+07	24	26	var.	−5·6v	M2	95	Betelgeux
HD 40657		06	00	03·3	−03	04	27	4·53	−0·1	K2	80	
61	μ	06	02	22·9	+09	38	51	4·12	1·2	A0	37	
62	χ^2	06	03	55·1	+20	08	18	4·63	−6·8	B2	1000	
67	ν	06	07	34·2	+14	46	06	4·42	−1·7	B3	170	
70	ξ	06	11	56·3	+14	12	31	4·48	−1·7	B3	170	

δ has long been classed as a variable, but is in fact an eclipsing star with a range of only 2·20 to 2·35 (period 5·73 days). η is also an eclipsing star with a range of only 0 m·2. π^5 is an ellipsoidal variable with a range of 0·1 mag.

Also above mag. 5:

		Mag.	Abs. mag.	Spectrum	Dist.
15		4·82	1·9	F2	41
23	m	5·00	−3·5	B1	470
25		4·95	−3·5	B1	490
49	d	4·80	1·0	A4	57
56		4·78	−2·2	K2	240
51		4·91	0·0	K1	87
69		4·95	−1·1	B5	160

VARIABLE STARS

	R.A. h	m	Dec. °	′	Range	Type	Period, d.	Spectrum
W	05	05·4	+01	11	5·9–7·7	Semi-reg.	212	N
S	05	29·0	−04	42	7·5–13·5	Mira	419·2	M
CK	05	30·3	+04	12	5·9–7·1	Semi-reg.	120	K
α	05	52	+07	24	0·1–0·9	Semi-reg.	2110	M
U	05	55·8	+20	10	4·8–12·6	Mira	372·4	M

DOUBLE STARS

	R.A. h	m	Dec. °	′	P.A. °	Sep. ″	Mags.	
π^3	04	49·8	+06	58	138	94·6	3·2, 8·7	
β	05	14·5	−08	12	202	9·5	0·1, 6·8	
ρ	05	15·3	+02	54	064	7·0	4·5, 8·3	
η	05	24·5	−02	24	AB 080	1·5	3·8, 4·8	
					AC 051	115·1	9·4	
δ	05	32·0	−00	18	359	52·6	2·2v, 6·3	
λ	05	35·1	+09	56	043	4·4	3·6, 5·5	
σ	05	38·7	−02	36	AB 137	0·2	4·0, 6·0	Binary, 170 y
					AB+C 238	11·4	10·3	
					AB+D 084	12·9	7·5	
					AB+E 061	42·6	6·5	

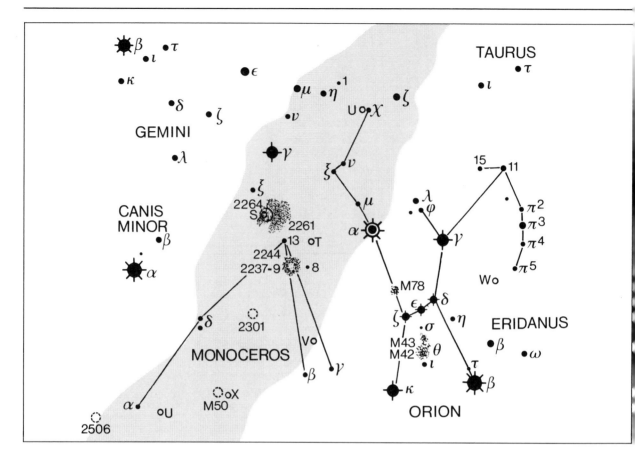

ORION *(continued)*

DOUBLE STARS *(continued)*

	R.A. h m	Dec. ° ′	P.A. °	Sep. ″	Mags.	
θ	05 35·3	−05 23	AB 031	8·8	6·7, 7·9	
			AC 132	12·8	5·1	
			AD 096	21·5	6·7	
ι	05 35·4	−05 55	141	11·3	2·8, 6·9	
ζ	05 40·8	−01 57	AB 162	2·4	1·9, 4·0	Binary, 1509 y
			AC 010	57·6	9·9	
μ	06 02·4	+09 39	023	0·4	4·4, 6·0	

OPEN CLUSTERS

M	NGC	R.A. h m	Dec. ° ′	Diameter ′	Mag.	No. of stars
	1981	05 35·2	−04 26	25	4·6	20
	2112	05 53·9	+00 24	11	9·1	5
	2175	06 09·8	+20 19	18	6·8	60
	2186	06 12·2	+05 27	4	8·7	30

NEBULÆ

M	NGC	R.A. h m	Dec. ° ′	Dimensions ′	Mag. of illuminating star	
42	1976	05 35·4	−05 27	66×60	5	Great Nebula
43	1982	05 35·6	−05 16	20×15	7	Extension of M.42
78	2068	05 46·7	+00 03	8×6	10	Nebula is mag. 8
	IC 434	05 41·0	−02 24	60×10	2	(ζ). Behind Horse's Head dark nebula. Barnard 33

PAVO

(Abbreviation: Pav).

One of the 'Southern Birds'. The brightest star, α, is somewhat isolated from the main pattern; the most celebrated object is κ, often called a Cepheid even though it is, strictly speaking, a W Virginis type variable.

Pavo includes 6 stars above the fourth magnitude:

α 1·94
β 3·42
δ 3·56
η 3·62
κ 3·9 (max)
ε 3·96

BRIGHTEST STARS

Star	R.A.			Dec.			Mag.	Abs. mag.	Spectrum	Dist.
	h	m	s	°	′	″				pc
η	17	45	43·8	−64	43	25	3·62	0·0	K1	45
π	18	08	34·6	−63	40	06	4·35	1·6	A	17
ξ	18	23	13·3	−61	29	38	4·36	−0·5	M1	94
ν	18	31	22·2	−62	16	42	4·64	−1·2	B8	150
ζ	18	43	02·0	−71	25	42	4·01	−0·1	K2	66
λ	18	52	12·8	−62	11	16	4·22v	−3·5	B1	350
κ	18	56	56·9	−67	14	01	var.	3·4v	F5v	23
ε	20	00	35·4	−72	54	38	3·96	0·6	A0	41
δ	20	08	43·3	−66	10	56	3·56	4·8	G5	5·7
α	20	25	38·7	−56	44	06	1·94	−2·3	B3	71
β	20	44	57·4	−66	12	12	3·42	1·2	A5	28
γ	21	26	26·6	−65	21	59	4·22	4·5	F6	86

Also above mag. 5:

	Mag.	Abs. mag.	Spectrum	Dist.
φ¹	4·76		F0	16
ρ	4·88	3·4	F5	24

VARIABLE STARS

	R.A.		Dec.		Range	Type	Period, d.	Spectrum
	h	m	°	′				
R	18	12·9	−63	37	7·5–13·8	Mira	229·8	M
λ	18	52·2	−62	11	3·4–4·3	Irreg.	−	B
κ	18	56·9	−67	14	3·9–4·7	W Virginis	9·09	F
T	19	50·7	−71	46	7·0–14·0	Mira	244·0	M
S	19	55·2	−59	12	6·6–10·4	Semi-reg.	386	M
Y	21	24·3	−69	44	5·7–8·5	Semi-reg.	233	N
SX	21	28·7	−69	30	5·4–6·0	Semi-reg.	50	M

DOUBLE STAR

	R.A.		Dec.		P.A.	Sep.	Mags.
	h	m	°	′	°	″	
ξ	18	23·2	−61	30	154	3·3	4·4, 8·6

GLOBULAR CLUSTER

M	NGC	R.A.		Dec.		Diameter	Mag.
		h	m	°	′	′	
	6752	19	10·9	−59	59	20·4	5·4

GALAXIES

M	NGC	R.A.		Dec.		Mag.	Dimensions	Type
		h	m	°	′		′	
	IC 4662	17	47·1	−64	38	11·4	2·2 × 1·4	Irreg.
	6684	18	49·0	−65	11	10·4	3·7 × 2·7	SB0
	6744	19	09·8	−63	51	9·0	15·5 × 10·2	SBb
	6753	19	11·4	−57	03	11·9	2·5 × 2·2	Sb

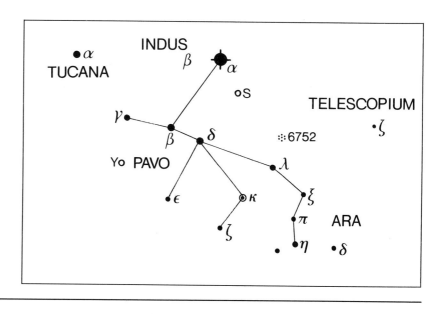

PEGASUS

(Abbreviation: Peg).

One of the most distinctive of the northern constellations. It commemorates the flying horse which the hero Bellerophon rode during an expedition to slay the fire-breathing Chimæra. The main stars of Pegasus make up a square; three of these are α, β and γ. The fourth is Alpheratz, which used to be included in Pegasus as δ Pegasi, but has been officially – and, frankly, illogically – transferred to Andromeda, as α Andromedæ. In Pegasus, excluding Alpheratz, there are nine stars above the fourth magnitude:

ε 2·38
β 2·4 (max)
α 2·49
γ 2·83v
η 2·94
ζ 3·40.
μ 3·48
θ 3·53
ι 3·76

BRIGHTEST STARS

Star		R.A.			Dec.			Mag.	Abs. mag.	Spectrum	Dist.	
		h	m	s	°	′	″				pc	
1		21	22	05·0	+19	48	16	4·08	0·0	K1	63	
2		21	29	56·8	+23	38	20	4·57	−0·5	M1	97	
8	ε	21	44	11·0	+09	52	30	2·38	−4·4	K2	160	Enif
9		21	44	30·5	+17	21	00	4·34	−4·5	G5	490	
10	κ	21	44	38·5	+25	38	42	4·13	2·1	F5	27	
24	ι	22	07	00·5	+25	20	42	3·76	3·4	F5	13	
29	π	22	09	59·1	+33	10	42	4·29	−0·6	F5	96	
26	θ	22	10	11·8	+06	11	52	3·53	1·4	A2	25	Biham
42	ζ	22	41	27·6	+10	49	53	3·40	0·0	B8·5	48	Homan
44	η	22	43	00·0	+30	13	17	2·94	−0·9	G2	53	Matar
47	λ	22	46	31·7	+23	33	56	3·95	−0·9	G8	33	
46	ξ	22	46	41·4	+12	10	22	4·19	3·8	F7	14	
48	μ	22	50	00·0	+24	36	06	3·48	0·2	K0	45	Sadalbari
53	β	23	03	46·3	+28	04	58	var.	−1·4v	M2	54	Scheat
54	α	23	04	45·5	+15	12	19	2·49	0·2	B9	31	Markab
55		23	07	00·1	+09	24	34	4·52	−0·5	M2	100	
62	τ	23	20	38·1	+23	44	25	4·60	1·2	A5	47	
68	υ	23	25	22·7	+23	24	15	4·40	2·4	F8	22	
70	q	23	29	09·1	+12	45	38	4·55	0·3	G8	70	
84	ψ	23	57	45·4	+25	08	29	4·66	−0·5	M3	110	
88	γ	00	13	14·1	+15	11	01	2·83v	−3·0	B2	150	Algenib

Also above mag. 5:

		Mag.	Abs. mag.	Spectrum	Dist.
22	υ	4·84	−0·3	K4	99
32		4·81	−0·2	B8	90
35		4·79	0·2	K0	77
43	o	4·79	1·2	A1	52
50	ρ	4·90	1·2	A1	55
56		4·76	−2·1	K0	240
72		4·98	−0·3	K4	110
78		4·93	0·2	K0	88
89	χ	4·80	−0·5	M2	120

VARIABLE STARS

	R.A.		Dec.		Range	Type	Period, d.	Spectrum
	h	m	°	′				
AG	21	51·0	+12	38	6·0–9·4	Z Andromedæ	830	WN+M
AW	21	52·3	+24	01	7·8–9·2	Algol	10·62	A+F
V	22	01·0	+06	07	7·0–15·0	Mira	302·3	M
TW	22	04·0	+28	21	7·0–9·2	Semi-reg.	956	M
RZ	22	05·9	+33	30	7·6–13·6	Mira	439·4	M
β	23	03·8	+28	05	2·3–2·8	Semi-reg.	38	M

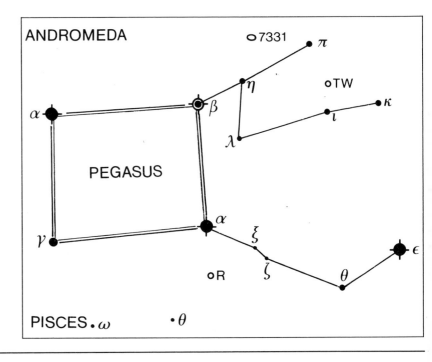

PEGASUS (continued)

VARIABLE STARS (continued)

	R.A.		Dec.		Range	Type	Period, d.	Spectrum
	h	m	°	'				
R	23	06·6	+10	33	6·9–13·8	Mira	378·0	M
W	23	19·8	+26	17	7·9–13·0	Mira	344·9	M
S	23	20·6	+08	55	7·1–13·8	Mira	319·2	M
Z	00	00·1	+25	53	7·7–13·6	Mira	325·5	M

DOUBLE STARS

	R.A.		Dec.		P.A.	Sep.	Mags.	
	h	m	°	'	°	"		
ε	21	44·2	+09	52	AB 325	81·8	2·4, 11·2	
					AC 320	142·5	9·4	
κ	21	44·6	+25	39	095	0·3	4·7, 5·0	Binary, 11·6 y
35	22	27·9	+04	42	AB 210	98·3	4·8, 9·8	
					AC 241	181·5	9·7	
37	22	30·0	+04	26	118	0·9	5·8, 7·1	Binary, 140 y
η	22	43·0	+30	13	339	90·4	2·9, 9·9	B is a close dble
β	23	03·8	+28	05	AB 211	108·5	2v, 11·6	
					AC 098	253·1	9·4	

GLOBULAR CLUSTER

M	NGC	R.A.		Dec.		Diameter	Mag.
		h	m	°	'	'	
15	7078	21	30·0	+12	10	12·3	6·3

GALAXIES

M	NGC	R.A.		Dec.		Mag.	Dimensions	Type
		h	m	°	'		'	
	7332	22	37·4	+23	48	11·8	4·2×1·3	E7
	7479	23	04·9	+12	19	11·0	4·1×3·2	SBb
	7814	00	03·3	+16	09	10·5	6·3×2·6	Sb

PERSEUS

(Abbreviation: Per).

A prominent constellation, containing the prototype eclipsing star Algol as well as the superb Sword-Handle cluster (H.VI.33–4). Mythologically, Perseus was the hero of one of the most famous of all legends; he killed the Gorgon, Medusa, and married Andromeda, daughter of Cepheus and Cassiopeia. The Gorgon's Head is marked by the winking 'Demon Star', Algol.

Perseus includes 12 stars above the fourth magnitude:

α 1·80
β 2·12 (max)
ζ 2·85
ε 2·89
γ 2·93
δ 3·01
ρ 3·2 (max)
η 3·76
ν 3·77
κ 3·80
ο 3·83
τ 3·95

BRIGHTEST STARS

Star		R.A.			Dec.			Mag.	Abs. mag.	Spectrum	Dist.	
		h	m	s	°	'	"				pc	
	φ	01	43	39·6	+50	41	20	4·07	−3·9	B1	350	
13	θ	02	44	11·9	+49	13	43	4·12	3·8	F7	13	
16		02	50	34·9	+38	19	07	4·23	0·6	F2	50	
15	η	02	50	41·8	+55	53	44	3·76	−4·4	K3	250	
17		02	51	30·8	+35	03	35	4·53	−0·3	K5	150	
18	τ	02	54	15·4	+52	45	45	3·95	0·3	G4	54	Kerb
22	π	02	58	45·6	+39	39	46	4·70	1·4	A2	44	
23	γ	03	04	47·7	+53	30	23	2·93	0·3	G8	34	
25	ρ	03	05	10·5	+38	50	25	var.	−0·5v	M4	60	
26	β	03	08	10·1	+40	57	21	2·12v	−0·2	B8	29	Algol
	ι	03	09	03·9	+49	36	49	4·05	3·7	G0	12	
27	κ	03	09	29·7	+44	51	27	3·80	0·2	K0	53	Misam
28	ω	03	11	17·3	+39	36	42	4·63	0·0	K0	180	
33	α	03	24	19·3	+49	51	40	1·80	−4·6	F5	190	Mirphak
34		03	29	22·0	+49	30	32	4·67	−2·3	B3	210	
35	σ	03	30	34·4	+47	59	43	4·35	−0·2	K3	71	
37	ψ	03	36	29·3	+48	11	34	4·23	−1·2	B5e	126	
39	δ	03	42	55·4	+47	47	15	3·01	−2·2	B5	100	
38	ο	03	44	19·1	+32	17	18	3·83	−4·4	B1	310	Ati
41	ν	03	45	11·6	+42	34	43	3·77	−2·0	F5	140	
44	ζ	03	54	07·8	+31	53	01	2·85	−5·7	B1	340	Atik
45	ε	03	57	51·1	+40	00	37	2·89	−3·7	B0·5	208	
46	ξ	03	58	57·8	+35	47	28	4·04	−5·4	O7	210	Menkib
47	λ	04	06	35·0	+50	21	05	4·29	0·2	B9	62	
48	υ	04	08	39·6	+47	42	45	4·04	−1·7	B3	140	Nembus
52	f	04	14	53·3	+40	29	01	4·71	−4·5	G5+A5	560	
51	μ	04	14	53·8	+48	24	33	4·14	−4·5	G0	460	
	b	04	18	14·6	+50	17	44	4·62v	var.	A2	55	

Also above mag. 5:

	Mag.	Abs. mag.	Spectrum	Dist.
12	4·91	4·2	F9	15
24	4·93	−0·1	K2	92
32	4·95	1·4	A2	51
40	4·97	0·5	B1	
54	4·93	0·3	G8	84
53	4·85	−1·7	B3	170

PERSEUS (continued)

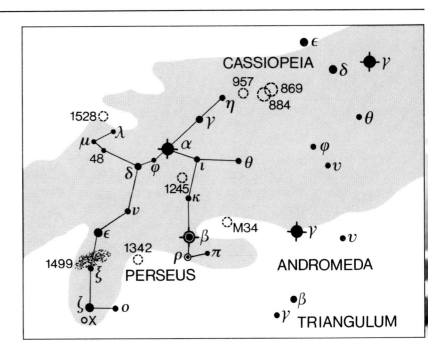

VARIABLE STARS

	R.A.		Dec.		Range	Type	Period, d.	Spectrum
	h	m	°	'				
U	01	59·6	+54	49	7·4–12·3	Mira	321·0	M
S	02	22·9	+58	35	7·9–11·5	Semi-reg.	Long	M
ρ	03	05·2	+38	50	3–4	Semi-reg.	33 to 55	M
β	03	08·2	+40	57	2·2–3·4	Algol	2·87	B+G
R	03	30·1	+35	40	8·1–14·8	Mira	210·0	M
X	03	55·4	+31	03	6·0–7·0	Irreg.	—	09·5 X-ray source
AW	04	47·8	+36	43	7·1–7·8	Cepheid	6·46	F–G

DOUBLE STARS

	R.A.		Dec.		P.A.	Sep.	Mags.	
	h	m	°	'	°	"		
η	02	50·7	+55	54	300	28·3	3·3, 8·5	
θ	02	44·2	+49	14	215	19·8	4·1, 9·9	Binary, 2720 y
γ	03	04·8	+53	30	326	57·0	2·9, 10·6	
ζ	03	54·1	+31	53	AB 208	12·9	2·9, 9·5	
					AC 286	32·8	11·3	
					AD 195	94·2	9·5	
					AE 185	120·3	10·2	
ε	03	57·9	+40	01	010	8·8	2·9, 8·1	

OPEN CLUSTERS

M	NGC	R.A.		Dec.		Diameter	Mag.	No. of stars	
		h	m	°	'	'			
	744	01	58·4	+55	29	11	7·9	20	
	869	02	19·0	+57	09	30	4·3	200	⎱ Sword-
	884	02	22·4	+57	07	30	4·4	150	⎰ Handle
	957	02	33·6	+57	32	11	7·6	30	
34	1039	02	42·0	+42	47	35	5·2	60	
	1245	03	14·7	+47	15	10	8·4	200	
	1444	03	49·4	+52	40	4	6·6	—	
	1513	04	10·0	+49	31	9	8·4	50	
	1528	04	15·4	+51	14	24	6·4	40	
	1545	04	20·9	+50	15	8	6·2	20	

PLANETARY NEBULA

M	NGC	R.A.		Dec.		Dimensions	Mag.	Mag. of central star	
		h	m	°	'	"			
76	650–1	01	42·4	+51	34	65×290	12·2	17	Little Dumbbell

PERSEUS *(continued)*

NEBULÆ

M	NGC	R.A.		Dec.		Dimensions	Mag. of illuminating star
		h	m	°	′	′	
	1333	03	29·3	+31	25	9×7	9·5 (Near dark nebula B.205)
	1499	04	00·7	+36	37	145×40	4 California Nebula

GALAXIES

M	NGC	R.A.		Dec.		Mag.	Dimensions	Type
		h	m	°	′		′	
	1003	02	39·3	+40	52	11·5	5·4×2·1	Sc
	1023	02	40·4	+39	04	9·5	8·7×3·3	E7p

PHŒNIX

(Abbreviation: Phe).

One of the 'Southern Birds'. It is not very distinctive, and Ankaa is the only bright star; there are however seven stars above the fourth magnitude:

α 2·39
β 3·31
γ 3·41
ζ 3·6 (max)
ε 3·88
κ 3·94
δ 3·95

BRIGHTEST STARS

Star	R.A.			Dec.			Mag.	Abs. mag.	Spectrum	Dist.	
	h	m	s	°	′	″				pc	
ι	23	35	04·4	−42	36	54	4·71	?	A	25?	
ε	00	09	24·6	−45	44	51	3·88	0·2	K0	23	
κ	00	26	12·1	−43	40	48	3·94	1·7	A3	19	
α	00	26	17·0	−42	18	22	2·39	0·2	K0	24	Ankaa
μ	00	41	19·5	−46	05	06	4·59	0·3	G8	70	
η	00	43	21·2	−57	27	48	4·36	0·0	A0	71	
β	01	06	05·0	−46	43	07	3·31	0·3	G8	40	
ζ	01	08	23·0	−55	14	45	var.	−0·2	B8	67	
γ	01	28	21·9	−43	19	06	3·41	−4·4	K5	280	
δ	01	31	15·0	−49	04	22	3·95	1·7	K0	28	
ψ	01	53	38·7	−46	18	09	4·41	−0·5	M4	96	

Also above mag. 5:

	Mag.	Abs. mag.	Spectrum	Dist.
λ¹	4·77	0·6	A0	130
ν	4·96	4·0	F8	15
HD 12055	4·83	2·6	G5	90

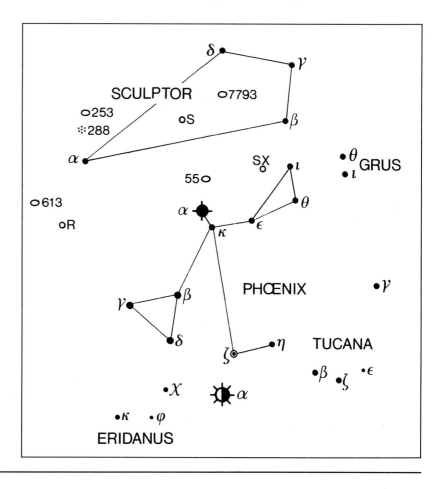

PHOENIX (continued)

VARIABLE STARS

	R.A. h m	Dec. ° ′	Range	Type	Period, d.	Spectrum
SX	23 46·5	−41 35	6·8–7·5	Delta Scuti	0·055	A
R	23 56·5	−49 47	7·5–14·4	Mira	267·9	M
S	23 53·1	−56 35	7·4–8·2	Semi-reg.	141	M
ζ	01 08·4	−55 15	3·9–4·4	Algol	1·67	B+B

DOUBLE STARS

	R.A. h m	Dec. ° ′	P.A. °	Sep. ″	Mags.
ξ	00 41·8	−56 30	253	13·2	5·8, 10·2
η	00 43·4	−57 28	217	19·8	4·4, 11·4
β	01 06·1	−46 43	346	1·4	4·0, 4·2

PICTOR

(Abbreviation: Pic).

An unremarkable constellation near Canopus, known originally under the cumbersome name of Equuleus Pictoris (the Painter's Easel). There are two stars above the fourth magnitude:

α 3·27

β 3·85

See chart for Carina

BRIGHTEST STARS

Star	R.A. h m s	Dec. ° ′ ″	Mag.	Abs. mag.	Spectrum	Dist. pc
β	05 47 17·1	−51 03 59	3·85	0·3	A5	24
γ	05 49 49·6	−56 10 00	4·51	0·0	K1	80
δ	06 10 17·9	−54 58 07	4·7v		B1	
α	06 48 11·4	−61 56 29	3·27	2·1	A5	16

VARIABLE STARS

	R.A. h m	Dec. ° ′	Range	Type	Period, d.	Spectrum
R	04 46·2	−49 15	6·7–10·0	Semi-reg.	164	M
S	05 11·0	−48 30	6·5–14·0	Mira	426·6	M
T	05 15·1	−46 55	7·9–14·4	Mira	200·6	M
δ	06 10·3	−54 58	4·7–4·9	Beta Lyræ	1·67	B

DOUBLE STARS

	R.A. h m	Dec. ° ′	P.A. °	Sep. ″	Mags.
ι	04 50·9	−53 28	058	12·3	5·6, 6·4
θ	05 24·8	−52 19	AB 152	0·2	6·9, 7·2
			AB+C 287	38·2	6·8
μ	06 32·0	−58 45	231	2·4	5·8, 9·0

PISCES

(Abbreviation: Psc).

A large but faint Zodiacal constellation; it now contains the Vernal Equinox. Its mythological associations are rather vague, but it may represent the fishes into which Venus and Cupid once changed themselves in order to escape from the monster Typhon. There are three stars above the fourth magnitude:

η 3·62

γ 3·69

α 3·79

BRIGHTEST STARS

Star		R.A. h m s	Dec. ° ′ ″	Mag.	Abs. mag.	Spectrum	Dist. pc	
4	β	23 03 52·5	+03 49 12	4·53	−0·4	B5p	101	
6	γ	23 17 09·7	+03 16 56	3·69	0·3	G8	48	
10	θ	23 27 57·9	+06 22 44	4·28	0·0	K1	72	
17	ι	23 39 56·9	+05 37 35	4·13	3·8	F7	13	
18	λ	23 42 02·6	+01 46 48	4·50	2·4	A7	26	
19	TX	23 46 23·3	+03 29 13	4·3–5·1	−2·0v	N	330	
28	ω	23 59 18·5	+06 51 48	4·01	0·8	F4	26	
30		00 01 57·5	−06 00 51	4·41	0·4	M3	30	
33		00 05 20·1	−05 42 27	4·61	0·0	K1	84	
63	δ	00 48 40·9	+07 35 06	4·43	−0·3	K5	88	
71	ε	01 02 56·5	+07 53 24	4·28	0·2	K0	66	
74	ψ	01 05 40·9	+21 28 24	4·7	−0·1, 0·2	B9·5, B9	120	
84	χ	01 11 27·1	+21 02 05	4·66	0·2	K0	75	
83	τ	01 11 39·5	+30 05 23	4·51	1·7	K0	29	
85	φ	01 13 44·8	+24 35 01	4·65	0·2	K0	74	
99	η	01 31 28·9	+15 20 45	3·62	0·3	G8	44	Alpherg
106	ν	01 41 25·8	+05 29 15	4·44	−0·2	K3	42	
110	ο	01 45 23·5	+09 09 28	4·26	0·2	K0	65	
111	ξ	01 53 33·3	+03 11 15	4·62	0·2	K0	77	
113	α	02 02 02·7	+02 45 49	3·79	1·4	A2	30	Al Rischa

Also above mag. 5:

		Mag.	Abs. mag.	Spectrum	Dist.
κ		4·94	2·2	A2	
27		4·86	0·2	G9	86
86	ζ	4·86	2·1	F6	33
90	υ	4·76	1·4	A2	47
98	μ	4·84	−0·3	K4	67
47	TV	4·8v	−0·5	M3	140

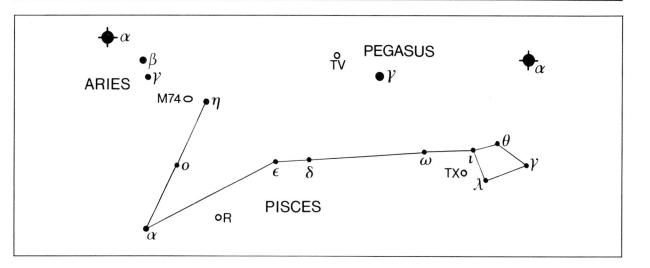

PISCES *(continued)*

VARIABLE STARS

	R.A. h m	Dec. ° '	Range	Type	Period, d.	Spectrum
TX	23 46·4	+03 29	6·9–7·7	Irreg.	–	N
TV	00 28·0	+17 24	4·6–5·4	Semi-reg.	70	M
Z	01 16·1	+25 46	7·0–7·9	Semi-reg.	144	N
R	01 30·6	+02 53	7·1–14·8	Mira	344·0	M

DOUBLE STARS

	R.A. h m	Dec. ° '	P.A. °	Sep. "	Mags.	
ζ	01 13·7	+07 35	063	23·0	5·6, 6·5	
ψ¹	01 05·6	+21 28	AB 159	30·0	5·6, 5·8	
			AC 123	92·6	11·2	
α	02 02·0	+02 46	279	1·9	4·2, 5·1	Binary, 933 y

GALAXIES

M	NGC	R.A. h m	Dec. ° '	Mag.	Dimensions '	Type
	470	01 19·7	+03 25	11·9	3·0×2·0	Sc
	474	01 20·1	+03 25	11·1	7·9×7·2	S0
	488	01 21·8	+05 15	10·3	5·2×4·1	Sb
	524	01 24·8	+09 32	10·6	3·2×3·2	E1
74	628	01 36·7	+15 47	9·2	10·2×9·5	Sc

PISCIS AUSTRALIS

(Abbreviation: PsA).

Also known as Piscis Austrinus. No specific mythological legends have been associated with it. It contains Fomalhaut, which is, incidentally, the most southerly of the first-magnitude stars to be visible from England (mag. 1·16), but there are no other stars above the fourth magnitude.

BRIGHTEST STARS

Star	R.A. h m s	Dec. ° ' "	Mag.	Abs. mag.	Spectrum	Dist. pc	
9 ι	21 44 56·7	−33 01 33	4·34	0·6	A0	42	
14 μ	22 08 22·8	−32 59 19	4·50	1·4	A2	42	
17 β	22 31 30·1	−32 20 46	4·29	0·6	A0	53	
18 ε	22 40 39·2	−27 02 37	4·17	−0·2	B8	75	
22 γ	22 52 31·4	−32 52 32	4·46	0·6	A0	59	
23 δ	22 55 56·8	−32 32 23	4·21	0·3	gG4	49	
24 α	22 57 38·9	−29 37 20	1·16	2·0	A3	6·7	Fomalhaut

Also above mag. 5:

	Mag.	Abs. mag.	Spectrum	Dist.
υ	4·99	−0·3	K5	110
15 τ	4·92	3·4	F5	11

VARIABLE STARS

	R.A. h m	Dec. ° '	Range	Type	Period, d.	Spectrum
S	22 03·8	−28 03	8·0–14·5	Mira	271·7	M
V	22 55·3	−29 37	8·0–9·0	Semi-reg.	148	M

PISCIS AUSTRALIS *(continued)*

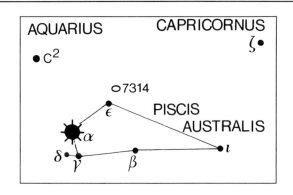

DOUBLE STARS

	R.A. h m	Dec. ° '	P.A. °	Sep. "	Mags.
η	22 00·8	−28 27	115	1·7	5·8, 6·8
β	22 31·5	−32 21	172	30·3	4·4, 7·9 (optical)
γ	22 52·5	−32 53	262	4·2	4·5, 8·0
δ	22 55·9	−32 32	244	5·0	4·2, 9·2

GALAXIES

M	NGC	R.A. h m	Dec. ° '	Mag.	Dimensions '	Type
	7172	22 02·0	−31 52	11·9	2·2×1·3	S
	7174	22 02·1	−31 59	12·6	1·3×0·7	S
	7314	22 35·8	−26 03	10·9	4·6×2·3	Sc Arp 14

PUPPIS

(Abbreviation: Pup).

The poop of the dismembered ship, Argo Navis. There are 12 stars above the fourth magnitude.

ζ 2·25
π 2·70
ρ 2·81v
τ 2·93
ν 3·17
σ 3·25
ξ 3·34
L² 3·4 (max)
c 3·59
a 3·73
κ 3·82
3 3·96

ρ is a variable with very small range (2·72–2·87).

See chart for Carina

BRIGHTEST STARS

Star	R.A. h m s	Dec. ° ' "	Mag.	Abs. mag.	Spectrum	Dist. pc	
ν	06 37 45·6	−43 11 45	3·17	−1·2	B8	75	
τ	06 49 56·1	−50 36 53	2·93	0·2	K0	25	
l	07 12 33·6	−46 45 34	4·49	2·6	F0	24	
L²	07 13 13·3	−45 10 59	var.	−3·1v	A0p	23	
π	07 17 08·5	−37 05 51	2·70	−0·3	K5	40	
υ¹	07 18 18·4	−36 44 03	4·66	−1·7	B3	180	
σ	07 29 13·8	−43 18 05	3·25	−0·3	K5	51	
HD 60532	07 34 03·1	−22 17 46	4·45	2·3	F7	26	
p	07 35 22·8	−28 22 10	4·64	0·1	B8	77	
f	07 37 22·0	−34 58 07	4·53	−0·2	B8	88	
m	07 38 17·9	−25 21 53	4·67	−0·1	B8	92	
k	07 38 49·7	−26 48 13	3·82	−0·7, −0·6	B8	110	
1	07 43 32·3	−28 24 40	4·59	−0·3	gK5	79	
3	07 43 48·4	−28 57 18	3·96	−7·5	A2	1700	
c	07 45 15·2	−37 58 07	3·59	−5·0	cK	530	
o	07 48 05·1	−25 56 14	4·50	−3·5	B1	390	
Q	07 48 20·2	−47 04 39	4·71	0·2	K0	72	
P	07 49 14·3	−46 22 24	4·11	−5·6	B0	820	
7 ξ	07 49 17·6	−24 51 35	3·34	−4·5	G3	230	Asmidiske
a	07 52 13·0	−40 34 33	3·73	0·3	G5	38	
J	07 53 03·7	−49 36 47	4·63	−3·6	B2	440	
11	07 56 51·5	−22 52 48	4·20	−2·0	F8	150	
V	07 58 14·3	−49 14 42	var.	−4·0v	B1 + B3	470	
232G	07 59 52·0	−18 23 58	4·61	1·7	A3	38	
ζ	08 03 35·0	−40 00 12	2·25	−7·1	O5·8	740	Suhail Hadar
ρ	08 07 32·6	−24 18 15	2·81v	−2·0	F6	92	Turais
16	08 09 01·5	−19 14 42	4·40	−1·1	B5	130	
19	08 11 16·2	−12 55 37	4·72	0·2	K0	45	
h¹	08 11 21·5	−39 37 07	4·45	−5·9	cK	850	
h²	08 14 02·8	−40 20 52	4·44	−0·1	gK2	81	
q	08 18 33·2	−36 39 34	4·45	0·5	A7	62	

Also above mag. 5:

	Mag.	Abs. mag.	Spectrum
Y	5·00	−0·2	G7
A	4·85		M3
d¹	4·91	−0·9	B3
r	4·77	−1·3	B3
W	4·94	1·8	M0

PUPPIS (continued)

VARIABLE STARS

	R.A. h m	Dec. ° '	Range	Type	Period, d.	Spectrum
L²	07 13·5	−44 39	2·6–6·2	Semi-reg.	140	M
Z	07 32·6	−20 40	7·2–14·6	Mira	499·7	M
VX	07 32·6	−21 56	7·7–8·5	Cepheid	3·01	F
X	07 32·8	−20 55	7·8–9·2	Cepheid	25·96	F–G
W	07 46·0	−41 12	7·3–13·6	Mira	120·1	M
AP	07 57·8	−40 07	7·1–7·8	Cepheid	5·08	F
V	07 58·2	−49 15	4·7–5·2	Beta Lyræ	1·45	B + B
AT	08 12·4	−36 57	7·5–8·4	Cepheid	6·66	F–G
RS	08 13·1	−34 35	6·5–7·6	Cepheid	41·39	F–G

DOUBLE STAR

	R.A. h m	Dec. ° '	P.A. °	Sep. "	Mags.
σ	07 29·2	−43 18	074	22·3	3·3, 9·4

OPEN CLUSTERS

M	NGC	R.A. h m	Dec. ° '	Diameter '	Mag.	No. of stars	
	2383	07 24·8	−20 56	6	8·4	40	
	2421	07 36·3	−20 37	10	8·3	70	
47	2422	07 36·6	−14 30	30	4·4	30	
	Mel 71	07 37·5	−12 04	9	7·1	80	
	Mel 72	07 38·4	−10 41	9	10·1	40	
	2432	07 40·9	−19 05	8	10·2	50	
	2439	07 40·8	−31 39	10	6·9	80	R Puppis. Asterism
46	2437	07 41·8	−14 49	27	6·1	100	
93	2447	07 44·6	−23 52	22	6·2	80	
	2451	07 45·4	−37 58	45	2·8	40	
	2477	07 52·3	−38 33	27	5·8	160	
	2479	07 55·1	−17 43	7	9·6	45	
	2489	07 56·2	−30 04	8	7·9	45	
	2509	08 00·7	−19 04	8	9·3	70	
	2527	08 05·3	−28 10	22	6·5	40	
	2533	08 07·0	−29 54	3·5	7·6	60	
	2539	08 10·7	−12 50	22	6·5	50	
	2546	08 12·4	−37 38	41	6·3	40	
	2567	08 18·6	−30 38	10	7·4	40	
	2571	08 18·9	−29 44	13	7·0	30	
	2580	08 21·6	−30 19	8	9·7	50	
	2587	08 23·5	−29 30	9	9·2	40	

PLANETARY NEBULÆ

M	NGC	R.A. h m	Dec. ° '	Dimensions "	Mag.	Mag. of central star
	2438	07 41·8	−14 44	66	10·1	17·7 In cluster NGC 2437
	2440	07 41·9	−18 13	14×32	10·8	14·3 Protoplanetary?

NEBULA

M	NGC	R.A. h m	Dec. ° '	Dimensions '	Mag. of illuminating star
	2467	07 52·5	−26 24	8×7	9·2 Gum 9

PYXIS

(Abbreviation: Pyx).

Also originally part of Argo. The only stars above the fourth magnitude are α (3·68) and β (3·97).

See chart for Carina

BRIGHTEST STARS

Star	R.A. h m s	Dec. ° ' "	Mag.	Abs. mag.	Spectrum	Dist. pc
β	08 40 06·1	−35 18 30	3·97	0·3	G4	46
α	08 43 35·5	−33 11 11	3·68	−4·4	B2	410
γ	08 50 31·9	−27 42 36	4·01	−0·3	K4	73
κ	09 08 02·8	−25 51 30	4·58	−0·4	gM0	97
θ	09 21 29·5	−25 57 55	4·72	−0·5	M1	100
λ	09 23 12·1	−28 50 02	4·69	0·3	gG7	76

Also above mag. 5:

	Mag.	Abs. mag.	Spectrum	Dist.
ζ	4·89	0·3	G4	75
δ	4·89	1·7	A3	42

PYXIS (continued)

VARIABLE STARS

	R.A. h m	Dec. ° '	Range	Type	Period, d.	Spectrum
TY	08 59·7	−27 49	6·9–7·5	Eclipsing	3·20	G + G
T	09 04·7	−32 23	6·3–14·0	Recurrent nova	–	Q Outbursts 1890, 1902, 1920, 1944, 1966
S	09 05·1	−23 05	8·0–14·2	Mira	206·4	M

DOUBLE STARS

	R.A. h m	Dec. ° '	P.A. °	Sep. "	Mags.
ζ	08 39·7	−29 34	061	52·4	4·9, 9·1
δ	08 55·5	−27 41	AB 268	23·8	4·9, 14·0
			CD 017	2·5	11·0, 11·0
ε	09 09·9	−30 22	A+BC 147	17·8	5·6, 10·5
			BC 088	0·3	10·5, 10·8
			AD 340	35·4	5·6, 13·5
κ	09 08·0	−25 52	263	2·1	4·6, 9·8

OPEN CLUSTERS

M	NGC	R.A. h m	Dec. ° '	Diameter '	Mag.	No. of stars
	2627	08 37·3	−29 57	11	8·4	60
	2658	08 43·4	−32 39	12	9·2	80

PLANETARY NEBULA

M	NGC	R.A. h m	Dec. ° '	Diameter "	Mag.	Mag. of central star
	2818	09 16·0	−36 28	38	13·0	13·0

GALAXY

M	NGC	R.A. h m	Dec. ° '	Mag.	Dimensions '	Type
	2613	08 33·4	−22 58	10·4	7·2 × 2·1	Sb

RETICULUM

(Abbreviation: Ret).

Originally Reticulum Rhomboidalis (the Rhomboidal Net). A small but quite distinctive constellation of the far south. There are two stars above the fourth magnitude:

α 3·35
β 3·85

BRIGHTEST STARS

Star	R.A. h m s	Dec. ° ' "	Mag.	Abs. mag.	Spectrum	Dist. pc
κ	03 29 22·7	−62 56 15	4·72	3·4	F5	20
β	03 44 12·0	−64 48 26	3·85	3·2	K0	17
δ	03 58 44·7	−61 24 01	4·56	−0·5	M2	97
˙γ	04 00 53·8	−62 09 34	4·51	?	Mb	?
α	04 14 25·5	−62 28 26	3·35	−2·1	G6	120
ε	04 16 28·9	−59 18 07	4·44	−0·3	gK5	21

Also above mag. 5:

	Mag.	Abs. mag.	Spectrum	Dist.
ι	4·97	−0·4	M0	120

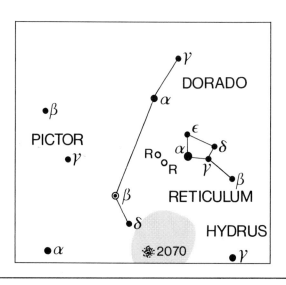

RETICULUM (continued)

VARIABLE STAR

	R.A. h m	Dec. ° ′	Range	Type	Period, d.	Spectrum
R	04 33·5	−63 02	6·5–14·0	Mira	278·3	M

DOUBLE STAR

	R.A. h m	Dec. ° ′	P.A. °	Sep. ″	Mags.
β	03 44·2	−64 48	following	1440	3·9, 8·1

GALAXIES

M	NGC	R.A. h m	Dec. ° ′	Mag.	Dimensions ′	Type
	1313	03 18·3	−66 30	9·4	8·5×6·6	SBd
	1559	04 17·6	−62 47	10·4	3·3×2·1	SBc

SAGITTA

(Abbreviation: Sge).

An original constellation; small though it is, it is quite distinctive. It has been identified with Cupid's bow, and also with the arrow used by Apollo against the one-eyed Cyclops. There are two stars above the fourth magnitude:

γ 3·47
δ 3·82

See chart for Aquila

BRIGHTEST STARS

Star	R.A. h m s	Dec. ° ′ ″	Mag.	Abs. mag.	Spectrum	Dist. pc
5 α	19 40 05·6	+18 00 50	4·37	−2·0	G0	190
6 β	19 41 02·8	+17 28 33	4·37	−2·1	G8	200
7 δ	19 47 23·0	+18 32 03	3·82	−2·4	M2	170
12 γ	19 58 45·3	+19 29 32	3·47	−0·3	K5	51

Also above mag. 5:

	Mag.	Abs. mag.	Spectrum	Dist.
8 ζ	5·00	1·7	A3	46

VARIABLE STARS

	R.A. h m	Dec. ° ′	Range	Type	Period, d.	Spectrum
U	19 18·8	+19 37	6·6–9·2	Algol	3·38	B–K
S	19 56·0	+16 38	5·3–6·0	Cepheid	8·38	F–G
X	20 05·1	+20 39	7·0–8·4	Semi-reg.	196	N
WZ	20 07·6	+17 42	7·0–15·5	Recurrent nova	−	Q Outbursts 1913, 1946, 1978

DOUBLE STAR

	R.A. h m	Dec. ° ′	P.A. °	Sep. ″	Mags.
ζ	19 49·0	+19 09	AB+C 311	8·6	5·5, 8·7
			AB 163	0·3	5·5, 6·2 Binary, 22·8 y

OPEN CLUSTER

M	NGC	R.A. h m	Dec. ° ′	Diameter ′	Mag.	No. of stars
	H.20	19 53·1	+18 20	7	7·7	15

GLOBULAR CLUSTER

M	NGC	R.A. h m	Dec. ° ′	Diameter ′	Mag.
71	6838	19 53·8	+18 47	7·2	8·3

PLANETARY NEBULÆ

M	NGC	R.A. h m	Dec. ° ′	Diameter ″	Mag.	Mag. of central star
	6879	20 10·5	+16 55	5	13·0	15
	IC 4997	20 20·2	+16 45	2	11·6	13 (v?)

SAGITTARIUS

(Abbreviation: Sgr).

The southernmost of the Zodiacal constellations, and not wholly visible from England. Mythologically it has been associated with Chiron, the wise centaur who was tutor to Jason and many others; but it would certainly be more logical to associate Chiron with Centaurus, and another version states that Chiron merely invented the constellation Sagittarius to help in guiding the Argonauts in their quest of the Golden Fleece. The centre of the Milky Way lies behind the star-clouds here, and the whole area is exceptionally rich; it abounds in Messier objects. It is worth commenting that the stars lettered α and β are relatively faint. There are 16 stars above the fourth magnitude:

ε	1·85
σ	2·02
ζ	2·59
δ	2·70
λ	2·81
π	2·89
γ	2·99
η	3·11
φ	3·17
τ	3·32
ξ²	3·51
o	3·77
μ	3·86v
ρ¹	3·93
β¹	3·93
α	3·97

μ is an Algol binary with a very small range (3·8 to 3·9).

BRIGHTEST STARS

Star		R.A. h	m	s	Dec. °	′	″	Mag.	Abs. mag.	Spectrum	Dist. pc	
3	X	17	47	33·4	−27	49	51	var.	−2·0	F7	200	
	W	18	05	01·1	−29	34	48	var.	var.	F8	400	
10	γ	18	05	48·3	−30	25	26	2·99	0·2	K0	36	Alnasr
HD	165634	18	08	04·8	−28	27	25	4·57	1·0	Gp	52	
13	μ	18	13	45·6	−21	03	32	3·86v	−7·1	B8	1200	Polis
	η	18	17	37·5	−36	45	42	3·11	−2·4	M3	130	
HD	167818	18	18	03·0	−27	02	33	4·65	−0·3	gK5	84	
19	δ	18	20	59·5	−29	49	42	2·70	−0·1	K2	25	Kaus Meridionalis
20	ε	18	24	10·2	−34	23	05	1·85	−0·3	B9	26	Kaus Australis
22	λ	18	27	58·1	−25	25	18	2·81	−0·1	K2	30	Kaus Borealis
27	φ	18	45	39·2	−26	59	27	3·17	−1·2	B8	75	
34	σ	18	55	15·7	−26	17	48	2·02	−2·0	B3	64	Nunki
37	ξ²	18	57	43·6	−21	06	24	3·51	0·0	K1	44	
38	ζ	19	02	36·5	−29	52	49	2·59	0·6	A2	24	Ascella
39	o	19	04	40·8	−21	44	30	3·77	0·3	gG8	43	
40	τ	19	06	56·2	−27	40	13	3·32	0·0	K1	40	
41	π	19	09	45·6	−21	01	25	2·89	−2·0	F2	95	Albaldah
44	ρ¹	19	21	40·2	−17	50	50	3·93	1·7	F0	28	
46	υ	19	21	43·5	−15	57	18	4·61v	3·0	F2	21	
	β¹	19	22	38·1	−44	27	32	3·93	−0·2	B8	67	Arkab
	β²	19	23	12·9	−44	47	59	4·29	0·6	F0	52	
	α	19	23	53·0	−40	36	58	3·97	−0·2	B8	36	Rukbat
52	h¹	19	36	42·3	−24	53	01	4·60	0·9	B9	55	
	ι	19	55	15·5	−41	52	06	4·13	0·2	K0	55	
58	ω	19	55	50·2	−26	17	58	4·70	5·2	dG5	11	
59		19	56	56·6	−27	10	12	4·52	−0·2	gK3	69	
	θ¹	19	59	44·0	−35	16	35	4·37	−2·3	B3	210	
62	c	20	02	39·4	−27	42	35	4·58	−0·5	M4	100	

Also above mag. 5:

		Mag.	Abs. mag.	Spectrum	Dist.
4		4·76	0·6	A0	21
1		4·98	0·2	K0	83
21		4·81	0·2	K0	57
32	ν¹	4·83	−6·0	K2	1400
36	ξ¹	5·00	0·6	A0	76
42	ψ	4·85	3·4	F5	21
43		4·90	−2·1	G8	210
60		4·83	0·3	G5	80

VARIABLE STARS

	R.A. h	m	Dec. °	′	Range	Type	Period, d.	Spectrum
X	17	47·6	−27	50	4·2–4·8	Cepheid	7·01	F
W	18	05·0	−29	35	4·3–5·1	Cepheid	7·59	F–G
VX	18	08·1	−22	13	6·5–12·5	Semi-reg.	732	M
RS	18	17·6	−34	06	6·0–6·9	Algol	2·41	B+A
Y	18	21·4	−18	52	5·4–6·1	Cepheid	5·77	F
RV	18	27·9	−33	19	7·2–14·8	Mira	317·5	M
U	18	31·9	−19	07	6·3–7·1	Cepheid	6·74	F–G
YZ	18	49·5	−16	43	7·0–7·7	Cepheid	9·55	F–G
UX	18	54·9	−16	31	7·6–8·4	Semi-reg.	100	M
ST	19	01·5	−12	46	7·6–16·0	Mira	395·1	S
T	19	16·3	−16	59	7·6–12·9	Mira	392·3	S
RY	19	16·5	−33	31	6·0–15	R Coronæ	−	Gp
R	19	16·7	−19	18	6·7–12·8	Mira	268·8	M
AQ	19	34·3	−16	22	6·6–7·7	Semi-reg.	200	N
RR	19	55·9	−29	11	5·6–14·0	Mira	334·6	M
RU	19	58·7	−41	51	6·0–13·8	Mira	240·3	M
RT	20	17·7	−39	07	6·0–14·1	Mira	305·3	M

DOUBLE STARS

	R.A. h	m	Dec. °	′	P.A. °	Sep. ″	Mags.	
21	18	25·3	−20	32	289	1·8	4·9, 7·4	
ζ	19	02·6	−29	53	320	0·3	3·3, 3·4	Binary, 21·2 y
η	18	17·6	−36	46	105	3·6	3·2, 7·8	
π	19	09·8	−21	01	AB 150	0·1	3·7, 3·7	
					AB+C 122	0·4	5·9	
β¹	19	22·6	−44	28	077	28·3	3·9, 8·0	Wide naked-eye pair with β²
κ²	20	23·9	−42	25	234	0·8	6·9, 6·9	

SAGITTARIUS (continued)

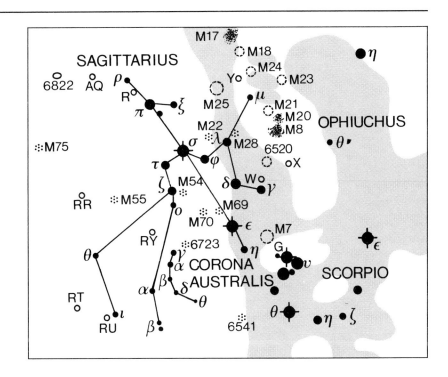

OPEN CLUSTERS

M	NGC	R.A.		Dec.		Diameter	Mag.	No. of stars	
		h	m	°	′	′			
	6469	17	52·9	−22	21	12	8·2	50	
23	6494	17	56·8	−19	01	27	5·5	150	
	6520	18	03·4	−27	54	6	7·6	60	In M.20
21	6531	18	04·6	−22	30	13	5·9	70	
	6530	18	04·8	−24	20	15	4·6	−	In M.20
	6546	18	07·2	−23	20	13	5·9	70	
	6568	18	12·8	−21	36	13	8·6	50	
24	−	18	16·9	−18	29	90	4·5	−	Star-cloud; not a true cluster
18	6613	18	19·9	−17	08	9	6·9	20	
25	IC 4725	18	31·6	−19	15	32	4·6	30	ν Sagittarii cluster
	6645	18	32·6	−16	54	10	8·5	40	
	6716	18	54·6	−19	53	7	6·9	20	

GLOBULAR CLUSTERS

M	NGC	R.A.		Dec.		Diameter	Mag.	
		h	m	°	′	′		
	6522	18	03·6	−30	02	5·6	8·6	
	6544	18	07·3	−25	00	8·9	8·2	
	6553	18	09·3	−25	54	8·1	8·2	
	6558	18	10·3	−31	46	3·7	−	
	6569	18	13·6	−31	50	5·8	8·7	
	6624	18	23·7	−30	22	5·9	8·3	H.I.50
28	6626	18	24·5	−24	52	11·2	6·9	
69	6637	18	31·4	−32	21	7·1	7·7	
	6638	18	30·9	−25	30	5·0	9·2	H.I.51
	6652	18	35·8	−32	59	3·5	8·9	
22	6656	18	36·4	−23	54	24·0	5·1	
54	6715	18	55·1	−30	29	9·1	7·7	
70	6681	18	43·2	−32	18	7·8	8·1	
55	6809	19	40·0	−30	58	19·0	6·9	
75	6864	20	06·1	−21	55	6·0	8·6	

PLANETARY NEBULÆ

M	NGC	R.A.		Dec.		Diameter	Mag.	Mag. of central star
		h	m	°	′	″		
	6567	18	13·7	−19	05	8	11·1	15·0
	6629	18	25·7	−23	12	15	11·6	12·8
	6644	18	32·6	−25	08	3	12·2	15·9
	6818	19	44·0	−14	09	17	9·9	13·0

SAGITTARIUS *(continued)*

NEBULÆ

M	NGC	R.A.		Dec.		Dimensions	Mag. of illuminating star	
		h	m	°	′			
20	6514	18	02·6	−23	02	29×27	7·6	Trifid Nebula
8	6523	18	03·8	−24	23	90×40	−	Lagoon Nebula
17	6618	18	20·8	−16	11	46×37	−	Omega Nebula

GALAXY

M	NGC	R.A.		Dec.		Mag.	Dimensions	Type
		h	m	°	′		′	
	6822	19	44·9	−14	48	9·3	10·2×9·5	Irreg. Barnard's Galaxy

SCORPIUS

(Abbreviation: Sco).

Alternatively, and less correctly, known as Scorpio. Mythologically it is usually associated with the scorpion which Juno caused to attack and kill the great hunter Orion. Note that Orion and Scorpius are now on opposite sides of the sky – placed there, it is said, so that the creature could do Orion no further damage!

Scorpius is one of the most magnificent of all constellations, and one of the few which gives at least a vague impression of the creature it is meant to represent. It is dominated by Antares, but the whole area is exceptionally rich. The 'sting', which includes Shaula – only just below the first magnitude – is to all intents and purposes invisible from England.

There are 20 stars above the fourth magnitude. Antares itself is very slightly variable (range 0·86–1·02). The leaders are:

α 0·96v
λ 1·63
θ 1·87
ε 2·29
δ 2·32
κ 2·41
β 2·64
υ 2·69
τ 2·82
σ 2·89v
π 2·89
ι¹ 3·03
μ¹ 3·04
G 3·21
η 3·33
μ² 3·57
ζ² 3·62
ρ 3·88
ω¹ 3·96
ν 4·00

σ is very slightly variable (2·82–2·90).

BRIGHTEST STARS

Star		R.A.			Dec.			Mag.	Abs. mag.	Spectrum	Dist.	
		h	m	s	°	′	″				pc	
1	b	15	50	58·6	−25	45	05	4·64	−1·7	B3	170	
2	A	15	53	36·6	−25	19	38	4·59	−2·1	B2·5	220	
5	ρ	15	56	53·0	−29	12	50	3·88	−2·5	B2	190	
6	π	15	58	51·0	−26	06	50	2·89	−3·5	B1	190	
7	δ	16	00	19·9	−22	37	18	2·32	−4·1	B0	170	Dschubba
	ξ	16	04	22·0	−11	22	24	4·16	2·2	F6	26	
8	β	16	05	26·1	−19	48	19	2·64	−3·7	B0·5+B2	250	Graffias
9	ω¹	16	06	48·3	−20	40	09	3·96	−3·5	B1	250	Jabhat al Akrab
10	ω²	16	07	24·2	−20	52	07	4·32	0·4	gG2	53	
14	ν	16	11	59·6	−19	27	38	4·00	0·0, −2·8	A0+B2	170	Jabbah
13	c²	16	12	18·1	−27	55	35	4·58	−2·1	B2·5	220	
19	o	16	20	38·0	−24	10	10	4·55	−2·1	A5	92	
20	σ	16	21	11·2	−25	35	34	2·89v	−4·4	B1	180	Alniyat
21	α	16	29	24·3	−26	25	55	0·96v	−4·7	M1	100	Antares
	N	16	31	22·8	−34	42	15	4·23	−2·5	B2	220	
23	τ	16	35	52·8	−28	12	58	2·82	−4·1	B0	240	
	H	16	36	22·4	−35	15	21	4·16	0·6	M0	55	
26	ε	16	50	09·7	−34	17	36	2·29	−0·1	K2	20	Wei
	μ¹	16	51	52·1	−38	02	51	3·04	−3·0	B1·5	160	
	μ²	16	52	20·0	−38	01	03	3·57	−3·0	B2	210	
	ζ¹	16	53	59·6	−42	21	44	4·73	−6·7	B1·5	780	
	ζ²	16	54	34·9	−42	21	41	3·62	−0·3	K5	50	
	η	17	12	09·0	−43	14	21	3·33	0·6	F2	21	
34	υ	17	30	45·6	−37	17	45	2·69	−5·7	B3	480	Lesath
35	λ	17	33	36·4	−37	06	14	1·63	−3·0	B2	84	Shaula
	Q	17	36	32·6	−38	38	07	4·29	0·2	gK0	58	
	θ	17	37	19·0	−42	59	52	1·87	−5·6	F0	280	Sargas
	κ	17	42	29·0	−39	01	48	2·41	−3·0	B2	120	Girtab
	ι¹	17	47	34·9	−40	07	37	3·03	−8·4	F2	1700	
	G	17	49	51·3	−37	02	36	3·21	−0·1	K2	46	

Also above mag. 5:

		Mag.	Abs. mag.	Spectrum	Dist.
22		4·79	−2·5	B2	260
ι²		4·81	−0·6	A2	
d		4·78	1·4	A2	44
k		4·87	−4·8	B1	
15	ψ	4·94	1·1	A0	40

VARIABLE STARS

	R.A.		Dec.		Range	Type	Period, d.	Spectrum
	h	m	°	′				
RT	17	03·5	−36	55	7·0–16·0	Mira	449·0	M
FV	17	13·7	−32	51	7·9–8·6	Algol	5·72	B
RY	17	50·9	−33	42	7·5–8·4	Cepheid	20·31	F–G
RR	16	55·6	−30	35	5·0–12·4	Mira	279·4	M
RS	16	56·6	−45	06	6·2–13·0	Mira	320·0	M
RV	16	58·3	−33	37	6·6–7·5	Cepheid	6·06	F–G
BM	17	41·0	−32	13	6·8–8·7	Semi-reg.	850	K
RU	17	42·4	−43	45	7·8–13·7	Mira	369·2	M

DOUBLE STARS

	R.A.		Dec.		P.A.	Sep.	Mags.	
	h	m	°	′	°	″		
2	15	53·6	−25	20	274	2·5	4·7, 7·4	
π	15	58·9	−26	07	132	50·4	2·9, 12·1	
ξ	16	04·4	−11	22	AB 040	0·8	4·8, 5·1	
					AC 051	7·6	7·3	
β	16	05·4	−19	48	AC 021	13·6	2·6, 4·9	A is a close dble
11	16	07·6	−12	45	257	3·3	5·6, 9·9	

SCORPIUS *(continued)*

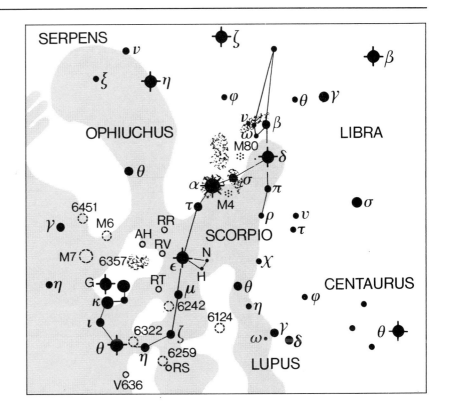

DOUBLE STARS *(continued)*

	R.A. h m	Dec. ° ′	P.A. °	Sep. ″	Mags.	
ν	16 12·0	−19 28	AB 003	0·9	4·3, 6·8	Binary, 45·7 y
			AC 337	41·1	6·4	
12	16 12·3	−28 25	073	4·0	5·9, 7·9	
σ	16 21·2	−25 36	273	20·0	2·9, 8·5	
α	16 29·4	−26 26	273	2·7	1·2, 5·4	Binary, 878 y

OPEN CLUSTERS

M	NGC	R.A. h m	Dec. ° ′	Diameter ′	Mag.	No. of stars	
	6124	16 25·6	−40 40	29	5·8	100	
	6178	16 35·7	−45 38	4	7·2	12	
	6192	16 40·3	−43 22	8	8·5	60	
	6231	16 54·0	−41 48	15	2·6	–	
	6242	16 55·6	−39 30	9	6·4	–	
	6259	17 00·7	−44 40	10	8·0	120	
	6268	17 02·4	−39 44	6	9·5	–	
	6281	17 04·8	−37 54	8	5·4	–	
	6383	17 34·8	−32 34	5	5·5	40	(Nebulosity)
	6400	17 40·8	−36 57	8	8·8	60	
6	6405	17 40·1	−32 13	15	4·2	50	Butterfly Cluster
	6416	17 44·4	−32 21	18	5·7	40	
	6451	17 50·7	−30 13	8	8·2	80	
7	6475	17 53·9	−34 49	80	3·3	80	

GLOBULAR CLUSTERS

M	NGC	R.A. h m	Dec. ° ′	Diameter ′	Mag.
80	6093	16 17·0	−22 59	8·9	7·2
4	6121	16 23·6	−26 32	26·3	5·9
	6388	17 36·3	−44 44	8·7	6·8

PLANETARY NEBULÆ

M	NGC	R.A. h m	Dec. ° ′	Diameter ″	Mag.	Mag. of central star
	6153	16 31·5	−40 15	25	11·5	–
	6302	17 13·7	−37 06	50	12·8	– Bug Nebula
	6337	17 22·3	−38 29	48	–	14·7

SCULPTOR

(Abbreviation: Scl).

Originally Apparatus Sculptoris. There is no star above the fourth magnitude.

See chart for Phœnix

BRIGHTEST STARS

Star	R.A. h	m	s	Dec. °	′	″	Mag.	Abs. mag.	Spectrum	Dist. pc
γ	23	28	49·3	−32	31	55	4·41	0·3	G8	47
β	23	32	58·0	−37	49	07	4·37	0·0	B9	77
δ	23	48	55·4	−28	07	49	4·51	0·6	A0	9·1
α	00	58	36·3	−29	21	27	4·31	−1·2	B8	130

Also above mag. 5:

	Mag.	Abs. mag.	Spectrum	Dist.
η	4·81		M5	

VARIABLE STARS

	R.A. h	m	Dec. °	′	Range	Type	Period, d.	Spectrum
Y	23	09·1	−30	08	7·5–9·0	Semi-reg.	300	M
S	00	15·4	−32	03	5·5–13·6	Mira	365·3	M
R	01	27·0	−32	33	5·8–7·7	Semi-reg.	370	N

DOUBLE STARS

	R.A. h	m	Dec. °	′	P.A. °	Sep. ″	Mags.	
δ	23	48·9	−28	08	AB 243	3·9	4·5, 11·5	
					AC 297	74·3	9·3	
ζ	00	02·3	−29	43	320	3·0	5·0, 13·0	
κ¹	00	09·3	−27	59	265	1·4	6·1, 6·2	
λ¹	00	42·7	−38	28	003	0·7	6·7, 7·0	
ε	01	45·6	−25	03	028	4·7	5·4, 8·6	Binary, 1192 y

GLOBULAR CLUSTER

M	NGC	R.A. h	m	Dec. °	′	Diameter ′	Mag.
	288	00	52·8	−26	35	13·8	8·1

GALAXIES

M	NGC	R.A. h	m	Dec. °	′	Mag.	Dimensions ′	Type
	IC 5332	23	34·5	−36	06	10·6	6·6×5·1	Sd
	7713	23	36·5	−37	56	11·6	4·3×2·0	SBd
	7755	23	47·9	−30	31	11·8	3·7×3·0	SBb
	7793	23	57·8	−32	35	9·1	9·1×6·6	Sd
	24	00	09·9	−24	58	11·5	5·5×1·6	Sb
	55	00	14·9	−39	11	8·2	32·4×6·5	SB
	134	00	30·4	−33	15	10·1	8·1×2·6	SBb
	253	00	47·6	−25	17	7·1	25·1×7·4	Scp
	300	00	54·9	−37	41	8·7	20·0×14·8	Sd
	613	01	34·3	−29	25	10·0	5·8×4·6	SBb

SCUTUM

(Abbreviation: Sct).

Originally Scutum Sobieskii or Clypeus Sobieskii (Sobieski's Shield). It has only one star brighter than the fourth magnitude (α 3·85), but is a rich area bordering Aquila, and contains the glorious 'Wild Duck' cluster M.11.

See chart for Aquila

BRIGHTEST STARS

Star	R.A. h	m	s	Dec. °	′	″	Mag.	Abs. mag.	Spectrum	Dist. pc
ζ	18	23	39·3	−08	56	03	4·68	0·2	K0	79
γ	18	29	11·7	−14	33	57	4·70	1·4	A2	45
α	18	35	12·1	−08	14	39	3·85	−0·2	K3	55
δ	18	42	16·2	−09	03	09	4·71v	1·3	F3	49
β	18	47	10·3	−04	44	52	4·22	−2·1	G5	150

Also above mag. 5:

	Mag.	Abs. mag.	Spectrum	Dist.
ε	4·90	−2·1	G8	250

VARIABLE STARS

	R.A. h	m	Dec. °	′	Range	Type	Period, d.	Spectrum
RZ	18	26·6	−09	12	7·3–8·8	Algol	15·19	B
R	18	47·5	−05	42	4·4–8·2	RV Tauri	140	G–K
S	18	50·3	−07	54	7·0–8·0	Semi-reg.	148	N

OPEN CLUSTERS

M	NGC	R.A. h	m	Dec. °	′	Diameter ′	Mag.	No. of stars	
	6664	18	36·7	−08	13	16	7·8	50	EV Scuti cluster
	6694	18	45·2	−09	24	15	8·0	30	
	6704	18	50·9	−05	12	6	9·2	30	
11	6705	18	51·1	−06	16	14	5·8	500	Wild Duck cluster

SCUTUM *(continued)*

GLOBULAR CLUSTER

M	NGC	R.A.		Dec.		Diameter	Mag.
		h	m	°	′	′	
	6712	18	53·1	−08	42	7·2	8·2

NEBULA

M	NGC	R.A.		Dec.		Dimensions	Mag. of illuminating star
		h	m	°	′	′	
	IC 1287	18	31·3	−10	50	4·4×3·4	5·5

SERPENS

(Abbreviation: Ser).

A curious constellation inasmuch as it is divided into two parts: Caput (the Head) and Cauda (the Body). It evidently represents the serpent with which Ophiuchus is struggling – and has had the worst of the encounter, since it has been pulled in half! Caput contains six stars above the fourth magnitude, and Cauda three:

Caput
α 2·65
μ 3·54
β 3·67
ε 3·71
δ 3·8 (combined)
γ 3·85

Cauda
η 3·26
θ 3·4 (combined)
ξ 3·54

See chart for Ophiuchus

BRIGHTEST STARS

Star		R.A.			Dec.			Mag.	Abs. mag.	Spectrum	Dist.	
		h	m	s	°	′	″				pc	
CAPUT												
13	δ	15	34	48·0	+10	32	21	3·80	1·7	F0	27	
21	ι	15	41	33·0	+19	40	13	4·52	1·2	A1	36	
24	α	15	44	16·0	+06	25	32	2·65	−0·1	K2	26	Unukalhai
28	β	15	46	11·2	+15	25	18	3·67	0·6	A2	37	
27	λ	15	46	26·5	+07	21	12	4·43	4·4	G0	11	
35	κ	15	48	44·3	+18	08	29	4·09	−0·5	M1	78	
32	μ	15	49	37·1	−03	25	49	3·54	0·6	A0	44	
37	ε	15	50	48·9	+04	28	40	3·71	1·6	A2	33	
41	γ	15	56	27·1	+15	39	42	3·85	3·7	F6	12	
CAUDA												
53	ν	17	20	49·4	−12	50	48	4·33	1·2	A1	42	
55	ξ	17	37	35·0	−15	23	55	3·54	1·7	F0	23	
56	ο	17	41	24·7	−12	52	31	4·26	1·4	A2	36	
57	ζ	18	00	28·7	−03	41	25	4·62	3·1	F3	22	
58	η	18	21	18·4	−02	53	56	3·26	1·7	K0	16	
63	θ	18	56	13·0	+04	12	13	4·06	2·1	A5	31	Combined
		18	56	14·5	+04	12	07	4·98	2·1	A5		mag. 3·4. Alya

Also above mag. 5:

		Mag.	Abs. mag.	Spectrum	Dist.
38	ρ	4·76	−0·3	K5	98
44	π	4·83	1·7	A3	42

VARIABLE STARS

	R.A.		Dec.		Range	Type	Period, d.	Spectrum
	h	m	°	′				
S	15	21·7	+14	19	7·0–14·1	Mira	368·6	M
τ⁴	15	36·5	+15	06	7·5–8·9	Irreg.	–	M
R	15	50·7	+15	08	5·1–14·4	Mira	356·4	M
U	16	07·3	+09	56	7·8–14·7	Mira	237·9	M
d	18	27·2	+00	12	4·9–5·9	?	?	G+A

DOUBLE STARS

	R.A.		Dec.		P.A.	Sep.	Mags.	
	h	m	°	′	°	″		
δ	15	34·8	+10	32	177	4·4	4·1, 5·2	Binary, 3168 y
β	15	46·2	+15	25	265	30·6	3·7, 9·9	
ν	17	20·8	−12	51	028	46·3	4·3, 8·3	
d	18	27·2	+00	12	318	3·8	5·3v, 7·6	
θ	18	56·2	+04	12	104	22·3	4·5, 4·5	

OPEN CLUSTERS

M	NGC	R.A.		Dec.		Diameter	Mag.	No. of stars
		h	m	°	′	′		
	6611	18	18·8	−13	47	7	6·0	In M.16
	6604	18	18·1	−12	14	2	7·0	30

GLOBULAR CLUSTER

M	NGC	R.A.		Dec.		Diameter	Mag.
		h	m	°	′	′	
5	5904	15	18·6	+02	05	17·4	5·8

NEBULA

M	NGC	R.A.		Dec.		Dimensions	
		h	m	°	′	′	
16	6611	18	18·8	−13	47	35×28	Eagle Nebula

SERPENS (continued)

GALAXY

M	NGC	R.A. h m	Dec. ° '	Mag.	Dimensions '	Type
	6118	16 21·8	−02 17	12·3	4·7×2·3	Sb

SEXTANS

(Abbreviation: Sxt).

A very obscure constellation, with no star as bright as the fourth magnitude.

See chart for Hydra

BRIGHTEST STAR

Star	R.A. h m s	Dec. ° ' "	Mag.	Abs. mag.	Spectrum	Dist. pc
15 α	10 07 56·2	−00 22 18	4·49	−1·1	B5	100

VARIABLE STAR

	R.A. h m	Dec. ° '	Range	Type	Period, d.	Spectrum
S	10 34·9	−00 20	8·2–13·5	Mira	261·0	M

DOUBLE STAR

	R.A. h m	Dec. ° '	P.A. °	Sep. "	Mags.
γ	09 52·5	−08 06	AB 067	0·6	5·5, 6·1 Binary, 75·6 y
			AC 325	35·8	12·0

GALAXIES

M	NGC	R.A. h m	Dec. ° '	Mag.	Dimensions '	Type
	2967	09 42·1	+00 20	11·6	3·0×2·9	Sc
	3115	10 05·2	−07 43	9·1	8·3×3·2	E6
	3166	10 13·8	+03 26	10·6	5·2×2·7	SBa
	3169	10 14·2	+03 28	10·4	4·8×3·2	Sb

TAURUS

(Abbreviation: Tau).

One of the brightest of the Zodiacal constellations. Mythologically it has been said to represent the bull into which Jupiter transformed himself when he wished to carry off Europa, daughter of the King of Crete.

Taurus includes the reddish first-magnitude star Aldebaran, and also the two most famous open clusters in the sky: the Pleiades and the Hyades. Altogether there are 16 stars above the fourth magnitude:

α	0·85v	27	3·63
β	1·65	γ	3·63
η	2·87	17	3·70
ζ	3·00	ξ	3·74
λ	3·3 (max)	δ¹	3·76
θ²	3·42	θ¹	3·85
ε	3·54	20	3·88
o	3·60	ν	3·91

Aldebaran has a very small range (0·78–0·93). β was formerly included in Auriga, as γ Aurigæ.

BRIGHTEST STARS

Star		R.A. h m s	Dec. ° ' "	Mag.	Abs. mag.	Spectrum	Dist. pc	
1	o	03 24 48·7	+09 01 44	3·60	0·3	G8	46	
2	ξ	03 27 10·1	+09 43 58	3·74	−0·1	B8p	30	
5		03 30 52·3	+12 56 12	4·11	−0·9	K0	97	
10		03 36 52·3	+00 24 06	4·28	4·0	F8	13	
17		03 44 52·5	+24 06 48	3·70	−1·9	B6	120	Electra (Pleiades)
19		03 45 12·4	+24 28 02	4·30	−0·9	B6	110	Taygete (Pleiades)
20		03 45 49·5	+24 22 04	3·88	−1·6	B7	120	Maia (Pleiades)
23		03 46 19·5	+23 56 54	4·18	−1·3	B6	120	Merope (Pleiades)
25	η	03 47 29·0	+24 06 18	2·87	−1·6	B7	73	Alcyone (Pleiades)
27		03 49 09·7	+24 03 12	3·63	−1·2	B8	90	Atlas (Pleiades)
28	BU	03 49 11·1	+24 08 12	var.	var.	B8p	29	Pleione (Pleiades)
35	λ	04 00 40·7	+12 29 25	var.	−1·7	B3	100	
38	ν	04 03 09·3	+05 59 21	3·91	1·2	A1	34	
37	A¹	04 04 41·7	+22 04 55	4·36	0·2	K0	60	
49	μ	04 15 32·0	+08 53 32	4·29	−1·7	B3	140	
54	γ	04 19 47·5	+15 37 39	3·63	0·2	K0	51	Hyadum Primus (Hyades)
61	δ¹	04 22 56·0	+17 32 33	3·76	0·2	K0	51	(Hyades)
65	κ	04 25 22·1	+22 17 38	4·22	2·4	A7	23	
68	δ³	04 25 29·3	+17 55 41	4·30	1·4	A2	38	(Hyades)
69	υ	04 26 18·4	+22 48 49	4·29	1·1	F0	42	(Hyades)
71		04 26 20·7	+15 37 06	4·49	2·6	F0	24	
73	π	04 26 36·4	+14 42 49	4·69	0·3	G8	72	
77	θ¹	04 28 34·4	+15 57 44	3·85	0·2	K0	54	(Hyades)
74	ε	04 28 36·9	+19 10 49	3·54	0·2	K0	45	Ain. (Hyades)
78	θ²	04 28 39·6	+15 52 15	3·42	0·5	A7	38	(Hyades)
86	ρ	04 33 50·8	+14 50 40	4·65	2·6	F0	26	
88	d	04 35 39·2	+10 09 39	4·25	2·4	A3	25	
87	α	04 35 55·2	+16 30 33	0·85	−0·3	K5	21	Aldebaran
90	c¹	04 38 09·4	+12 30 39	4·27	2·1	A5	27	
92	σ²	04 39 16·4	+15 55 05	4·68	2·1	A5	33	
94	τ	04 42 14·6	+22 57 25	4·28	−1·7	B3	150	

TAURUS *(continued)*

BRIGHTEST STARS *(continued)*

Star		R.A. h m s			Dec. ° ′ ″			Mag.	Abs. mag.	Spectrum	Dist. pc	
102	ι	05	03	05·7	+21	35	24	4·64	2·4	A7	28	
112	β	05	26	17·5	+28	36	27	1·65	−1·6	B7	40	Al Nath. = γ Aurigæ
119	CE	05	32	12·7	+18	35	39	4·3v	−4·8v	M2	290	
123	ζ	05	37	38·6	+21	08	33	3·00	−3·0	B2	150	Alheka
136		05	53	19·6	+27	36	44	4·58	0·4	B9·5	64	

Also above mag. 5:

		Mag.	Abs. mag.	Spectrum	Dist.
47		4·84	0·3	G5	81
50	ω	4·94	2·3	A5	
52	φ	4·95	0·0	K1	90
64	δ²	4·80	2·4	A7	30
75		4·97	−0·1	K2	100
104		4·92	5·0	G4	13
109		4·94	0·3	G8	85
111		4·98	4·0	F8	16
114		4·88	−1·7	B3	210
126		4·86	−2·3	B3	250
132		4·86	0·3	G8	73
134		4·91	0·3	B9	110
139		4·82	−5·7	B1	1100

VARIABLE STARS

	R.A. h m		Dec. ° ′		Range	Type	Period, d.	Spectrum	
BU	03	49·2	+24	08	4·8–5·5	Irreg.	–	Bp	Pleione
λ	04	00·7	+12	29	3·3–3·8	Algol	3·95	B+A	
T	04	22·0	+19	32	8·4–13·5	T Tauri	Irreg.	G–K	
R	04	28·3	+10	10	7·6–14·7	Mira	323·7	M	
HU	04	38·3	+20	41	5·9–6·7	Algol	2·06	A	
ST	05	45·1	+13	35	7·8–8·6	W Virginis	4·03	F–G	
TU	05	45·2	+24	25	5·9–8·6	Semi-reg.	190	N	
SU	05	49·1	+19	04	9·1–16·0	R Coronæ	–	G0p	

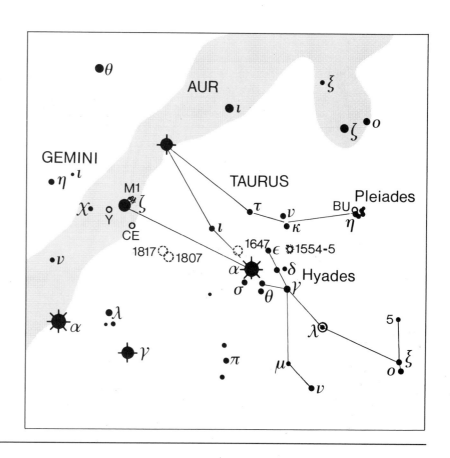

TAURUS (continued)

DOUBLE STARS

	R.A.		Dec.		P.A.	Sep.	Mags.	
	h	m	°	'	°	"		
φ	04	20·4	+27	21	250	52·1	5·0, 8·4	
χ	04	22·6	+25	38	024	19·4	5·5, 7·6	
66	04	23·9	+09	28	265	0·1	5·8, 5·9	Binary, 51·6 y
κ+67	04	25·4	+22	18	173	339	4·2, 5·3	
θ	04	28·7	+15	32	346	337·4	3·4, 3·8	
σ	04	39·3	+15	55	193	431·2	4·7, 5·1	
126	05	41·3	+16	32	238	0·3	5·3, 5·9	

OPEN CLUSTERS

M	NGC	R.A.		Dec.		Diameter	Mag.	No. of stars	
		h	m	°	'	'			
45	1432/5	03	47·0	+24	07	110	1·2	300+	Pleiades
	–	04	27	+16		330	1	200+	Hyades
	1647	04	46·0	+19	04	45	6·4	200	
	1746	05	03·6	+23	49	42	6·1	20	
	1807	05	10·7	+16	32	17	7·0	20	Asterism?
	1817	05	12·1	+16	42	16	7·7	60	

PLANETARY NEBULA

M	NGC	R.A.		Dec.		Diameter	Mag.	Mag. of central star
		h	m	°	'	"		
	1514	04	09·2	+30	47	114	10	9·4

NEBULÆ

M	NGC	R.A.		Dec.		Dimensions	Mag. of illuminating star
		h	m	°	'	'	
	1554–5	04	21·8	+19	32	var.	9v Hind's Variable Nebula (T Tauri)
1	1952	05	34·5	+22	01	6×4	16 Crab Nebula: SNR

TELESCOPIUM

(Abbreviation: Tel).

A small constellation with only one star above the fourth magnitude: α (3·51).

See chart for Ara

BRIGHTEST STARS

Star	R.A.			Dec.			Mag.	Abs. mag.	Spectrum	Dist.
	h	m	s	°	'	"				pc
ε	18	11	13·6	−45	57	15	4·53	0·3	G5	57
α	18	26	58·2	−45	58	06	3·51	−2·9	B3	180
ζ	18	28	49·7	−49	04	15	4·13	0·3	gG8	51

Also above mag. 5:

	Mag.	Abs. mag.	Spectrum	Dist.
λ	5·00	−0·8	B9	140
L	4·90	0·2	G9	74
ξ	4·94	−0·5	M2	120

VARIABLE STARS

	R.A.		Dec.		Range	Type	Period, d.	Spectrum
	h	m	°	'				
BL	19	06·6	−51	25	7·7–9·8	Eclipsing	778·1	F+M
RR	20	04·2	−55	43	6·5–16·5	Z Andromedæ	–	F5p
R	20	14·7	−46	58	7·6–14·8	Mira	461·9	M

PLANETARY NEBULA

M	NGC	R.A.		Dec.		Diameter	Mag.
		h	m	°	'	"	
	IC 4699	18	18·5	−45	59	10	11·9

TRIANGULUM

(Abbreviation: Tri).

A small but original constellation – and its main stars really do form a triangle! There are two stars above the fourth magnitude:

β 3·00

α 3·41

See chart for Andromeda

BRIGHTEST STARS

Star		R.A.			Dec.			Mag.	Abs. mag.	Spectrum	Dist.	
		h	m	s	°	'	"				pc	
2	α	01	53	04·8	+29	34	44	3·41	2·2	F6	18	Rasalmo- thallah
4	β	02	09	32·5	+34	59	14	3·00	0·3	A5	35	
9	γ	02	17	18·8	+33	50	50	4·01	0·6	A0	46	

Also above mag. 5:

		Mag.	Abs. mag.	Spectrum	Dist.
6		4·94	0·3	G5	85
8	δ	4·87	4·4	G0	10

TRIANGULUM (continued)

VARIABLE STAR

	R.A.		Dec.		Range	Type	Period, d.	Spectrum
	h	m	°	'				
R	02	37·0	+34	16	5·4–12·6	Mira	266·5	M

DOUBLE STARS

	R.A.		Dec.		P.A.	Sep.	Mags.
	h	m	°	'	°	"	
6	02	12·4	+30	18	071	3·9	5·3, 6·9
ι	02	15·9	+33	21	240	2·3	5·4, 7·0

GALAXIES

M	NGC	R.A.		Dec.		Mag.	Dimensions	Type
		h	m	°	'		'	
33	598	01	33·9	+30	39	5·7	62×39	Sc
	925	02	27·3	+33	35	10·0	9·8×6·0	SBc

TRIANGULUM AUSTRALE

(Abbreviation: TRA).

This 'triangle' also merits its name. There are 4 stars above the fourth magnitude:

α 1·92
β 2·85
γ 2·89
δ 3·85

BRIGHTEST STARS

Star	R.A.			Dec.			Mag.	Abs. mag.	Spectrum	Dist.	
	h	m	s	°	'	"				pc	
γ	15	18	54·5	−68	40	46	2·89	0·6	A0	28	
ε	15	36	43·1	−66	19	02	4·11	0·2	K0	44	
β	15	55	08·4	−63	25	50	2·85	3·0	F5	10	
δ	16	15	26·2	−63	41	08	3·85	−2·1	G2	110	
α	16	48	39·8	−69	01	39	1·92	−0·1	K2	17	Atria

Also above mag. 5:

	Mag.	Abs. mag.	Spectrum	Dist.
ζ	4·91	4·4	G0	12

VARIABLE STARS

	R.A.		Dec.		Range	Type	Period, d.	Spectrum
	h	m	°	'				
X	15	14·3	−70	05	8·1–9·1	Irreg.	–	N
R	15	19·8	−66	30	6·4–6·9	Cepheid	3·39	F–G
S	16	01·2	−63	47	6·1–6·8	Cepheid	6·32	F
U	16	07·3	−62	55	7·5–8·3	Cepheid	2·57	F

DOUBLE STARS

	R.A.		Dec.		P.A.	Sep.	Mags.
	h	m	°	'	°	"	
ε	15	36·7	−66	19	218	83·2	4·1, 9·5
ι	16	28·0	−64	03	016	29·6	5·3, 10·3

OPEN CLUSTER

M	NGC	R.A.		Dec.		Diameter	Mag.	No. of stars
		h	m	°	'	'		
	6025	16	03·7	−60	30	12	5·1	60

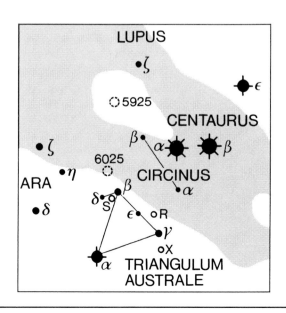

TUCANA

(Abbreviation: Tuc).

The dimmest of the 'Southern Birds', but graced by the presence of the glorious globular cluster 47 Tucanæ – inferior only to ω Centauri. The only star above the fourth magnitude is α (2·86), but the combined magnitude of β¹ and β² is 3·7.

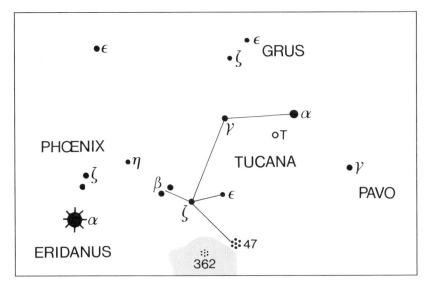

BRIGHTEST STARS

Star	R.A. h m s	Dec. ° ′ ″	Mag.	Abs. mag.	Spectrum	Dist. pc	
α	22 18 30·1	−60 15 35	2·86	−0·2	K3	35	
δ	22 27 19·9	−64 58 00	4·48	−0·2	B8	76	
γ	23 17 25·6	−58 14 08	3·99	0·6	F0	45	
ε	23 59 54·9	−65 34 38	4·50	0·0	B8·5	67	
ζ	00 20 04·4	−64 52 30	4·23	5·0	G0	7·1	
β {	00 32 32·7	−62 57 30	4·37	−0·2	B8 }	33	Combined
{	00 32 33·6	−62 57 57	4·53		A2 }		mag. 3·7

Also above mag. 5:

	Mag.	Abs. mag.	Spectrum	Dist.
η	5·00		A2	

VARIABLE STARS

	R.A. h m	Dec. ° ′	Range	Type	Period, d.	Spectrum
T	22 40·6	−61 33	7·7–13·8	Mira	250·8	M
S	00 23·1	−61 40	8·2–15·0	Mira	240·7	M
U	00 57·2	−75 00	8·0–14·8	Mira	259·5	M

DOUBLE STARS

	R.A. h m	Dec. ° ′	P.A. °	Sep. ″	Mags.	
δ	22 27·3	−64 58	282	6·9	4·5, 9·8	
β¹	00 31·5	−62 58	169	27·1	4·4, 4·8	
β²	00 31·6	−62 58	295	0·6	4·8, 6·0	Binary, 44·4 y
κ	01 15·8	−68 53	336	5·4	5·1, 7·3	

GLOBULAR CLUSTERS

M	NGC	R.A. h m	Dec. ° ′	Diameter ′	Mag.	
	104	00 24·1	−72 05	30·9	4·0	47 Tucanæ
	362	01 03·2	−70 51	12·9	6·6	

GALAXY

	R.A. h m	Dec. ° ′	Mag.	Dimensions ′
Small Cloud of Magellan	00 53	−72 50	2·3	280 × 160

URSA MAJOR

(Abbreviation: UMa).

The most famous of all northern constellations; circumpolar in England and the northern United States. Mythologically it represents Callisto, daughter of King Lycaon of Arcadia. Her beauty surpassed that of Juno, which so infuriated the goddess that she ill-naturedly changed Callisto into a bear. Years later Arcas, Callisto's son, found the bear while out hunting, and was about to shoot it when Jupiter intervened, swinging both Callisto and Arcas – also transformed into a bear – up to the sky: Callisto as Ursa Major, Arcas as Ursa Minor. The sudden jolt explains why both bears have tails stretched out to decidedly un-ursine length!

The seven main stars of Ursa Major are often called the Plough; sometimes King Charles' Wain, and, in America, the Big Dipper. One of the Plough stars is Mizar, the most celebrated naked-eye double in the sky since it makes a pair with Alcor; Mizar is itself a compound system, and the two main components are easily separable with a small telescope.

Ursa Major contains nineteen stars above the fourth magnitude:

ε 1·77
α 1·79
η 1·86
ζ 2·09
β 2·37
γ 2·44
ψ 3·01
μ 3·05
ι 3·14
θ 3·17
δ 3·31
o 3·36
λ 3·45
ν 3·48
κ 3·60
h 3·67
χ 3·71
ξ 3·79
υ 3·80v

Several of the Plough stars have alternative proper names; thus η may also be called Benetnasch, while γ may be Phekda or Phecda. However, Mizar is the only star whose proper name is generally used.

BRIGHTEST STARS

Star		R.A. h	m	s	Dec. °	'	"	Mag.	Abs. mag.	Spectrum	Dist. pc	
1	o	08	30	15·8	+60	43	05	3·36	−0·9	G4	71	Muscida
4	π²	08	40	12·9	+64	19	40	4·60	−0·1	K2	87	Ta Tsun
9	ι	08	59	12·4	+48	02	29	3·14	2·4	A7	15	Talita
10		09	00	38·3	+41	46	57	3·97	3·4	F5	14	
12	κ	09	03	37·5	+47	09	23	3·60	−0·6	A0	28	
15	f	09	08	52·2	+51	36	16	4·48	+2·1	A0	29	
14	τ	09	10	55·0	+63	30	49	4·67	0·5	F6+A5	70	
23	h	09	31	31·7	+63	03	42	3·67	1·7	F0	25	
25	θ	09	32	51·3	+51	40	38	3·17	2·2	F6	14	
24	d	09	34	28·8	+69	49	49	4·56	3·0	G2	21	
26		09	34	49·4	+52	03	05	4·50	1·4	A2	42	
29	υ	09	50	59·3	+59	02	19	3·80v	1·9	F2	26	
30	φ	09	52	06·3	+54	03	41	4·59	0·9	A3	13	
33	λ	10	17	05·7	+42	54	52	3·45	0·6	A2	37	Tania Borealis
34	μ	10	22	19·7	+41	29	58	3·05	−0·4	M0	48	Tania Australis
48	β	11	01	50·4	+56	22	56	2·37	1·2	A1	19	Merak
50	α	11	03	43·6	+61	45	03	1·79	0·2	K0	23	Dubhe
52	ψ	11	09	39·7	+44	29	54	3·01	0·0	K1	37	
53	ξ	11	18	10·9	+31	31	45	3·79	4·9	G0	7·7	Alula Australis
54	ν	11	18	28·7	+33	05	39	3·48	−0·2	K3	46	Alula Borealis
63	χ	11	46	03·0	+47	46	45	3·71	0·2	K0	37	Alkafzah
64	γ	11	53	49·7	+53	41	41	2·44	0·6	A0	23	Phad
69	δ	12	15	25·5	+57	01	57	3·31	1·7	A3	20	Megrez
77	ε	12	54	01·7	+55	57	35	1·77v	0·2	A0	19	Alioth
79	ζ	13	23	05·5	+54	55	31	2·09	0·4, 2·1	A2+A6	18	Mizar
80	g	13	25	13·4	+54	59	17	4·01	2·1	A5	25	Alcor
83		13	40	44·1	+54	40	54	4·66	−0·5	M2	100	
85	η	13	47	32·3	+49	18	48	1·86	−1·7	B3	33	Alkaid

Also above mag. 5:

	Mag.	Abs. mag.	Spectrum	Dist.
8 ρ	4·76		M0	
13 σ²	4·80	3·1	F7	22
18	4·83	2·1	A5	34
55	4·78	1·4	A2	42
56	4·99	−2·1	G8	260
78	4·93	3·0	F2	24

VARIABLE STARS

	R.A. h	m	Dec. °	'	Range	Type	Period, d.	Spectrum
X	08	40·8	+50	08	8·0–14·8	Mira	248·8	M
W	09	43·8	+55	57	7·9–8·6	W UMa	0·33	F+F
R	10	44·6	+68	47	6·7–13·4	Mira	301·7	M
TX	10	45·3	+45	34	7·1–8·8	Algol	3·06	B+F
VY	10	45·7	+67	25	5·9–6·5	Irreg.	–	N
VW	10	59·0	+69	59	6·8–7·7	Semi-reg.	125	M
ST	11	27·8	+45	11	7·7–9·5	Semi-reg.	81	M
CF	11	53·0	+37	43	8·5–12	Flare	–	B
Z	11	56·5	+57	52	6·8–9·1	Semi-reg.	196	M
RY	12	20·5	+61	19	6·7–8·5	Semi-reg.	311	M
T	12	36·4	+59	29	6·6–13·4	Mira	256·5	M
S	12	43·9	+61	06	7·0–12·4	Mira	225·0	S

DOUBLE STARS

	R.A. h	m	Dec. °	'	P.A. °	Sep. "	Mags.	
ι	08	59·2	+48	02	100	1·8	3·1, 10·2	Binary, 818 y
κ	09	03·6	+47	09	258	0·1	4·2, 4·4	Binary, 70 y
σ²	09	10·4	+67	08	000	3·4	4·8, 8·2	Binary, 1067 y
φ	09	52·1	+54	04	188	0·2	5·3, 5·4	Binary, 105·5 y
α	11	03·7	+61	45	283	0·7	1·9, 4·8	Binary, 44·7 y
ξ	11	18·2	+31	32	060	1·3	4·3, 4·8	Binary, 59·8 y
ν	11	18·5	+33	06	147	7·2	3·5, 9·9	
78	13	00·7	+56	22	057	1·5	5·0, 7·4	Binary, 116 y
ζ	13	23·9	+54	56	AB 152	14·4	2·3, 4·0	
					AC 071	708·7	2·1, 4·0	

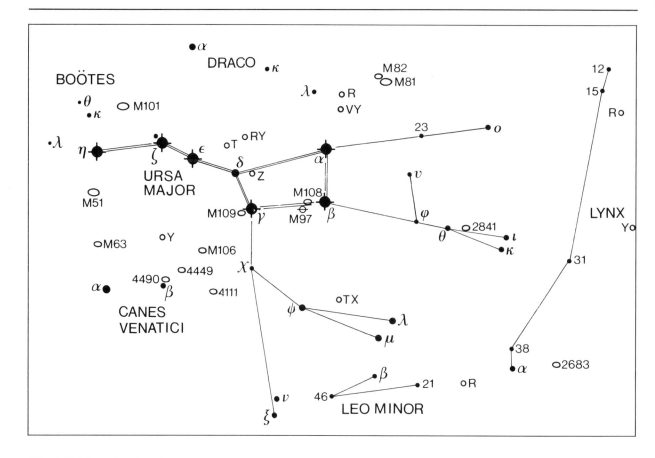

URSA MAJOR *(continued)*

GALAXIES

M	NGC	R.A. h	 m	Dec. °	 '	Mag.	Dimensions '	Type
	2681	08	53·5	+51	19	10·3	3·8×3·5	Sa
	2685	08	55·6	+58	44	11·0	5·2×3·0	Sbp
	2768	09	11·6	+60	02	10·0	6·3×2·8	E5
	2787	09	19·3	+69	12	10·8	3·4×2·3	Sap
	2841	09	22·0	+50	58	9·3	8·1×3·8	Sb
	2976	09	47·3	+67	55	10·1	4·9×2·5	Scp
	2985	09	50·4	+72	17	10·5	4·5×3·4	Sb
81	3031	09	55·6	+69	04	6·9	25·7×14·1	Sb
82	3034	09	55·8	+69	41	8·4	11·2×4·6	Pec.
	3077	10	03·3	+68	44	9·8	4·6×3·6	E2p
	3079	10	02·0	+55	41	10·6	7·6×1·7	Sb
	3184	10	18·3	+41	25	9·7	6·9×6·8	Sc
	3198	10	19·9	+45	33	10·4	8·3×3·7	Sc
	3310	10	38·7	+53	30	10·9	3·6×3·0	SBc
	3359	10	46·6	+63	13	10·4	6·8×4·3	SBc
	3610	11	18·4	+58	47	10·7	3·2×2·5	E2p
	3631	11	21·0	+53	10	10·4	4·6×4·1	Sc
	3675	11	26·1	+43	35	10·9	5·9×3·2	Sb
	3687	11	28·0	+29	31	12·6	2·0×2·0	Sb
	3718	11	32·6	+53	04	10·5	8·7×4·5	SBap
	3726	11	33·3	+47	02	10·4	6·0×4·5	Sc
	3877	11	46·1	+47	30	11·6	5·4×1·5	Sb
	3898	11	49·2	+56	05	10·8	4·4×2·6	Sb
	3945	11	53·2	+60	41	10·6	5·5×3·6	SBa
	3949	11	53·7	+47	52	11·0	3·0×1·8	Sb
	3953	11	53·8	+52	20	10·1	6·6×3·6	Sb
	3998	11	57·9	+55	27	10·6	3·1×2·5	E2p
	4026	11	59·4	+50	58	11·7	5·1×1·4	S0
	4036	12	01·4	+61	54	10·6	4·5×2·0	E6
	4041	12	02·2	+62	08	11·1	2·8×2·7	Sc
	4051	12	03·2	+44	32	10·3	5·0×4·0	Sc
	4062	12	04·1	+31	54	11·2	4·3×2·0	Sb
	4088	12	05·6	+50	33	10·5	5·8×2·5	Sc
	4096	12	06·0	+47	29	10·6	6·5×2·0	Sc
	4100	12	06·2	+49	35	11·5	5·2×1·9	Sb

URSA MAJOR (continued)

GALAXIES (continued)

M	NGC	R.A.		Dec.		Mag.	Dimensions	Type	
		h	m	°	'		'		
	4605	12	40·0	+61	37	11·0	5·5×2·3	SBcp	
	5308	13	47·0	+60	58	11·3	3·5×0·8	S0	
	5322	13	49·3	+60	12	10·0	5·5×3·9	E2	
101	5457	14	03·2	+54	21	7·7	26·9×26·3	Sc	Pinwheel
	5475	14	05·2	+55	45	12·4	2·2×0·6	Sa	

PLANETARY NEBULA

M	NGC	R.A.		Dec.		Diameter	Mag.	Mag. of central star	
		h	m	°	'	"			
97	3587	11	14·8	+55	01	194	12·0	15·9	Owl Nebula

URSA MINOR

(Abbreviation: UMi).

The north polar constellation. Mythologically it represents Arcas, son of Callisto (see Ursa Major). There are three stars above the fourth magnitude:

α 1·99v
β 2·08
γ 3·05

BRIGHTEST STARS

Star		R.A.			Dec.			Mag.	Abs. mag.	Spectrum	Dist.	
		h	m	s	°	'	"				pc	
2		01	01	31	+85	59	24	4·52	0·0	K0	77	
1	α	02	31	50·4	+89	15	51	1·99v	−4·6	F8	208	Polaris
4	π²	08	40	12·9	+64	19	40	4·60	−0·1	K2	87	
5		14	27	31·4	+75	41	45	4·25	−0·3	K4	79	
7	β	14	50	42·2	+74	09	19	2·08	−0·3	K4	29	Kocab
13	γ	15	20	43·6	+71	50	02	3·05	−1·1	A3	69	Pherkad Major
16	ζ	15	44	03·3	+77	47	40	4·32	1·7	A3	33	Alifa
22	ε	16	45	57·8	+82	02	14	4·23	0·3	G5	61	
23	δ	17	32	12·7	+86	35	11	4·36	1·2	A1	44	Yildun

Also above mag. 5:

		Mag.	Abs. mag.	Spectrum	Dist.	
4		4·82	−0·2	K3	90	
21	η	4·95	2·6	F0	28	Alasco

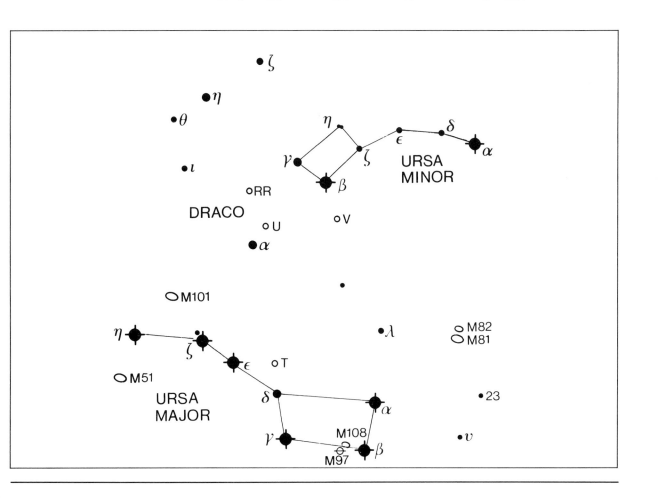

URSA MINOR *(continued)*

VARIABLE STARS

	R.A. h	m	Dec. °	'	Range	Type	Period, d.	Spectrum
T	13	34·7	+73	26	8·1–15·0	Mira	313·9	M
V	13	38·7	+74	19	7·4–8·8	Semi-reg.	72	M
U	14	17·3	+66	48	7·4–12·7	Mira	326·5	M
RR	14	57·6	+65	56	6·0–6·5	Semi-reg.?	40?	M
S	15	29·6	+78	38	7·7–12·9	Mira	326·2	N
R	16	30·0	+72	17	8·8–11·0	Semi-reg.	324	M

DOUBLE STAR

	R.A. h	m	Dec. °	'	P.A. °	Sep. "	Mags.
α	02	31·8	+89	16	218	18·4	2·0, 9·0

VELA

(Abbreviation: Vel).

The Sails of the dismembered ship Argo. There are 14 stars above the fourth magnitude:

γ 1·78
δ 1·96
λ 2·21v
κ 2·50
μ 2·69
N 3·13
φ 3·54
ψ 3·60
o 3·62v
c 3·75
p 3·84
b 3·84
q 3·85
a 3·91

δ and κ make up the 'False Cross' with ε and ι Carinæ.

See chart for Carina

BRIGHTEST STARS

Star	R.A. h	m	s	Dec. °	'	"	Mag.	Abs. mag.	Spectrum	Dist. pc	
γ	08	09	31·9	−47	20	12	1·78	−4·1	WC7	160	Regor
e	08	37	38·6	−42	59	21	4·14	−5·8	A9	950	
o	08	40	17·6	−52	55	19	3·62v	−1·7	B3	120	
δ	08	44	42·2	−54	42	30	1·96	0·6	A0	21	Koo She
b	08	40	37·6	−46	38	55	3·84	−8·4	F2	1900	
a	08	46	01·7	−46	02	30	3·91	−0·6	A0	75	
W	09	00	05·4	−41	15	14	4·45	2·4	F8	20	
c	09	04	09·2	−47	05	02	3·75	−0·1	K2	57	
λ	09	07	59·7	−43	25	57	2·21v	−4·4	K5	150	Al Suhail al Wazn
k²	09	15	45·1	−37	24	47	4·62	2·5	F3	22	
κ	09	22	06·8	−55	00	38	2·50	−3·0	B2	120	Markeb
ψ	09	30	41·9	−40	28	00	3·60	1·9	F2	19	
N	09	31	13·3	−57	02	04	3·13	−0·3	K5	45	
M	09	36	49·7	−49	21	18	4·35	2·1	dA5	28	
m	09	51	40·7	−46	32	52	4·58	0·3	gG6	46	
φ	09	56	51·7	−54	34	03	3·54	−6·0	B5	770	
q	10	14	44·1	−42	07	19	3·85	1·4	A2	31	
HD 89682	10	19	36·8	−55	01	46	4·57	−5·9	cK	910	
HD 89890	10	20	55·4	−56	02	36	4·50	−2·3	B3	210	
p	10	37	18·0	−48	13	32	3·84	2·0	F4	23	
x	10	39	18·3	−55	36	12	4·28	−2·1	G2	150	
μ	10	46	46·1	−49	25	12	2·69	0·3	G5	30	
i	11	00	09·2	−42	13	33	4·39	0·6	A2	52	

Also above mag. 5:

	Mag.	Abs. mag.	Spectrum	Dist.
HD 74272	4·77	−2·3	A3	260
HD 92036	4·89	−0·3	K5	93

VARIABLE STARS

	R.A. h	m	Dec. °	'	Range	Type	Period, d.	Spectrum
AH	08	12·0	−46	39	5·5–5·9	Cepheid	4·23	F
AI	08	14·1	−44	34	6·4–7·1	Delta Scuti	0·11	A–F
RZ	08	37·0	−44	07	6·4–7·6	Cepheid	20·40	G
T	08	37·7	−47	22	7·7–8·3	Cepheid	4·64	F
SW	08	43·6	−47	24	7·4–9·0	Cepheid	23·47	K
SX	08	44·9	−46	21	8·0–8·6	Cepheid	9·55	G
CV	09	00·6	−51	33	6·5–7·3	Algol	6·89	B+B
SY	09	12·4	−43	47	7·6–8·1	Semi-reg.	63	M
RW	09	20·3	−49	31	7·8–12·0	Mira	451·7	M
V	09	22·3	−55	58	7·2–7·9	Cepheid	4·37	F
S	09	33·2	−45	13	7·7–9·5	Algol	5·93	A+K
U	09	33·2	−45	31	7·9–8·2	Semi-reg.	37	M
Z	09	52·9	−54	11	7·8–14·8	Mira	421·6	M
SV	10	44·9	−56	17	7·9–9·1	Cepheid	14·10	F–G

DOUBLE STARS

	R.A. h	m	Dec. °	'	P.A. °	Sep. "	Mags.
γ	08	09·5	−47	20	AB 220	41·2	1·9, 4·2
					AC 151	62·3	8·2
					AD 141	93·5	9·1
					DE 146	1·8	12·5
b	08	40·6	−46	39	058	37·5	3·8, 10·2

VELA (continued)

DOUBLE STARS (continued)

	R.A. h m	Dec. ° ′	P.A. °	Sep. ″	Mags.
δ	08 44·7	−54 43	AB 153	2·6	2·1, 5·1
			AC 061	69·2	11·0
			CD 102	6·2	13·5
μ	10 46·8	−49 25	055	2·3	2·7, 6·4 Binary, 116 y

OPEN CLUSTERS

M	NGC	R.A. h m	Dec. ° ′	Diameter ′	Mag.	No. of stars	
	2547	08 10·7	−49 16	20	4·7	80	
	IC 2391	08 40·2	−53 04	50	2·5	30	o Velorum cluster
	IC 2395	08 41·1	−48 12	8	4·6	40	
	2669	08 44·9	−52 58	12	6·1	40	
	2670	08 45·5	−48 47	9	7·8	30	
	IC 2488	09 27·6	−56 59	15	7·4	70	
	2910	09 30·4	−52 54	5	7·2	30	
	2925	09 33·7	−53 26	12	8·3	40	
	2972	09 40·3	−50 20	4	9·9	25	
	3033	09 48·8	−56 25	5	8·8	50	
	3228	10 21·8	−51 43	18	6·0	15	

GLOBULAR CLUSTER

M	NGC	R.A. h m	Dec. ° ′	Diameter ′	Mag.
	3201	10 17·6	−46 25	18·2	6·7

PLANETARY NEBULA

M	NGC	R.A. h m	Dec. ° ′	Diameter ″	Mag.	Mag. of central star
	3132	10 07·7	−40 26	47	8·2	10·1

VIRGO

(Abbreviation: Vir).

A very large constellation, representing Astræa – the goddess of justice, daughter of Jupiter and Themis. The 'bowl' of Virgo abounds in faint galaxies. Of the 9 stars down to the fourth magnitude, Spica is an eclipsing variable over a small range (0·91–1·01), and γ (Arich) used to be one of the most spectacular binary pairs in the sky, though it is closing up and will have become very difficult to separate by the end of the 20th century. The leading stars are:

α	0·98v
γ	2·75
ε	2·83
ζ	3·37
δ	3·38
β	3·61
109	3·72
μ	3·88
η	3·89

BRIGHTEST STARS

Star		R.A. h m s	Dec. ° ′ ″	Mag.	Abs. mag.	Spectrum	Dist. pc	
3	ν	11 45 51·5	+06 31 45	4·03	−0·5	M1	51	
5	β	11 50 41·6	+01 45 53	3·61	3·6	F8	10	Zavijava
8	π	12 00 52·3	+06 36 51	4·66	1·7	A3	37	
9	o	12 05 12·5	+08 43 59	4·12	0·3	G8	34	
15	η	12 19 54·3	−00 40 00	3·89	1·4	A2	32	Zaniah
26	χ	12 39 14·7	−07 59 45	4·66	−0·1	K2	78	
29	γ	12 41 39·5	−01 26 57	2·75	2·6	F0+F0	11	Arich
43	δ	12 55 36·1	+03 23 51	3·38	−0·5	M3	45	Minelauva
47	ε	13 02 10·5	+10 57 33	2·83	0·2	G9	32	Vindemiatrix
51	θ	13 09 56·9	−05 32 20	4·38	1·2	A1	43	
61		13 18 24·2	−18 18 21	4·74	5·1	G6	8·4	
67	α	13 25 11·5	−11 09 41	0·98v	−3·5	B1	79	Spica
74		13 31 57·8	−06 15 21	4·69	−0·5	gM3	110	
79	ζ	13 34 41·5	−00 35 46	3·37	1·7	A3	340	Heze
93	τ	14 01 38·7	+01 32 40	4·26	1·7	A3	32	
98	κ	14 12 53·6	−10 16 25	4·19	−0·2	K3	71	
99	ι	14 16 00·8	−06 00 02	4·08	0·7	F6	22	Syrma
100	λ	14 19 06·5	−13 22 16	4·52	2·3	A0	20	Khambalia
107	μ	14 43 03·5	−05 39 30	3·88	1·9	F3	26	
109		14 46 14·9	+01 53 34	3·72	0·6	A0	38	
110		15 02 53·9	+02 05 28	4·40	0·2	K0	65	

Also above mag. 5:

		Mag.	Abs. mag.	Spectrum	Dist.
2	ξ	4·85		A3	
16		4·96	0·2	K0	67
30	ρ	4·88	0·6	A0	63
40	ψ	4·80	−0·5	M3	120
60	σ	4·80	−0·5	M2	100
69		4·76	0·0	K1	32
78		4·94	0·9	A2	26

VARIABLE STARS

	R.A. h m	Dec. ° ′	Range	Type	Period, d.	Spectrum
X	12 01·9	+09 04	7·3–11·2	?	−	F
SS	12 25·3	+00 48	6·0–9·6	Mira	354·7	N
R	12 38·5	+06 59	6·0–12·1	Mira	145·6	M

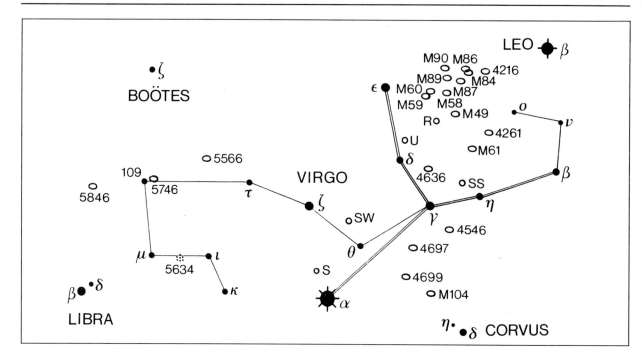

VIRGO *(continued)*

VARIABLE STARS *(continued)*

	R.A. h m	Dec. ° '	Range	Type	Period, d.	Spectrum
U	12 51·1	+05 33	7·5–13·5	Mira	206·8	M
S	13 33·0	−07 12	6·3–13·2	Mira	377·4	M
RS	14 27·3	+04 41	7·0–14·4	Mira	352·8	M

DOUBLE STARS

	R.A. h m	Dec. ° '	P.A. °	Sep. "	Mags.	
17	12 22·5	+05 18	337	20·0	6·6, 9·4	
γ	12 41·7	−01 27	287	3·0	3·5, 3·5	Binary, 171·4 y
θ	13 09·9	−05 32	343	7·1	4·4, 9·4	
73	13 32·0	−18 44	183	0·1	6·7, 6·9	
84	13 43·1	+03 32	229	2·9	5·5, 7·9	
τ	14 01·6	+01 33	290	80·0	4·3, 9·6	
φ	14 28·2	−02 14	110	4·8	4·8, 9·3	

GALAXIES

M	NGC	R.A. h m	Dec. ° '	Mag.	Dimensions '	Type	
	4216	12 15·9	+13 09	10·0	8·3×2·2	Sb	
	4261	12 19·4	+05 49	10·3	3·9×3·2	E2	
61	4303	12 21·9	+04 28	9·7	6·0×5·5	Sc	
84	4374	12 25·1	+12 53	9·3	5·0×4·4	E1	
	4429	12 27·4	+11 07	10·2	5·5×2·6	S0	
	4438	12 27·8	+13 01	10·1	9·3×3·9	Sap	
	4442	12 28·1	+09 48	10·5	4·6×1·9	E5p	
86	4406	12 26·2	+12 57	9·2	7·4×5·5	E3	
49	4472	12 29·8	+08 00	8·4	8·9×7·4	E4	
87	4486	12 30·8	+12 24	8·6	7·2×6·8	E1	Virgo A
	4699	12 49·0	−08 40	9·6	3·5×2·7	Sa	
	4535	12 34·3	+08 12	9·8	6·8×5·0	SBc	
	4546	12 35·5	−03 48	10·3	3·5×1·7	E6	
	4527	12 34·1	+02 39	11·3	6·3×2·3	Sb	
	4536	12 34·5	+02 11	11·0	7·4×3·5	Sc	
	4546	12 35·5	−03 48	11·3	3·5×1·7	E6	
89	4552	12 35·7	+12 33	9·8	4·2×4·2	E0	
90	4569	12 36·8	+13 10	9·5	9·5×4·7	Sb	
58	4579	12 37·7	+11 49	9·8	5·4×4·4	Sb	
104	4594	12 40·0	−11 37	8·3	8·9×4·1	Sb	Sombrero Hat
	4596	12 39·9	+10 11	10·5	3·9×2·8	SBa	
59	4621	12 42·0	+11 39	9·8	5·1×3·4	E3	
60	4649	12 43·7	+11 33	8·8	7·2×6·2	E1	
	4654	12 44·0	+13 08	10·5	4·7×3·0	Sc	
	4636	12 42·8	+02 41	9·6	6·2×5·0	E1	

VIRGO (continued)

GALAXIES (continued)

M	NGC	R.A.		Dec.		Mag.	Dimensions	Type
		h	m	°	′		′	
	4660	12	44·5	+11	11	11·9	2·8×1·9	E5
	4697	12	48·6	−05	48	9·3	6·0×3·8	E4
	4699	12	49·0	−08	40	9·6	3·5×2·7	Sa
	4753	12	52·4	−01	12	9·9	5·4×2·9	Pec.
	4762	12	52·9	+11	14	10·2	8·7×1·6	SB0
	4856	12	59·3	−15	02	10·4	4·6×1·6	SBa
	5247	13	38·1	−17	53	10·5	5·4×4·7	Sb
	5363	13	56·1	+05	15	10·2	4·2×2·7	Ep
	5364	13	56·2	+05	01	10·4	7·1×5·0	SB+p
	5068	13	18·9	−21	02	10·8	6·9×6·3	SBc
	5850	15	07·1	+01	33	11·7	4·3×3·9	SBb

VOLANS

(Abbreviation: Vol).

Originally Piscis Volans. A small constellation, intruding into Carina. There are four stars above the fourth magnitude:

γ 3·6 (combined)
β 3·77
ζ 3·95
δ 3·98

See chart for Carina

BRIGHTEST STARS

Star	R.A.			Dec.			Mag.	Abs. mag.	Spectrum	Dist.	
	h	m	s	°	′	″				pc	
γ {	07	08	42·3	−70	29	50	5·7	2·8	dF4 }	23	Combined
{	07	08	45·0	−70	29	57	3·8	0·9	G8 }		mag. 3·6
δ	07	16	49·8	−67	57	27	3·98	−5·4	F8	730	
ζ	07	41	49·3	−72	36	22	3·95	0·2	K0	54	
ε	08	07	55·9	−68	37	02	4·35	−1·1	B5	120	
β	08	25	44·3	−66	08	13	3·77	−0·1	K2	59	
α⁻	09	02	26·9	−66	23	46	4·00	2·1	A5	24	

VARIABLE STAR

	R.A.		Dec.		Range	Type	Period, d.	Spectrum
	h	m	°	′				
S	07	29·8	−73	23	7·7–13·9	Mira	395·8	M

DOUBLE STARS

	R.A.		Dec.		P.A.	Sep.	Mags.
	h	m	°	′	°	″	
γ²	07	08·8	−70	30	300	13·6	4·0, 5·9
ζ	07	41·8	−72	36	116	16·7	4·0, 9·8
ε	08	07·9	−68	37	024	6·1	4·4, 8·0
κ	08	19·8	−71	31	AB 057	65·0	5·4, 5·7
					BC 030	37·7	8·5
θ	08	39·1	−70	23	108	45·0	5·3, 10·3

GALAXY

M	NGC	R.A.		Dec.		Mag.	Dimensions	Type
		h	m	°	′		′	
	2442	07	36·4	−69	32	11·2	6·0×5·5	SBb

VULPECULA

(Abbreviation: Vul).

Originally Vulpecula et Anser, the Fox and Goose – nowadays the goose has disappeared (possibly the fox has eaten it). Vulpecula contains no star above magnitude 4·4, and is notable only because of the presence of the Dumbbell Nebula and the fact that several novæ have appeared within the boundaries of the constellation.

See chart for Cygnus

BRIGHTEST STARS

Star		R.A.			Dec.			Mag.	Abs. mag.	Spectrum	Dist.
		h	m	s	°	′	″				pc
6	α	19	28	42·2	+24	39	54	4·44	0·0	M0	26
13		19	53	27·5	+24	04	47	4·58	−0·6	A0	110
15		20	01	05·9	+27	45	13	4·64	1·7	A0	18
23		20	15	49·1	+27	48	51	4·52	−0·2	K3	88

Also above mag. 5:

	Mag.	Abs. mag.	Spectrum	Dist.
12	4·95	−1·7	B3	210

VARIABLE STARS

	R.A.		Dec.		Range	Type	Period, d.	Spectrum
	h	m	°	′				
RS	19	17·7	+22	26	6·9–7·6	Algol	4·46	B+A
Z	19	21·7	+25	34	7·4–9·2	Algol	2·45	B+A
U	19	36·6	+20	20	6·8–7·5	Cepheid	7·99	F–G
T	20	51·5	+28	15	5·4–6·1	Cepheid	4·44	F–G
SV	19	51·5	+27	28	6·7–7·7	Cepheid	45·03	F–K
R	21	04·4	+23	49	7·0–14·3	Mira	136·4	M

DOUBLE STARS

	R.A.		Dec.		P.A.	Sep.	Mags.
	h	m	°	′	°	″	
2 (ES)	19	17·7	+23	02	127	1·8	5·4, 9·2
α −8	19	28·7	+24	40	028	413·7	4·4, 5·8

VULPECULA *(continued)*

OPEN CLUSTERS

M	NGC	R.A. h	m	Dec. °	′	Diameter ′	Mag.	No. of stars
	6823	19	43·1	+23	18	12	7·1	30
	6830	19	51·0	+23	04	12	7·9	20
	6885	20	12·0	+26	29	7	5·7	30
	6940	20	34·6	+28	18	31	6·3	60

PLANETARY NEBULA

M	NGC	R.A. h	m	Dec. °	′	Dimensions ″	Mag.	Mag. of central star	
27	6853	19	59·6	+22	43	350×910	7·6	13·9	Dumbbell Nebula

Dome of the Russian 236-inch reflector, the world's largest single-mirror telescope.

TELESCOPES AND OBSERVATORIES

OBSERVATORIES

The oldest observatory building now standing is the Chomsong-dae, in Kyongju, South Korea, AD 632.

The oldest observatory to be associated with a national body is at Leiden University, Holland, dating from 1632.

The oldest national observatory was that at Copenhagen, Denmark. Unfortunately, the original buildings were destroyed by fire.

The first national British observatory is the Royal Greenwich Observatory, founded in 1675 by order of King Charles II. The original observatory in Greenwich Park, some buildings of which were designed by Wren, is now a museum (Flamsteed House) and the research equipment was shifted to Herstmonceux, Sussex as well as La Palma in the Canary Islands. It has unfortunately been announced that the R.G.O. is to be transferred to Cambridge, and Herstmonceux closed.

The world's highest observatory is at Boulder, Colorado (University of Denver High Altitude Observatory) at 4297 m. It was opened in 1973. The main telescope is the 60·48 cm Ealing Beck reflector. Only slightly lower is the Mauna Kea Observatory in Hawaii, close to the summit of a volcano which is (we hope!) extinct; altitude 4205 m. The main telescopes are the UKIRT (United Kingdom Infra-Red Telescope) (375 cm), the CFH or Canada–France–Hawaii reflector (366 cm), the University of Hawaii reflector (224 cm) and the NASA infra-red telescope (305 cm). At this level one is above 40 per cent of the atmosphere and 90 per cent of the atmospheric water vapour, making the site excellent for infra-red work.

The world's lowest observatory is at Homestake Mine, South Dakota, at a depth of 1·5 km below ground level. The 'telescope' is a 100 000 gallon tank of 'cleaning fluid' to trap solar neutrinos. The experiment was started in 1965, and is still continuing.

The only flying observatory is the KAO (Kuiper Airborne Observatory) which is really a modified Lockheed C–141 aircraft carrying a 91 cm reflector. At peak altitude it flies above 85 per cent of the atmosphere. The telescope is installed in an open cavity recessed into the port side of the aircraft, immediately ahead of the wing.

The first space telescope should be the 238 cm due to be launched from the Shuttle in 1989. It is not easy to decide whether this equipment should be classified as a telescope or as an observatory!

TELESCOPES

The world's largest optical telescope is the 600 cm reflector at Mount Semirodriki, near Zelenchukskaya in the Caucasus (USSR), altitude 2080 m. The 70-tonne mirror was completed in 1974, and regular observations began on 7 Feb 1976. The weight of the full assembly is 840 tonnes. The only other telescope over 500 cm aperture is the Hale Reflector on Palomar Mountain, Cal., USA (508 cm), completed in 1948.

The world's largest refractor is the 101·6 cm telescope at Yerkes Observatory, USA. (The largest object-glass ever made was the 124·5 cm lens shown at the Paris Exhibition of 1900, but it was never used seriously.)

The first telescope whose existence can be definitely proved was made by H. Lippershey (Holland) in 1608, but it is strongly suspected that some telescopes had been made before this. **The first reflector** was made by Newton, and presented to the Royal Society in 1671, so that the actual construction was a year or two before this, possibly 1668 or 1669.

The first multiple-mirror telescope is the MMT on Mount Hopkins, Arizona. It uses six 183 cm mirrors in conjunction, so that the total light-grasp is equal to a single 442 cm mirror. It has proved to be very successful, and further MMTs are being planned, for Mauna Kea and elsewhere.

The largest speculum metal mirror ever made for an optical telescope was 182·9 cm in diameter. It was made by the third Earl of Rosse for his observatory at Birr Castle, Ireland in 1845. He later made a second mirror for the same material. The main mirror is now in the Science Museum, South Kensington, and the telescope tube is on display at Birr.

The world's largest infra-red telescope is the UKIRT, on Mauna Kea. It was made in Sheffield, and has proved to be so good that it is used for optical research as well as for infra-red.

The world's largest telescope designed specially for microwave observations is the James Clerk Maxwell Telescope, on Mauna Kea. It has a 'dish' 15 metres in diameter, made up of 276 separate panels which have been fitted together to form the correct shape.

The largest telescope ever set up in Britain was the INT or Isaac Newton Telescope, with a mirror 244 cm in diameter (slightly less than the figure usually quoted). It was completed in 1967, but has now been moved to the new Canary Island observatory – El Observatorio des Roques de los Muchachos (the Observatory of the Rocks of the Boys) on La Palma (2423 m altitude). It has a new 256 cm mirror. The 4·2-metre William Herschel Telescope on La Palma was brought into full operation in 1987.

The largest balloon-borne telescope in the world is the 101·6 cm reflector designed and used by Dr. R. Joseph, Imperial College, University of London.

200-inch Reflector. (Patrick Moore)

The largest solar telescope is at Kitt Peak, USA. It uses a 203 cm heliostat, and has an inclined tunnel 146 m long. It was completed in 1962. Though designed for solar work, it is also capable of being used for other types of observation as well – an initially unexpected bonus.

PLANETARIA

The ancestor of the modern planetarium is the Gottorp Globe, made by H. Busch in Denmark about 1654–64. It was 4 m in diameter. The audience sat inside it; the stars were painted on the inside of the globe.

The first modern-type planetarium (due to Dr. W. Bauersfelt of the Zeiss optical works) was opened at the Deutsches Museum in Bonn in 1923.

The world's largest planetarium is at Moscow, 25·15 m.

The first British planetarium was at Madame Tussaud's, Baker Street, London; the dome is 21·33 m in diameter. The planetarium was opened in 1958. **The first Irish planetarium** was at Armagh, opened in 1968.

SOME OF THE WORLD'S GREAT OBSERVATORIES

Location	Name	Latitude °	'	"	Longitude °	'	"	Altitude metres
Aarhus, Denmark	Ole Rømer Obs.	+56	07	40·0	E 10	11·8		50
Alma Ata, USSR	Mountain Obs. of Academy of Sciences	+43	11	16·9	E 76	57·4		1450
Ann Arbor, Michigan	Univ. of Michigan	+42	16	48·7	W 83	43·8		282
Arcetri (Florence), Italy	Astrophysical Obs.	+43	45	44·7	E 11	15·3		184
Armagh, N. Ireland	Armagh Obs.	+54	21	11·1	W 06	38·9		64
Athens, Greece	National Obs.	+37	58	19·7	E 23	43·0		110
Auckland, N. Zealand	Auckland Public Obs.	−36	54	28·0	E 174	46·6		80
Barcelona, Spain	Fabra Obs.	+41	24	59·3	E 02	07·6		415
Beirut, Lebanon	American Univ. Obs.	+33	54	22·0	E 35	28·2		38
Belgrade, Yugoslavia	Obs. of Academy of Science	+44	48	13·2	E 20	30·8		253
Berlin, E. Germany	Wilhelm Förster Obs.	+52	28	30·0	E 13	25·5		40
Berlin-Treptow, E. Germany	Archenhold Obs.	+52	29	07·0	E 13	28·6		38
Berne, Switzerland	Astr. Inst. of Univ.	+46	57	12·7	E 07	25·7		563
Bloemfontein, S. Africa	Boyden Obs.	−29	02	18·0	E 26	24·3		1387
Bochum, Germany	Astronomical Station	+51	27	54·8	E 07	13·4		132
Bogota, Colombia	National Obs.	+04	35	55·2	W 74	04·9		2640
Bologna, Italy	University Obs.	+44	29	52·8	E 11	21·1		84
Bombay, India	Government Obs.	+18	53	36·2	E 72	48·9		14
Bonn, W. Germany	University Obs.	+50	43	45·0	E 07	05·8		62
Bordeaux, France	Obs. of Univ. of Bordeaux	+44	50	07	E 00	31·6		73
Boulder, Colorado	Sommers-Bausch Obs.	+40	00	13·0	W 105	15·7		1648
Bucharest, Rumania	National Obs.	+44	24	49·4	E 26	05·8		83
Budapest, Hungary	Konkoly Obs.	+47	29	58·6	E 18	57·9		474
Buenos Aires, Argentina	Naval Obs.	−34	37	18·3	W 58	21·3		6
Byurakan, Armenia	Astronomical Obs. of Academy of Sciences	+40	20	07·0	E 44	17·5		1500
Cambridge, England	University Obs.	+52	12	51·6	E 00	05·7		28
Cambridge, Mass.	Harvard College Obs.	+42	22	47·6	W 71	07·8		24
Canberra, Australia	Mount Stromlo Obs.	−35	19	16·0	E 149	00·3		768
Cape, South Africa	S. African Astr. Obs.	−33	56	02·5	E 18	28·6		10
Caracas, Venezuela	Cagigal Obs.	+10	30	24·3	W 66	55·7		1042
Castel Gandolfo, Italy	Vatican Obs.	+41	44	47·4	E 12	39·1		450
Catania, Sicily	Astrophysical Obs.	+37	31	42·0	E 15	04·7		193
Cincinnati, Ohio	Cincinnati Obs.	+39	08	19·8	W 84	25·3		247
Copenhagen, Denmark	Urania Obs.	+55	41	19·2	E 12	32·3		10
Cordoba, Argentina	National Obs.	−31	25	16·4	W 64	11·8		434
Cracow, Poland	University Obs.	+50	03	52·0	E 19	57·6		221
Crimea, USSR	Crimean Astro. Obs.	+44	43	42·0	E 34	01·0		550
Dublin, Eire	Dunsink Obs.	+53	23	13·1	W 06	20·3		86
Edinburgh, Scotland	Royal Obs.	+55	55	30·0	W 03	11·0		146
Flagstaff, Arizona	US Naval Obs.	+35	11	00·0	W 111	44·4		2310
Flagstaff, Arizona	Lowell Obs.	+35	12	06·0	W 111	39·8		2210
Fort Davis, Texas	McDonald Obs.	+30	40	17·7	W 104	0·13		2081
Göttingen, Germany	University Obs.	+51	31	48·2	E 09	56·6		161
Greenbelt, Maryland	Goddard Research Obs.	+39	01	11·5	W 76	49·6		49
Groningen, Holland	Kapteyn Astron. Lab.	+53	13	13·8	E 06	33·4		4
Haleakala, Hawaii	Univ. of Hawaii	+20	42	22·0	W 156	15·4		3054
Hamburg, W. Germany	Bergedorf Obs.	+53	28	46·9	E 10	14·4		41
Hartebeespoort, S. Africa	Republic Obs. Annexe	−25	46	22·4	E 27	52·6		1220
Helsinki, Finland	University Obs.	+60	09	42·3	E 24	57·3		33
Helwan, Egypt	Helwan Obs.	+29	51	31·1	E 31	20·5		115
Herstmonceux, England	Royal Greenwich Obs.	+50	52	18·0	E 00	20·3		34
Jena, E. Germany	Karl Schwarzschild Obs.	+50	58	51·0	E 11	42·8		331
Johannesburg, S. Africa	Republic Obs.	−26	10	55·3	E 28	04·5		1806
Juvisy, France	Flammarion Obs.	+48	41	37·0	E 02	22·3		92
Kodaikanal, India	Astrophysical Obs.	+10	13	50·0	E 77	28·1		2343
Kyoto, Japan	Kwasan Obs.	+34	59	40·8	E 135	47·6		234
Leyden, Holland	University Obs.	+52	09	19·8	E 04	29·0		6
Los Angeles, Calif.	Griffith Obs.	+34	06	46·8	W 118	18·1		357
Lund, Sweden	Royal University Obs.	+55	41	51·6	E 13	11·2		34
Madrid, Spain	Astronomical Obs.	+40	24	30·0	W 03	41·3		655
Mauna Kea, Hawaii	Univ. of Hawaii	+19	49	34·0	W 155	28·3		4205
Meudon, France	Obs. of Physical Astr.	+48	48	18·0	E 02	13·9		162
Milan, Italy	Brera Obs.	+45	27	59·2	E 09	11·5		120
Mill Hill, London, Eng.	Univ. of London Obs.	+51	36	46·3	W 00	14·4		82
Minneapolis, Minnesota	Univ. of Minnesota Obs.	+44	58	40·0	W 93	14·3		260
Montevideo, Uruguay	National Obs.	−34	54	33·0	W 56	12·7		24

Onion Dome, old Royal Observatory, Greenwich. (Patrick Moore)

Location	Name	Latitude °	"	Longitude °	"	Altitude metres
Montreal, Quebec, Canada	McGill Univ. Obs.	+45 30	20·0	W 73	34·7	57
Moscow, USSR	Sternberg Inst. Obs.	+55 45	19·8	E 37	34·2	166
Mount Hamilton, Calif.	Lick Obs.	+37 20	25·3	W 121	38·7	1283
Mount Semirodriki, Caucasus, USSR	Zelenchukskaya	+43 49	32·0	E 41	35·4	973
Mount Wilson, Calif.	Hale Obs.	+34 12	59·5	W 118	03·6	1742
Nanking, China	Purple Mountain Obs.	+32 03	59·9	E 118	49·3	367
Naples, Italy	Capodimonte Obs.	+40 51	45·7	E 14	15·4	164
Nice, France	Nice Obs.	+43 43	17·0	E 07	18·0	376
Ondřejov, Czecho-slovakia	Astrophysical Obs.	+49 54	38·1	E 14	47·0	533
Ottawa, Ontario, Canada	Dominion Obs.	+45 23	38·1	W 75	43·0	87
Palermo, Sicily	University Astron. Obs.	+38 06	43·6	E 13	21·5	72
Palomar, California	Hale Obs.	+33 21	22·4	W 116	51·8	1706
Pic du Midi, France	Obs. of University of Toulouse	+42 56	12·0	E 00	08·5	2862
Pittsburgh, Penn.	Allegheny Obs.	+40 20	58·1	W 80	01·3	370
Potsdam, E. Germany	Astrophysical Obs.	+52 22	56·0	E 13	04·0	107
Pulkovo, USSR	Astron. Obs. of Academy of Sciences	+59 46	18·5	E 30	19·6	75
Quito, Equador	National Obs.	−00 14	00·0	W 78	29·6	2908
Richmond Hill, Ontario, Canada	David Dunlap Obs.	+43 51	46·0	W 79	25·3	244
Rio de Janeiro, Brazil	National Obs.	−22 53	42·2	W 43	13·4	33
St. Andrews, Scotland	University Obs.	+56 20	12·0	W 02	48·9	30
Santiago, Chile	Cerro Tololo Obs.	−33 23	50·0	W 70	32·9	860
Sâo Paulo, Brazil	Astronomical and Geophysical Inst.	−23 39	06·9	W 46	37·4	800

The Lovell Telescope at Jodrell Bank, Cheshire (Patrick Moore)

Location	Name	Latitude °		"	Longitude °		"	Altitude metres
Siding Spring, NSW, Australia	Siding Spring Obs.	−31	16	37·3	E	149	04·0	1165
Stockholm, Sweden	Saltsjöbaden Obs.	+59	16	18·0	E	18	18·5	55
Sutherland, CP, South Africa	S. African Astronomical Observatories	−32	22	46·5	E	20	48·6	1830
Sydney, Australia	Government Obs.	−33	51	41·1	E	151	12·3	44
Tokyo, Japan	Tokyo Obs. at Mitaka	+35	40	21·4	E	139	32·5	59
Tonantzintla, Mexico	National Astro. Obs.	+19	01	57·9	W	98	18·8	2150
Toruń, Poland	Copernicus Univ. Obs.	+53	05	47·7	E	18	33·3	90
Tucson, Arizona	Catalina Obs.	+32	25	00·7	W	110	43·9	2510
Tucson, Arizona	Kitt Peak Nat. Obs.	+31	57	30·3	W	111	35·7	2064
Uccle, Belgium	Royal Obs.	+50	47	55·0	E	04	21·5	105
Uppsala, Sweden	Univ. Astron. Obs.	+59	51	29·4	E	17	37·5	21
Utrecht, Holland	Sonnenborgh Obs.	+52	05	09·6	E	05	07·8	14
Victoria, BC, Canada	Dominion Astro. Obs.	+48	31	15·7	W	123	25·0	229
Vienna, Austria	University Obs.	+48	13	55·1	E	16	20·3	240
Washington, DC, USA	US Naval Obs.	+38	55	14·0	W	77	03·9	86
Wellington, N. Zealand	Carter Obs.	−41	17	03·9	E	174	45·9	129
Williams Bay, Wisconsin, USA	Yerkes Obs.	+42	34	13·4	W	88	33·4	334

RADIO OBSERVATORIES

Location	Name	Latitude °		"	Longitude °		"	Altitude metres
Arcetri (Florence), Italy	Astrophysical Obs.	+43	45	14·4	E	11	15·3	184
Arecibo, Puerto Rico	Arecibo Obs., Cornell University	+18	20	36·6	W	66	45·2	496
Bochum, Germany	Radio Telescope Stn.	+51	25	43·0	E	07	11·5	160
Boulder, Colorado	High Altitude Obs.	+40	04	42·0	W	105	16·5	1692
Cambridge, England	Mullard Radio Astron. Obs.	+52	09	45·0	E	00	02·4	26
Crimea, USSR	Crimean Astrophys. Obs.	+44	43	42·0	E	34	01·0	550
Delaware, Ohio	Ohio State Obs.	+40	15	04·7	W	83	02·9	282
Eschweiler, W. Germany	Stockert (Bonn Univ.)	+50	34	14·0	E	06	43·4	435
Goldstone, California	Jet Propulsion Lab.	+35	23	34·2	W	116	50·9	1038
Green Bank, W. Virginia	Nat. Radio Astron. Obs.	+38	26	17·0	W	79	50·2	823
Harestua, Norway	Univ. of Oslo Obs.	+60	12	30·0	E	10	45·5	585
Jodrell Bank, England	Nuffield Rad. Ast. Lab.	+53	14	11·0	W	02	18·4	70
Kitt Peak, Arizona	Nat. Radio Ast. Obs.	+31	57	11·0	W	111	36·8	1920
Nançay, France	Rad. Obs. of Nançay	+47	22	48·0	E	02	11·8	150
Nederhorst den Berg, Holland	Rad. Astr. Obs.	+52	14	03·0	E	05	04·6	0
Parkes, NSW, Australia	Australian Nat. Rad. Obs.	−33	00	00·4	E	148	15·7	392
Pulkovo, USSR	Astron. Obs. Acad. Sciences	+59	46	05·5	E	30	19·4	70
Richmond Hill, Ontario, Canada	David Dunlap Obs.	+43	51	44·0	W	79	25·2	244
St. Michel, France	Nat. Centre of Scientific Research	+43	55	00·0	E	05	42·5	614
Tokyo, Japan	Tokyo Obs. at Mitaka	+35	40	18·2	E	139	32·4	70
Washington, DC, USA	Radio Astron. Obs. National Lab.	+38	49	16·6	W	77	01·6	30

The I.N.T. or Isaac Newton Telescope, now installed in its new dome at La Palma in the Canary Islands. Photographed by Patrick Moore in 1986 soon after the telescope had been transferred to La Palma from its old site at Herstmonceux in Sussex.

THE WORLD'S LARGEST TELESCOPES

The first really giant telescope was built by William Herschel in 1789; it had a 49 in (124·5 cm) mirror. This was surpassed in 1845 by Lord Rosse's 72 in (189·9 cm), which was actually the largest in the world until 1917, when the Hooker reflector at Mount Wilson was completed. Of course, these two early giants are now dismantled, but both were used for outstanding work. The current list of giant telescopes is as follows:

REFLECTORS

Aperture cm	Observatory	Date of completion
600	Mount Semirodriki, USSR	1976
508	Palomar, California, USA (Hale reflector)	1948
420	La Palma	1987
401	Cerro Tololo, Chile	1970
401	Kitt Peak, Arizona, USA	1970
389	Siding Spring, NSW (Anglo-Australian Telescope or AAT)	1974
381	Mount Stromlo, Canberra, Australia	1972
381	La Cilla, Chile	1975
375	UKIRT, Mauna Kea, Hawaii	1979
366	CFH (Canada–France–Hawaii), Mauna Kea	1970
305	Lick, California	1959
305	NASA Infra-red Telescope, Mauna Kea	1978
272	McDonald, Fort Davis, Texas, USA	1968
264	Crimea, USSR	1960
260	Byurakan, Armenia, USSR	1976
256	Isaac Newton Telescope (INT), La Palma (original 1967 mirror was 244 cm)	1983
254	Mount Wilson, California, USA	1917
254	Irénée du Pont Telescope, Las Campanas, Chile	1980
224	University of Hawaii Telescope, Mauna Kea	1970

This list excludes the MMT or Multiple-Mirror Telescope at Mount Hopkins, Arizona, USA which uses 6 mirrors and can equal the light-grasp of a single 442 cm mirror—which would make it the third largest in the world.

REFRACTORS

40·0	101·6	Yerkes, Williams Bay, USA	1897
36·0	91·4	Lick, California	1888
32·7	83·1	Meudon, France	1893
32·0	81·3	Potsdam, East Germany	1899
30·0	76·2	Allegheny, Pittsburgh	1914
30·0	76·2	Nice, France	1880

The Yerkes Refractor, with a 40-inch (101·6-cm) object-glass, photographed by Patrick Moore in 1987. It is the largest refractor in the world, and is in regular use. The Yerkes Observatory is at Williams Bay, some distance from Chicago (USA).

GAMMA-RAY ASTRONOMY

The ultra-short, highly-penetrating gamma-rays are difficult to detect. Spark chambers carried in balloons and artificial satellites are used, though care must be taken to distinguish gamma-rays from the much more plentiful cosmic rays.

The first discrete gamma-ray source identified was in 1969 (in Sagittarius).

The most powerful gamma-ray source in the sky is the Vela pulsar. This is also the faintest optical object ever detected! The second most powerful source is the Crab Nebula, which was also the first gamma-ray source to be optically identified.

Gamma-rays come from the Milky Way, and may be due to collisions between cosmic rays and hydrogen gas, though discrete sources have also been identified. Gamma-rays are not abundant. Altogether, about a million gamma-ray photons have been collected – the same number as the photons of visible light received from a bright star, such as Vega, in one second.

Very high energy gamma-rays (wavelengths a million million times shorter than that of light) may be studied from Earth, because of the luminous effects caused by the polarization of atoms in the atmosphere by gamma-rays. **The main instrument used for this work** is the 10-metre reflector at Mount Hopkins, in Arizona.

The first interplanetary gamma-ray detectors were carried on Apollos 15 and 16; they were designed to study the background gamma-ray emission from the Milky Way.

The first gamma-ray satellites were SAS II (1972) and COS B (1975). SAS II failed after six months but COS B operated until 1982.

Gamma-ray bursts have also been identified. They come from neither the Sun nor the Earth, but their origin is still obscure. Explanations range from flare stars to neutron stars, White Dwarfs, and even (less plausibly) comets hitting neutron stars! At present it must be admitted that gamma-ray astronomy is still in a very early stage.

Dome of the 200 in Hale reflector on Palomar Mountain (Patrick Moore)

'INVISIBLE' ASTRONOMY

Until comparatively modern times, astronomers were limited to studying radiations in the visible range of the electromagnetic spectrum. Then, in 1931, came the detection of radio waves from the Milky Way, and by now it is possible to examine almost all wavelengths, from long radio waves down to the very short gamma-rays.

RADIO ASTRONOMY

The first detection of radio waves from the sky was due to the American radio engineer K. Jansky in 1931. The discovery was fortuitous; he was investigating 'static' on behalf of the Bell Telephone Company with a home-made aerial nicknamed the 'Merry-go-Round' (part of the mounting was made from a dismantled Ford car). Jansky's first paper was published in 1932. **The first international radio telescope** was made by the American amateur Grote Reber, whose first paper appeared in 1940; Reber's telescope was a 9·5-metre 'dish'.

Radio waves from the Sun were first detected in 1942 by a British team led by J. S. Hey; originally the effects were thought to be due to German jamming of British radar! In February 1946 it was found that the giant sunspot of that month was a strong radio source.

At Jodrell Bank, meteor work began in 1945, when radar was used to measure meteor trails. (**Radio echoes from the Moon** were also detected in 1945, independently by an American team led by J. H. de Witt and a Hungarian team led by Z. Bay.) The first Jodrell Bank radio telescope was a fixed 66-metre paraboloid, due to B. Lovell (now Sir Bernard Lovell).

The first discrete radio source detected beyond the Solar System was Cygnus A, in 1946. In 1948 M. Ryle and F. G. Smith (now Sir Francis Graham-Smith) detected Cassiopeia A. **The first optical identifications of radio sources beyond the Solar System** were made in 1949; Taurus A (the Crab Nebula), Virgo A (the galaxy M.87) and Centaurus A (the galaxy NGC 5128). **The first planet found to be a radio source** was Jupiter, in 1955, by two Americans, B. F. Burke and K.

Franklin – again fortuitously. **The first quasar** was identified in 1963 because of its radio emissions, and the **first pulsar** by Jocelyn Bell-Burnell, at Cambridge, in 1967.

The Jodrell Bank 250-ft radio telescope came into operation in 1957. In 1987, at a special ceremony, it was re-named The Lovell Telescope.

Today, **the largest steerable radio astronomy** 'dish' is at the Max Planck Institute, Bonn, West Germany; it was completed in 1971, and has a dish 100 metres in diameter. **The largest non-steerable dish** is at Arecibo, Puerto Rico; it is 304·8 metres in diameter, and was completed in 1963. **The largest array** is the VLA or Very Large Array, 80 km west of Sococco in New Mexico; it is Y-shaped, each arm being 20·9 km long, with 27 movable antennæ, each 25 metres across. It was completed in 1981.

MICROWAVE ASTRONOMY

The microwave region of the electromagnetic spectrum extends between 1 mm and 30 cm, longer than infra-red but shorter than radio radiation.

The microwave background at 3 degrees above absolute zero was detected in 1964 by A. Penzias and R. Wilson at the Bell Telephone Laboratories in Holmdel, New Jersey, fortuitously when they were calibrating a 7·35 cm wavelength receiver built for satellite communications – at first the effect was thought to be due to pigeon droppings in the equipment! The background radiation had been predicted theoretically by R. Dicke. There had been an earlier prediction of background radiation (1940: Alpher, Herman and Gamow) but this idea had not been followed up at the time.

The largest microwave telescope is the James Clerk Maxwell Telescope at the Mauna Kea Observatory in Hawaii. It has a 15-metre dish, made up of 276 panels adjusted to form the correct curve, and is protected by a membrane for most of the observing time. It was completed in 1987.

INFRA-RED ASTRONOMY

The discovery of infra-red radiation from the Sun was made in 1801 by William Herschel, by placing a thermometer beyond the red end of the solar spectrum. It extends from a micron (0·001 mm) to several hundreds of microns, beyond which comes the microwave region.

The first major survey of the infra-red sky was undertaken in the 1960s by G. Neugebauer and R. Leighton; they discovered 6000 discrete sources.

The infra-red astronomical satellite IRAS was planned in 1974, and launched on 25 January 1983; the reflector was of 60 cm aperture, enclosed in a cooling vessel containing 500 litres of liquid helium, holding the temperature not far above absolute zero (−273 °C). IRAS operated for 300 days, and could have picked up the radiation from a speck of dust several kilometres away. There were eight staggered rows of detectors, and observations were made in four main bands (12, 25, 60 and 100 microns). Among the discoveries were several comets (the first on 26 April 1983, Comet IRAS–Araki–Alcock); the comet passed within 3 000 000 km of Earth, and was found to have a dust tail. Comet Tempel 2 also showed a dust-tail. The asteroid Phæthon was discovered on 11 October 1983. A ring of material in the Solar System was also discovered in the asteroid region, and infra-red excesses were found for many stars, including Vega, Fomalhaut and Beta Pictoris. **The brightest infra-red source beyond the Solar System** is the irregular galaxy M.82.

ULTRA-VIOLET ASTRONOMY

The ultra-violet region of the electromagnetic spectrum extends between 2 and 380 nanometres. (One nanometre is equal to one thousand-millionth of a millimetre.)

The existence of ultra-violet radiation from the Sun was demonstrated in 1801 by J. Ritter, by producing a spectrum with a prism and noting the darkening of paper soaked in sodium chloride held in the region beyond the violet. This was possible because ultra-violet radiation with a wavelength of between 300 and 400 nm can penetrate the Earth's atmosphere.

The first ultra-violet spectrograph was launched on 28 June 1946, from White Sands (New Mexico), by a V2 rocket. It crashed, and the film was lost, but on 10 October 1946 a V2 soared to 88 km, and R. Tousey recorded the first recoverable solar ultra-violet spectra from above the atmosphere.

Ultra-violet observations were subsequently made from the American satellites of the OAO (Orbiting Astronomical Observatory) series. OAO 1 (8 April 1966) failed. OAO 2 (7 December 1968) operated until 13 February 1973; it had an orbit extending between 770 and 780 km above the ground. Its 11 ultra-violet telescopes viewed 1930 objects. Discoveries included a hydrogen cloud round a comet (Tago–Sato–Kosaka), magnetic fields of stars, and ultra-violet radiation from a supernova. OAO 3 (the Copernicus satellite) was launched on 21 August 1972 (orbit 748×740 km) and discovered many new sources.

The most successful ultra-violet satellite so far is the International Ultra-Violet Explorer (IUE), launched on 26 January 1978, carrying a 45-cm aperture telescope which can feed one of two spectrographs (one in the 190–320 nm range, the other 115–200 nm). Though its planned life expectancy was 3 years, it was still functioning in 1988. It has examined the ultra-violet spectra of all the principal bodies of the Solar System, including Halley's Comet and some asteroids; observations have been made of the interstellar medium, stars of all kinds and also external galaxies.

The region shorter than 90 nm is termed the 'extreme ultra-violet'. Few EUV sources have been detected as yet, but a small EUV telescope was carried on the Apollo–Soyuz mission in 1975.

X-RAY ASTRONOMY

The first observations of X-rays from the sky were made on 5 August 1948, when R. Burnright of the US Naval Research Laboratory detected solar X-rays from the darkening of a photographic emulsion carried on a V2 rocket. On 29 September 1948 H. Friedman and Hulburt, also using a V2, detected intense X-radiation from the Sun. In 1956 Friedman's team recorded results which could have been celestial X-rays, but they could not be sure, as they always had the Sun in their field of view.

In 1959 R. Giacconi and his colleagues published a paper in which they predicted X-rays from very hot stars and supernova remnants. They referred particularly to the Crab Nebula.

On 24 October 1961 a rocket was launched from White Sands to search for X-rays from the Moon. It was thought that these could be due to incident X-rays striking the lunar surface and causing X-ray fluorescence, together with X-rays due to the surface being struck by energetic electrons from the solar wind. The equipment failed, as the protective covers failed to open. A second rocket – an Aerobee – was launched on 18 June 1962. No lunar X-rays were found, but a discrete source was detected, later found to be Scorpius X-1. On 12 October 1962 a new launch recorded X-rays from the Crab Nebula, with an intensity 15 per cent of those from Scorpius X-1, and also obtained a good value for the position of Scorpius X-1. By 1966, 30 sources had been identified, including the first galaxy, M.87.

The first X-ray source to be optically identified was the Crab Nebula (1962). Scorpius X-1 was optically identified in March 1966 by Japanese observers, using the Tokyo 178 cm reflector; the possibility had been suggested by Minoru Odo.

The first X-ray satellite, Uhuru, was launched from Kenya on 12 December 1970. It discovered 300 sources. In 1971 Uhuru results showed that Centaurus X-3 is a binary system; the intensity of the X-radiation varied rapidly in a period of 4·8 seconds, suggesting a neutron star, and a period of 2·087 days indicated a binary companion eclipsing the neutron star. The larger component was then optically identified. In 1971 a second binary, Hercules X-1, was identified.

The first X-ray nova, Centaurus X-

4, was discovered in May 1969 by an X-ray detector carried in a Vela satellite. An X-ray nova seen in December 1974, near the known source Centaurus A (though certainly unconnected with it) reached its maximum on Christmas Day, and was inevitably nicknamed CenXmas; another, in May 1975 close to the Crab Nebula (again, unconnected with it) was equally inevitably nicknamed Fresh Crab. X-ray novæ seem to be binaries with low-mass optical companions.

A few stars are known to be X-ray sources – notably the variable SS Cygni (detected at X-ray wavelengths on 14 June 1978) and the massive binary X Persei.

The most successful British X-ray satellite to date has been Ariel 5, launched in October 1974; it operated for five years.

The Einstein Observatory (HEAO 2) was launched on 13 November 1978, and operated until 1980, observing 5 per cent of the entire sky. It carried the first large X-ray telescope, consisting of four nested paraboloids and four nested hyperboloids, with an outer diameter of 58 cm. X-rays were detected from sources of various kinds, including stars, galaxies, clusters of galaxies, and quasars (previously, only three X-ray quasars had been known).

The first X-ray satellite launched by the European Space Agency was Exosat (26 May 1983). It has a $356 \times 19\,581$ km orbit, with an inclination of 72 degrees. This very eccentric polar orbit means that it remains in sunlight for most of the time, maintaining the sensors at stable temperatures.

THE HISTORY OF ASTRONOMY

To give every date of importance in the history of astronomy would be a mammoth undertaking. What I have therefore tried to do is to make a judicious selection, separating our purely space-research advances and discoveries.

It is impossible to say just when astronomy began, but even the earliest men capable of coherent thought must have paid attention to the various objects to be seen in the sky, so that it may be fair to say that astronomy is as old as *Homo sapiens*. Among the earliest peoples to make systematic studies of the stars were the Mesopotamians, the Egyptians and the Chinese, all of whom drew up constellation patterns. (There have also been suggestions that the constellations we use as a basis today were first worked out in Crete, but this is speculation only.) It seems that some constellation-systems date back to 3000 BC, probably earlier, but of course all dates in these very ancient times are uncertain.

The first essential among ancient civilizations was the compilation of a good calendar. Probably the first reasonably accurate value of the length of the year (365 days) was given by the Egyptians. (The first recorded monarch of all Egypt was Menes, who seems to have reigned around 3100 BC; he was eventually killed by a hippopotamus – possibly the only sovereign ever to have met with such a fate!) They paid great attention to the star Sirius (Sothis), because its 'heliacal rising', or date when it could first be seen in the dawn sky, gave a reliable clue to the time of the annual flooding of the Nile, upon which the Egyptian economy depended. The Pyramids are, of course, astronomically aligned, and arguments about the methods by which they were constructed still rage as fiercely as ever.

Obviously the Egyptians had no idea of the scale of the universe, and they believed the flat Earth to be all-important. So too did the Chinese, who also made observations. It has been maintained that a conjunction of the five naked-eye planets recorded during the reign of the Emperor Chuan Hsü refers to either 2449 or 2446 BC. There is also the legend of the Court Astronomers, Hsi and Ho, who were executed in 2136 BC (or, according to some authorities, 2159 BC) for their failure to predict a total solar eclipse; since the Chinese believed eclipses to be due to attacks on the Sun by a hungry dragon, this was clearly a matter of extreme importance! However, this legend is discounted by modern scholars.

The earliest data-collectors were the Assyrians; all students of ancient history know of the Library of Ashurbanipal (668–626 BC). This included the 'Venus Tablet', discovered by Sir Henry Layard and deciphered in 1911 by F. X. Kugler. It claims that when Venus appears, 'rains will be in the heavens'; when it returns after an absence of three months 'hostility will be in the land; the crops will prosper'. Early attempts at drawing up tables of the movements of the Moon and planets may well date from pre-Greek times, largely for astrological reasons; until relatively modern times astrology was regarded as a true science, and all the ancient astronomers (even Ptolemy) were also astrologers.

Babylonian astronomy continued well into Greek times, and some of the astronomers, such as Naburiannu (about 500 BC) and Kidinnu (about 380 BC) may have made great advances; but we know relatively little about them, and reliable dating begins only with the rise of Greek science.

Little progress was made in the following centuries, though there were some interesting Indian writings (Aryabhāta, 5th century AD), and in AD 570 Isidorus, Bishop of Seville, was the first to draw a definite distinction between astronomy and astrology. The revival of astronomy was due to the Arabs. In 813 Al-Ma'mūn founded the Baghdad school of astronomy, and various star catalogues were drawn up, the most notable being that of Al-Sūfī (born about 903). During this period two supernovæ were observed by Chinese astronomers; the star of 1006 (in Lupus) and 1054 (in Taurus, the remnant of which is today seen as the Crab Nebula).

The improved 'Alphonsine Tables' of planetary motions were published in 1270 by order of Alphonso X of Castile. In 1433 Ulūgh Beigh set up an elaborate observatory at Samarkand, but unfortunately he was a firm believer in astrology, and was told that his eldest son Abdallatif was destined to kill him. He therefore banished his son, who duly returned at the head of an army and had Ulūgh Beigh murdered. This marked the end of the Arab school of astronomy, and subsequent developments were mainly European. Some of the important dates in the history of astronomy are as follows:

1543 Publication of Copernicus' book *De Revolutionibus Orbium Cœlestium*. This sparked off the 'Copernican revolution' which was not really complete until the publication of Newton's *Principia* in 1687.
1572 Tycho Brahe observed the supernova in Cassiopeia.
1576–96 Tycho worked at Hven, drawing up the best star catalogue of pre-telescopic times.
1600 Giordano Bruno burned at the stake in Rome, partly because of his defence of the theory that the Earth revolves round the Sun.
1603 Publication of Johann Bayer's star catalogue, *Uranometria*.
1604 Appearance of the last supernova to be observed in our Galaxy (Kepler's Star, in Ophiuchus).
1608 Invention of the telescope, by Lippershey in Holland. (Telescopes may well have been invented earlier, but we have no definite proof.)
1609 First telescopic lunar map, drawn by Thomas Harriot. **Serious** telescopic work begun by Galileo, who made a series of spectacular discoveries in 1609–10 (phases of Venus, satellites of Jupiter, stellar nature of the Milky Way). **Publication**

Design of Tycho Brahe's observatory at Uraniborg, on Hven in the Baltic, used from 1576 to 1596. (Science Photo Lib.)

of Kepler's first two Laws of Planetary Motion.

1618 Publication of Kepler's third Law of Planetary Motion.

1627 Publication by Kepler of improved planetary tables (the Rudolphine Tables).

1631 First transit of Mercury observed, by Gassendi (following Kepler's prediction of it).

1632 Publication of Galileo's *Dialogue*, which amounted to a defence of the Copernican system. In 1633 he was condemned by the Inquisition in Rome, and forced into a completely hollow recantation. **Founding** of the first official observatory (the tower observatory at Leiden in Holland).

1637 Founding of the first national observatory (Copenhagen, Denmark).

1638 Identification of the first variable star (Mira Ceti, by Phocylides Holwarda in Holland).

1639 First transit of Venus observed (by Horrocks and Crabtree, in England, following Horrocks' prediction of it).

1647 Publication of Hevelius' map of the Moon.

1651 Publication of Riccioli's map of the Moon, introducing the modern-type lunar nomenclature.

1655 Discovery of Saturn's main satellite, Titan, by C. Huygens, who announced the correct explanation of Saturn's ring system in the same year.

1656 Founding of the second Copenhagen Observatory.

1659 Markings on Mars seen for the first time (by Huygens).

1663 First description of the principle of the reflecting telescope, by the Scottish mathematician James Gregory.

1665 Newton's pioneer experiments on light and gravitation, carried out at Woolsthorpe in Lincolnshire while Cambridge University was temporarily closed because of the Plague.

1666 First observation of the Martian polar caps, by G. D. Cassini.

1667 Founding of the Paris Observatory, with Cassini as Director. (It was virtually in action by 1671.)

1668 First reflector made, by Newton. (This is the probable date. It was presented to the Royal Society in 1671, and still exists.)

1675 Founding of the Royal Greenwich Observatory. **Velocity** of light measured, by O. Rømer (Denmark).

1676 First serious attempt at cataloguing the southern stars, by Edmond Halley from St. Helena.

1685 First astronomical observations made from South Africa (Father Guy Tachard, at the Cape).

1687 Publication of Newton's *Principia*, finally proving the truth of the theory that the Sun is the centre of the Solar System.

1704 Publication of Newton's other major work, *Opticks*.

1705 Prediction of the return of a comet, by Halley (for 1758).

1723 Construction of the first really good reflecting telescope (a 6-in, by Hadley).

1725 Publication of the final version of the star catalogue by Flamsteed, drawn up at Greenwich. (Publication was posthumous.)

1728 Discovery of the aberration of light, by James Bradley.

1750 First extensive catalogue of the southern stars, by Lacaille at the Cape. (His observations extended from 1750 to 1752.) **Wright's** theory of the origin of the Solar System.

1758 First observation of a comet at a predicted return (Halley's Comet, discovered on 25 Dec by Palitzsch. Perihelion occurred in 1759). **Principle** of the achromatic refractor discovered by Dollond. (Previously described, by Chester More Hall in 1729, but his basic theory was erroneous, and his discovery had been forgotten.)

1761 Discovery of the atmosphere of Venus, during the transit of that year, by M. V. Lomonosov in Russia.

1762 Completion of a new star catalogue by

James Bradley; it contained the measured positions of 60 000 stars.

1767 Founding of the *Nautical Almanac*, by Nevil Maskelyne.

1769 Observations of the transit of Venus made from many stations all over the world, including Tahiti (the expedition commanded by James Cook).

1774 First recorded astronomical observation by William Herschel.

1779 Founding of Johann Schröter's private observatory at Lilienthal, near Bremen.

1781 Publication of Charles Messier's catalogue of clusters and nebulæ. **Discovery** of the planet Uranus, by William Herschel.

1783 First explanation of the variations of Algol, by Goodricke. (Algol's variability had been discovered by Montanari in 1669.)

1784 First Cepheid variable discovered; δ Cephei itself, by Goodricke.

1786 First reasonably correct description of the shape of the Galaxy given, by William Herschel.

1789 Completion of Herschel's great reflector, with a mirror 49 in (124·5 cm) in diameter and a focal length of 40 ft (12·2 m).

1796 Publication of Laplace's 'Nebular Hypothesis' of the origin of the Solar System.

1799 Great Leonid meteor shower, observed by W. Humboldt.

1800 Infra-red radiation from the Sun detected by W. Herschel.

1801 First asteroid discovered (Ceres, by Piazzi at Palermo).

1802 Second asteroid discovered (Pallas, by Olbers). **Existence** of binary star systems established by W. Herschel. **Dark** lines in the solar spectrum observed by W. H. Wollaston.

1804 Third asteroid discovered (Juno, by Harding).

1807 Fourth asteroid discovered (Vesta, by Olbers).

1814–18 Founding of the Calton Hill Observatory, Edinburgh.

1815 Fraunhofer's first detailed map of the solar spectrum (324 lines).

1820 Foundation of the Royal Astronomical Society.

1821 Arrival of F. Fallows at the Cape, as Director of the first observatory in South Africa. **Founding** of the Paramatta Observatory by Sir Thomas Brisbane, Governor of New South Wales. (This was the first Australian observatory. It was dismantled in 1847.)

1822 First calculated return of a short-period comet (Encke's, recovered by Rümker at Paramatta).

1824 First telescope to be mounted equatorially, with clock drive (the Dorpat refractor, made by Fraunhofer).

1827 First calculation of the orbit of a binary star (ξ Ursæ Majoris, by Savary).

1829 Completion of the Royal Observatory at the Cape.

1834–8 First really exhaustive survey of the southern stars, carried out by John Herschel at Feldhausen (Cape).

1835 Second predicted return of Halley's Comet.

1837 Publication of the famous lunar map by Beer and Mädler. **Publication** of the first good catalogue of double stars (W. Struve's *Mensuræ Micrometricæ*).

1838 First announcement of the distance of a star (61 Cygni, by F. W. Bessel).

1839 Pulkova Observatory completed.

1840 First attempt to photograph the Moon (by J. W. Draper).

1842 Important total solar eclipse, from which it was inferred that the corona and prominences are solar rather than lunar. First attempt to photo-

graph totality (by Majocci), though he recorded only the partial phase.

1843 First daguerreotype of the solar spectrum obtained (by Draper).

1844 Founding of the Harvard College Observatory (first official observatory in the United States). The 15 in refractor was installed in 1847.

1845 Completion of Lord Rosse's 72 in reflector at Birr Castle, and the discovery with it of the spiral forms of galaxies ('spiral nebulæ'). **Daguerreotype** of the Sun taken by Fizeau and Foucault, in France. **Discovery** of the fifth asteroid (Astræa, by Hencke).

1846 Discovery of Neptune, by Galle and D'Arrest at Berlin, from the prediction by Le Verrier. The large satellite of Neptune (Triton) was discovered by W. Lassell in the same year.

1850 First photograph of a star (Vega, from Harvard College Observatory). Castor was also photographed, and the image was extended, though the two components were not shown separately. **Discovery** of Saturn's Crêpe Ring (Bond, at Harvard).

1851 First photograph of a total solar eclipse (by Berkowski). **Schwabe's** discovery of the solar cycle established by W. Humboldt.

1857 Clerk Maxwell proved that Saturn's rings must be composed of discrete particles. **Founding** of the Sydney Observatory. **First** good photograph of a double star (Mizar, with Alcor, by Bond, Whipple and Black).

1858 First photograph of a comet (Donati's, photographed by Usherwood).

1859 Explanation of the absorption lines in the solar spectrum given by Kirchhoff and Bunsen. **Discovery** of the Sun's differential rotation (by Carrington).

1860 Total solar eclipse. Final demonstration that the corona and prominences are solar rather than lunar.

1861–2 Publication of Kirchhoff's map of the solar spectrum.

1862 Construction of the first great refractors, including the Newall 25 in made by Cooke. (It was for many years at Cambridge, and is now in Athens.) **Discovery** of the Companion of Sirius (by Clark, at Washington). **Completion** of the *Bonner Durchmusterung*.

1863 Secchi's classification of stellar spectra published.

1864 Huggins' first results in his studies of stellar spectra. **First** spectroscopic examination of a comet (Tempel's, by Donati). **First** spectroscopic proof that 'nebulæ' are gaseous (by Huggins). **Founding** of the Melbourne Observatory. (The 'Great Melbourne Reflector' completed 1869.)

1866 Association between comets and meteors established (by G. V. Schiaparelli). **Great Leonid** meteor shower. **Announcement** by J. Schmidt of an alteration in the lunar crater Linné. (Though the reality of change is now discounted, regular lunar observation dates from this time.)

1867 Studies of 'Wolf-Rayet' stars by Wolf and Rayet, at Paris.

1868 First description of the method of observing the solar prominences at times of non-eclipse (independently by Janssen and Lockyer). **Publication** of a detailed map of the solar spectrum, by A. Ångström.

1870 First photograph of a solar prominence (by C. Young).

1872 First photograph of the spectrum of a star (Vega, by H. Draper).

1874 Transit of Venus; solar parallax redetermined. (Another transit occurred in 1882, but the overall results were disappointing.) **Founding** of observatories at Meudon (France) and Adelaide (Australia).

1876 First use of dry gelatine plates in stellar photography; spectrum of Vega photographed by

Huggins.

1877 Discovery of the two satellites of Mars (by Hall, at Washington). Observations of the 'canals' of Mars (by Schiaparelli, at Milan).

1878 Publication of the elaborate lunar map by J. Schmidt (from Athens). **Completion** of the Potsdam Astrophysical Observatory.

1879 Founding of the Brisbane Observatory.

1880 First good photograph of a gaseous nebula (M.42, by Draper).

1882 Gill's classic photograph of the Great Comet of 1882, showing so many stars that the idea of stellar cataloguing by photography was born.

1885 Founding of the Tokyo Observatory. (An earlier naval observatory in Tokyo had been established in 1874.) **Supernova** in M.31, the Andromeda Galaxy (S Andromedæ). This was the only recorded extragalactic supernova to reach the fringe of naked-eye visibility until 1987.

1886 Photograph of M.31 (the Andromeda Galaxy) by Roberts, showing spiral structure. (A better photograph was obtained by him in 1888.)

1887 Completion of the Lick 36 in refractor.

1888 Publication of J. L. E. Dreyer's *New General Catalogue* of clusters and nebulæ. **Vogel's** first spectrographic measurements of the radial velocities of stars.

1889 Spectrum of M.31 photographed by J. Scheiner, from Potsdam. **Discovery** at Harvard of the first spectroscopic binaries (ζ Ursæ Majoris and β Aurigæ). **First** photographs of the Milky Way taken by E. E. Barnard.

1890 Foundation of the British Astronomical Association. **Unsuccessful** attempts to detect radio waves from the Sun, by Edison. (Sir Oliver Lodge was equally unsuccessful in 1896.) **Publication** of the Draper Catalogue of stellar spectra.

1891 Completion of the Arequipa southern station of Harvard College Observatory. **Spectroheliograph** invented by G. E. Hale. **First** photographic discovery of an asteroid (by Max Wolf, from Heidelberg).

1892 First photographic discovery of a comet (by E. E. Barnard).

1893 Completion of the 28 in Greenwich refractor.

1894 Founding of the Lowell Observatory at Flagstaff, in Arizona.

1896 Publication of the first lunar photographic atlas (Lick). **Founding** of the Perth Observatory. **Completion** of the Meudon 33 in (83 cm) refractor. **Completion** of the new Royal Observatory at Blackford Hill, Edinburgh. **Discovery** of the predicted Companion of Procyon, by Schaeberle.

1897 Completion of the Yerkes Observatory.

1898 Discovery of the first asteroid to come within the orbit of Mars (433 Eros, discovered by Witt at Berlin).

1899 Spectrum of the Andromeda Galaxy (M.31) photographed by Scheiner.

1900 Publication of Burnham's catalogue of 1290 double stars. **Horizontal** refractor, of 49 in aperture, focal length 197 ft (60 m), shown at the Paris Exhibition. (It was never used for astronomical research.)

1905 Founding of the Mount Wilson Observatory (California).

1908 Giant and dwarf stellar divisions described by E. Hertzsprung (Denmark). **Completion** of the Mount Wilson 60 in reflector. **Fall** of the Siberian meteorite.

1912 Studies of short-period variables in the Small Magellanic Cloud, by Miss H. Leavitt, leading on to the period-luminosity law of Cepheids.

1913 Founding of the Dominion Astrophysical Observatory, Victoria (British Columbia). **H. N. Russell's** theory of stellar evolution announced.

1915 W. S. Adams' studies of Sirius B, leading to the identification of White Dwarf stars.

1917 Completion of the 100 in Hooker reflector at Mount Wilson (the largest until 1948).

1918 Studies by H. Shapley leading him to the first accurate estimate of the size of the Galaxy.

1919 Publication of Barnard's catalogue of dark nebulæ.

1920 The Red Shifts in the spectra of galaxies announced by V. M. Slipher.

1923 Proof given (by E. Hubble) that the **galaxies** are true independent systems rather than parts of our Milky Way system. **Invention** of the spectrohelioscope, by Hale.

1925 Establishment of the Yale Observatory at Johannesburg. (It was finally dismantled in 1952, its work done.)

1927 Completion of the Boyden Observatory at Bloemfontein, South Africa.

1930 Discovery of Pluto, by Clyde Tombaugh at Flagstaff. **Invention** of the Schmidt camera, by Bernhard Schmidt (Estonia).

1931 First experiments by K. Jansky at Holmdel, New Jersey, with an improvised aerial, leading on to the founding of radio astronomy. Jansky published his first results in 1932, and in 1933 found that the radio emission definitely came from the Milky Way.

1932 Discovery of carbon dioxide in the atmosphere of Venus (by T. Dunham).

1933–5 Completion of the David Dunlap Observatory near Toronto (Canada).

1937 First intentional radio telescope built (by Grote Reber); it was a 'dish' 31 ft (9·4 m) in diameter.

1938 New (and correct) theory of stellar energy proposed by H. Bethe and, independently, by G. von Weizsäcker.

1942 Solar radio emission detected by M. H. Hey and his colleagues (27–28 Feb). The emission had previously been attributed to intentional jamming by the Germans!.

1944 Suggestion, by H. C. van de Hulst, that interstellar hydrogen must emit radio waves at a wavelength of 21.1 cm.

1945 Thermal radiation from the Moon detected at radio wavelengths (by R. H. Dicke).

1945–6 First radar contact with the Moon, by Z. Bay (Hungary) and independently by the US Army Signal Corps Laboratory.

1946 Work at Jodrell Bank begun (radar reflections from the Giacobinid meteor trails, 10 Oct). **Beginning** of radio astronomy in Australia (solar work by a team led by E. G. Bowen). **Identification** of the radio source Cygnus A by Hey, Parsons and Phillips.

1947–8 Photoelectric observations of variable stars in the infra-red carried out by Lenouvel, using a Lallemand electronic telescope.

1948 Completion of the 200 in Hale reflector at Palomar (USA). **Identification** of the radio source Cassiopeia A, by M. Ryle and F. G. Smith.

1949 Identification of further radio sources; Taurus A (the Crab Nebula), Virgo A (M.87), and Centaurus A (NGC 5128). These were the first radio sources beyond the Solar System to be identified with optical objects.

1950 M.31 (the Andromeda Galaxy) detected at radio wavelengths by M. Ryle, F. G. Smith and B. Elsmore. **Funds** for the building of the great Jodrell Bank radio telescope obtained by Sir Bernard Lovell.

1951 Discovery by H. Ewen and E. Purcell of the 21 cm emission from interstellar hydrogen, thus confirming van de Hulst's prediction. **Optical** identification of Cygnus A and Cassiopeia A (by Baade and Minkowski, using the Palomar reflector, from the positions given by Smith).

1952 W. Baade's announcement of an error in the Cepheid luminosity scale, showing that the galaxies are about twice as remote as had been previous-ly thought. **Electronic** images of Saturn and θ Orionis obtained by Lallemand and Duchesne (Paris). **Tycho's** supernova of 1572 identified at radio wavelengths by Hanbury Brown and Hazard.

1953 I. Shklovskii explains the radio emission from the Crab Nebula as being due to synchroton radiation.

1955 Completion of the 250 ft radio 'dish' at Jodrell Bank. **First** detection of radio emissions from Jupiter (by Burke and Franklin). **Construction** of a radio interferometer by M. Ryle, and also the completion of the 2nd Cambridge catalogue of radio sources. (The 3rd Cambridge catalogue was completed in 1959.)

1958 Observations of a red event in the lunar crater Alphonsus, by N. A. Kozyrev (Crimean Astrophysical Observatory, USSR). **Venus** detected at radio wavelengths (by Mayer).

1959 Radar contact with the Sun (Eshleman, at the Stanford Research Institute, USA).

1960 Aperture synthesis method developed by M. Ryle and A. Hewish.

1961 Completion of the Parkes radio telescope, 330 km west of Sydney.

1962 Thermal radio emission detected from Mercury, by Howard, Barrett and Haddock, using the 85 ft radio telescope at Michigan. **First** radar contact with Mercury (Kotelnikov, USSR). **First X-ray** source detected (in Scorpius). **Sugar Grove** fiasco; the US attempt to build a 600 ft fully steerable radio 'dish'. Work had been begun in 1959, and when discontinued had cost $96 000 000.

1963 Announcement, by P. van de Kamp, of a planet attending Barnard's Star. **Identification** of quasars (M. Schmidt, Palomar).

1965–6 Identification of the 3 °K microwave radiation, as a result of theoretical work by Dicke and experiments by Penzias and Wilson.

1967 Completion of the 98 in Isaac Newton reflector at Herstmonceux. **Identification** of the first pulsar, CP 1919, by Miss Jocelyn Bell at Cambridge.

1968 Identification of the Vela pulsar (Large, Vaughan and Mills).

1969 First optical identification of a pulsar; the pulsar in the Crab Nebula by Cocke, Taylor and Disney at the Steward Observatory, USA.

1970 Completion of the 100 m radio 'dish' at Bonn (Germany). **Completion** of the large reflectors for Kitt Peak (Arizona) and Cerro Tololo (Chile); each 158 in (401 cm) aperture. First large reflector to be erected on Mauna Kea, Hawaii; an 88 in (224 cm).

1973 Opening of the Sutherland station of the South African Astronomical Observatories.

1974 Completion of the 153 in (389 cm) reflector at the Siding Spring Observatory, Australia.

1976 Completion of the 236 in (600 cm) reflector at Mount Semirodriki (USSR).

1977 Optical identification of the Vela pulsar (at Siding Spring). **Discovery** of Chiron (by C. Kowal, USA). **Discovery** of the rings of Uranus.

1978 Completion of the new Russian underground neutrino telescope. **Reported discovery** of a satellite of Pluto (J. Christy, USA). **Discovery** of the first satellite of an asteroid (Herculina). **Rings of Uranus** recorded from Earth (Matthews, Neugebauer, Nicholson). **Discovery** of X-rays from SS Cygni (HEAO 1).

1979 Official opening of the observatory at La Palma. **Pluto** and Charon recorded separately (Bonneau and Foy, Mauna Kea, thereby confirming Charon's independent existence). **First comet** observed to hit the Sun.

1980 Discovery of the first scintar (SS 433).

1981 Five asteroids contacted by radar from Arecibo, including two Apollos (Apollo itself, and Quetzalcoatl).

1982 Discovery of the remote quasar, PKS 2000–330 (Wright and Jauncey, Parkes). **Recovery** of Halley's Comet.

1983 Discovery of the fastest-vibrating pulsar, PKS 1937+215 in Vulpecula: period 1·557806449022 milliseconds – twenty times shorter than the Crab pulsar. It spins 642 times per second.

1984 Isaac Newton Telescope installed on La Palma.

1986 Return of Halley's Comet.

1987 Completion of the William Herschel telescope on La Palma. Completion of the James Clerk Maxwell telescope on Mauna Kea. Supernova seen in the Large Cloud of Magellan.

ASTRONOMY AND SPACE RESEARCH

It is no longer possible to separate what may be called 'pure' astronomy from space research. The Space Age began on 4 October 1957, with the launching of Russia's first artificial satellite, Sputnik 1; less than a decade later Neil Armstrong and Edwin Aldrin stepped out on to the surface of the Moon, and there seems little doubt that elaborate space-stations will be in orbit during the 1980s. To talk of a lunar base before AD 2000 is no longer in the least far-fetched.

The following list has been restricted to the more 'astronomical' events. Full lists of all lunar and planetary probes are not given, because they will be found in the appropriate sections elsewhere in this book.

Pre-1957

c. 150 Lucian of Samosata's *True History* about a journey to the Moon – possibly the first of all science-fiction stories.

1232 Military rockets used in a battle between the Chinese and the Mongols.

1865 Publication of Jules Verne's novel *From the Earth to the Moon.*

1881 Early rocket design by N. I. Kibaltchitch. (Unwisely, he made the bomb used to kill the Czar of Russia, and was predictably executed.)

1891 Public lecture about space-flight by the eccentric German inventor Hermann Ganswindt.

1895 First scientific papers about space-flight by K. E. Tsiolkovskii. He published important papers in 1903 and subsequent years. The Russians refer to him as 'the father of space-flight'.

1919 Monograph, *A Method of Reaching Extreme Altitudes*, published by R. H. Goddard in America. This included a suggestion of sending a small vehicle to the Moon, and adverse Press comments made Goddard disinclined to expose himself to further ridicule.

1924 Publication of *The Rocket into Interplanetary Space*, by H. Oberth. This was the first truly scientific account of space-research techniques.

1926 First liquid-propelled rocket launched, by Goddard.

1927 Formation of the German rocket group, *Verein für Raumschiffart*.

1931 First European firing of a liquid-propelled rocket (Winkler, in Germany).

1937 First rocket tests at the Baltic research station at Peenemünde. One of the leaders of the

Yuri Gagarin, the first man in space (April 1961).

team was Wernher von Braun, who died in 1977.
1942 First firing of the A4 rocket (better known as the V2) from Peenemünde. 1944–5 many V2s fell upon Southern England.
1945 White Sands proving ground established in New Mexico. **Idea of synchronous** artificial satellites for communications purposes proposed by Arthur C. Clarke.
1949 First step-rocket fired from White Sands; it reached an altitude of almost 400 km. **Rocket** testing ground established at Cape Canaveral, Florida.
1955 Announcement of the US 'Vanguard' project for launching artificial satellites.

The Space Age

1957 4 Oct; launching of the first artificial satellite, *Sputnik 1* (USSR).
1958 First successful US artificial satellite (*Explorer 1*). Instruments carried in it were responsible for the detection of the Van Allen radiation zones surrounding the Earth.
1959 First lunar probes; *Lunas 1, 2* and *3* (all USSR). *Luna 1* by-passed the Moon, *Luna 2* crash-landed there, and *Luna 3* went on a round trip, sending back pictures of the Moon's far side.
1960 First television weather satellite (*Tiros 1*, USA).
1961 First attempted Venus probe (USSR); contact with it lost. **First manned space-flight** (Yuri Gagarin, USSR). **First manned US** space-flight (A. Shepard; sub-orbital).
1962 First American to orbit the Earth (J. Glenn). **First** British-built satellite (*Ariel 1*, launched from Cape Canaveral). **First** transatlantic television pictures relayed by satellite (*Telstar*). **First** attempted Mars probe (USSR; contact lost). **First** successful planetary probe: *Mariner 2* to Venus.
1963 First occasion when two manned spacecraft were in orbit simultaneously (Nikolayev and Popovich, USSR). **First space-woman**: Valentina

Tereshkova-Nikolayeva (USSR).
1964 First good close-range photographs of the Moon (*Ranger 7*, USA).
1965 First 'space-walk' (A. Leonov, USSR). **First** successful Mars probe (*Mariner 4*, USA).
1966 First soft landing on the Moon by an automatic probe (*Luna 9*, USSR). **First** landing of a probe on Venus (*Venera 3*, USSR), though contact with it was lost. **First** soft landing of an American probe on the Moon (*Surveyor 1*). **First** circum-lunar probe (*Luna 10*, USSR). **First** really good close-range lunar pictures (*Orbiter 1*, USA).
1967 First soft landing of an unmanned probe on Venus (*Venera 4*, USSR).
1968 First recovery of a circum-lunar probe (*Zond 5*, USSR). **First** manned Apollo orbital flight (*Apollo 7*; Schirra, Cunningham, Eisele, USA). **First** manned flight round the Moon; *Apollo 8* (Borman, Lovell, Anders, USA).
1969 First testing of the lunar module in orbit round the Moon (*Apollo 10*; Stafford, Cernan, Young, USA). **21 July**, first lunar landing (*Apollo 11*; N. Armstrong, E. Aldrin, USA).
1970 First Chinese and Japanese artificial satellites.
1971 First capsule landed on Mars, from the USSR probe *Mars 2*.
1971–2 First detailed pictures of Martian volcanoes, obtained from the probe *Mariner 9*, which entered orbit round the planet. (USA)
1972 End of the Apollo programme, with *Apollo 17* (Cernan, Schmitt, Evans, USA).
1973–4 Operational 'life' of the US space-station Skylab, manned by 3 successive three-man crews, and from which much pioneer astronomical work was carried out.
1973 First close-range information from Jupiter (including pictures) obtained from the fly-by probe *Pioneer 10*. (*Pioneer 11* repeated the experiments in 1974.) *Pioneer 10* was also the first probe to escape from the Solar System, while

Pioneer 11 was destined to be the first probe to bypass Saturn (in 1979). (USA)
1974 First pictures of the cloud-tops of Venus from close range, from the two-planet probe *Mariner 10*; the probe then encountered Mercury, and sent back the first pictures of the cratered surface.
1975 First pictures received from the surface of Venus, from the Russian probes *Venera 9* and *Venera 10*.
1976 First successful soft landings on Mars (*Vikings 1* and *2*) sending back direct pictures and information from the surface of the planet. (USA)
1977 Launching of *Voyagers 1* and *2* to the outer planets. (USA)
1978 Launch of two Pioneer probes to Venus – the first American attempts to put a vehicle into orbit round the planet and to land capsules there. **Launching** of the X-ray 'Einstein Observatory', which operated successfully for over 2 years.
1979 Fly-by of Jupiter by *Voyagers 1* and *2* (USA). **Decay** of *Skylab* in the Earth's atmosphere (11 July).
1980 *Voyager 1* fly-by of Saturn, obtaining data from Titan and other satellites during its pass. (USA)
1981 Successful *Voyager 2* pass of Saturn. (USA)
1982 Landings of *Veneras 13* and *14* on Venus, obtaining improved pictures and data (USSR). **Longest** space-mission undertaken by the Russians (2 months in orbit; A. Berezevoy and V. Lebedev in *Salyut 7*).
1983 Launch of IRAS (Infra-Red Astronomical Satellite).
1986 *Voyager 2* fly-by of Uranus. Shuttle disaster, with the destruction of the *Challenger*. Five probes to Halley's Comet; two Japanese, two Russian, and one European (Giotto).
1987 Longest space mission (326 days) completed by Yuri Romanenko, on the space-station Mir.

ASTRONOMERS

Selecting a limited number of astronomers for short biographical notes may be somewhat invidious. However, the list given here includes most of the great pioneers and researchers. No astronomers still living at the time of writing are included. All dates are A.D. unless otherwise stated.

Abul Wafa, Mohammed. 959–88. Last of the famous Baghdad school of astronomers. He wrote a book called *Almagest*, a summary of Ptolemy's great work also called the *Almagest* in Arabic.

Adams, John Couch, 1819–92. English astronomer, *b.* Lidcot, Cornwall. He graduated brilliantly from Cambridge in 1843, but had already formulated a plan to search for a new planet by studying the perturbations of Uranus. By 1845 his results were ready, but no quick search was made, and the actual discovery was due to calculations by U. Le Verrier. Later he became Director of the Cambridge Observatory, and worked upon lunar acceleration, the orbit of the Leonid meteor shower, and upon various other investigations.

Airy, George Biddell, 1801–92. English astronomer. Born in Northumberland, he graduated from Cambridge 1823; and was Professor of Astronomy there 1826–35. On becoming Astronomer Royal (1835–81) he totally reorganized Greenwich Observatory and raised it to its present eminence. He re-equipped the Observatory and ensured that the best use was made of its instruments; it is ironical that he is probably best remembered for his failure to instigate a prompt search for Neptune when receiving Adams' calculations.

Aitken, Robert Grant, 1864–1951. American astronomer, *b.* Jackson, Cal. In 1895 he joined the staff of Lick Observatory, and specialized in double star work; he discovered 31 000 new pairs, and wrote a standard book on the subject. From 1930 until his retirement in 1935 he was Director of the Lick Observatory.

Albategnius. *c.* 850–929. This is the Latinized form of the name of the Arab prince Al Battani. *b.* Batan, Mesopotamia; he drew up improved tables of the Sun and Moon, and found a more accurate value for the precession of the equinoxes. His *Movements of the Stars* enabled Hevelius, in the 17th century, to discover the secular variation in the Moon's motion. Albategnius was also a pioneer mathematician; in trigonometry, he introduced the use of sines.

Alfonso X. 1223–84. King of Castile. At Toledo he assembled many of the leading astronomers of the world, and drew up the famous Alphonsine Tables, which remained the standard for three centuries.

Alhazen (Abu Ali al Hassan). 987–1038. Arab mathematician, *b.* Basra. He went to Cairo, where he made his observations and also wrote the first important book on optics since the time of Ptolemy.

Al-Ma'mŭn, Abdalla. ?-833. Often referred to as Almanon. He was Caliph of Baghdad, son of Harun al Raschid; he collected and translated many Greek and Persian works, and built a major observatory in 829.

Al-Sŭfi. 903–86. A Persian nobleman, who compiled an invaluable catalogue of 1018 stars, giving their approximate positions, magnitudes and colours.

Anaxagoras. 500–428 BC. *b.* Clazomenæ, Ionia. In Athens he became a friend of Pericles, and it was because of this friendship that he was merely banished, rather than being condemned to death, for teaching that the Moon contains plains, valleys and mountains, while the Sun is a blazing stone larger than the Peloponnesus (the peninsula upon which Athens stands).

Anaximander. *c.* 611–547 BC. Greek philosopher, *b.* Miletus. He believed the Earth to be a cylinder, suspended freely in the centre of a spherical universe. He attempted to draw up a map of the world, and introduced the gnomon into Greece.

Anaximenes. *c.* 585–525 BC. Greek philosopher, *b.* Miletus. He believed the Sun to be hot because of its quick motion round the Earth, that the stars were too remote to send us detectable heat, and that the stars were fastened on to a crystal sphere.

Ångström, Anders. 1814–74. Swedish physicist, who graduated from Uppsala. He mapped the solar spectrum, and was the first to examine the spectra of auroræ. The Ångström unit (100-millionth part of a centimetre) is named in his honour.

Antoniadi, Eugenios. 1870–1944. Greek astronomer, who spent most of his life in France and became a naturalized Frenchman. He worked mainly at the Juvisy Observatory (with Camille Flammarion) and at Meudon, nr Paris, where he used the 83 cm refractor to make classic observations of the planets. Before the Space Age, his maps of Mars and Mercury were regarded as the standard works. He died in Occupied France during World War II.

Apian, Peter Bienewitz, 1495–1552. *b.* Leisnig, Saxony. Became professor of mathematics at Ingolstädt. He observed five comets, and was the first to note that their tails always point away from the Sun. The 1531 comet is known to be Halley's, and his observations of it enabled Edmond Halley to identify it with the comets of 1607 and 1682.

Apollonius. *c.* 250–200 BC. *b.* Perga, Asia Minor, but lived in Alexandria. An expert mathematician, he was one of the first to develop the theory of epicycles to represent the movement of the Sun, Moon and planets.

Arago, François Jean Dominique. 1786–1853. Director of the Paris Observatory from 1830. He made many important contributions, including a recognition of the importance of photography in astronomy. He made an exhaustive study of the great total solar eclipse of 1842, and maintained (correctly!) that the Sun is wholly gaseous.

Argelander, Friedrich Wilhelm August. 1799–1875. German astronomer, who became Director of the Bonn Observatory 1836. Here he drew up his atlas of the northern heavens (the *Bonn Durchmusterung*), containing the positions of 324 198 stars down to the ninth magnitude. This standard work was published in 1863.

Aristarchus. *c.* 310–250 BC. Greek astronomer, *b.* Samos. One of the first (quite possibly the very first) to maintain that the Earth moves round the Sun, and he also tried to measure the relative distances of the Sun and Moon by a method which was sound in theory, though inaccurate in practice.

Aristotle (384–322 BC) believed in a finite, spherical universe. He further developed the theory of concentric spheres, and gave the first practical proofs that the Earth cannot be flat.

Baade, Wilhelm Heinrich Walter. 1893–1959. German astronomer. In 1920, while assistant at Hamburg Observatory, he discovered the unique asteroid 944 Hidalgo. In 1931 he went to America, and joined the staff of Mount Wilson. In 1952 his work

Eugenios Antoniadi. (Patrick Moore)

Wilhelm Heinrich Walter Baade. (Patrick Moore)

upon the two classes of 'Cepheid' short-period variables enabled him to show that the galaxies are approximately twice as remote as had previously been thought.

Bailey, Solon Irving. 1854–1931. American astronomer, b. New Hampshire. He joined the Harvard staff in 1879, and was for many years in charge of the Harvard southern station at Arequipa, Peru. His studies of globular clusters led to the discovery of 'cluster variables', now known as RR Lyræ stars.

Barnard, Edward Emerson. 1857–1923. American astronomer, b. Nashville, Tennessee. He was self-taught, but joined the staff at Lick Observatory in 1888, moving to Yerkes in 1897. He was a renowned comet-hunter; he discovered the fifth satellite of Jupiter (Amalthea) and the swift-moving star in Ophiuchus now called Barnard's Star. He also specialized in studies of dark nebulæ.

Barrow, Isaac. 1630–77. English mathematician. He made various important contributions to science, but is perhaps best known because in 1669 he resigned his post as Lucasian Professor at Cambridge so that his pupil Isaac Newton could succeed him.

Bayer, Johann. 1572–1625. German astronomer; a lawyer by profession and an amateur in science. He is remembered for his 1603 star catalogue, in which he introduced the system of allotting Greek letters to the stars in each constellation–the system still in use today.

Beer, Wilhelm. 1797–1850. A Berlin banker, who set up a private observatory and collaborated with Mädler in the great map of the Moon published in 1837–8. This map remained the standard for many years. Beer was the brother of Meyerbeer, the famous composer.

Belopolsky, Aristarch. 1854–1934. Russian astronomer, who went from Moscow to the Pulkova Observatory in 1888. He became Director of the Observatory in 1916, but resigned in 1918. He specialized in spectroscopic astronomy and in studies of variable stars.

Bessel, Friedrich Wilhelm, 1784–1846. German astronomer. He went to Lilienthal as assistant to Schröter, but in 1810 became Director of the Königsberg Observatory, retaining the post until his death. He determined the position of 75 000 stars by reducing Bradley's observations; was the first to obtain a parallax value for a star (61 Cygni, in 1838), and predicted the positions of the then-unknown companions of Sirius and Procyon.

Biela, Wilhelm von. 1782–1856. Austrian army officer, and amateur astronomer, who is remembered for his discovery (in 1826) of the now-defunct periodical comet which bears his name.

Bode, Johann Elert. 1747–1826. German astronomer. b. Hamburg. Appointed director of Berlin Observatory 1772. In the same year he drew attention to the 'law' of planetary distances which had been discovered by Titius of Wittenberg; rather unfairly, perhaps, this is known as Bode's Law. He published a star catalogue, did much to popularize astronomy, and for 50 years edited the *Berlin Astronomisches Jahrbuch.*

Bond, George Phillips. 1825–65. Son of W. C. Bond. b. Massachusetts. In 1859 succeeded his father as Director of the Harvard Observatory. He was a pioneer of planetary and cometary photography, and was the first to assert upon truly scientific principles that Saturn's rings could not be solid.

Bond, William Cranch. 1789–1859. American astronomer, b. Maine. He began his career as a watchmaker, but his fame as an amateur astronomer led to his appointment as Director of the newly-founded Harvard Observatory. In 1848 he discovered Saturn's satellite Hyperion, and in 1850 he discovered Saturn's Crêpe Ring. He was also a pioneer of astronomical photography.

Bouvard, Alexis. 1767–1843. A shepherd boy, born in a hut at Chamonix. He went to Paris, taught

himself mathematics, and was appointed assistant to Laplace. He made contributions to lunar theory and drew up tables of the motions of the outer planets, as well as discovering several comets.

Bradley, James. 1692–1762. English astronomer (Astronomer Royal, 1742–62). He was educated in Gloucestershire, and entered the Ministry, becoming Vicar of Bridstow in 1719; in 1721 he went to Oxford as Professor of Astronomy, and remained there until his appointment to Greenwich, mainly on the recommendation of his close friend Halley. He discovered the aberration of light and the nutation of the Earth's axis, but his greatest work was his catalogue of the positions of 60 000 stars.

Brorsen, Theodor. 1819–95. Danish astronomer, who discovered several comets, and in 1854 made the first scientific observations of the Gegenschein.

Brown, Ernest William. 1866–1938. English astronomer, who graduated from Cambridge and then went to USA. His chief work was on lunar theory, and his tables of the Moon's motion are still recognized as the standard.

Burnham, Sherburne Wesley. 1838–1921. American astronomer, who began as an amateur and then went successively to Lick (1888) and Yerkes (1897). He specialized in double star work, and discovered over 1300 new pairs. His *General Catalogue of Double Stars* remains a standard reference work.

Campbell, William Wallace, 1862–1938. American astronomer, b. Ohio. He joined the staff at Lick Observatory in 1891, and was Director from 1900 until his retirement in 1930. His main work was in spectroscopy; he discovered 339 spectroscopic binaries (among them Capella), and determined the radial velocities of stars and of 125 nebulæ as well as carrying out spectroscopic observations of the planets.

Cannon, Annie Jump. 1863–1941. Outstanding American woman astronomer, b. Delaware. In 1896 she joined the staff at Harvard College Observatory, where she worked unceasingly on the classification of stellar spectra; the present system is due largely to her. She also discovered five novæ and over 300 variable stars. From 1938 she was William Cranch Bond Astronomer.

Carrington, Richard Christopher. 1826–75. English amateur astronomer, who had his Observatory at Redhill, Surrey. He concentrated upon the Sun, and made many contributions, including the first observation of a solar flare and the independent discovery of Spörer's Law concerning the distribution of sunspots throughout a cycle.

Cassini, Giovanni Domenico. 1625–1712. Italian astronomer; Professor of Astronomy, Bologna 1650–69, when he went to Paris as the first Director of the Observatory there. He discovered 4 of Saturn's satellites as well as the main division in the rings; he drew up new tables of Jupiter's satellites, made pioneer observations of Mars, and made the first reasonably good measurement of the distance of the Sun.

Cassini, Jacques J. 1677–1756. Son of G. D. Cassini; b. Paris. He succeeded his father as Director of the Paris Observatory. He confirmed Halley's discovery of the proper motions of certain stars, and played an important part in measuring an arc of meridian from Dunkirk to the Pyrenees in order to determine the figure of the Earth.

Challis, James. 1803–62. English astronomer; Professor of Astronomy at Cambridge from 1836. He accomplished much useful work, but, unfortunately, is remembered as the man who failed to discover Neptune before the success by Galle and D'Arrest at Berlin.

Charlier, Carl Vilhelm Ludwig. 1862–1934. Swedish cosmologist; Professor of Astronomy at Lund from 1897. He accomplished outstanding work

Friedrich Wilhelm Bessel. (Patrick Moore)

with regard to the distribution of stars in our Galaxy.

Christie, William Henry Mahoney. 1845–1922. English astronomer (Astronomer Royal 1881–1910). He modernized Greenwich Observatory, and fully maintained its great reputation, achieved under Airy.

Clairaut, Alexis Claude. 1713–65. French mathematical genius, who published his first important paper at the age of 12. He studied the motion of the Moon, and worked out the perihelion passage of Halley's Comet in 1759 to within a month of the actual date.

Clavius, Christopher Klau. 1537–1612. German Jesuit mathematical teacher, who laid down the calendar reform of 1582 at the request of Pope Gregory.

Copernicus, Nicolaus. 1473–1543. The Latinized name of Mikołaj Kopernik, b. Toruń, Poland. He entered the Church, and became Canon of Frombork. He had a varied career, including medicine and also the defence of his country against the Teutonic Knights, but is remembered for his great book *De Revolutionbus Orbium Cœlestium*, finally published during the last days of his life (he had previously withheld it because he was well aware of Church opposition). It was this book which revived the heliocentric theory according to which the Earth moves round the Sun, and sparked off the 'Copernican revolution' which came to its end with the work of Newton more than a century later.

Curtis, Heber Doust. 1872–1942. American astronomer, who worked at Lick, Allegheny and Michigan Observatories. He was an outstanding spectroscopist, and in 1920 took part in the 'Great Debate' with Shapley about the size of the Galaxy and the status of the resolvable nebulæ; Curtis was wrong about the size of the Galaxy, but correct in maintaining that the spiral nebulæ were independent galaxies. He also played a major rôle in the establishment of the McMath-Hulbert Observatory, renowned for its solar research.

D'Arrest, Heinrich Ludwig. 1822–75. German astronomer, b. Berlin. While Assistant at the Berlin Observatory, he joined Galle in the successful search for Neptune. From 1857 he worked at Copenhagen Observatory. He specialized in comet and asteroid work, and also published improved positions for about 2000 nebulæ.

Darwin, George Howard. 1845–1912. Son of Charles Darwin. From 1883 he was Professor of Astronomy at Cambridge, and drew up his famous though now rejected tidal theory of the origin of the planets. He was knighted in 1906.

Dawes, William Rutter. 1799–1868. English clergyman, and a keen-eyed amateur observer who specialized in observations of the Sun, planets and double stars. He discovered Saturn's Crêpe Ring independently of Bond.

Delambre, Jean-Baptiste Joseph. 1749–1822. French astronomer, best remembered for his work on the history of the science but also a skilled computer of planetary tables.

De la Rue, Warren. 1815–89. English astronomer, born in Guernsey. He was a pioneer of astronomical photography; in 1852 he obtained the first good photographs of the Moon, and in 1857 of the Sun. His photographs of the total solar eclipse of 1860 finally proved that the prominences are solar rather than lunar.

Delaunay, Charles. 1816–72. French astronomer, who specialized in studies of the Moon's motion. He became Director of the Paris Observatory in 1870, but was drowned in a boating accident two years later.

Democritus. c. 460–360 BC. Greek philosopher, b. Abdera, Thrace. He adopted Leucippus' atomic theory, and was the first to claim that the Milky Way is made up of stars.

Descartes, René. 1596–1650. French astronomer, author of the theory that matter originates as vortices in an all-pervading ether. He also made great improvements in optics. His books were published in Holland, but he died in Sweden.

De Sitter, Willem. 1872–1934. Dutch astronomer and cosmologist, b. Friesland; he went to the Cape, and from 1908 was Professor of Astronomy at Leiden. He studied the motions of Jupiter's satellites and also the rotation of the Sun, but is best remembered for his pioneer work in relativity theory. The 'De Sitter universe', finite but unbounded, was calculated to be 2000 million light-years in radius and to contain 80 000 million galaxies.

Deslandres, Henri Alexander. 1853–1948. French astronomer (originally an Army officer); from 1907 Director of the Meudon Observatory, and from 1927 Director of the Paris Observatory also. He was a pioneer spectroscopist, and developed the spectroheliograph independently of Hale.

Dollond, John. 1706–61. English optician, who re-invented the achromatic lens in 1758 and thus improved refractors beyond all recognition.

Donati, Giovanni Battista. 1826–73. Italian astronomer who discovered the great comet of 1858, and was the first to obtain the spectrum of a comet (Tempel's of 1864). From 1859 Director of the observatory at Florence, in 1872 he was largely responsible for the creation of the now-celebrated observatory at Arcetri.

Dreyer, John Louis Emil. 1852–1926. Danish astronomer, b. Copenhagen, who went to Ireland as astronomer to Lord Rosse at Birr Castle and became Director of the Armagh Observatory in 1882. He was a great astronomical historian, but is best remembered for his *New General Catalogue* of Clusters and Nebulæ (the N.G.C.) still regarded as a standard work. In 1916 he retired from Armagh and went to Oxford, where he lived for the rest of his life.

Dyson, Frank Watson. 1868–1939. English astronomer (Astronomer Royal, 1910–33). A great administrator as well as an energetic observer of eclipses; he also carried out important work in the field of astrophysics and stellar motions.

Eddington, Arthur Stanley. 1882–1945. English astronomer, b. Kendal. After working at Cambridge and Greenwich, he was appointed Professor of Astronomy at Cambridge, 1913. He was a pioneer of the theory of the evolution and constitution of the stars, and an outstanding relativist; in 1919 he confirmed Einstein's prediction of the displacement of star positions near the eclipsed Sun. He was knighted in 1930. In addition to his outstanding work, Eddington was one of the best of all writers of popular scientific books, and was a splendid broadcaster.

Einstein, Albert. 1879–1955. German Jew, whose name will be remembered as long as Newton's; in 1905 he laid down the Special Theory of Relativity, and from 1915–17 he developed the General Theory. In 1933 he left Germany, fearing persecution of the Jews, and settled in USA.

Elger, Thomas Gwyn. 1838–97. English amateur astronomer; he was first Director of the Lunar Section of the British Astronomical Association, and in 1895 published an excellent outline map of the Moon.

Empedocles of Agrigentum (c. 490–450 BC) believed the Sun to be a reflection of fire, but is credited with being the first to maintain that light has a finite velocity.

Encke, Johann Franz. 1791–1865. German astronomer; from 1825, Director of the Berlin Observatory. He was responsible for compiling the star maps which enabled Galle and D'Arrest to locate Neptune. In 1818 he computed the orbit of a faint comet, and successfully predicted its return; this

Nicolaus Copernicus. (Patrick Moore)

Johann Franz Encke. (Patrick Moore)

David Gill. (Patrick Moore)

was Encke's Comet, which has the shortest period of any known comet (3·3 years).

Eratosthenes. c. 276–196 BC. Greek philosopher, b. Cyrene; he became Librarian at Alexandria, and made a remarkably accurate measurement of the circumference of the Earth.

Eudoxus. c. 408–355 BC. Greek astronomer, b. Cnidus. He went to Athens, and attended lectures by Plato. Finally he settled in Sicily. He developed the theory of concentric spheres–the first truly scientific attempt to explain the movements of the celestial bodies.

Euler, Leonhard, 1707–83. Brilliant Swiss mathematician, b. Basle. He pioneered studies of the lunar theory, the movements of planets, comets, and the tides. He lost his sight in 1766, but this did not stop him from working; he undertook the complicated calculations mentally!

Fabricius, David. 1564–1617. Dutch minister and amateur astronomer, who observed Mira Ceti in 1596 (though without recognizing it as a variable) and made pioneer telescopic observations, notably of the Sun. In 1617 he announced from the pulpit that he knew the identity of a member of his congregation who had stolen one of his geese – and he was presumably correct, since he was assassinated before he could divulge the name of the culprit!

Fabricius, Johann. 1587–1616. Son of David Fabricius, and also a pioneer observer of the Sun by telescopic means; he discovered sunspots independently of Galileo and Scheiner. He died a year before his father.

Fallows, Fearon. 1789–1831. English astronomer, b. Cumberland. He went to South Africa in 1821 as the first Director of the Cape Observatory. He established the observatory, working under almost incredible difficulties, but the primitive living conditions undermined his health. The reduction of his Cape observations was undertaken by Airy.

Fauth, Philipp Johann Heinrich. 1867–1943. German astronomer, who compiled a large map of the Moon. Unfortunately he believed in the absurd theory that the Moon is ice-covered, and this influenced all his work.

Ferguson, James. 1710–76. Scottish popularizer of astronomy, who began life as a shepherd-boy but whose books gained great influence. He was also one of the first to suggest an evolutionary origin of the Solar System.

Flammarion, Camille. 1842–1925. French astronomer, renowned both for his observations of Mars and for his popular books. He set up his own observatory at Juvisy, and founded the Société Astronomique de France.

Flamsteed, John. 1646–1720. English astronomer (Astronomer Royal, 1675–1720), though at first the title was 'unofficial'). His main work was the compilation of a new star catalogue, the final version of which was published posthumously. Flamsteed was also Rector of Burstow, Surrey.

Fleming, Wilhelmina. 1857–1911. Scottish woman astronomer, who emigrated to America and worked at Harvard College Observatory, where she was in charge of the famous Draper star-catalogue. She discovered 10 novæ and 222 variable stars.

Fontana, Francisco. 1585–1656. Italian amateur (a lawyer by profession); one of the earliest telescopic observers. He left sketches of Mars and Venus, though the 'markings' which he recorded were certainly illusory.

Fowler, Alfred. 1868–1940. English astronomer, whose spectroscopic work in connection with the Sun, stars and comets was of great importance.

Franklin-Adams, John. 1843–1912. English businessman who took up astronomy as a hobby at the age of 47, and compiled a photographic chart of the stars which is still regarded as a standard work.

Fraunhofer, Joseph von. 1787–1826. Outstanding German optical worker, orphaned in early childhood and rescued from poverty by the Elector of Bavaria. He joined the Physical and Optical Institute of Munich, and was Director from 1823. He invented the diffraction grating, constructed the best lenses in the world, and studied the dark lines in the solar spectrum (the 'Fraunhofer Lines'). He made the Dorpat refractor for Struve (the first telescope to be clock-driven) and also the Königsberg heliometer. His comparatively early death was a tragedy for science.

Galilei, Galileo. 1564–1642. The first great telescopic observer–and also the true founder of experimental mechanics. He worked successively at Pisa, Padua and Florence. The story of his remarkable telescopic discoveries (including the satellites of Jupiter, the phases of Venus and the gibbous aspect of Mars, the starry nature of the Milky Way and many more), and of how his defence of the Copernican theory brought him into conflict with the Church, is one of the most famous in scientific history. He was condemned by the Inquisition in 1633, and was kept a virtual prisoner in his villa at Arcetri; in his last years he also lost his sight.

Galle, Johann Gottfried. 1812–1910. German astronomer, best remembered as being the first (with D'Arrest) to locate Neptune in 1846. He discovered three comets, and in 1872, while director of the Breslau Observatory, was the first to use an asteroid for measuring solar parallax.

Gassendi, Pierre. 1592–1655. French mathematician and astronomer. In 1631 he made the first of all observations of a transit of Mercury.

Gauss, Karl Friedrich. 1777–1855. German mathematical genius. In 1801 he calculated the orbit of the first asteroid, Ceres, from a few observations, and enabled Olbers to recover it in the following year. He invented the 'method of least squares', known to every mathematician.

Gill, David. 1843–1914. Scottish astronomer. In 1877 he used observations of Mars to redetermine the solar parallax, and in 1879 went to South Africa as HM Astronomer at the Cape. It was his photograph of the comet of 1882 which showed him the importance of mapping the sky photographically – since his plate showed many stars as well as the comet. He was also deeply involved in cataloguing the southern stars. He was knighted in 1900.

Goldschmidt, Hermann. 1802–66. German astronomer, who settled in Paris. Using small telescopes poked through his attic window, he discovered 14 asteroids between 1852–61.

Goodacre, Walter. 1856–1938. English amateur astronomer, who published an excellent map of the Moon in 1910.

Goodricke, John. 1764–86. Born of English parents in Holland. He was a deaf-mute, with a brilliant brain. It was he who found that Algol is an eclipsing binary rather than true variable, and he also discovered the fluctuations of the intrinsic variable δ Cephei.

Gould, Benjamin Apthorp. 1824–96. American astronomer, who founded the *Astrophysical Journal*. From Cordoba Observatory, Argentina, he compiled the *Uranimetria Argentina*, the first major catalogue of the southern stars.

Green, Charles. 1735–71. English astronomer who went with Captain Cook to study the 1769 transit of Venus. He died on the return voyage.

Gregory, James. 1638–75. Scottish mathematician. In 1663 he described the principle of the reflecting telescope, but never actually made one.

Grimaldi, Francesco Maria. 1618–63. Italian Jesuit, who made observations of the Moon used in the lunar map compiled by his friend Riccioli. Grimaldi discovered the refraction of light.

Gruithuisen, Franz von Paula. 1744–1852. German astronomer; from 1826 Professor of Astronomy, Munich. He was an assiduous observer of the Moon and planets, but his vivid imagination tended to discredit his work; at one stage he even reported the discovery of artificial structures on the Moon.

He also proposed the impact theory of lunar crater formation.

Gum, Colin. 1924–60. Australian astronomer, who carried out work of vital importance in the surveying of southern radio sources. The famous 'Gum Nebula' in Vela/Puppis is named after him. He was killed in a skiing accident at Zermatt in Switzerland.

Hadley, John. 1682–1743. English astronomer; friend of Bradley. He made the first really good reflecting telescope (6 in aperture) in 1723, and 1731 constructed his 'reflecting quadrant', which replaced the astrolabe and the cross-staff in navigation.

Hale, George Ellery. 1868–1938. American astronomer. A pioneer solar observer, who invented the spectroheliograph and discovered the magnetic fields of sunspots. In 1897 he became Director of Yerkes Observatory, and transferred to Mount Wilson in 1905, where he master-minded the building of the 60 in and 100 in reflectors, as well as the Yerkes refractor. He was mainly responsible for the building of the Palomar 200 in reflector, unfortunately not completed in his lifetime.

Hall, Asaph. 1829–1907. American astronomer, noted for his planetary work. At Washington, in 1877, he discovered the two satellites of Mars. From 1896 Professor of Astronomy at Harvard.

Halley, Edmond. 1656–1742. English astronomer (Astronomer Royal, 1720–42). Though best known for his prediction of the return of the great comet which now bears his name, Halley accomplished much other valuable work; he catalogued the southern stars from St Helena, studied star-clusters and nebulæ, and discovered the proper motions of some of the bright stars. Even more importantly, he was responsible for the writing of Newton's *Principia* and personally financed its publication.

Harding, Karl Ludwig. 1765–1834. German astronomer, who was at first assistant to Schröter and then was appointed Professor of Astronomy at Göttingen. In 1804 he discovered the third asteroid, Juno.

Harriot, Thomas. 1560–1621. English scholar, once tutor to Sir Walter Raleigh. He compiled the first telescopic map of the Moon, and completed it some months before Galileo began his work.

Harrison, John. 1693–1776. English clockmaker, who invented the marine chronometer which revolutionized navigation. Several of his original chronometers are now on display in London.

Hartmann, Johannes Franz. 1865–1936. German astronomer. Director Göttingen Observatory 1909–21, when he went to Argentina to superintend the National Observatory there. His important work was connected with stellar and nebular radial velocities, in the course of which he discovered interstellar absorption lines in the spectrum of δ Orionis.

Hay, William Thompson. 1888–1949. 'Will Hay' was probably the only skilled amateur astronomer who was by profession a stage and screen comedian! In 1933 he discovered the famous white spot on Saturn—the most prominent ever seen on that planet.

Heis, Eduard. 1806–77. German astronomer; Professor at Münster from 1852. He was a leading authority on the Zodiacal Light, meteors and variable stars, and published a valuable star catalogue. He was renowned for his keen eyesight, and is said to have counted 19 naked-eye stars in the Pleiades.

Hencke, Karl Ludwig. 1793–1866. German amateur astronomer—postmaster at Driessen. In 1845, after 15 years' search, he discovered the fifth asteroid, Astræa.

Henderson, Thomas. 1798–1844. Scottish astronomer. 1823–3 HM Astronomer at the Cape. While there, he made the measurements which enabled him to measure the parallax of α Centauri. In 1834

he became the first Astronomer Royal for Scotland.

Heraclides of Pontus (c. 388–315 BC) declared that the apparent daily rotation of the sky is due to the real rotation of the Earth. He also discovered that Mercury and Venus revolve round the Sun, not round the Earth.

Heraclitus of Ephesus (born c. 544 BC) took fire to be the principal element, and maintained that the diameter of the Sun was about one foot!.

Herschel, Friedrich Wilhelm (always known as William Herschel). 1738–1822. Probably the greatest observer of all time. He was born in Hanover, but spent most of his life in England. He was the best telescope maker of his day, and in 1781 became famous by his discovery of the planet Uranus. He made innumerable discoveries of double stars, clusters and nebulæ; he found that many doubles are physically-associated or binary systems, and he was the first to give a reasonable idea of the shape of the Galaxy. He was knighted in 1816, and received every honour that the scientific world could bestow. George III appointed him King's Astronomer (not Astronomer Royal).

Herschel, Caroline. 1750–1848. William Herschel's sister and constant assistant in his astronomical work. She discovered 8 comets.

Herschel, John Frederick William. 1792–1871. William Herschel's son. He graduated from Cambridge 1813 and from 1832–8 took a large telescope to the Cape to make the first really systematic observation of the southern heavens. He discovered 3347 double stars and 525 nebulæ, and may be said to have completed his father's pioneering work.

Hertzsprung, Ejnar. 1873–1967. Danish astronomer, who worked successively at Frederiksberg, Copenhagen, Göttingen, Mount Wilson and Leiden (Director Leiden Observatory from 1935). In 1905 he discovered the giant and dwarf subdivisions of late-type stars, and this led on to the compilation of H-R or Hertzsprung-Russell Diagrams, which are of fundamental importance in astronomy.

Hevelius. 1611–87. The Latinized name of Johannes Hewelcke of Danzig (now Gdańsk). From his private observatory he drew up a catalogue of 1500 stars, and observed planets, the Moon and comets, using the unwieldy long-focus, small-aperture refractors of his day. His observatory was burned down in 1679, but he promptly constructed another. His original map of the Moon has been lost; tradition says that the copper engraving was melted down and made into a teapot after his death.

Hind, John Russell. 1823–95. English astronomer, who discovered 11 asteroids, the 1848 nova in Ophiuchus, and his 'variable nebula' round T Tauri. He also computed many cometary orbits, and from 1853 was superintendent of the *Nautical Almanac.*

Hipparchus. Fl. 140 BC. Great Greek astronomer, who lived in Rhodes. He drew up a star catalogue, later augmented by Ptolemy. Among his many discoveries was that of precession; he also constructed trigonometric tables. Unfortunately all his original works have been lost.

Hooke, Robert. 1653–1703. English scientific genius, contemporary with (though no friend of!) Newton. He built various astronomical instruments, and made some useful observations, including sketches of lunar craters.

Horrocks, Jeremiah, 1619–41. English astronomer who, with his friend Crabtree, was the first to observe a transit of Venus (1639). He also worked on lunar theory. His early death was a great tragedy for science.

Hubble, Edwin Powell. 1889–1953. American astronomer, who served in the Army during world War I and was also a boxing champion. In 1923, using the Mount Wilson 100 in reflector, he discovered short-period variables in the Andromeda

Friedrich Wilhelm Herschel. (Patrick Moore)

Edwin Hubble. (Patrick Moore)

Ejnar Hertzsprung and Henry Norris Russell. (Patrick Moore)

Milton Humason. (Patrick Moore)

Johannes Kepler. (Patrick Moore)

Spiral, and proved the Spiral to be an independent galaxy. He also established the velocity/distance relationship known as Hubble's Law.

Huggins, William. 1824–1910. Pioneer English spectroscopist, who had his private observatory at Tulse Hill, nr London. Pioneer of stellar spectroscopy; he established that the irresolvable nebulæ are gaseous; he was the first to determine stellar radial motions by means of the Doppler shifts in their spectral lines, and he carried out important solar and planetary work. He was knighted in 1897.

Humason, Milton La Salle. 1891–1972. b. Minnesota. He was mainly self-taught, but joined the staff of Mount Wilson Observatory in 1920, and from then on worked closely with Hubble, studying the forms, spectra, radial motions and nature of the galaxies; he also photographed the spectra of supernovæ in external systems. In 1919 he carried out a photographic search for a trans-Neptunian planet at the request of W. H. Pickering, who had made independent calculations similar to Lowell's. Humason took several plates, but failed to locate the planet. When the plates were re-examined years later, after Pluto had been discovered at Flagstaff, it was found that Humason had recorded the planet twice – but once the image was masked by a star, and on the other occasion it fell on a flaw in the plate!

Huygens, Christiaan. 1629–95. Dutch astronomer; probably the best telescopic observer of his time. He discovered Saturn's brightest satellite (Titan) in 1655, and was the first to realize that the curious appearance of the planet was due to a system of rings. He was also the first to see markings on Mars. His activities extended into many fields of science; in particular, he invented the pendulum clock.

Innes, Robert Thorburn Ayton. 1861–1933. Scottish astronomer, who emigrated first to Australia (becoming a wine merchant) and then went to South Africa, as director of the Observatory at Johannesburg. He specialized in double star work, discovering more than 1500 new pairs; he also discovered Proxima Centauri, the nearest star beyond the Sun.

Janssen, Pierre Jules César. 1824–1907. French astronomer, who specialized in solar work (in 1870 he escaped from the besieged city of Paris by balloon to study a total eclipse). Independently of Lockyer, he discovered the means of observing the Sun's chromosphere and prominences without waiting for an eclipse. From 1876 he was Director of the Meudon Observatory, and in 1904 published an elaborate solar atlas, containing more than 8000

photographs. The square at the entrance to the Meudon Observatory is still called the Place Janssen, and his statue is to be seen there.

Jansky, Karl Guthe. 1905–49. American radio engineer, of Czech descent. He joined the Bell Telephone Laboratories, and was using an improvised aerial to investigate problems of static when he detected radio waves which he subsequently showed to come from the Milky Way. This was, in fact, the beginning of radio astronomy; but for various reasons Jansky paid little attention to it after 1937, and virtually abandoned the problem.

Jeans, James Hopwood. 1877–1946. English astronomer. He elaborated the plausible but now rejected theory of the tidal origin of the planets, but his major work was in connection with stellar constitution, in which he made notable advances. He was also an expert writer of popular scientific books, and was famous as a lecturer and broadcaster.

Jones, Harold Spencer. 1890–1960. English astronomer (Astronomer Royal 1933–55). A Cambridge graduate, who was HM Astronomer at the Cape from 1923 until his appointment to Greenwich. From the Cape he carried out much important work, mainly in connection with star catalogues and stellar radial velocities. While Astronomer Royal he redetermined the solar parallax by means of the world-wide observations of Eros, and published several excellent popular books as well as technical papers. He played a major rôle in the removal of the main equipment from Greenwich to the new site at Herstmonceux, in Sussex, and himself transferred to Herstmonceux in 1948, though it was not until 1958 that the move was completed. He was knighted in 1943.

Kant, Immanuel. 1724–1804. German philosopher, remembered astronomically for proposing a theory of the origin of the Solar System which had some points of resemblance to Laplace's later Nebular Hypothesis.

Kapteyn, Jacobus Cornelius. 1851–1922. Dutch astronomer and cosmologist. His most celebrated discovery was that of 'star-streaming'.

Kepler, Johannes. 1571–1630. German astronomer, b. Württemberg. He was the last assistant to Tycho Brahe, and after Tycho's death used the mass of observations to establish his three Laws of Planetary Motion. He observed the 1604 supernova, and also several comets, as well as making improvements to the refracting telescope, but his main achievements were theoretical. He ranks with Copernicus and Galileo as one of the main figures in the story of the 'Copernican revolution'.

Kirch, Gottfried. 1639–1710. German astronomer; Director of the Berlin Observatory from 1705. He was one of the earliest of systematic observers of comets, star-clusters and variable stars. In 1686 he discovered the variability of χ Cygni.

Kirchhoff, Gustav Robert. 1824–87. Professor of physics at Heidelberg. One of the greatest of German physicists, who explained the dark lines in the Sun's spectrum. His great map of the solar spectrum was published from Berlin in 1860.

Kirkwood, Daniel. 1814–95. American astronomer; an authority on asteroids and meteors. He drew attention to gaps in the asteroid belt, known today as the Kirkwood Gaps; they are due to the gravitational influence of Jupiter.

Kuiper, Gerard P. 1905–73. Dutch-American astronomer, who made notable advances in planetary and lunar work and was deeply involved with the programmes of sending probes beyond the Earth. The first crater to be identified on Mercury from *Mariner 10* was named in his honour.

Kulik, Leonid. 1883–1942. Russian scientist, trained as a forester, who achieved fame because of his work in meteorite research. In particular, he led several expeditions to study the Tunguska object of 1908. He died in a German prison camp in 1942.

Lacaille, Nicolas Louis de. 1713–62. French astronomer, who went to the Cape to draw up the first good southern-star catalogue.

Lagrange, Joseph Louis de. 1736–1813. French mathematical genius, and author of the classic *Mécanique Analytique*. He wrote numerous astronomical papers, dealing, among other topics, with the Moon's libration and the stability of the Solar System.

Laplace, Pierre Simon. 1749–1827. French mathematician who made great advances in dynamical astronomy. In 1796 he wrote *Système du Monde*, in which he outlined his Nebular Hypothesis of the origin of the planets. Discarded in its original form, but modern theories have many points of resemblance to it.

Lassell, William. 1799–1880. English astronomer, who would have taken part in the hunt for Neptune but for a sprained ankle. He discovered Triton, Neptune's larger satellite, and (independently of Bond) Hyperion, the 7th satellite of Saturn, as well as two satellites of Uranus (Ariel and Umbriel). He set up a 24 in reflector in Malta, and with it discovered 600 nebulæ.

Leavitt, Henrietta Swan. 1868–1921. American woman astronomer, best remembered for her observations of Cepheids in the Small Magellanic Cloud (1912), based on photographs taken in S. America; these led on to the discovery of the vital period-luminosity law for Cepheids. She also discovered 4 novæ, several asteroids, and over 2400 variable stars.

Lemaître, Georges. 1894–1966. Belgian priest, who was a leading mathematician; from 1927, Professor at Louvain University. His most important paper, leading to what is now called the 'Big Bang' theory of the universe, appeared in 1927, but did not become well-known until publicized by Eddington three years later. During World War I Lemaître served in the Belgian Army, and won the Croix du Guerre.

Le Monnier, Pierre Charles. 1715–99. French astronomer who was concerned in star cataloguing. He observed the planet Uranus several times, but did not check his observations, and missed the chance of a classic discovery. It was said that he never failed to quarrel with anyone whom he met!

Le Verrier, Urbain Jean Joseph. 1811–77. French astronomer, whose calculations led in 1846 to the discovery of Neptune. He was an authority on meteors, and in 1867 computed the orbit of the Leonids. He also developed solar and planetary theory, and believed in the existence of a planet (Vulcan) closer to the Sun than Mercury–now known to be a myth. He was forced to resign the Directorship of the Paris Observatory in 1870 because of his irritability, but was reinstated on the death by drowning of his successor, Delaunay.

Lexell, Anders John. 1740–84. Finnish astronomer, *b*. Abö. He became Professor of Mathematics at St Petersburg. He discovered the periodical comet of 1770 (now lost), and was one of the first to prove that the object discovered by Herschel in 1781 was a planet rather than a comet.

Lindsay, Eric Mervyn. 1907–74. Irish astronomer, Director of the Armagh Observatory from 1936 until his death. His main work was in connection with the Magellanic Clouds and with quasars. He had close connections with the Boyden Observatory in South Africa (where he had previously been assistant astronomer) and forged close links between it, Harvard, Dunsink (Dublin) and Armagh. He was an excellent lecturer on popular astronomy, and founded the Armagh Planetarium in 1966.

Lockyer, Joseph Norman. 1836–1920. English astronomer, and an independent discoverer of the method of studying the solar chromosphere and prominences at time of non-eclipse. He was knighted in 1897. He founded the Norman Lockyer Observatory at Sidmouth in Devon, which still exists even though no astronomical work is now carried on there.

Lohrmann, Wilhelm Gotthelf. 1796–1840. German land surveyor, who began an elaborate lunar map but was unable to complete it owing to ill-health. The map was completed 40 years later by Julius Schmidt.

Lomonosov, Mikhail. 1711–65. Russian astronomer; he was also termed 'the founder of Russian literature'. His father was a fisherman. In 1735 he went to the University of St Petersburg, and then to Marburg in Germany to study chemistry. On his return to Russia in 1741 he insulted some of his colleagues at the St Petersburg Academy and was imprisoned for several months, during which time he wrote two of his most famous poems. However, he later became Professor of Chemistry at St Petersburg, and in 1746 became a Secretary of State. He drew up the first accurate map of the Russian Empire, described a 'solar furnace', and investigated electrical phenomena. He also studied auroræ. In 1761 he observed the transit of Venus, and rightly concluded that Venus has a considerable atmosphere. His most important contribution was his championship of the Copernican theory and of Newton's theories, neither of which had really taken root in Russia before Lomonosov's work.

Lowell, Percival. 1855–1916. American astronomer, who founded the Lowell Observatory at Flagstaff, Arizona, in 1894. He paid great attention to Mars, and believed the 'canals' to be artificial waterways. His calculations led to the discovery of the planet Pluto, though the planet was not actually found until 1930 – by Clyde Tombaugh, at the Lowell Observatory. Lowell himself was a great astronomer who did much for science, and it is regrettable that he is today remembered mainly because of his erroneous theories about the Martian canals.

Lyot, Bernard. 1897–1953. Great French astronomer; Director of the Meudon Observatory. He made many advances in instrumental techniques, and invented the coronagraph, which enables the inner corona to be studied at times of non-eclipse. He died suddenly while taking part in an eclipse expedition to Africa.

Maclear, Thomas. 1794–1879. Irish astronomer, who in 1833 succeeded Henderson as HM Astronomer at the Cape. He made an accurate measure-

Gerard P. Kuiper. (Patrick Moore)

Johann Heinrich von Mädler. (Patrick Moore)

ment of an arc of meridian as well as verifying Henderson's parallax of α Centauri; he also studied comets and nebulæ. He was knighted in 1860.

Mädler, Johann Heinrich von. 1794–1874. German astronomer, who was the main observer in the great lunar map by himself and Beer, published in 1837–8, a map which remained the standard for several decades. In 1840 he left his Berlin home to become Director of the Dorpat Observatory in Estonia. He erroneously believed that η Tauri (Alcyone) was the star lying at the centre of the Galaxy. He retired in 1865 and spent his last years in Hanover.

Maraldi, Giacomo Filippo. 1665–1729. Italian astronomer; nephew of G. D. Cassini. He was renowned for his observations of the planets, particularly Mars, and assisted his uncle at the Paris Observatory.

Maskelyne, Nevil. 1732–1811. English astronomer (Astronomer Royal 1765–1811). Educated at Cambridge; he then went to St Helena, at the suggestion of Bradley, to observe the transit of Venus, and decided to make a serious study of navigation. During his régime as Astronomer Royal he founded the *Nautical Almanac.*

Méchain, Pierre François Andre. 1744–1805. French astronomer, who discovered 8 comets 1781–99.

Menzel, Donald H. 1901–76. American astronomer, celebrated for his research into problems of the Sun and planets as well as in stellar studies. He was also an excellent lecturer, and a skilled writer of popular books.

Messier, Charles. 1730–1817. French astronomer, interested mainly in comets. Though he discovered 13 comets, he is remembered chiefly because of his catalogue of star-clusters and nebulæ, published in 1781.

Michell, John. 1725–93. English clergyman, and an amateur astronomer who made the first suggestion that many double stars may be physically associated or binary systems.

Milne, Edward Arthur. 1896–1950. English astronomer, who graduated from Cambridge and then went successively to Manchester and Oxford. He made important contributions to astrophysics, and developed his theory of 'kinematic relativity', which was for a time regarded as an alternative to general Einsteinian relativity.

Minkowski, Rudolf. 1895–1976. German astronomer, who went to Mount Wilson in 1935 and remained there. He was one of the leading authorities on novæ and planetary nebulæ, and after the war became a pioneer in the new science of radio astronomy. His studies of rapidly-moving gases in radio galaxies led to the rejection of the 'colliding galaxies' theory.

Montanari, Geminiano. 1633–87. Italian astronomer, who worked at Bologna and then at Padua. In 1669 he discovered the variability of Algol.

Nevill, Edmund Neison. 1851–1940. English astronomer, who published an important book and map concerning the Moon in 1876; he wrote under the name of Neison. He was Director of the Natal Observatory at Durban in South Africa 1882–1910, returning to England when the Observatory was closed.

Newcomb, Simon. 1835–1909. American astronomer, for some years head of the American Nautical Almanac office. His chief work was in mathematical astronomy, to which he made valuable contributions. He is also remembered as the man who proved to his own satisfaction that no heavier-than-air machine could ever fly!

Newton, Isaac. 1643–1727. Probably the greatest of all astronomers. To list all his contributions here would be pointless; suffice to say that his *Principia,* published in 1687, has been described as the greatest mental effort ever made by one man. In addition to his scientific work, he sat briefly in Parliament, and served as Master of the Mint. He was knighted in 1705, and on his death was buried in Westminster Abbey.

Olbers, Heinrich Wilhelm Matthias. 1758–1840. German doctor, who was a skilled amateur astronomer and established his private observatory in Bremen. He discovered two of the first four asteroids (Pallas and Vesta) and rediscovered the first (Ceres); he carried out important work in connection with cometary orbits, and discovered a periodical comet which has a period of 69·5 years, and last returned in 1956. Olbers also wrote about his celebrated paradox: 'Why is it dark at night'?

Parmenides of Elea (second half of the 6th century BC) believed the stars to be of compressed fire, but agreed that the Earth was spherical, and in equilibrium because it was equidistant from all points on the sphere representing the universe.

Perrine, Charles Dillon. 1867–1951. American astronomer, who discovered two of Jupiter's satellites as well as nine comets. He worked at the Lick Observatory until 1909, when he became Director of the Cordoba Observatory in Argentina, where he constructed a 30 in reflector and made many observations of southern galaxies. He also planned a major star catalogue, but was politically unpopular, and after a narrow escape from assassination he retired (1936).

Peters, Christian Heinrich Friedrich. 1813–90. Danish astronomer, who emigrated to America in 1848. He discovered 48 asteroids.

Piazzi, Giuseppe. 1746–1826. Italian astronomer, who became Director of the Palermo Observatory in Sicily. During the compilation of a star catalogue he discovered the first asteroid, Ceres (on 1 Jan 1801, the 1st day of the new century).

Pickering, Edward Charles. 1846–1919. American astronomer; for 43 years, from 1876, Director of the Harvard College Observatory. He concentrated upon photometry, variable stars, and above all stellar spectra; in the famous *Draper Catalogue,* the stars were classified according to their spectra. During his régime the Harvard Observatory was modernized, and a southern outstation was set up at Arequipa in Peru.

Pickering, William Henry. 1858–1938. Brother of E. C. Pickering, who worked with him at Harvard and also served for a while as astronomer-in-charge of the Arequipa out-station. In 1898 he discovered Saturn's 9th satellite, Phœbe. He made extensive studies of the Moon and Mars, mainly from the Harvard station in Jamaica which was set up in 1900. Independently of Lowell, he calculated the position of the planet Pluto.

Plutarch. c. 46–120. Greek biographer, mentioned here because of his authorship of *De Facie in Orbe Lunæ* – On the Face in the Orb of the Moon – in which he claims that the Moon is a world of mountains and valleys.

Pond, John. 1767–1836. English astronomer (Astronomer Royal, 1811–35). Though an excellent and painstaking astronomer, Pond was handicapped by ill-health during the latter part of his régime at Greenwich, and was eventually asked to resign. He tried unsuccessfully to obtain star-distances by the parallax method.

Pons, Jean Louis. 1761–1831. French astronomer, whose first post at an observatory (Marseilles) was that of caretaker! He was self-taught and concentrated on hunting for comets; he found 36 in all, and ended his career as Director of the Museum Observatory in Florence.

Proctor, Richard Anthony. 1837–88. English astronomer, who was an excellent cosmologist but is best known for his many popular books. In 1881 he emigrated to America, and remained there for the rest of his life. Proctor paid considerable attention to the planets, and constructed a map of Mars.

Ptolemy (Claudius Ptolemæus). c. 120–180. The 'Prince of Astronomers', who lived and worked in

Alexandria. Nothing is known about his life, but his great work has come down to us through its Arab translation (the *Almagest*). Ptolemy's star catalogue was based on that of Hipparchus but with many contributions of his own; he also brought the geocentric system to its highest state of perfection, so that it is always known as the Ptolemaic theory. He constructed a reasonable map of the Mediterranean world, and even showed Britain, though it is true that he joined Scotland on to England in a back-to-front position.

Purbach, Georg von. 1423–61. Austrian astronomer, who became a professor at Vienna in 1450. He founded a new school of astronomy, compiled tables, and began to write an *Epitome of Astronomy* based on Ptolemy's *Almagest*. After Purbach's premature death, the book was completed by his friend and pupil Regiomontanus.

Pythagoras. *c.*572–500 BC. The great Greek geometer, mentioned here because he was one of the very first to maintain that the Earth is spherical rather than flat. He seems also to have studied the movements of the planets.

Ramsden, Jesse. 1735–1800. English maker of astronomical instruments. His meridian circles were the first to be lit through the hollow axis.

Rayet, Georges Antoine. 1839–1906. French astronomer; in 1867, with Wolf, drew attention to the Wolf-Rayet stars, which have bright lines in their spectra. He went from Paris to Bordeaux, and became Director of the Observatory there.

Redman, Richard Oliver. 1905–75. English astronomer, who graduated from Cambridge. He made extensive studies of the Sun, stellar velocities, galactic rotation and the photometry of galaxies. In 1937 he went to the Radcliffe Observatory, Pretoria, and designed the spectrograph for the 74 in reflector. In 1947 he returned to Cambridge as Professor of Astrophysics and Director of the Observatories. Many programmes were carried through, and Redman also devoted much time and energy in the planning and construction of the 153 in Anglo-Australian telescope at Siding Spring.

Regiomontanus. 1436–76. The Latinized name of Johann Müller, Purbach's pupil. He completed the *Epitome of Astronomy*, and at Nürnberg set up a printing press, publishing the first printed astronomical ephemerides. He died in Rome, where he had been invited to help in reforming the calendar.

Rhæticus, Georg Joachim. 1514–76. German astronomer, who became Professor of Astronomy, at Wittenberg in 1536. An early convert to the Copernican system, he visited Copernicus at Frombork, and persuaded him to send his great book for publication.

Riccioli, Giovanni Battista. 1598–1671. Italian Jesuit astronomer, who taught at Padua and Bologna. He was a pioneer telescopic observer, and drew up a lunar map, inaugurating the system of nomenclature which is still in use. Oddly enough, he never accepted the Copernican system!

Robinson, Romney. 1792–1882. Irish astronomer, who was Director of the Armagh Observatory from 1823 to his death. He published the Armagh catalogue of over 5000 stars, and made many other contributions; he also invented the cup anemometer. It is on record that when the railway company planned to build a line' close to Armagh, Robinson managed to have it diverted, since he maintained that he trains would shake his telescopes!

Rømer, Ole. 1644–1710. Danish astronomer. In 1675 he used the eclipse times of Jupiter's satellites to make an accurate measurement of the velocity of light. In 1681 he became Director of the Copenhagen Observatory. Among his numerous inventions are the transit instrument and the meridian circle.

Rosse, 3rd Earl of. 1800–67. Irish amateur astronomer, who in 1845 completed the building of a 72 in reflector and erected it at his home at Birr Castle. The 72 in with its metal mirror, was much the largest telescope ever built up to that time. His greatest discovery was that many of the galaxies are spiral in form.

Rosse, 4th Earl of. 1840–1908. Continued his father's work, and was also the first man to measure the tiny quantity of heat coming from the Moon. After his death the telescope was not used again, though the tube, between the stone walls which served as an observatory, may still be seen at Birr Castle, and a museum has been erected on the site.

Rowland, Henry Augustus. 1848–1901. American scientist; Professor of Physics in Baltimore from 1876. His great map of the solar spectrum was published in 1895–7; it showed 20 000 absorption lines.

Russell, Henry Norris. 1877–1957. American astronomer; Director, Princeton Observatory from 1908. He devoted much of his energy to studies of stellar constitution and evolution, and independently of Hertzsprung he discovered the giant and dwarf sub-divisions of stars of late spectral type. This led on to the H-R or Hertzsprung-Russell Diagram, in which luminosity (or the equivalent) is plotted against spectral type.

Rutherfurd, Lewis Morris. 1816–92. American barrister, who gave up his profession to devote himself to astronomy. A pioneer in astronomical photography, his pictures of the Moon were outstanding; his ruled solar gratings for solar spectra were the best of their time.

Scheiner, Christoph. 1575–1650. German Jesuit, who was for some time a professor of mathematics in Rome. He discovered sunspots independently of his contemporaries, and wrote a book, *Rosa Ursina*, which contains solar drawings and observations for the years 1611–25. He was unfriendly towards Galileo, and played a rather discreditable part in the events leading up to Galileo's trial and condemnation.

Schiaparelli, Giovanni Virginio. 1835–1910. Italian astronomer, who graduated from Turin and became Director of the Brera Observatory in Milan. He discovered the connection between meteors and comets, but his most famous work was in connection with the planets. It was he who first drew attention to the 'canal network' on Mars, in 1877.

Schlesinger, Frank. 1871–1943. American astronomer, *b.* New York. His main work was in connection with stellar parallaxes. He was Director successively of the Yale and Allegheny Observatories, and was responsible for the Yale 'southern station' in Johannesburg. His major works, *General Catalogue of Parallaxes* and its supplement, dealt with more than 2000 stars. He also pioneered the determination of star positions by using wide-angle cameras.

Schmidt, Julius (actually Johann Friedrich Julius). 1825–84. German astronomer, who became Director of the Athens Observatory in 1858 and spent most of his life in Greece. He concentrated upon lunar work, producing an elaborate map (based on Lohrmann's early work) and making great improvements in selenography. He drew attention to the alleged change in the lunar crater Linné, in 1866. He also discovered the outburst of the recurrent nova T Coronæ, in 1866.

Schönfeld, Eduard. 1828–91. German astronomer, who collaborated with Argelander in preparing the *Bonn Durchmusterung* and later extended it to the southern hemisphere.

Schröter, Johann Hieronymus. 1745–1816. Chief magistrate of Lilienthal, nr Bremen. He set up a private observatory, and made outstanding observations of the Moon and planets. Unfortunately many of his notebooks were lost in 1813, with the destruction of his observatory by the invading French troops.

Third Earl of Rosse. (Patrick Moore)

Giovanni Virginio Schiaparelli. (Patrick Moore)

Schwabe, Heinrich. 1789–1875. German apothecary, who became a noted amateur astronomer concentrating on the Sun. His great discovery was that of the 11-year sunspot cycle.

Schwarzschild, Karl. 1873–1916. German astronomer, who worked successively at Vienna, Göttingen and (as Observatory Director) Potsdam. His early work dealt with photometry, but he was also a pioneer of theoretical astrophysics. Military service during the first war broke his health and led to his premature death.

Secchi, Angelo. 1818–78. Italian Jesuit astronomer; one of the great pioneers of stellar spectroscopy, classifying the stars into four types (a system superseded later by that of Harvard). He was also an authority in solar work, and his planetary observations were equally outstanding.

Seyfert, Carl. 1911–60. American astronomer, who concentrated upon studies of galaxies. In 1942 he drew attention to those galaxies with very condensed nuclei, now always termed Seyfert galaxies.

Shapley, Harlow. 1885–1972. Great American astronomer, who began his main work at Princeton under H. N. Russell. In 1914 he advanced the pulsation theory of Cepheid variables, and was soon able to use the variables in globular clusters to give the first accurate picture of the shape and size of the Galaxy. In 1921 he became Director of the Harvard College Observatory. In later years he concentrated upon studies of galaxies and upon the international aspect of astronomy. He was also an excellent lecturer, and a skilled writer of popular books.

Smyth, William Henry. 1788–1865. English naval officer, rising to the rank of Admiral, who in 1830 established a private observatory at Bedford and made numerous observations. He is best remembered for his famous book, *Cycle of Celestial Objects*.

Smyth, Piazzi (actually Charles Piazzi). 1819–1900. Son of Admiral Smyth. Astronomer Royal for Scotland from 1844 until his death. He was a skilled astronomer who carried out much valuable work, including spectroscopic examinations of the Zodiacal Light, but also an eccentric who wrote a large and totally valueless volume about the significance of the Great Pyramid!

Sosigenes. Greek astronomer, who flourished about 46 BC. He was entrusted by Julius Cæsar with the reform of the calendar. Nothing is known about his life.

South, James. 1785–1867. English amateur astronomer, who founded a private observatory in Southwark and collaborated with John Herschel in studies of double stars. In 1822 he observed an occultation of a star by Mars, and the virtually instantaneous disappearance convinced him that the Martian atmosphere must be extremely tenuous.

Spörer, Friedrich Wilhelm Gustav. 1822–95. German astronomer, who joined the staff at Potsdam Observatory. He concentrated upon the Sun, and discovered the variation in latitude of spot zones over the course of a solar cycle (Spörer's Law).

Struve, Friedrich Georg Wilhelm. 1793–1864. German astronomer, *b*. Altona. He went to Dorpat in Estonia (then, as now, controlled by Russia) and in 1818 became Director of the Observatory. Using the 9 in Fraunhofer refractor – the first telescope to be clock-driven–he began his classic work on double stars. In 1839 he went to Pulkova, to become director of the new observatory set up by Tsar Nicholas. Here he continued his double-star work, and his *Mensuræ Micrometricæ* gives details of over 3000 pairs. Struve also measured the parallax of Vega; his value was announced in 1840.

Struve, Otto (Wilhelm). 1819–1905. Son of F. G. W. Struve; *b*. Dorpat. He became assistant to his father, accompanying him to Pulkova. He continued his father's work, and became a leading authority on double stars. He succeeded to the directorship of Pulkova Observatory in 1861, retiring in 1889 and returning to Germany.

Struve, Karl Hermann. 1854–1920. Son of Otto Struve; *b*. Pulkova, later becoming assistant to his father. His main work was in connection with planetary satellites. In 1895 he went to Königsberg, and in 1904 became Director of the Berlin Observatory, which was reorganized by him and transferred to Babelsberg during his term of office.

Struve, Gustav Wilhelm Ludwig. 1858–1920. Son of Otto Struve, and brother of Karl. He too was born at Pulkova and acted as assistant to his father. He went to Dorpat in 1886, and from 1894 was Director of the Kharkov Observatory. He was concerned mainly with statistical astronomy and with the motion of the Sun.

Struve, Otto. 1897–1963. Son of Gustav; often known as Otto Struve II. He was born in Kharkov, and fought during World War I; joined the White Army under Wrangel and Denikin, and after their defeat reached Constantinople, where he worked as a labourer. Finally he was offered a post at the Yerkes Observatory, where he arrived in 1921. He spent the rest of his life in America; in 1932 he became Director at Yerkes, after which he founded the McDonald Observatory in Texas and was its director from 1939–47, when he became Chairman of the Department of Astronomy at Chicago. In 1959 he began a new career as the first Director of the National Radio Astronomy Observatory, but ill-health forced his resignation in 1962. He was a brilliant astrophysicist, dealing mainly with spectroscopic binaries, stellar rotation and interstellar matter; he was also an author of popoular books. He is (so far!) the last of the famous Struve astronomers. It is notable that all four were in succession awarded the Gold Medal of the Royal Astronomical Society–a sequence unique in astronomical history.

Swift, Lewis. 1820–1913. American astronomer who specialized in hunting for comets and nebulæ. He found 13 comets (including the Great Comet of 1862) and 900 nebulæ.

Tempel, Ernst Wilhelm. 1821–89. German astronomer, who became Director of the Arcetri Observatory. In 1859 he discovered the nebula in the Pleiades; he also discovered 6 asteroids and several comets, including the comet of 1865–6 which is associated with the Leonid meteors.

Thales. *c*. 624–547 BC. The first of the great Greek philosophers. He believed the Earth to be flat, and floating in an ocean, but he was a pioneer mathematician and observer, and successfully predicted the eclipse of 585 BC. which stopped the war between the Lydians and the Medes.

Timocharis. Fl. *c*. 280 BC. Greek astronomer, who made some accurate measurements of star positions; one of these (of Spica) enabled Hipparchus, 150 years later, to demonstrate the precession of the equinoxes.

Turner, Herbert Hall. 1861–1930. English astronomer who played an important rôle in the organization and preparation of the International Astrographic Chart. In 1903 he discovered Nova Geminorum.

Tycho Brahe. 1546–1601. The great Danish observer–probably the best of pre-telescopic times. He studied the supernova of 1572 and from 1576 to 1596 worked at his observatory at Hven, an island in the Baltic, making amazingly accurate measurements of star positions and the movements of the planets, particularly Mars. His observatory – Uraniborg – became a scientific centre, but Tycho was haughty and tactless (during his student days he had part of his nose sliced off in a duel, and made himself a replacement out of gold, silver and wax!), and after quarrels with the Danish Court he left Hven and went to Prague as Imperial Mathematician to the Holy Roman Emperor,

Tycho Brahe. (Patrick Moore)

Rudolph II. Here he was joined by Kepler, who acted as his assistant. When Tycho died, Kepler came into possession of the Hven observations, and used them to prove that the Earth moves round the Sun—something which Tycho himself could never accept.

Van Maanen, Adriaan. 1884–1947. Dutch astronomer, who emigrated to America and joined the Mount Wilson staff in 1912. He specialized in stellar parallaxes and proper motions, and accomplished much valuable work, though his alleged detection of movements in the spiral arms of galaxies later proved to be erroneous. He also discovered the white dwarf still known as Van Maanen's Star.

Vogel, Hermann Carl. 1842–1907. German astronomer, born and educated in Leipzig. He went to Potsdam in 1874, and concentrated upon stellar spectroscopy, pioneering research into spectroscopic binaries. In 1883 he published the first catalogue of stellar spectra.

Walther, Bernard. 1430–1504. (Often spelled 'Walter'.) German amateur astronomer, who lived in Nürnberg; he financed Regiomontanus' equipment, and carried on the work when Regiomontanus died. He was a very accurate observer, whose measurements of star and planetary positions were of great value to later astronomers.

Wargentin, Pehr Vilhelm. 1717–83. Swedish astronomer, and Director of the Stockholm Observatory. His best work was in the preparation of extremely accurate tables of Jupiter's satellites.

Webb, Thomas William, 1806–85. Vicar of Hardwicke in Herefordshire. He was an excellent observer, but is best remembered for his book *Celestial Objects for Common Telescopes*, which remains a classic.

Wilkins, Hugh Percy. 1896–1960. Welsh amateur astronomer (by profession a Civil Servant) who concentrated upon lunar observation, and produced a 300 in map of the Moon. He was for many years Director of the Lunar Section of the British Astronomical Association.

Wolf, Maximilian Franz Joseph Cornelius. 1863–1932. (Better known as Max Wolf.) German astronomer, who was born and lived in Heidelberg. He studied comets, and discovered his periodical comet in 1884; he was the first to hunt for asteroids photographically, and discovered well over 1000, and he also carried out research into dark nebulæ.

Wright, Thomas. 1711–85. *b.* nr Durham. Trained as a clockmaker, though he afterwards taught mathematics. He is remembered for his book published in 1750, in which he suggested that the Galaxy is disk-shaped. He also believed Saturn's rings to be composed of small particles.

Xenophanes. *c.* 570–478 BC. Greek philosopher, *b.* Colophon. His astronomical theories sound strange today; an infinitely thick flat Earth, a new Sun each day, and celestial bodies which—apart from the Moon—were made of fire!

Zach, Franz Xavier von. 1754–1832. Hungarian baron, who became renowned as an amateur astronomer. He published tables of the Sun and Moon, and was one of the chief organizers of the 'Celestial Police' who banded together to hunt for the supposed planet between Mars and Jupiter. He became Director of the Seeberg Observatory at Gotha, and did much for international co-operation among astronomers.

Zöllner, Johann Carl Friedrich. 1834–82. German astronomer, born in Leipzig, becoming Professor Astronomy there in 1874. He carried out pioneer spectroscopic work, observing the forms of solar prominences; he invented the polarizing photometer, and was the first to suggest that the spectral types of stars represent an evolutionary sequence.

Zwicky, Fritz. 1898–1974. Swiss astronomer; *b.* Bulgaria, but remained a Swiss citizen throughout his life. He graduated from Zürich, and in 1925 went to the California Institute of Technology, where he remained permanently, becoming Professor of Astrophysics from 1942 until his retirement in 1968. He became famous for his studies of galaxies and intergalactic matter; he predicted the existence of neutron stars (1934) and even black holes. He discovered many supernovæ in external galaxies, and masterminded a catalogue of compact galaxies. He was also active in the development of astronomical instrumentation, and was a pioneer worker with Schmidt telescopes. He received the Gold Medal of the Royal Astronomical Society in 1973.

GLOSSARY

Aberration of starlight. As light does not move infinitely fast, but at a rate of practically 300 000 km/s, and as the Earth is moving round the Sun at an average velocity of 28 km/s, the stars appear to be shifted slightly from their true positions. The best analogy is to picture a man walking along in a rainstorm, holding an umbrella. If he wants to keep himself dry, he will have to slant the umbrella forward; similarly, starlight seems to reach us 'from an angle'. Aberration may affect a star's position by up to 20·5 seconds of arc.

Absolute magnitude. The **apparent magnitude** that a star would have if it could be observed from a standard distance of 10 **parsecs** (32·6 light-years).

Achromatic object-glass. An **object-glass** which has been corrected so as to eliminate **chromatic aberration** or false colour as much as possible.

Aerolite. A **meteorite** whose main composition is stony.

Albedo. The reflecting power of a planet or other non-luminous body. The Moon is a poor reflector; its albedo is a mere 7% on average.

Altazimuth mounting for a telescope. A mounting on which the telescope may swing freely in any direction.

Altitude. The angular distance of a celestial body above the horizon.

Ångström unit. One hundred-millionth part of a centimetre.

Aphelion. The furthest distance of a planet or other body from the Sun in its orbit.

Apogee. The furthest point of the Moon from the Earth in its orbit.

Apparent magnitude. The apparent brightness of a celestial body. The lower the magnitude, the brighter the object: thus the Sun is approximately −27, the Pole Star +2, and the faintest stars detectable by modern techniques around +26.

Areography. The official name for 'the geography of Mars'.

Asteroids. One of the names for the minor planet swarm.

Astrograph. An astronomical telescope designed specially for astronomical photography.

Astrolabe. An ancient instrument used to measure the altitudes of celestial bodies.

Astronomical unit. The mean distance between the Earth and the Sun. It is equal to 149 598 500 km.

Aurora. Aurorae are 'polar lights'; Aurora Borealis (northern) and Aurora Australis (southern). They occur in the Earth's upper atmosphere, and are caused by charged particles emitted by the Sun.

Azimuth. The bearing of an object in the sky, measured from north (0°) through east, south and west.

Baily's beads. Brilliant points seen along the edge of the Moon just before and just after a total solar eclipse. They are caused by the sunlight shining through valleys at the Moon's limb.

Barycentre. The centre of gravity of the Earth-Moon system. Because the Earth is 81 times as massive as the Moon, the barycentre lies well inside the Earth's globe.

Binary star. A stellar system made up of two stars, genuinely associated, and moving round their common centre of gravity. The revolution periods range from millions of years for very widely-separated visual pairs down to less than half an hour for pairs in which the components are almost in contact with each other. With very close pairs, the components cannot be seen separately, but may be detected by spectroscopic methods.

Black hole. A region round a very small, very massive collapsed star from which not even light can escape.

BL Lacertæ objects. Variable objects which are powerful emitters of infra-red radiation, and appear to be very luminous and remote. Their nature is uncertain; they may be associated with quasars.

Bode's law. A mathematical relationship linking the distances of the planets from the Sun. It may or may not be genuinely significant. Strictly speaking it should be called Titius' Law, since it was discovered by J. D. Titius some years before J. E. Bode popularized it in 1772.

Bolide. A brilliant exploding meteor.

Bolometer. An instrument used to measure small quantities of heat radiation.

Carbon stars. Red stars of spectral types R and N with unusually carbon-rich atmospheres.

Cassegrain reflector. A reflecting telescope in which the secondary mirror is convex; the light is passed back through a hole in the main mirror. Its main advantage is that it is more compact than the Newtonian reflector.

Celestial sphere. An imaginary sphere surrounding the Earth, whose centre is the same as that of the Earth's globe.

Cepheid. A short-period **variable star**, very regular in behaviour; the name comes from the prototype star, Delta Cephei. Cepheids are astronomically important because there is a definite law linking their variation periods with their real luminosities, so that their distances may be obtained by sheer observation.

Chromatic aberration. A defect in all lenses, due to the fact that light is a mixture of all wavelengths – and these wavelengths are refracted unequally, so that false colour is produced round a bright object such as a star. The fault may be reduced by making the lens a compound arrangement, using different kinds of glasses.

Chromosphere. That part of the Sun's atmosphere which lies above the bright surface or photosphere.

Circumpolar star. A star which never sets. For instance, Ursa Major (the Great Bear) is circumpolar as seen from England; Crux Australis (the Southern Cross) is circumpolar as seen from New Zealand.

Cluster variables. An obsolete name for the stars now known as RR Lyræ variables.

Cœlostat. An optical instrument making use of two mirrors, one of which is fixed, while the other is movable and is mounted parallel to the Earth's axis; as the Earth rotates, the light from the star (or other object being observed) is caught by the rotatable mirror and is reflected in a fixed direction on to the second mirror. The result is that the eyepiece of the instrument need not move at all.

Collapsar. The end product of a very massive star, which has collapsed and has surrounded itself with a **black hole**.

Colour index. The difference between a star's visual magnitude and its photographic magnitude. The redder the star, the greater the positive value of the colour index; bluish stars have negative colour indices. For stars of type Ao, colour index is zero.

Colures. Great circles on the celestial sphere.

Conjunction. (1) A planet is said to be in conjunction with a star, or with another planet, when the two bodies are apparently close together in the sky. (2) For the inferior planets, Mercury and Venus, inferior conjunction occurs when the planet is approximately between the Earth and the Sun; superior conjunction, when the planet is on the far side of the Sun and the three bodies are again lined up. Planets beyond the Earth's orbit can never come to inferior conjunction, for obvious reasons!

Corona. The outermost part of the Sun's atmosphere, made up of very tenuous gas. It is visible with the naked eye only during a total solar eclipse.

Coronagraph. A device used for studying the inner **corona** at times of non-eclipse.

Cosmic rays. High-velocity particles reaching the Earth from outer space. The heavier cosmic-ray particles are broken up when they enter the upper atmosphere.

Cosmogony. The study of the origin and evolution of the universe.

Cosmology. The study of the universe considered as a whole.

Counterglow. The English name for the sky-glow more generally called by its German name of the **Gegenschein.**

Culmination. The maximum altitude of a celestial body above the horizon.

Dawes' limit. The practical limit for the resolving power of a telescope; it is 4·56/d, where d is the aperture of the telescope in inches.

Day, sidereal. The interval between successive meridian passages, or **culminations**, of the same star: 23h 56m 4s·091.

Day, solar. The mean interval between successive meridian passages of the Sun: 24h 3m 56s·555 of mean sidereal time. It is longer than the sidereal day because the Sun seems to move eastward against the stars at an average rate of approximately one degree per day.

Declination. The angular distance of a celestial body north or south of the celestial equator. It corresponds to latitude on the Earth.

Dewcap. An open tube fitted to the upper end of a refracting telescope. Its rôle is to prevent condensation upon the object-glass.

Dichotomy. The exact half-phase of the Moon or an **inferior planet.**

Diffraction grating. A device used for splitting up light; it consists of a polished metallic surface upon which thousands of parallel lines are ruled. It may be regarded as an alternative to the prism.

Direct motion. Movement of revolution or rotation in the same sense as that of the Earth.

Doppler effect. The apparent change in wavelength of the light from a luminous body which is in motion relative to the observer. With an approaching object, the wavelength is apparently shortened, and the spectral lines are shifted to the blue end of the spectral band; with a receding body there is a red shift, since the wavelength is apparently lengthened.

Double star. A star made up of two components – either genuinely associated (binary systems) or merely lined up by chance (optical pairs).

Driving clock. A mechanism for driving a telescope round at a rate which compensates for the axial rotation of the Earth, so that the object under observation remains fixed in the field of view.

Dwarf novæ. A term sometimes applied to the U Geminorum (or SS Cygni) variable stars.

Earthshine. The faint luminosity of the night side of the Moon, frequently seen when the Moon is in its crescent phase. It is due to light reflected on to the Moon from the Earth.

Eclipse, lunar. The passage of the Moon through the shadow cast by the Earth. Lunar eclipses may be either total or partial. At some eclipses, totality may last for approximately 1¾ hours, though most are shorter.

Eclipse, solar. The blotting-out of the Sun by the Moon, so that the Moon is then directly between the Earth and the Sun. Total eclipses can last for over 7 minutes under exceptionally favourable circumstances. In a partial eclipse, the Sun is incompletely covered. In an annular eclipse, exact alignment occurs when the Moon is in the far part of its orbit, and so appears smaller than the Sun; a ring of sunlight is left showing round the dark body of the Moon. Strictly speaking, a solar 'eclipse' is the **occultation** of the Sun by the Moon.

Eclipsing variable (or Eclipsing Binary). A **binary star** in which one component is regularly **occulted** by the other, so that the total light which we receive from the system is reduced. The prototype eclipsing variable is Algol (Beta Persei).

Ecliptic. The apparent yearly path of the Sun among the stars. It is more accurately defined as the projection of the Earth's orbit on to the celestial sphere.

Electron. Part of an atom; a fundamental particle carrying a negative electric charge.

Electron density. The number of 'free' (unattached) electrons in unit volume of space.

Elongation. The angular distance of a planet from the Sun, or of a satellite from its primary planet.

Ephemeris. A table showing the predicted positions of a celestial body such as a comet, asteroid or planet.

Epoch. A date chosen for reference purposes in quoting astronomical data.

Equator, celestial. The projection of the Earth's equator on to the **celestial sphere**.

Equatorial mounting for a telescope. A mounting in which the telescope is set up on an axis which is parallel with the axis of the Earth. This means that one movement only (east to west) will suffice to keep an object in the field of view.

Equinox. The equinoxes are the two points at which the **ecliptic** cuts the **celestial equator**. The vernal equinox or First Point of Aries now lies in the constellation of Pisces; the Sun crosses it about 21 March each year. The autumnal equinox is known as the First Point of Libra; the Sun reaches it about 22 September yearly.

Escape velocity. The minimum velocity which an object must have in order to escape from the surface of a planet, or other celestial body, without being given any extra impetus.

Evection. An inequality in the Moon's motion, due to slight changes in the shape of the lunar orbit.

Exosphere. The outermost part of the Earth's atmosphere.

Extinction. The apparent reduction in brightness of a star or planet when low down in the sky, so that more of its light is absorbed by the Earth's atmosphere. With a star 1° above the horizon, extinction amounts to 3 magnitudes.

Eyepiece (or Ocular). The lens, or combination of lenses, at the eye-end of a telescope. It is responsible for magnifying the image of the object under study. With a positive eyepiece (for instance, a Ramsden, Orthoscopic or Monocentric) the image plane lies between the eyepiece and the object-glass (or main mirror); with a negative eyepiece (such as a Huyghenian or Tolles) the image plane lies inside the eyepiece. A Barlow lens is concave, and is mounted in a short tube which may be placed between the eyepiece and the object-glass (or mirror). It increases the effective focal length of the telescope, thereby providing increased magnification.

Faculæ. Bright, temporary patches on the surface of the sun.

Filar micrometer. A device used for measuring very small angular distances as seen in the eyepiece of a telescope.

Finder. A small, wide-field telescope attached to a larger one, used for sighting purposes.

Fireball. A very brilliant **meteor**.

Flares, solar. Brilliant eruptions in the outer part of the Sun's atmosphere. Normally they can be detected only by spectroscopic means (or the equivalent), though a few have been seen in integrated light. They are made up of hydrogen, and emit charged particles which may later reach the Earth, producing magnetic storms and displays of auroræ. Flares are generally, though not always, associated with sunspot groups.

Flare stars. Faint Red Dwarf stars which show sudden, short-lived increases in brilliancy, due possibly to intense flares above their surfaces.

Flash spectrum. The sudden change-over from dark to bright lines in the Sun's spectrum, just before the onset of totality in a **solar eclipse**. The phenomenon is due to the fact that at this time the Moon has covered up the bright surface of the Sun, so that the chromosphere is shining 'on its own'.

Flocculi. Patches of the Sun's surface, observable with spectroscopic equipment. They are of two main kinds: bright (calcium) and dark (hydrogen).

Forbidden lines. Lines in the spectrum of a celestial body which do not appear under normal conditions, but may be seen in bodies where conditions are exceptional.

Fraunhofer lines. The dark absorption lines in the spectrum of the Sun.

Galaxies. Systems made up of stars, nebulæ, and interstellar matter. Many, though by no means all, are spiral in form.

Galaxy, the. The system of which our Sun is a member. It contains approximately 100 000 million stars, and is a rather loose spiral.

Gamma-rays. Radiation of extremely short wavelength.

Gauss. Unit of measurement of a magnetic field. (The Earth's field, at the surface, is on average about 0·3 to 0·6 gauss.)

Gegenschein. A faint sky-glow, opposite to the Sun and very difficult to observe. It is due to thinly-spread interplanetary material.

Geodesy. The study of the shape, size, mass and other characteristics of the Earth.

Gibbous phase. The phase of the Moon or planet when between half and full.

Globules. Small dark patches inside gaseous nebulæ. They may be embryo stars.

Gnomon. In a sundial, the gnomon is a pointer whose function is to cast the Sun's shadow on to the dial. The gnomon always points to the celestial pole.

Great circle. A circle on the surface of a sphere whose plane passes through the centre of that sphere.

Green Flash. Sudden, brief green light seen as the last segment of the Sun disappears below the horizon. It is purely an effect of the Earth's atmosphere. Venus has also been known to show a Green Flash.

Gregorian reflector. A telescope in which the secondary mirror is concave, and placed beyond the focus of the main mirror. The image obtained is erect. Few Gregorian telescopes are in use nowadays.

H.I and H.II regions. Clouds of hydrogen in the Galaxy. In H.I regions the hydrogen is neutral; in H.II regions the hydrogen is ionized, and the presence of hot stars will make the cloud shine as a nebula.

Halo, galactic. The spherical-shaped cloud of stars round the main part of the Galaxy.

Heliacal rising. The rising of a star or planet at the same time as the Sun, though the term is generally used to denote the time when the object is first detectable in the dawn sky.

Herschelian reflector. An obsolete type of telescope in which the main mirror is tilted, thus removing the need for a secondary mirror.

Hertzsprung-Russell diagram (usually known as the H-R Diagram). A diagram in which stars are plotted according to their spectral types and their **absolute magnitudes**.

Horizon. The great circle on the celestial sphere which is everywhere 90 degrees from the observer's zenith.

Hour angle (of a celestial object). The time which has elapsed since the object crossed the meridian. If RA = right ascension of the object and LST = the local sidereal time, then Hour Angle = LST − RA.

Hour circle. A great circle on the **celestial sphere**, passing through both celestial poles. The zero hour circle coincides with the observer's meridian.

Hubble's constant. The rate of increase in the recession of a galaxy with increased distance from the Earth.

Inferior planets. Mercury and Venus, whose distances from the Sun are less than that of the Earth.

Infra-red radiation. Radiation with wavelength longer than that of visible light (approximately 7500 Ångströms).

Interferometer, stellar. An instrument for measuring star diameters. The principle is based upon light-interference.

Ion. An atom which has lost or gained one or more of its planetary electrons, and so has respectively a positive or negative charge.

Ionosphere. The region of the Earth's atmosphere lying above the stratosphere.

Irradiation. The effect which makes very brilliant bodies appear larger than they really are.

Julian day. A count of the days, starting from 12 noon on 1 January 4713 BC. Thus 1 January 1977 was Julian Day 2 443 145. (The name 'Julian' has nothing to do with Julius Cæsar! The system was invented in 1582 by the mathematician Scaliger, who named it in honour of his father, Julius Scaliger.)

Kepler's laws of planetary motion. These were laid down by Johannes Kepler, from 1609 to 1618. They are: (1) The planets move in elliptical orbits, with the Sun occupying one focus. (2) The radius vector, or imaginary line joining the centre of the planet to the centre of the Sun, sweeps out equal areas in equal times. (3) With a planet, the square of the sidereal period is proportional to the cube of the mean distance from the Sun.

Kiloparsec. One thousand **parsecs** (3260 light-years).

Latitude, celestial. The angular distance of a celestial body from the nearest point on the **ecliptic**.

Libration. The apparent 'tilting' of the Moon as seen from Earth. There are three librations: latitudinal, longitudinal and diurnal. The overall effect is that at various times an observer on Earth can see a total of 59% of the total surface of the Moon, though, naturally, no more than 50% at any one moment!

Light-year. The distance travelled by light in one year: 9·4607 million million kilometres.

Local group. A group of more than two dozen galaxies, one member of which is our own **Galaxy**. The largest member of the Local Group is the Andromeda Spiral, M.31.

Longitude, celestial. The angular distance of a celestial body from the **vernal equinox**, measured in degrees eastward along the ecliptic.

Lunation. The interval between successive new

moons: 29d 12h 44m. (Also known as the Synodic Month.)

Magnetosphere. The region of the magnetic field of a planet or other body. In the Solar System, only the Earth, Jupiter, Mercury, Saturn and Uranus are known to have detectable magnetospheres, and Neptune probably has.

Main Sequence. A band along an **H-R Diagram**, including most normal stars except for the giants.

Maksutov telescope. An astronomical telescope involving both mirrors and lenses.

Mass. The quantity of matter that a body contains. It is not the same as 'weight'.

Mean sun. An imaginary sun travelling eastward along the celestial equator, at a speed equal to the average rate of the real Sun along the **ecliptic**.

Megaparsec. One million **parsecs**.

Meridian, celestial. The great circle on the **celestial sphere** which passes through the **zenith** and both celestial poles.

Meteor. A small particle, friable in nature and usually smaller than a sand grain, moving round the Sun, and visible only when it enters the upper atmosphere and is destroyed by friction. Meteors may be regarded as cometary debris.

Meteorite. A larger object, which may fall to the ground without being destroyed in the upper atmosphere. A meteorite is fundamentally different from a **meteor**. Meteorites are not associated with comets, but may be closely related to asteroids.

Micrometeorite. A very small particle of interplanetary material, too small to cause a luminous effect when it enters the Earth's upper atmosphere.

Micrometer. A measuring device, used together with a telescope to measure very small angular distances – such as the separations between the components of double stars.

Micron. One-thousandth of a millimetre. The usual symbol is μ.

Month. (1) Anomalistic: the interval between successive **perigee** passages of the Moon (27·55 days). (2) Sidereal: the revolution period of the Moon with reference to the stars (27·32 days). (3) Synodical: the interval between successive new moons (29·53 days). (4) Nodical or Draconitic: the interval between successive passages of the Moon through one of its nodes (27·21 days). (5) Tropical: the time taken for the Moon to return to the same celestial longitude (about 7 seconds shorter than the sidereal month).

Nadir. The point on the celestial sphere directly below the observer.

Nebula. A cloud of gas and dust in space. Galaxies were once known as 'spiral nebulæ' or 'extragalactic nebulæ'.

Neutrino. A fundamental particle with little or no mass and no electric charge.

Neutron. A fundamental particle with no electric charge, but a mass practically equal to that of a **proton**.

Neutron star. The remnant of a very massive star which has exploded as a **supernova**. Neutron stars send out rapidly-varying radio emissions, and are therefore called 'pulsars'. Only two (the Crab and Vela pulsars) have as yet been identified with optical objects.

Newtonian reflector. A reflecting telescope in which the light is collected by a main mirror, reflected on to a smaller flat mirror set at an angle of 45°, and thence to the side of the tube.

Nodes. The points at which the orbit of the Moon, a planet or a comet cuts the plane of the **ecliptic**; south to north (Ascending Node) or north to south (Descending Node).

Nova. A star which suddenly flared up to many times its normal brilliancy, remaining bright for a relatively short time before fading back to obscurity.

Nutation. A slow, slight 'nodding' of the Earth's axis, due to the gravitational pull of the Moon on the Earth's equatorial bulge.

Object-glass (or Objective). The main lens of a refracting telescope.

Objective prism. A small prism placed in front of the **object-glass** of a telescope. It produces small-scale spectra of the stars in the field of view.

Obliquity of the ecliptic. The angle between the **ecliptic** and the celestial equator: 23° 26′ 45″.

Occultation. The covering-up of one celestial body by another.

Opposition. The position of a planet when exactly opposite to the Sun in the sky; the Sun, the Earth and the planet are then approximately lined up.

Orbit. The path of a celestial object.

Orrery. A model showing the Sun and the planets, capable of being moved mechanically so that the planets move round the Sun at their correct relative speeds.

Parallax, trigonometrical. The apparent shift of an object when observed from two different directions.

Parsec. The distance at which a star would have a parallax of one second of arc: 3·26 **light-years**, 206 265 **astronomical units**, or 30·857 million million kilometres.

Penumbra. (1) The area of partial shadow to either side of the main cone of shadow cast by the Earth. (2) The lighter part of a sunspot.

Perigree. The position of the Moon in its orbit when closest to the Earth.

Perihelion. The position in orbit of a planet or other body when closest to the Sun.

Perturbations. The disturbances in the orbit of a celestial body produced by the gravitational effects of other bodies.

Phases. The apparent changes in shape of the Moon and the inferior planets from new to full. Mars may show a **gibbous phase**, but with the other planets there are no appreciable phases as seen from Earth.

Photoelectric cell. An electronic device; light falling on the cell produces an electric current, the strength of which depends upon the intensity of the light.

Photoelectric photometer. A **photoelectric cell** used together with a telescope for measuring the magnitudes of celestial bodies.

Photometer. An instrument used to measure the intensity of light from any particular source.

Photometry. The measurement of the intensity of light.

Photon. The smallest 'unit' of light.

Photosphere. The bright surface of the Sun.

Planetary nebula. A small, dense, hot star surrounded by a shell of gas. The name is ill-chosen, since planetary nebulæ are neither planets nor nebulæ!

Planetoid. An **asteroid** or minor planet.

Poles, celestial. The north and south points of the celestial sphere.

Populations, stellar. Two main types of star regions: I (in which the brightest stars are hot and white), and II (in which the brightest stars are old Red Giants).

Position angle. The apparent direction of one object with reference to another, measured from the north point of the main object through east, south and west.

Precession. The apparent slow movement of the celestial **poles**. This also means a shift of the celestial equator, and hence of the equinoxes; the vernal **equinox** moves by 50″ of arc yearly, and has moved out of Aries into Pisces. Precession is due to the pull of the Moon and Sun on the Earth's equatorial bulge.

Prime meridian. The meridian on the Earth's surface which passes through the Airy Transit Circle at Greenwich Observatory. It is taken as longitude 0°.

Prominences. Masses of glowing gas rising from the surface of the Sun. They are made up chiefly of hydrogen.

Proper motion, stellar. The individual movement of a star on the celestial sphere.

Proton. A fundamental particle with a positive electric charge. The nucleus of the hydrogen atom is made up of a single proton.

Quadrant. An ancient astronomical instrument used for measuring the apparent positions of celestial bodies.

Quadrature. The position of the Moon or a planet when at right-angles to the Sun as seen from the Earth.

Quantum. The amount of energy possessed by one photon of light.

Quasar. A very remote, superluminous object. Quasars are now known to be the cores of very active galaxies, though the source of their energy is still a matter for debate.

Radial Velocity. The movement of a celestial body toward or away from the observer; positive if receding, negative if approaching.

Radiant. The point in the sky from which the meteors of any particular shower seem to radiate.

Regression of the nodes. The nodes of the Moon's orbit move slowly westward, making one complete revolution in 18·6 years. This regression is caused by the gravitational pull of the Sun.

Retardation. The difference in the time of moonrise between one night and the next.

Retrograde motion. Orbital or rotational movement in the sense opposite to that of the Earth's motion.

Reversing layer. The gaseous layer above the Sun's **photosphere.**

Right ascension. The angular distance of a celestial body from the vernal equinox, measured eastward. It is usually given in hours, minutes and seconds of time, so that the right ascension is the time-difference between the **culmination** of the vernal **equinox** and the culmination of the body.

Roche limit. The distance from the centre of a planet within which a second body would be broken up by the planet's gravitational pull. Note, however, that this would be the case only for a body which had no appreciable gravitational cohesion.

Saros. The period after which the Earth, Moon and Sun return to almost the same relative positions: 18 years 11·3 days. The saros may be used in eclipse prediction, since it is usual for an eclipse to be followed by a similar eclipse exactly one saros later.

Schmidt camera (or Schmidt telescope). An instrument which collects its light by means of a spherical mirror; a correcting plate is placed at the top of the tube. It is a purely photographic instrument.

Schwarzschild radius. The radius that a body must have if its **escape velocity** is to be equal to the velocity of light.

Scintillation. Twinkling of a star; it is due to the Earth's atmosphere. Planets may also show scintillation when low in the sky.

Secular acceleration of the Moon. The apparent speeding-up of the Moon in its orbit as measured over a long period of time, caused by the gradual slowing of the Earth's rotation (by 0·000 000 02 second per day).

Selenography. The study of the surface of the Moon.

Sextant. An instrument used for measuring the altitude of a celestial object.

Seyfert galaxies. Galaxies with relatively small, bright nuclei and weak spiral arms. Some of them are strong radio emitters.

Sidereal period. The revolution period of a planet round the Sun, or of a satellite round its primary planet.

Sidereal time. The local time reckoned according to the apparent rotation of the **celestial sphere**. When the vernal **equinox** crosses the observer's **meridian**, the sidereal time is 0 hours.

Solar wind. A flow of atomic particles streaming out constantly from the Sun in all directions.

Solstices. The times when the Sun is at its maximum **declination** of approximately 23½ degrees; around 22 June (summer solstice, with the Sun in the northern hemisphere of the sky) and 22 December (winter solstice, Sun in the southern hemisphere).

Specific gravity. The density of any substance, taking that of water as 1. For instance, the Earth's specific gravity is 5·5, so that the Earth 'weighs' 5·5 times as much as an equal volume of water would do.

Spectroheliograph. An instrument used for photographing the Sun in the light of one particular wavelength only. The visual equivalent of the spectroheliograph is the spectrohelioscope.

Spectroscopic binary. A binary system whose components are too close together to be seen individually, but which can be studied by means of spectroscopic analysis.

Speculum. The main mirror of a reflecting telescope.

Spherical aberration. Blurring of a telescopic image; it is due to the fact that the lens (or mirror) does not bring the light-rays falling on its edge and on its centre to exactly the same focal point.

Superior planets. All the planets lying beyond the orbit of the Earth in the Solar System (that is to say, all the principal planets apart from Mercury and Venus).

Supernova. A colossal stellar outburst, involving (1) the total destruction of the white dwarf member of a binary system, or (2) the collapse of a very massive star.

Synodic period. The interval between successive **oppositions** of a **superior planet.**

Syzygy. The position of the Moon in its orbit when new or full.

Tektites. Small, glassy objects found in a few localized parts of the Earth. Nobody is yet certain whether or not they come from the sky!

Terminator. The boundary between the day- and night-hemispheres of the Moon or a planet.

Thermocouple. An instrument used for measuring very small amounts of heat.

Transit. (1) The passage of a celestial body across the observer's meridian. (2) The projection of Mercury or Venus against the face of the Sun.

Transit instrument. A telescope mounted so that it can move only in **declination**; it is kept pointing to the meridian, and is used for timing the passages of stars across the meridian. Transit instruments were once the basis of all practical timekeeping. The Airy transit instrument at Greenwich is accepted as the zero for all longitudes on the Earth.

Troposphere. The lowest part of the Earth's atmosphere; its top lies at an average height of about 11 km. Above it lies the stratosphere; and above the stratosphere come the ionosphere and the exosphere.

Twilight. The state of illumination when the Sun is below the horizon by less than 18 degrees.

Umbra. (1) The main cone of shadow cast by the Earth. (2) The darkest part of a sunspot.

Van Allen zones. Zones of charged particles around the Earth. There are two main zones; the outer (made up chiefly of **electrons**) and the inner (made up chiefly of **protons**).

Variable stars. Stars which change in brilliancy over short periods. They are of various types.

Variation. An inequality in the Moon's motion, due to the fact that the pull of the Sun on the Moon is not constant for all positions in the lunar orbit.

White dwarf. A very small, very dense star which has used up its nuclear energy, and is in a very late stage of its evolution.

Widmanstätten patterns. If an iron **meteorite** is cut, polished and then etched with acid, characteristic figures of the iron crystals appear; these are the Widmanstätten patterns.

Wolf-Rayet stars. Very hot, greenish-white stars which are surrounded by expanding gaseous envelopes. Their spectra show bright (emission) lines.

Year. (1) Sidereal: the period taken for the Earth to complete one journey round the Sun (365·26 days). (2) Tropical: the interval between successive passages of the Sun across the vernal equinox (365·24 days). (3) Anomalistic: the interval between successive perihelion passages of the Earth (365·26 days; slightly less than 5 minutes longer than the sidereal year, because the position of the perihelion point moves along the Earth's orbit by about 11 seconds of arc every year). (4) Calendar: the mean length of the year according to the Gregorian calendar (365·24 days, or 365d 5h 49m 12s).

Zenith. The observer's overhead point (altitude 90°).

Zenith distance. The angular distance of a celestial object from the **Zenith.**

Zodiac. A belt stretching round the sky, 8° to either side of the **ecliptic**, in which the Sun, Moon and principal planets are to be found at any time. (Pluto is the only planet which can leave the Zodiac, though many asteroids do so.)

Zodiacal light. A cone of light rising from the horizon and stretching along the **ecliptic**; visible only when the Sun is a little way below the horizon. It is due to thinly-spread interplanetary material near the main plane of the Solar System.

INDEX